**Ruthenium Complexes**

# Ruthenium Complexes

Photochemical and Biomedical Applications

*Edited by*
*Alvin A. Holder*
*Lothar Lilge*
*Wesley R. Browne*
*Mark A.W. Lawrence*
*Jimmie L. Bullock Jr.*

WILEY-VCH

**Editors**

***Prof. Alvin A. Holder***
Old Dominion University
Department of Chemistry and Biochemistry
4541 Hampton Blvd.
VA
United States

***Prof. Lothar Lilge***
University of Toronto
Princess Margaret Cancer Centre
101 College Street
M5G 1L7 ON
Canada

***Prof. Wesley R. Browne***
University of Groningen
Stratingh Institute of Chemistry
Nijenborgh 4
9747 AG Groningen
Netherlands

***Dr. Mark A.W. Lawrence***
Old Dominion University
Department of Chemistry and Biothchnology
4541 Hampton Blvd.
VA
United States

***Jimmie L. Bullock Jr.***
Old Dominion University
Department of Chemistry and Biochemistry
4541 Hampton Blvd.
VA
United States



**Library of Congress Card No.:**
applied for

**British Library Cataloguing-in-Publication Data**
A catalogue record for this book is available from the British Library.

**Bibliographic information published by the Deutsche Nationalbibliothek**
The Deutsche Nationalbibliothek lists this publication in the Deutsche Nationalbibliografie; detailed bibliographic data are available on the Internet at <http://dnb.d-nb.de>.



**Print ISBN:** 978-3-527-33957-0
**ePDF ISBN:** 978-3-527-69520-1
**ePub ISBN:** 978-3-527-69524-9
**Mobi ISBN:** 978-3-527-69521-8
**oBook ISBN:** 978-3-527-69522-5

**Cover Design** Grafik-Design Schulz, Fußgönheim, Germany
**Typesetting** SPi Global, Chennai, India
**Printing and Binding** C.O.S. Printers Pte Ltd Singapore

Printed on acid-free paper

10 9 8 7 6 5 4 3 2 1

Dedicated to Karen with admiration, affection, and respect!!

Dear Karen, we will miss you for
your class, humour, and knowledge!!
Selah
R.I.P.

# Contents

**About the Editors** *xv*
**Preface** *xvii*
**Acknowledgments** *xix*

**Section I Introduction** *1*

1 **Karen J. Brewer (1961–2014): A Bright Star that Burned Out Far Too Soon** *3*
*Seth C. Rasmussen*
1.1 Introduction *3*
1.2 Early Years *4*
1.3 Graduate Studies and Clemson University *6*
1.4 Postdoctoral Research and the University of California, Berkeley *11*
1.5 Washington State University: Beginning an Independent Career *13*
1.6 Move to Virginia Tech *15*
1.7 Collaboration with Brenda Winkel and the Study of Metal-DNA Interactions *16*
1.8 A Return to Where It All Started: Photochemical $H_2$ Production *18*
1.9 A Career Cut Tragically Short *19*
1.10 Karen's Legacy *20*
Acknowledgments *20*
References *20*

2 **Basic Coordination Chemistry of Ruthenium** *25*
*Mark A. W. Lawrence, Jimmie L. Bullock, and Alvin A. Holder*
2.1 Coordination Chemistry of Ruthenium *25*
2.1.1 The Element *25*
2.1.2 Stereochemistry and Common Oxidation States *26*
2.1.2.1 Ruthenium in Low Oxidation States *27*
2.1.2.2 Chemistry of Ruthenium(II) and (III) *31*
2.1.2.3 Higher Oxidation States of Ruthenium *36*
2.1.3 Conclusion *37*
References *37*

**Section II Artificial Photosynthesis** *43*

**3 Water Oxidation Catalysis with Ruthenium** *45*
*Andrea Sartorel*
3.1 Introduction *45*
3.1.1 Energy Issue and Energy from the Sun *45*
3.1.2 Photosynthesis and Solar Fuels *46*
3.1.3 Water Oxidation *48*
3.1.4 Artificial Water Oxidation *49*
3.2 Ruthenium in Water Oxidation Catalyst *50*
3.2.1 Ruthenium Oxide *50*
3.2.2 Molecular Ruthenium WOC *52*
3.2.2.1 Meyer's Blue Dimer *53*
3.2.2.2 The Ru-Hbpp Catalyst *54*
3.2.2.3 Single-Site Ru-WOCs *55*
3.2.2.4 Heptacoordinated Ru Intermediates *56*
3.2.3 Polyoxometalates: The Bridge Between Metal Oxides and Coordination Complexes *57*
3.3 Conclusions and Perspectives *60*
References *61*

**4 Ruthenium- and Cobalt-Containing Complexes and Hydrogenases for Hydrogen Production** *67*
*Michael J. Celestine, Raj K. Gurung, and Alvin A. Holder*
4.1 Introduction *67*
4.2 (A) Ruthenium- and Cobalt-Containing Complexes for Hydrogen Production *68*
4.2.1 Nonbridged Systems *68*
4.2.2 Bridged Systems *70*
4.3 (B) Ruthenium(II)-Containing Complexes and Hydrogenases for Hydrogen Generation in Aqueous Solution *77*
4.3.1 Hydrogenases *77*
4.3.2 Hydrogenases with Ruthenium(II) Complexes *78*
4.4 Conclusions *84*
References *85*

**Section III Applications in Medicine** *89*

**5 Ligand Photosubstitution Reactions with Ruthenium Compounds: Applications in Chemical Biology and Medicinal Chemistry** *91*
*Samantha L. Hopkins and Sylvestre Bonnet*
5.1 Introduction *91*
5.2 Caging and Uncaging Biologically Active Ligands with a Nontoxic Ruthenium Complex *92*
5.3 Caging Cytotoxic Ruthenium Complexes with Organic Ligands *96*

5.4 Low-Energy Photosubstitution *100*
5.4.1 Introduction *100*
5.4.2 Modulating Ru Photophysics by Ligand Modulation *100*
5.4.3 Upconversion (UC) *105*
5.4.3.1 Triplet–Triplet Annihilation Upconversion *105*
5.4.3.2 Upconverting Nanoparticles (UCNPs) *106*
5.4.3.3 Two-Photon Absorption (TPA) Photosubstitution *109*
5.5 Conclusions *110*
References *111*

**6 Use of Ruthenium Complexes as Photosensitizers in Photodynamic Therapy** *117*
*Lothar Lilge*
6.1 Introduction *117*
6.2 The Basics of Photodynamic Therapy *118*
6.2.1 Singlet Oxygen Production *120*
6.2.2 Other Radical Production *120*
6.2.3 PDT Dose Definition *120*
6.2.3.1 PDT Dosimetry *In Vitro* *122*
6.2.3.2 PDT Dosimetry *In Vivo* *124*
6.2.3.3 Oxygen Consumption Model *125*
6.2.3.4 *In Vivo* Tissue Response Models *125*
6.2.4 PDT and Immunology *126*
6.3 Status of Ru Photosensitizing Complexes *126*
6.3.1 Photostability for Ru-PS Complexes *128*
6.3.2 Long Wavelength Activation of Ru(II)-PS Complexes *128*
6.4 Issues to Be Considered to Further Develop Ru-Based Photosensitizers *129*
6.4.1 Subcellular Localization *130*
6.4.2 Ruthenium Complex Photosensitizers and the Immune Response *131*
6.5 Future Directions for Ru-PS Research *131*
6.6 Conclusion *132*
References *132*

**7 Photodynamic Therapy in Medicine with Mixed-Metal/Supramolecular Complexes** *139*
*Jimmie L. Bullock and Alvin A. Holder*
7.1 Introduction *139*
7.2 Platinum and Rhodium Centers as Bioactive Sites *140*
7.2.1 Platinum(II)-Based Chemotherapeutics *140*
7.2.2 Rhodium(III) as a Bioactive Site *141*
7.3 Supramolecular Complexes as DNA Photomodification Agents *142*
7.4 Mixed-Metal Complexes as Photodynamic Therapeutic Agents *143*
7.4.1 Photosensitizers with a Ru(II) Metal Center Coupled to Pt(II) Bioactive Sites *143*

7.4.1.1 Binuclear Complexes with Ru(II) and Pt(II) Metal Centers with Bidentate Ligands *143*
7.4.1.2 Binuclear and Trinuclear Complexes with Ru, Pt with Tridentate Ligands *146*
7.4.2 Photosensitizers with a Ru(II) Metal Center Coupled to Rh(III) Bioactive Sites *147*
7.4.2.1 Trinuclear Complexes with Ru(II), Rh(III), and Ru(II) Metal Centers *147*
7.4.2.2 Binuclear Complexes with Ru(II) and Rh(III) Metal Centers *149*
7.4.3 Photosensitizers with a Ru(II) Metal Cenetr Coupled to Other Bioactive Sites *150*
7.4.3.1 Binuclear Complexes with Ru(II) and Cu *150*
7.4.3.2 Binuclear Complexes with Ru(II) and Co(III) Metal Centers *151*
7.4.3.3 Binuclear Complexes with Ru (II) and V(IV) Metal Centers *151*
7.4.3.4 Applications of Ru(II) Metal Centers in Nanomedicine *152*
7.5 Summary and Conclusions *155*
Abbreviations *156*
References *157*

**8 Ruthenium Anticancer Agents En Route to the Tumor: From Plasma Protein Binding Agents to Targeted Delivery** *161*
*Muhammad Hanif and Christian G. Hartinger*
8.1 Introduction *161*
8.2 Protein Binding $Ru^{III}$ Anticancer Drug Candidates *163*
8.2.1 $Ru^{III}$ Anticancer Drug Candidates Targeting Primary Tumors *163*
8.2.2 Antimetastatic $Ru^{III}$ Compounds *165*
8.3 Functionalization of Macromolecular Carrier Systems with Ru Anticancer Agents *166*
8.3.1 Proteins as Delivery Vectors for Organometallic Compounds *166*
8.3.2 Polymers and Liposomes as Delivery Systems for Bioactive Ruthenium Complexes *168*
8.3.3 Dendrimers *169*
8.4 Hormones, Vitamins, and Sugars: Ruthenium Complexes Targeting Small Molecule Receptors *169*
8.5 Peptides as Transporters for Ruthenium Complexes into Tumor Cells and Cell Compartments *173*
8.6 Polynuclear Ruthenium Complexes for the Delivery of a Cytotoxic Payload *174*
8.7 Summary and Conclusions *175*
Acknowledgments *175*
References *176*

**9 Design Aspects of Ruthenium Complexes as DNA Probes and Therapeutic Agents** *181*
*Madeleine De Beer and Shawn Swavey*
9.1 Introduction *181*
9.2 Physical Interaction to Disrupt DNA Structure *181*

9.2.1 Irreversible Covalent Binding *182*
9.2.2 Intercalation *184*
9.2.3 Additional Noncovalent Binding Interactions *185*
9.3 Biological Consequences of Ru-Complex/DNA Interactions *186*
9.4 Effects of Ru Complexes on Topoisomerases and Telomerase *191*
9.5 Summary and Conclusions *196*
References *197*

**10 Ruthenium-Based Anticancer Compounds: Insights into Their Cellular Targeting and Mechanism of Action** *201*
*António Matos, Filipa Mendes, Andreia Valente, Tânia Morais, Ana Isabel Tomaz, Philippe Zinck, Maria Helena Garcia, Manuel Bicho, and Fernanda Marques*
10.1 Introduction *201*
10.2 Cellular Uptake *204*
10.3 DNA and DNA-Related Cellular Targets *205*
10.4 Targeting Signaling Pathways *207*
10.5 Targeting Enzymes of Specific Cell Functions *207*
10.6 Targeting Glycolytic Pathways *209*
10.7 Macromolecular Ruthenium Conjugates: A New Approach to Targeting *211*
10.8 Conclusions *214*
References *215*

**11 Targeting cellular DNA with Luminescent Ruthenium(II) Polypyridyl Complexes** *221*
*Martin R. Gill and Jim A. Thomas*
11.1 Introduction *221*
11.1.1 DNA-Binding Modes of Small Molecules *222*
11.1.2 Metal Complexes and DNA *223*
11.2 $[Ru(bpy)_2(dppz)]^{2+}$ and the DNA "Light-Switch" Effect *224*
11.3 Cellular Uptake of RPCs and Application as DNA-Imaging Agents *226*
11.3.1 Mononuclear Complexes *226*
11.3.2 Dinuclear Complexes *228*
11.3.3 Cyclometalated Systems *228*
11.4 Alternative Techniques to Assess Cellular Uptake and Localization *231*
11.5 Toward Theranostics: luminescent RPCs as Anticancer Therapeutics *232*
11.6 Summary and Conclusions *234*
References *235*

**12 Biological Activity of Ruthenium Complexes With Quinoline Antibacterial and Antimalarial Drugs** *239*
*Jakob Kljun and Iztok Turel*
12.1 Introduction *239*

12.2 Antibacterial (Fluoro)quinolones *240*
12.2.1 Quinolones and Their Interactions with Metal Ions *241*
12.2.2 Ruthenium and Quinolones *241*
12.2.3 Ruthenium and HIV Integrase Inhibitor Elvitegravir *245*
12.3 Antibacterial 8-Hydroxyquinolines *246*
12.3.1 Mode of Action of 8-Hydroxyquinoline Agents *246*
12.3.2 Ruthenium and 8-Hydroxyquinolines *247*
12.4 Antimalarial 4-Aminoquinolines *248*
12.4.1 Mechanism of Action of Antimalarial Quinoline Agents *248*
12.5 Metallocene Analogues of Chloroquine *249*
12.6 Conclusions *252*
References *252*

**13 Ruthenium Complexes as NO Donors: Perspectives and Photobiological Applications** *257*
*Loyanne C.B. Ramos, Juliana C. Biazzotto, Juliana A. Uzuelli, Renata G. de Lima, and Roberto S. da Silva*
13.1 Introduction *257*
13.2 Photochemical Processes of Some Nitrogen Oxide Derivative–Ruthenium Complexes *258*
13.2.1 Metal-Ligand Charge-Transfer Photolysis of $\{Ru\text{-}NO\}^6$ *258*
13.2.2 Nitrosyl Ruthenium Complexes: Visible-Light Stimulation *261*
13.3 Photobiological Applications of Nitrogen Oxide Compounds *265*
13.3.1 Photovasorelaxation *265*
References *268*

**14 Trends and Perspectives of Ruthenium Anticancer Compounds (Non-PDT)** *271*
*Michael A. Jakupec, Wolfgang Kandioller, Beatrix Schoenhacker-Alte, Robert Trondl, Walter Berger, and Bernhard K. Keppler*
14.1 Introduction *271*
14.2 Ruthenium(III) Compounds *272*
14.2.1 NAMI-A *273*
14.2.1.1 Biotransformation *273*
14.2.1.2 Antimetastatic Activity *274*
14.2.1.3 Mode of Action *274*
14.2.1.4 Clinical Studies and Perspectives *275*
14.2.2 KP1019/NKP-1339 *276*
14.2.2.1 Tumor Targeting Mediated by Plasma Proteins *276*
14.2.2.2 Activation by Reduction *277*
14.2.2.3 Mode of Action *278*
14.2.2.4 Clinical Studies and Perspectives *281*
14.3 Organoruthenium(II) Compounds *282*
14.3.1 Ruthenium(II)–Arene Compounds in Preclinical Development *282*
14.3.1.1 Organoruthenium Complexes Bearing Bioactive Ligand Scaffolds *284*

14.3.1.2 Cytotoxic Organoruthenium Complexes without Activation by Aquation *285*
References *286*

**15 Ruthenium Complexes as Antifungal Agents** *293*
*Claudio L. Donnici, Maria H. Araujo, and Maria A. R. Stoianoff*
15.1 Introduction *293*
15.2 Antifungal Activity Investigations of Ruthenium Complexes *304*
15.2.1 Ruthenium Complexes with Activity against Several Pathogenic Fungi Species: Dinuclear, Trinuclear, and Tetranuclear ruthenium Polydentate Polypyridil ligands, Heterotrimetallic di-Ruthenium-Mono-Palladium Complexes, Dinuclear bis-β-Diketones and Pentadithiocarbamate Ligands *304*
15.2.2 Aromatic and Heteroaromatic Ligands in Ru Monometallic Centers (Pyridine, Phenantroline, Terpyridine, Quinoline, and Phenazine) *305*
15.2.3 Schiff bases, Thiosemicarbazones, and Chalcones *307*
15.2.3.1 Schiff bases (Tetradentate Salen Like, Tridentate, and bidentate) *307*
15.2.3.2 Thiosemicarbazones *309*
15.2.3.3 Chalcone Derivatives *310*
15.2.4 Other ligands (Dithio-Naphtyl-Benzamide, Arylazo, Catecholamine, Organophosphorated, Hydridotris(pyrazolyl)borate and Bioactive Azole Ligands) *310*
15.3 Conclusion *312*
References *313*

**Index** *319*

## About the Editors

**Alvin A. Holder** is an associate professor at Old Dominion University in Norfolk, USA. He graduated from the University of the West Indies (UWI), Mona Campus, Jamaica, with a BSc (special chemistry) in 1989 and acquired his PhD in inorganic chemistry in 1994 with Prof. Tara P. Dasgupta. He was a faculty member at the University of the West Indies, Cave Hill Campus, Barbados, and an assistant professor in chemistry at the University of Southern Mississippi. His current research involves transition metal chemistry and he has published more than 650 articles and several textbooks and book chapters. In 2012, he was awarded an NSF Career Award.

**Lothar Lilge** is a Senior Scientist at the Princess Margaret Cancer Centre and holds a professorship at the University of Toronto, Canada. He obtained his Diploma in physics from the Johann Wolfgang Goethe University in Frankfurt, Germany, and his PhD in biophysics from the Westfaehlische Wilhelms University in Muenster, Germany. Additional training was provided through the Wellman Laboratories of Photomedicine at Massachusetts General Hospital, Boston, USA, and during a post-doc at McMaster University in Hamilton, Canada. His work is focused on photodynamic therapy including the use of ruthenium-based photosensitizers and optical spectroscopy for diagnostic and risk assessment among a range of other biophotonic application in medicine.

**Wesley R. Browne** is an associate professor at Stratingh Institute for Chemistry at the University of Groningen, The Netherlands, since 2013. He completed his PhD at Dublin City University, Ireland, with Prof. J. G. Vos in 2002, followed by a post-doc under the joint guidance of Prof. J. G. Vos and Prof. J. J. McGarvey, Queens University Belfast, UK. Between 2003 and 2007 he was a postdoctoral research fellow in the group of Prof. B. L. Feringa at the University of Groningen. He was appointed assistant professor in 2008. His current research interests include transition-metal-based oxidation catalysis, electrochromic materials, and responsive surfaces. He is an advisory board member for the European Journal of Inorganic Chemistry, Particle & Particle Characterization (both Wiley) and Chemical Communications (RSC). He has (co-)authored over 150 research papers, reviews, and book chapters.

**Mark A. W. Lawrence** was a post-doctoral fellow at Old Dominion University in Norfolk, USA, in the group of Prof. A. Holder. He received his BSc degree in 2006 and his PhD degree in inorganic-physical chemistry in 2011 from the University of the West Indies (UWI), Mona Campus, Jamaica, with Prof. Tara P. Dasgupta. His research interests include synthesis of hydrazones and functionalized pyridyl benzothiazoles, their transition metal complexes and application to catalysis and biological processes.

**Jimmie L. Bullock Jr** is a PhD student at the University of Kentucky in Lexington, USA, in the department of Chemistry. He received his BSc from Longwood University, Farmville, USA, and his MS in biological inorganic chemistry from Old Dominion University, Norfolk, USA, in 2013 and 2016, respectively. His research interests include studying activation of signaling pathways induced by non-platinum-based chemotherapeutic agents and synthesis of lanthanide sensor molecules.

# Preface

Ruthenium, a second-row transition metal, continues to attract much attention in scientific research, as it possesses a vast array of novel applications and properties. The enormous chemistry of ruthenium, much of which remains untapped, has been and continues to be investigated by numerous researchers. One such person was an icon, Prof. Karen J. Brewer. Karen, as she was affectionately called by many of her friends and research students, is being honored for her contribution to research on ruthenium with this textbook.

Ruthenium and its compounds are also paramount in catalysis and medicine, so it is not surprising that its biological activities and coordination chemistry remain very active areas of research. Ruthenium-containing complexes have long been known to be well suited for biological applications, and have long been studied as replacements to popular platinum-based drugs.

The textbook entitled "Ruthenium Complexes: Photochemical and Biomedical Applications" focuses on the uses and application of ruthenium-containing complexes in medicine and renewable energy. This title is unique as it discusses potential applications of ruthenium complexes in solving some of the world's foremost problems. While the biological application of ruthenium-containing complexes has been known for years, their application as photosensitizers in the emerging field of photodynamic therapy, also known as photochemotherapy, is of special interest. Photodynamic therapy can be utilized to treat a wide range of medical conditions including macular degeneration and malignant cancers. Ruthenium-containing photosensitizers have been shown to be especially active in the latter, with often minimal dark toxicity. Light-activated ruthenium-containing complexes are also gaining much attention as molecular catalysts in artificial photosynthesis for the production of hydrogen gas in aqueous media, after water oxidation.

Our goal at the outset was to capture the full vibrancy of the biological and coordination chemistry of this very important element called ruthenium and, in this way, to reflect the insight and enthusiasm of the honoree, Karen. To do so, we have divided this textbook into three sections with 15 chapters: (1) Introduction (Chapters 1–2), (2) Artificial Photosynthesis (Chapters 3 and 4), and (3) Applications in Medicine (Chapters 5–15). As such, we invited experts in each of these areas to complete this project by contributing a chapter. The chapters are as follows: Chapter 1: Karen J. Brewer (1961–2014): A Bright Star that Burned Out Far Too Soon; Chapter 2: Basic Coordination Chemistry of Ruthenium;

Chapter 3: Water Oxidation Catalysis with Ruthenium; Chapter 4: Ruthenium- and Cobalt-Containing Complexes and Hydrogenases for Hydrogen Production; Chapter 5: Ligand Photosubstitution Reactions with Ruthenium Compounds: Applications in Chemical Biology and Medicinal Chemistry; Chapter 6: Use of Ruthenium Complexes as Photosensitizers in Photodynamic Therapy; Chapter 7: Photodynamic Therapy in Medicine with Mixed-Metal/Supramolecular Complexes; Chapter 8: Ruthenium Anticancer Agents En Route to the Tumor: From Plasma Protein Binding Agents to Targeted Delivery; Chapter 9: Design Aspects of Ruthenium Complexes as DNA Probes and Therapeutic Agents; Chapter 10: Ruthenium-Based Anticancer Compounds: Insights into Their Cellular Targeting and Mechanism of Action; Chapter 11: Targeting cellular DNA with Luminescent Ruthenium(II) Polypyridyl Complexes; Chapter 12: Biological Activity of Ruthenium Complexes With Quinoline Antibacterial and Antimalarial Drugs; Chapter 13: Ruthenium Complexes as NO Donors: Perspectives and Photobiological Applications; Chapter 14: Trends and Perspectives of Ruthenium Anticancer Compounds (Non-PDT); and Chapter 15: Ruthenium Complexes as Antifungal Agents.

It has been our good fortune to work with so many exceptionally talented contributors from all over the world in compiling a textbook that we believe will be a valuable resource for graduate students, young investigators, and more senior scholars in the field of biological and coordination chemistry. We thank all the contributors for their hard work and their willingness to assist us whenever requested.

July 20, 2017

Alvin Holder
Co-Editor
Norfolk, Virginia, USA

## Acknowledgments

This is my acknowledgment which is based on the influence of Professor Karen Brewer on my life as she taught me how to carry out good and sensible chemistry with osmium(II), ruthenium(II), and rhodium(III) complexes. Photodynamic therapeutic studies with pUC18 and pBluescript DNA plasmids and Vero cells were the order of the day! This research catalyzed my career and research in the USA.

The task of working with so many gifted authors has been a real treat for me. The project also presented many challenges. We would not have made it to the finish line without the assistance of so many colleagues. We have not lost any Soldados on this journey. Thank God!!

In an Invited Plenary Talk: 251st ACS National Meeting and Exposition, March 13–17, 2016, San Diego, California. Abstract # INOR-1141. Title: "Light that pleases the world in science: The Karen Brewer's effect on my academic career." Author: *Alvin A. Holder*; I learnt about the seven (7) Ps from Professor Mark Richter and the Brew Crew, who attended the ACS conference. They are as follows:

The seven (7) Ps

Proper
Prior
Planning
Prevents
Piss
Poor
Performance

Credit for the seven (7) Ps must be given to my deceased former postdoctoral mentor, Professor Karen J. Brewer. She was a great Lady, who believed in "Family First"!! Please see http://www.chem.vt.edu/media/karen-brewer-obituary.pdf.

**R.I.P.**

I would like to thank the National Science Foundation (NSF) for a National Science Foundation CAREER Award as this material is based upon work supported by the National Science Foundation under CHE-1431172 (formerly CHE – 1151832). I would also like to thank Old Dominion University's Faculty Proposal Preparation Program (FP3), and also for the Old Dominion University start-up package that allowed for the successful completion of this work. Full gratitude to Professor Karen Brewer (R.I.P.), Professor Brenda Winkel, Professor Larry Taylor, Dr Myra Gordon, the research group (The Brew Crew), and all at Virginia Tech.

Personally, I would like to thank Dr. Anne Brennführer, Dr. Eva-Stina Müller, Ramprasad Jayakumar, Anne, Claudia Nussbeck, Dr. Eva-Stina Müller, and Samnaa Srinivas.

Alvin A. Holder
Co-Editor

# Section I

# Introduction

# 1

# Karen J. Brewer (1961–2014): A Bright Star that Burned Out Far Too Soon

*Seth C. Rasmussen*

*North Dakota State University, Department of Chemistry and Biochemistry, Fargo, ND 58108-6050, USA*

## 1.1 Introduction

Over the span of her career, first at Washington State University (WSU) and later at Virginia Tech, Karen J. Brewer (Figure 1.1) earned international acclaim as a prolific and pioneering researcher in the photochemistry and photophysics of multimetallic complexes [1, 2]. Ranging from synthesis of new multimetallic complexes to the study of their ground- and excited-state properties, her contributions aimed to elucidate the effect of the specific assembly of such complexes on their respective spectroscopic and electrochemical properties. In the process, Karen studied the application of complexes to molecular photovoltaics, solar $H_2$ production, artificial photosynthesis, electrocatalysis, Pt-based DNA binders, and photodynamic therapy [1–6]. Publishing her first paper in 1985, she accumulated over 125 peer-reviewed research publications in her career, which have in turn garnered over 3000 citations to date [1, 2], and her research pace was as active as ever at the time of her premature death in 2014 (Figure 1.2) [1].

**Figure 1.1** Karen J. Brewer (1961–2014) in the Spring of 2014. (Courtesy of Virginia Tech.)

*Ruthenium Complexes: Photochemical and Biomedical Applications*, First Edition.
Edited by Alvin A. Holder, Lothar Lilge, Wesley R. Browne, Mark A.W. Lawrence, and Jimmie L. Bullock Jr.

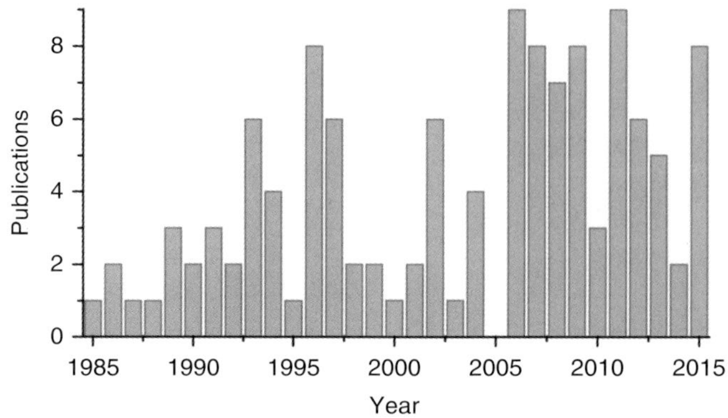


**Figure 1.2** Publications per year from 1985 to 2015.

Although known specifically for her various research contributions, Karen was also an award-recognized educator. She was comfortable teaching chemistry at all levels, from first-year students in general chemistry to graduate students in special topics classes such as electrochemistry and the photophysics of transition metal complexes. Her enthusiasm in the classroom was infectious and she inspired students to change not only their view of chemistry but, in some cases, their major to chemistry [1].

For many, including this author, Karen will be remembered most for her role as mentor and role model. She had tremendous impact on everyone who transitioned through her research laboratory, from undergraduates to postdocs. Throughout her career, Karen was a strong advocate for women and minorities in chemistry and was a role model and mentor for many female students and researchers [1, 2, 5]. Her passion in the promotion of chemistry as a career choice for women was most evident in her extensive outreach efforts to K-12 students. Throughout her career, she regularly visited primary and secondary school classrooms and hosted students in her laboratories at Virginia Tech [1, 2, 5]. In the process, Karen provided a real-life role model for young girls and others with aspirations to work in the physical sciences [1, 5].

Over the years, Karen received significant recognition for her collective efforts in research, teaching, and outreach. This included a College of Arts and Sciences Diversity Award in 1996, shortly after arriving at Virginia Tech [1, 5], as well as various teaching awards [3] and a Popular Mechanics Breakthrough Innovator Award in 2010, which she shared with collaborator Dr Brenda Winkel [2–5]. Most recently, Virginia Tech recognized her outreach efforts with the 2014 Alumni Award for Outreach Excellence [1–4], which she shared with Dr Shamindri Arachchige, Virginia Tech instructor of chemistry and a former postdoctoral researcher from her research group [1, 5, 7].

## 1.2 Early Years

Karen Sonja Jenks was born on June 27, 1961, in Wiesbaden, Germany to parents Gerda and Henry Jenks [3, 4]. As the daughter of a career military man,

**Figure 1.3** Karen in kindergarten at age 5. (Courtesy of Elise Naughton and the Brewer family.)

Karen moved frequently in her youth (Figure 1.3) [2–4], which provided her the opportunity to see much of the United States and the world as a young girl [3, 4]. The family ultimately settled in Lancaster, South Carolina in 1974, where Karen graduated with honors from Lancaster High School in 1979 (Figure 1.4) [3, 4].

Karen then attended Wofford College in Spartanburg, South Carolina [2–4, 6] on a Reserve Officers' Training Corps (ROTC) scholarship [8]. It was an interesting time to attend Wofford College, as it had formerly been an all-male school and had transitioned to a coeducational institution only 3 years before she began her studies there [8]. Karen soon decided that the military was not what she wanted to do with her career and enrolled in Wofford's K-12 education program, where she was involved in teacher training at the middle school level [8]. Her father had instilled a love of learning and teaching [8] and this probably influenced her decision. Ultimately, however, she developed an interest in chemistry and she finished her undergraduate studies in the chemistry program. While at Wofford she also participated in women's basketball and became a member of both Alpha Phi Omega and the American Chemical Society during her senior year [2, 6]. The Wofford chemistry faculty thought highly of Karen as a student [9] and she received her BS degree in chemistry in 1983 [2–4, 6]. After the completion of her undergraduate studies, she married Ralph Gary Brewer (who went by Gary), with

**Figure 1.4** Karen in her senior year of high school at age 17. (Courtesy of Elise Naughton and the Brewer family.)

their wedding held on the same day as their Wofford graduation ceremonies on Sunday, May 15, 1983. Following her marriage, Karen was known both personally and professionally as Karen Jenks Brewer.

## 1.3 Graduate Studies and Clemson University

Karen then entered the Chemistry graduate program at Clemson University in the fall of 1983, where she began working under the supervision of Dr John D. Petersen (b. 1947, PhD University of California, Santa Barbara 1975) (Figures 1.5 and 1.6). Notable coworkers during her time in the Petersen group included Ronald Ruminski (Professor, University of Colorado Colorado Springs; PhD University of New Mexico 1980; Petersen postdoc 1981–1984) [10], Wyatt Rorer Murphy, Jr (Professor, Seton Hall University; PhD University of North Carolina at Chapel Hill 1984; Petersen postdoc 1984–1986) [11], and D. Brent MacQueen (PhD Clemson 1989) [12].

Karen began her research in the Petersen laboratory by joining ongoing efforts to develop bimetallic complexes capable of converting radiation to usable chemical potential energy. The basic design of such species included

**Figure 1.5** John D. Petersen (b. 1947). (Courtesy of John Petersen.)

three components (Figure 1.5): (i) a strongly absorbing, but photochemically unreactive, metal center (antenna complex); (ii) a second metal center capable of undergoing a useful chemical reaction from a nonspectroscopic excited state (reactive complex); and (iii) a bridging ligand (BL) that both couples the two metal fragments and facilitates intramolecular energy transfer between the two metal centers [12, 13]. While others had previously studied electron transfer in bimetallic complexes utilizing primarily monodentate BLs (Figure 1.7), the Petersen group focused on the application of bidentate BLs (Figure 1.8) as a method to increase stability of the bimetallic species during excitation. Karen's first publication was as fourth author on a 1985 paper published in *Coordination Chemistry Reviews* that presented this design and discussed the optimization of the three basic components [13].

Karen's research initially focused on evaluating the effect of bidentate BLs such as dpp on the photophysics of Ru(II)-based antenna complexes. This resulted in the publication of her initial first-author paper in 1986, which reported the synthesis of $[Ru(dpp)_3]^{2+}$ along with its photophysical and electrochemical properties [14]. The conclusion of this work was that in comparison to $[Ru(bpy)_3]^{2+}$, the dpp analogue exhibited similar electronic absorption and emission spectra, as well as a similar luminescence quantum yield. As such, the application of dpp should allow the tethering of Ru(II)-based antenna complexes of reactive metal centers without the loss of the desired photophysical properties [12, 14].

The ultimate focus of the majority of her graduate work was the potential application of bimetallic complexes to the photochemical elimination of molecular hydrogen. These efforts began with an intermolecular sensitization study using $Fe(bpy)_2(CN)_2$ as the donor and the dihydride species $[Co(bpy)(PEt_2Ph)_2H_2]ClO_4$ in order to evaluate the relative energy levels of the visible light accessible excited state of the Fe(II) antenna and the reactive

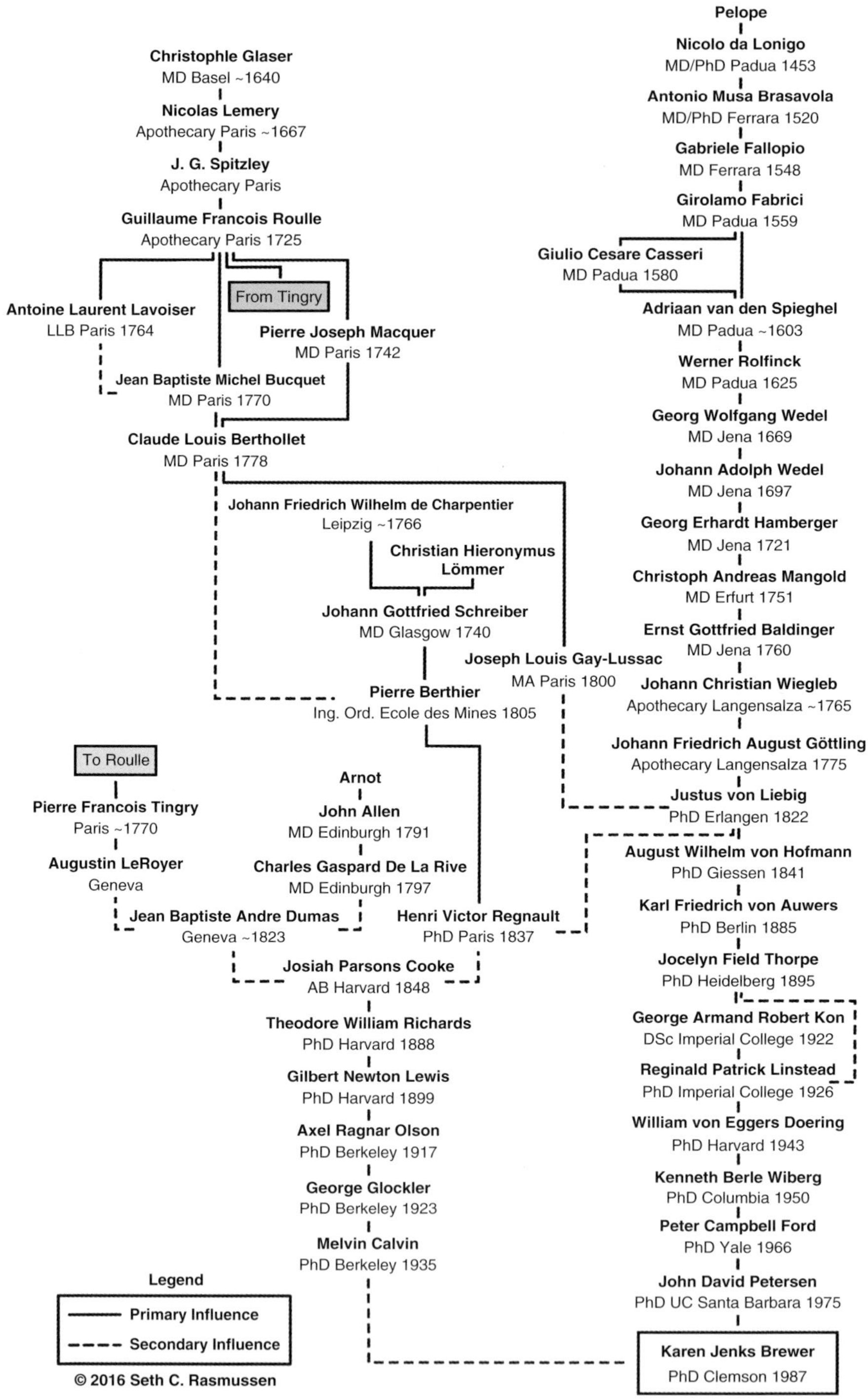


**Figure 1.6** Karen Brewer's academic genealogy.

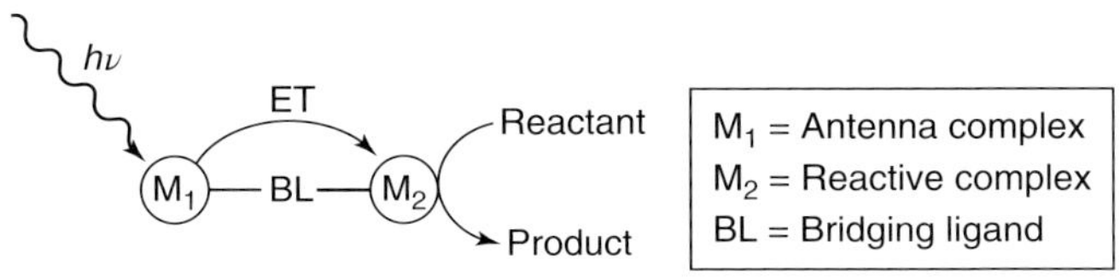


**Figure 1.7** Basic design of bimetallic complexes capable of efficient photochemical processes.

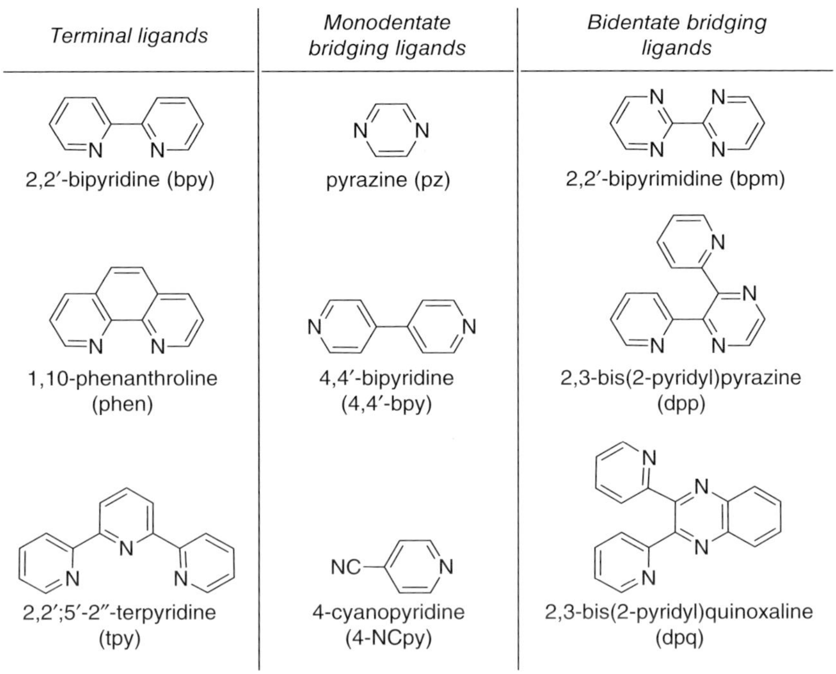


**Figure 1.8** Polypyridyl ligands referenced throughout the chapter.

state of the cobalt dihydride model reactive fragment [12, 15]. It was found that when either $Fe(bpy)_2(CN)_2$ or $[Co(bpy)(PEt_2Ph)_2H_2]^+$ was irradiated alone at 577 nm, no reaction occurred. Irradiation of a mixture of $Fe(bpy)_2(CN)_2$ and $[Co(bpy)(PEt_2Ph)_2H_2]^+$ at 577 nm, however, would result in the production of $H_2$. The hydrogen generated was due to the initial excitation of the Fe(II) complex, followed by energy transfer to the Co(III) dihydride and resulting in loss of molecular hydrogen. A limiting quantum yield of 0.13 mol/einstein was obtained for hydrogen production, which is within the experimental uncertainty of that obtained for the direct irradiation of the Co(III) complex at 436 nm [12, 15]. These results demonstrated that visible light most certainly could be used to drive the photoelimination reaction of a transition metal dihydride.

A particular challenge in the choice of hydrogen production for the chemical reaction of the reactive complex was the necessity of working with molecular hydrogen. Here, the Co(III) dihydride was generated via reaction of $[Co(bpy)(PEt_2Ph)_2]^+$ with $H_2$ and, of course, the photoelimination reaction resulted in the production of $H_2$. The highly flammable nature of hydrogen was an extra challenge at Clemson in the 1980s as smoking was still allowed in University buildings at that time, including in the chemistry research laboratories. Karen recounted on multiple occasions stories of people who would walk into the laboratory with lit cigarettes while she was working with hydrogen, even though she had posted clear warning signs of the danger on the door. It probably speaks to her care while working with sensitive materials that such events never resulted in fire or worse.

Karen then continued to study the effect of bidentate BLs on antenna complexes through the investigation of a series of monometallic and bimetallic Fe(II) cyano complexes containing either dpp or bpm BLs (Figure 1.5) [12, 16]. The bimetallic systems revealed that either bpm or dpp could successfully bridge two first-row transition-metal centers and the electrochemical data indicated good communication between the metal centers, supporting the model of possible energy transfer across the BL. The results also indicated that dpp may yield more promising results than bpm in the preparation of mixed-metal bimetallic complexes as it significantly lowers the steric crowding without reducing the communication between metal centers.

This investigation of BL effects on antenna complexes was then continued with bimetallic and tetrametallic Ru(II) complexes via dpp [12, 17]. The various tetrametallic complexes utilized the $[Ru(dpp)_3]^{2+}$ core reported in her previous 1986 paper [14], which were then capped with three $Ru(bpy)_2{}^{2+}$, $Ru(phen)_2{}^{2+}$, or $Ru(tpy)Cl^+$ fragments. The results of the bimetallic systems were consistent with that previously observed for the analogous iron complexes. For the tetrametallic complexes, however, it was found that they were unstable when more than two electrons/mol are removed from the complexes. Still, the complexes allowed an overall predictability of the chromophore-based excited-state energies, as well as the ability to modulate the ground- and excited-state redox potentials by variation of metal center, BL, and terminal ligand. As a result, it was concluded that this structural motif provided significant promise for the further design and synthesis of heteropolymetallic complexes with specifically controlled ground- and excited-state properties [17].

Lastly, the ability to apply the previously designed multimetallic frameworks to heterometallic complexes was successfully demonstrated by the production of mixed-metal Ru(II)—Fe(II) bimetallic and tetrametallic complexes [12]. In these cases, the initial monometallic Ru(II) species $[(bpy)_2Ru(dpp)]^{2+}$ or $[Ru(dpp)_3]^{2+}$ were capped with $Fe(CN)_4{}^{2-}$ fragments. The synthesis of a bpm-bridged mixed Fe(II)—Co(II) bimetallic complex was also investigated.

By 1986, Karen had amassed a sizable amount of research and could reasonably have begun the preparation and defense of a relatively strong dissertation. However, that same year Petersen was awarded an Alexander von Humboldt research fellowship to spend a year at the Universitat Regensburg as a guest professor, were he began working in the laboratory of Arnt Vogler in December of 1986 [18]. At this point, Karen was the senior student in the Petersen group and he did not have a postdoc at the time, as Rory Murphy had left in August of 1986 to accept a faculty position at Seton Hall University [11]. As such, he really needed to convince Karen to stay and supervise the group at Clemson until his return. As an incentive to wait to defend until after he returned from Germany, Petersen agreed to allow Karen to pursue some of her own research ideas while he was away.

Karen took advantage of this and began to develop ideas that she was interested in pursuing as part of her future career. These efforts resulted in the production and study of the Ir(III) complex $[Ir(dpp)_2Cl_2]^+$ [12]. The complex exhibited absorption in the visible regime, room temperature luminescence, as well as simple, stepwise electrochemical behavior. Overall, however, her interest in this species was the ultimate possibility of replacing the chlorides with hydride to produce a new reactive center for hydrogen production which could then be incorporated into multimetallic frameworks. Karen's development and synthesis of this complex was a critical moment in her career and this compound became the seed from which she grew the majority of her later independent research.

Petersen then returned to Clemson in August of 1987 [18] and Karen successfully defended her PhD dissertation entitled "The development of novel Fe, Co, Ir, and Ru complexes for the capture and utilization of solar energy" that same month before a graduate committee comprised of Petersen, Dr Kilian Dill, Dr James C. Fanning, and Dr Darryl D. DesMarteau [12]. Petersen commonly viewed Karen as the most successful PhD student of his career, stating in 2009 [6]:

"Karen was the most driven and most successful of the Ph.D. students I trained in chemistry."

Particularly after pursuing her own research ideas during Petersen's absence in Germany, Karen was eager to apply for faculty positions and begin her own independent career. However, an opportunity to work as a postdoc under Nobel Laureate Melvin Calvin (1911–1997) presented itself just as Karen finished and that was not an experience that she could pass up.

## 1.4 Postdoctoral Research and the University of California, Berkeley

After receiving her PhD from Clemson in 1987 [2–4, 6], Karen moved across the country to begin a postdoctoral fellowship with Melvin Calvin (PhD University of Minnesota 1935) (Figure 1.9) at the University of California, Berkeley [2–4]. Melvin Calvin was awarded the 1961 Nobel Prize in Chemistry for his work on the mechanistic study of photosynthesis, with an emphasis on the path of carbon during photosynthetic $CO_2$ reduction, and his research had a strong focus on the electronic, photoelectric, and photochemical properties of metal chelate complexes [19]. As such, the focus and efforts of the Calvin group was synergetic with Karen's previous work under Petersen. Although she stayed at Berkeley for only 1 year, she accumulated three publications from her contributions to the group [20–22].

Her efforts in the Calvin group focused on metal chelates of macrocyclic tetradentate ligands, particularly 1,4,8,11-tetraazacyclotetradecane (cyclam)

**Figure 1.9** Melvin Calvin. (Photo courtesy of Lawrence Berkeley National Laboratory.)

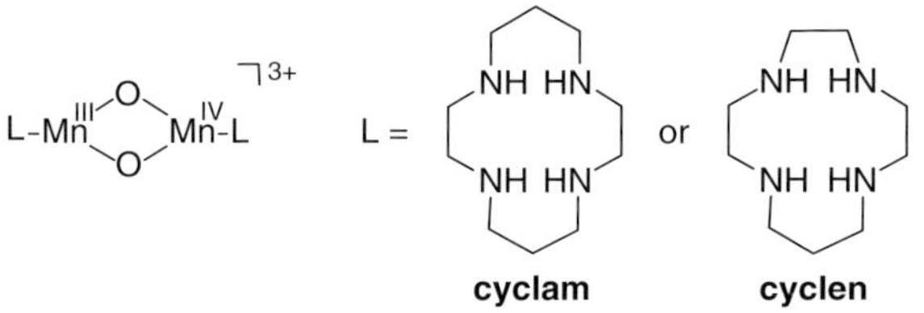


**Figure 1.10** Mixed-valence manganese μ-oxo dimers of macrocyclic tetradentate ligands.

and 1,4,7,10-tetraazacyclododecane (cyclen) [20–22]. The majority of this work focused on mixed-valence manganese $\mu$-oxo dimers (Figure 1.10) as potential homogeneous, multiple-electron redox catalysts [20, 21]. Of particular interest was the possibility of using such complexes for the catalytic oxidation of water to molecular oxygen. The electrochemistry of these complexes revealed them to be remarkably stable to both oxidation and reduction [20, 21] and that the redox potentials could be tuned by the choice of macrocyclic ligands [20]. In addition, it was found that cyclic voltammetry in the presence of water exhibited an additional process not observed in dry $CH_3CN$, which was ultimately determined to be consistent with the oxidation of water at the electrode following the electrochemical oxidation of the Mn(III)—Mn(IV) dimer to the Mn(IV)—Mn(IV) species [20]. It was later found that these species could be generated electrochemically via the oxidation of $Mn(SO_3CF_3)_2$ in a water/acetonitrile solution in the presence of the ligand, which revealed that the oxygens of the $\mu$-oxo bridge came from water [21].

## 1.5 Washington State University: Beginning an Independent Career

Karen then moved to Pullman, Washington to join the Department of Chemistry at WSU as an assistant professor in the fall of 1988 [2, 6]. There she joined Professors John Hunt and Scot Wherland as part of the Inorganic Chemistry Division. At the time she joined the department, Karen became the second woman on the chemistry faculty [1, 5]. Karen had been assigned research space in one of the two new additions to Fulmer Hall, which were still in the final stages of construction at the time of her arrival [23]. As a result, she was assigned temporary laboratory space on the first floor of Fulmer Hall and it was there that she started to build her research group. The first student to join her research group was a third-year chemistry undergraduate, Seth Rasmussen, who had responded to a work-study position Karen had posted for a research assistant early in the fall of 1988. A second student, Eugene Yi, joined the group shortly thereafter.

During the spring of 1989, the laboratory transitioned to its permanent space in the ground floor of the new Synthesis Unit of Fulmer Hall and it was here that the Brewer Group really began to take shape and grow. Karen's personality and enthusiasm for research continued to draw undergraduates to the group and Karen recruited her first PhD student, Mark M. Richter (B.A. Gustavus Adolphus College 1989) in the summer of 1989. Thus, by the spring of 1990, the group had already grown to an appreciable size (Figure 1.11).

Karen's research at WSU began as an extension of the $[Ir(dpp)_2Cl_2]^+$ complex that she prepared during her final year at Clemson, with the synthesis

**Figure 1.11** Brewer Group, Spring 1990 (L to R): Eric Kimble (undergrad), Bob Williamson (undergrad), Seth Rasmussen (undergrad), Keith Blomgren (undergrad), Karen Brewer (Asst. Prof.), Sumner Jones (undergrad), Mark Richter (grad student), and Jon Bridgewater (undergrad). (Courtesy of Hannah Rodgers and the Brewer Group.)

and characterization of a series of Ir(III) and Rh(III) complexes of the type $[Ir(BL)_2Cl_2]^+$ and $[Rh(BL)_2Br_2]^+$ [24]. Here, the polypyridyl ligands consisted of one of four bidentate BLs, including the previously discussed bpm, dpp, and dpq (Figure 1.8), as well as the benzo-annulated analogue of dpq, 2,3-bis(2-pyridyl)benzoquinoxaline (dpb). Although her initial interest in such complexes back in 1986–87 was as potential reactive centers for hydrogen production, her focus had shifted to the application of these complexes as potential electrocatalysts for $CO_2$ reduction [24]. This new interest in both electrocatalysts and $CO_2$ reduction was most certainly a result of her time in the Calvin laboratory.

All of the Rh(III) and Ir(III) complexes successfully catalyzed the reduction of $CO_2$ to formate ($HCO_2^-$):

$$CO_2 + 2e^- + H^+ \rightarrow HCO_2^- \tag{1.1}$$

Current efficiencies for these catalytic reductions ranged from ca. 19 to 70%, with the Rh(III) complexes providing higher efficiencies than the analogous Ir(III) species [24]. These results confirmed that the application of the BLs did not significantly alter the desired catalytic activity of the metal previously observed in $[Rh(bpy)_2(OTf)_2]^+$ (where OTf = trifluoromethylsulfonate). The application of the BLs thus provided the promise of developing multimetal systems that could ultimately allow the potential photochemical reduction of $CO_2$.

In addition to this initial work on group 9 complexes, efforts began in the study of osmium compounds of various BLs as new Os(II)-based antenna complexes for the eventual production of multimetallic complexes. These efforts began with the synthesis and study of a series of complexes of the form $[Os(bpy)_2BL](PF_6)_2$, where BL consisted of dpp, dpq, or dpb [25]. This was then followed by the preparation and study of the $Ru(bpy)_2^{2+}$ and $Os(bpy)_2^{2+}$ capped bimetallic complexes $[(bpy)_2Os(BL)Ru(bpy)_2](PF_6)_4$ [26] and $[(bpy)_2Os(BL)Os(bpy)_2](PF_6)_4$ [27]. In general, the Os-based metal-to-ligand charge-transfer (MLCT) transitions occur at lower energy than the analogous Ru-based transitions and intervalence charge-transfer studies indicated that the metal–metal coupling is enhanced in the Os/Os systems relative to the Os/Ru mixed-metal analogues [26, 27]. Synthesis of the bis-BL species $Os(BL)_2Cl_2$ then allowed the production of the trimetallic complexes $[\{(bpy)_2Ru(BL)\}_2OsCl_2](PF_6)_4$ [28].

Work also continued on the initial Rh and Ir complexes, with efforts focused on the addition of $Ru(bpy)_2^{2+}$ moieties to the two BLs of the central complexes. Purification of the resulting trimetallic complexes was challenging, such that although these efforts dated back to early 1990, the first report of these complexes was not until 1993. This report detailed the synthesis and characterization of a series of complexes of the type $[\{(bpy)_2Ru(BL)\}_2IrCl_2]^{5+}$, where BL included dpp, dpq, and dpb [29].

At the same time, the group began the investigation of multimetallic complexes of the terdentate BL 2,3,5,6-tetrakis(2-pyridyl)pyrazine (tpp, Figure 1.12). This began with the synthesis of $Ir(tpp)Cl_3$ and $[Ru(tpy)(tpp)](PF_6)_2$, before the ultimate synthesis of the mixed-metal bimetallic species $[(tpy)Ru(tpp)IrCl_3](PF_6)_2$ [30]. It is perhaps not surprising that this study provided evidence that the

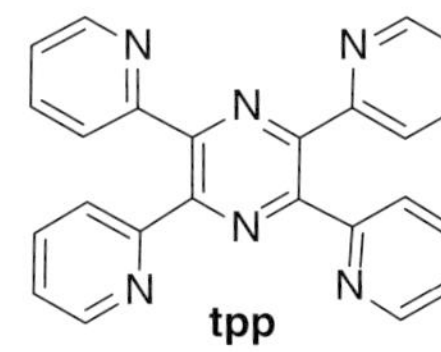


**Figure 1.12** The terdentate bridging ligand 2,3,5,6-tetrakis(2-pyridyl)pyrazine.

increased tridenate chelate of the tpp ligand provides excellent communication between the MLCT states centered on different metals.

While building her research group at WSU, this time in Pullman also brought changes to her nonacademic life. In the fall of 1989, Karen and Gary were looking forward to a trip to Hawaii that December to attend the International Chemical Congress of Pacific Basin Societies (Pacifichem 1989), with plans of exotic drinks and time on the beach. The details of these plans changed abruptly just a few weeks before the trip, however, when Karen discovered that she was pregnant, thus removing exotic drinks (at least those of the alcoholic kind) from the table. Their first child, Nicole Brewer, was born during the summer of 1990.

Karen's time in Pullman was somewhat short-lived, however, as she accepted an offer to move to Virginia Polytechnic Institute and State University (Virginia Tech) after only 4 years at WSU. Her choice to move her career was the result of several factors, including moving back closer to both her and Gary's families. In addition, she felt somewhat limited by what she perceived as a male-dominated culture within the department at WSU and thought things might be better at Virginia Tech. Even so, she left on good terms with her WSU colleagues, particularly Roger Willett with whom she continued to collaborate for years to come [31–34].

## 1.6 Move to Virginia Tech

Karen thus moved to Blacksburg, Virginia to become an assistant professor in the Department of Chemistry at Virginia Tech in the fall of 1992 [1–5]. Three of her graduate students from WSU, Sharon Molnar, Lisa Vogler, and Glen Jensen, moved to Virginia with her to become the core of her new research group there [35]. Mark Richter, who was close to finishing at the time, remained at WSU to finish writing his dissertation and was the only of Karen's graduate students to graduate from WSU. At the time that she joined Virginia Tech, she became one of two women on the faculty in the department [8] and was only the third woman ever hired in chemistry [1, 5].

At Virginia Tech, Karen's group continued to work on the synthesis, characterization, and photophysics/photochemistry of various multimetallic complexes. This included further work on the Rh- and Ir-based trimetallic complexes initially developed at WSU [36–38], as well as the eventual application of the Ir trimetallics as electrocatalysts for $CO_2$ reduction [39]. In comparison to the monometallic analogues, the trimetallics were found to be much better electrocatalysts. Not only were they found to be extremely selective toward the electrocatalytic reduction of $CO_2$ to CO but they also gave nearly quantitative current efficiencies and appeared quite robust under the electrocatalytic conditions [39].

In only 3 short years, Karen was promoted to associate professor in 1995 and she became the second woman granted tenure in the Department of Chemistry [1]. It was also about this same time that Karen and Gary welcomed another addition to their family with the birth of their second daughter, Kaitlyn Brewer. In the end, however, Karen's marriage did not last, eventually resulting in separation and divorce. Following the divorce, Karen took on the primary responsibility to continue raising Nicole and Kaitlyn. It has been said that Karen viewed her greatest achievement to be raising her two daughters [1].

Karen continued to be successful at Virginia Tech, not only in her research but also as an exceptional teacher and as a champion of science via numerous outreach activities. She was finally promoted to the rank of full professor in 2005 [1–4].

## 1.7 Collaboration with Brenda Winkel and the Study of Metal-DNA Interactions

Another landmark moment in her career occurred shortly after she arrived at Virginia Tech in August of 1992, when she met Brenda Winkel Shirley (now Brenda Winkel, Figure 1.13) at new faculty orientation [8, 40–42]. Like Karen, Brenda was a new faculty member at Virginia Tech and had just joined

**Figure 1.13** Karen J. Brewer and collaborator Brenda Winkel in 2005. (Courtesy of Virginia Tech.)

the Department of Biological Sciences and upon meeting her, Karen innocently asked her if she knew anything about DNA [42]. With initial training in chemistry and biochemistry, followed by a PhD in Genetics, Brenda was perhaps the perfect possible collaborator for the potential application of the inorganic species being developed in the Brewer laboratory to biological systems. The study of the interaction of multimetallic species with DNA then eventually became a major focus in the Brewer Group [8].

These efforts began by looking at ways to overcome some of the limitations of the widely used cancer drug cisplatin [*cis*-$Pt(NH_3)_2Cl_2$] [40, 41]. As such, early efforts focused on the potential DNA binding of Pt bimetallic complexes of the form $[(bpy)_2M(dpb)PtCl_2]Cl_2$ (where M=Ru(II) or Os(II)) [43–45]. In fact, both complexes were shown to successfully bind to DNA and exhibited binding thought to be primarily intrastrand cross-linking in nature, similar to cisplatin [44, 45]. In contrast to cisplatin, however, the bimetallic $Ru^-$ and Os-based complexes exhibited a higher percentage of interstrand cross-links. These studies ruled out a purely intercalative mode of interaction and strongly suggested that these complexes form covalent bonds to DNA through the platinum metal site [45]. These bimetallic species then formed the basis of a new class of complexes that interact with DNA in a manner that is similar to but still observably different from the known anticancer agent cisplatin. In addition, the higher water solubility of the bimetallic complexes made them attractive as potential chemotherapeutic agents [41, 45].

Study of the Pt-based bimetallics was then expanded to include a greater diversity of BLs beyond the initially studied dpb [46]. In addition, the previous $Ru(bpy)_2^{2+}$ moiety was replaced with a Ru center capped with the terdentate ligand tpy, that is, $Ru(tpy)Cl^+$. The chloride could then be substituted with monodentate phosphine ligands for greater structural and chemical diversity [47, 48]. As with the previous complexes, all of these species avidly bind DNA and occupy binding sties consistent with the covalent attachment of Pt [46, 48].

In addition to the successful Pt-based complexes, it was shown that previously studied Rh-based trimetallic complexes could successfully be used to photocleave DNA [49, 50]. In the absence of molecular oxygen, the DNA plasmid irradiated ($\lambda \geq 475$ nm) in the presence of $[\{(bpy)_2Ru(dpp)\}_2RhCl_2]^{5+}$ exhibited conversion of the supercoiled DNA to the nicked form. While the mode of the DNA photocleavage was unclear, the cleavage observed was consistent with reactivity arising from the photogenerated Rh(II) site. This work was then expanded by tuning of the antenna species, including the use of $Os(bpy)_2^{2+}$ and $Ru(tpy)Cl^+$ fragments [50]. The photodynamic action of these complexes was reported to inhibit cell replication after exposure to light while displaying no impact on cell replication in the dark [51, 52].

These two approaches were then combined with the production of the tetrametallic complex, $[\{(bpy)_2Ru(dpp)\}_2Ru(dpp)PtCl_2]^{6+}$ [53]. This complex utilized the central $[Ru(dpp)_3]^{2+}$ developed by Karen during her graduate work with Petersen [12, 14], coupled to two $Ru(bpy)_2^{2+}$ antenna complexes and the Pt center for DNA binding. As a result, this complex was shown to display both Pt-based DNA binding and Ru-based DNA photocleavage.

These efforts continued with the preparation and study of additional complexes, with an overall emphasis on photodynamic therapy. In particular, efforts focused on the production of materials that absorb efficiently within the therapeutic window of red light [54–56]. Ultimately, these efforts included an additional collaboration with John Robertson in biomedical engineering [8], as well as efforts to take their research into a clinical testing phase [40]. This also included a partnership with Theralase Technologies Inc., a developer of light sources capable of light treatment on deep tissue [40, 41]. It was reported in 2010 that when irradiated by lasers developed by Theralase Technologies, the complexes produced by the Brewer Group had been proved effective in destroying breast cancer cells in preclinical trials [57].

## 1.8 A Return to Where It All Started: Photochemical $H_2$ Production

As previously discussed, the photochemical process that was the focus of Karen's graduate work was that of $H_2$ production [12]. As such, it seems only natural that at some point Karen would return to this area of photochemical research. Thus, efforts began to apply the previously reported Rh-based trimetallic complex $[\{(bpy)_2Ru(dpp)\}_2RhCl_2]^{5+}$ to the photocatalytic production of $H_2$ from $H_2O$, which were first reported in 2007 [58]. In this initial communication, it was reported that the complex produced hydrogen catalytically when excited at 470 nm in the presence of dimethylaniline as a sacrificial electron donor, with a quantum yield of ca. 0.01. Following photolysis of this mixture for 4 h, 8.2 μmol of $H_2$ was produced representing 30 turnovers of the catalytic system with no reduction in the rate of $H_2$ production [58]. It was then shown in a follow-up report that the $RhBr_2$ analogue exhibited enhanced catalytic performance to generate 10.9 μmol of $H_2$ (38 turnovers) under the same conditions [59].

These initial efforts were then followed by a more expansive study in which the antenna complexes were varied via the capping of the $[Rh(dpp)_2X_2]^+$ core with either $Ru(bpy)_2{}^{2+}$, $Os(bpy)_2{}^{2+}$, $Ru(phen)_2{}^{2+}$, $Ru(tpy)Cl^+$, or $Os(tpy)Cl^+$ [60]. For comparison, the Ir-based complex $[\{(bpy)_2Ru(dpb)\}_2IrCl_2]^{5+}$ was also evaluated and all but the Ir complex and the $Os(tpy)Cl^+$-capped Rh complex demonstrated photocatalytic activity. Although the previously studied $[\{(bpy)_2Ru(dpp)\}_2RhBr_2]^{5+}$ continued to provide the best catalytic performance [59, 60], the current study did illustrate that the use of phen terminal ligands in place of bpy had only minor effects on the photocatalytic activity.

In efforts to further tune the activity of the best photocatalyst, the potential effects of complex counterion ($PF_6{}^-$ vs $Br^-$) was investigated, as well as the effect of pH and buffer acid [61]. While the use of $Br^-$ counterions does provide water solubility, this also resulted in a reduction in catalyst activity. In terms of pH, it was found that values near the $pK_a$ of the amine used for the electron donor gave a notable increase in $H_2$ production [61].

A shift to a Pt-based catalyst was then reported in 2011 with the tetrametallic species $[\{(phen)_2Ru(dpp)\}_2Ru(dpq)PtCl_2](PF_6)_6$ [62]. In contrast to the previous Rh-based complexes, this showed enhanced activity with the production of 21.1 μmol of $H_2$ and 115 turnovers of the catalytic system when irradiated at 470 nm for 5 h.

About the same time, however, significant advances had been made in the applications of the previous champion photocatalyst, $[\{(bpy)_2Ru(dpp)\}_2RhBr_2]^{5+}$ [63]. Through careful tuning of reaction conditions, it was found that $H_2$ production was enhanced by the choice of solvent (DMF), increased electron-donor concentration, increased headspace, and purging of the system of $H_2$. Using such optimized conditions, 420 turnovers of the catalyst were achieved over 50 h, with a $\Phi = 0.023$. As a result, over 20 mL (810 μmol) of $H_2$ could be photocatalytically produced over a period of 19.5 h. This was then further improved by replacing the terminal bpy ligands with 4,7-diphenyl-1,10-phenanthroline ($Ph_2$phen) to increase the maximum $\Phi$ to 0.073. As a result, this allowed the production of 44 ml (1640 μmol) of $H_2$ in 20 h, corresponding to a catalyst turnover of 610 [64].

## 1.9 A Career Cut Tragically Short

For all of her professional career, Karen suffered from various health issues. By her teens, she had already been diagnosed with Crohn's disease [8], a relapsing systemic inflammatory disease that mainly affects the gastrointestinal tract, but can also include other health effects such as immune disorders [65]. Even during her graduate studies, her health was always on the edge, but she never complained or used it as an excuse [18]. Her constant battle with Crohn's resulted in various surgeries and a significant number of hospital stays throughout her life [8], and such health issues too often caused her to unexpectedly change plans or cancel speaking engagements. At one point, her struggles with her health even led to her needing a reduced teaching load while she worked to get her health under control [35]. The fact that she was able to accomplish so much during her career while dealing with such a handicap makes her successes just that much more impressive.

Unrelated to her Crohn's, Karen also had issues with her knees such that she finally had to give up basketball while she was attending Wofford [35]. As she got older, her knees continued to give her issues until she finally underwent knee-replacement surgery in June of 2012 [66]. Tragically, she did not recover from the surgery as expected, and it was eventually discovered that she was suffering from a serious infection related to the surgery. Treatment with intravenous antibiotics was not able to cure the infection, which ultimately required the removal of the knee replacement in order to treat the infection at the source in July of 2013 [35, 66]. As a result, she had to spend over a month without a physical knee while the infection was dealt with. It has been proposed that her various surgeries and hospital stays throughout her life led to such an antibiotic-resistant infection. Once the infection was dealt with, she underwent surgery once again in August of 2013 to replace the knee. In August of the following year, however, it was found that the infection had returned, thus requiring another knee surgery [66]. It was sometime in the months following this final surgery that it is believed that she developed a blood clot, resulting in a pulmonary embolism. As a result, Karen died unexpectedly in Blacksburg, Virginia, on October 24, 2014, at the age of 53 [2–5]. Funeral services were held in Blacksburg on October 31, 2014, and she was laid to rest at Memorial Gardens of the New River Valley [3, 5].

## 1.10 Karen's Legacy

While there is no doubt that Karen left an extensive record of discoveries and top-notch research, which will be forever remembered and cited by anyone working in the field of multimetallic coordination chemistry, her true legacy is all of those that she trained over the years. As anyone who ever spent time in her group can confirm, the Brewer Group was family and every student or postdoc was far more than just the means to yet another publication. In the process, she did her best to instill in them everything that made her the scientist and teacher that she was. The success of this can be seen in the large number of Brewer Group members who remained in academia and are now faculty members, instructors, and research advisors themselves. This includes Seth Rasmussen (North Dakota State University), Mark Richter (Missouri State University), Sharon Molnar (West Virginia State University), Shawn Swavey (University of Dayton), Alvin Holder (Old Dominion University), Matthew Mongelli (Kean University), Matthew Milkevitch (Philadelphia University), Shamindri Arachchige (Virginia Tech), Michael Jordan (Oklahoma Baptist University), Avijita Jain (Indiana University of Pennsylvania), Jessica and Travis White (Ohio University), and I am sure even more that I have missed. All of these former members of the Brewer Group will then pass on aspects of Karen to their own students, thus guaranteeing that she will continue to live on in spirit and never be forgotten.

## Acknowledgments

I thank everyone who shared stories of Karen with me during the preparation of this biography. In particular, I thank John Petersen, Sharon Molnar, Brenda Winkel, Theodore Canterbury, Joseph Merola, David Whisnant, Elise Naughton, and Hannah Rodgers for resources, specific information, and/or photographs. Lastly, I thank the Department of Chemistry and Biochemistry at North Dakota State University, which continues to support my historical pursuits.

## References

1 Good, L.S., Ed. *Elements, The Alumni Magazine of the Department of Chemistry at Virginia Tech.* 2014 Fall Issue, p 3.
2 Ainsworth, S.J. (2014) *Chem. Eng. News*, **92**, 34.
3 Obituaries - Brewer, Karen Jenks. *The Roanoke Times*, October 30, 2014.
4 Dr. Karen (Jenks) Brewer (1961–2014). *The Lancaster News*, November 7, 2014.

5 Bushey, R. *Virginia Tech News*, http://www.vtnews.vt.edu/articles/2014/10/103014-science-brewermemoriam.html (accessed February 15, 2016).
6 Brewer, K.J. LinkedIn profile; https://www.linkedin.com/in/karenjbrewer (accessed October 2015).
7 Owczarski, M. and Bushey, R. *Virginia Tech News*, http://www.vtnews.vt.edu/articles/2014/04/042114-facstaffaward-arachchigebrewer.html (accessed February 15, 2016).
8 Lovegrove, R. (2014) *Virginia Tech College of Science Annual Magazine*, **1**, 12–13.
9 Whisnant, D.M. (2016) *Personal communication.*
10 Ruminski, R. (2016) *Personal communication.*
11 Murphy, W.R. (2016) *Personal communication.*
12 Brewer, K. S. J. PhD Dissertation, Clemson University, 1987.
13 Petersen, J.D., Murphy, W.R. Jr., Sahai, R., Brewer, K.J., and Ruminski, R.R. (1985) *Coord. Chem. Rev.*, **64**, 261–272.
14 Brewer, K.J., Murphy, W.R. Jr., Spurlin, S.R., and Petersen, J.D. (1986) *Inorg. Chem.*, **25**, 882–884.
15 Brewer, K.J., Murphy, W.R. Jr., Moore, K.J., Eberle, E.C., and Petersen, J.D. (1986) *Inorg. Chem.*, **25**, 2470–2472.
16 Brewer, K.J., Murphy, W.R. Jr., and Petersen, J.D. (1987) *Inorg. Chem.*, **26**, 3376–3379.
17 Murphy, W.R. Jr., Brewer, K.J., Gettliffe, G., and Petersen, J.D. (1989) *Inorg. Chem.*, **28**, 81–84.
18 Petersen, J.D. (2016) *Personal communication.*
19 Calvin, M. (1964) *Nobel Lectures, Chemistry 1942–1962*, Elsevier Publishing Company, Amsterdam.
20 Brewer, K.J., Liegeois, A., Otvos, J.W., Calvin, M., and Spreer, L.O. (1988) *J. Chem. Soc., Chem. Commun.*, 1219–1220.
21 Brewer, K.J., Calvin, M., Lumpkin, R.S., Otvos, J.W., and Spreer, L.O. (1989) *Inorg. Chem.*, **28**, 4446–4451.
22 Scott, B., Brewer, K.J., Spreer, L.O., Craig, C.A., Otvos, J.W., Calvin, M., and Taylor, S. (1990) *J. Coord. Chem.*, **21**, 307–313.
23 New Chemistry Complex to be Among Best. *Hilltopics*, October 1987.
24 Rasmussen, S.C., Richter, M.M., Yi, E., Place, H., and Brewer, K.J. (1990) *Inorg. Chem.*, **29**, 3926–3932.
25 Richter, M.M. and Brewer, K.J. (1991) *Inorg. Chim. Acta*, **180**, 125–131.
26 Richter, M.M. and Brewer, K.J. (1992) *Inorg. Chem.*, **31**, 1594–1598.
27 Richter, M.M. and Brewer, K.J. (1993) *Inorg. Chem.*, **32**, 2827–2834.
28 Richter, M.M. and Brewer, K.J. (1993) *Inorg. Chem.*, **32**, 5762–5768.
29 Bridgewater, J.S., Vogler, L.M., Molnar, S.M., and Brewer, K.J. (1993) *Inorg. Chim. Acta*, **208**, 179–188.
30 Vogler, L.M., Scott, B., and Brewer, K.J. (1993) *Inorg. Chem.*, **32**, 898–903.
31 Willett, R.D., Wang, Z., Molnar, S., Brewer, K., Landee, C.P., Turnbull, M.M., and Zhang, W. (1993) *Mol. Cryst. Liq. Cryst. Sci. Technol., Sect. A*, **233**, 277–282.
32 Chen, X., Long, G.y., Willett, R.D., Hawks, T., Molnar, S., and Brewer, K. (1996) *Acta Crystallogr., Sect. C: Cryst. Struct. Commun.*, **C52**, 1924–1928.

**33** Wang, Z., Willett, R.D., Molnar, S., and Brewer, K.J. (1996) *Acta Crystallogr., Sect. C: Cryst. Struct. Commun.*, **C52**, 581–583.
**34** Chen, X., Willett, R.D., Hawks, T., Molnar, S., and Brewer, K.J. (1996) *J. Chem. Crystallogr.*, **26**, 261–265.
**35** Molnar, S.M. (2016) *Personal communication.*
**36** Molnar, S.M., Nallas, G., Bridgewater, J.S., and Brewer, K.J. (1994) *J. Am. Chem. Soc.*, **116**, 5206–5210.
**37** Molnar, S.M., Jensen, G.E., Vogler, L.M., Jones, S.W., Laverman, L., Bridgewater, J.S., Richter, M.M., and Brewer, K.J. (1994) *J. Photochem. Photobiol., A*, **80**, 315–322.
**38** Nallas, G.N.A., Jones, S.W., and Brewer, K.J. (1996) *Inorg. Chem.*, **35**, 6974–6980.
**39** Nallas, G.N.A. and Brewer, K.J. (1996) *Inorg. Chim. Acta*, **253**, 7–13.
**40** Doss, C. (2006) *Virginia Tech Research Magazine*, http://www.research.vt.edu/resmag/ResearchMagJan06/light.html (accessed April 5, 2016).
**41** Doss, C. (2005) *Virginia Tech College of Science Magazine*, **1** (Fall), 2–5.
**42** Winkel, B. (2016) *Personal communication.*
**43** Milkevitch, M., Brauns, E., and Brewer, K.J. (1996) *Inorg. Chem.*, **35**, 1737–1739.
**44** Milkevitch, M., Shirley, B.W., and Brewer, K.J. (1997) *Inorg. Chim. Acta*, **264**, 249–256.
**45** Milkevitch, M., Storrie, H., Brauns, E., Brewer, K.J., and Shirley, B.W. (1997) *Inorg. Chem.*, **36**, 4534–4538.
**46** Williams, R.L., Toft, H.N., Winkel, B., and Brewer, K.J. (2003) *Inorg. Chem.*, **42**, 4394–4400.
**47** Swavey, S., Fang, Z., and Brewer, K.J. (2002) *Inorg. Chem.*, **41**, 2598–2607.
**48** Fang, Z., Swavey, S., Holder, A., Winkel, B., and Brewer, K.J. (2002) *Inorg. Chem. Commun.*, **5**, 1078–1081.
**49** Swavey, S. and Brewer, K.J. (2002) *Inorg. Chem.*, **41**, 6196–6198.
**50** Holder, A.A., Swavey, S., and Brewer, K.J. (2004) *Inorg. Chem.*, **43**, 303–308.
**51** Storrie, B., Holder, A., and Brewer, K.J. (2006) *Proc. SPIE*, **6139**, 61391G/1–61391G/7.
**52** Holder, A.A., Zigler, D.F., Tarrago-Trani, M.T., Storrie, B., and Brewer, K.J. (2007) *Inorg. Chem.*, **46**, 4760–4762.
**53** Miao, R., Mongelli, M.T., Zigler, D.F., Winkel, B.S.J., and Brewer, K.J. (2006) *Inorg. Chem.*, **45**, 10413–10415.
**54** Wang, J., Higgins, S.L.H., Winkel, B.S.J., and Brewer, K.J. (2011) *Chem. Commun.*, **47**, 9786–9788.
**55** Higgins, S.L.H., Tucker, A.J., Winkel, B.S.J., and Brewer, K.J. (2012) *Chem. Commun.*, **48**, 67–69.
**56** Higgins, S.L.H. and Brewer, K.J. (2012) *Angew. Chem., Int. Ed.*, **51**, 11420–11422.
**57** Doss, C. *Virginia Tech News.* http://www.vtnews.vt.edu/articles/2010/01/2010-45.html (accessed April 3, 2016).
**58** Elvington, M., Brown, J., Arachchige, S.M., and Brewer, K.J. (2007) *J. Am. Chem. Soc.*, **129**, 10644–10645.

59 Arachchige, S.M., Brown, J., and Brewer, K.J. (2008) *J. Photochem. Photobiol. A*, **197**, 13–17.

60 Arachchige, S.M., Brown, J.R., Chang, E., Jain, A., Zigler, D.F., Rangan, K., and Brewer, K.J. (2009) *Inorg. Chem.*, **48**, 1989–2000.

61 Rangan, K., Arachchige, S.M., Brown, J.R., and Brewer, K.J. (2009) *Energy Environ. Sci.*, **2**, 410–419.

62 Knoll, J.D., Arachchige, S.M., and Brewer, K.J. (2011) *ChemSusChem*, **4**, 252–261.

63 Arachchige, S.M., Shaw, R., White, T.A., Shenoy, V., Tsui, H.-M., and Brewer, K.J. (2011) *ChemSusChem*, **4**, 514–518.

64 White, T.A., Higgins, S.L.H., Arachchige, S.M., and Brewer, K.J. (2011) *Angew. Chem., Int. Ed.*, **50**, 12209–12213.

65 Baumgart, D.C. and Sandborn, W.J. (2012) *The Lancet*, **380**, 1590–1605.

66 Canterbury, T. (2016) *Personal communication.*

# 2

# Basic Coordination Chemistry of Ruthenium

*Mark A. W. Lawrence, Jimmie L. Bullock, and Alvin A. Holder*

*Old Dominion University, Department of Chemistry and Biochemistry, 4541 Hampton Boulevard, Norfolk, VA 23529-0126, USA*

## 2.1 Coordination Chemistry of Ruthenium

### 2.1.1 The Element

Ruthenium [1], element 44, has an electronic configuration: [Kr] $4d^7$ $5s^1$ and an average atomic mass of 101.07 g $mol^{-1}$, a melting point of 2334 °C, and a density of 12.27 g $cm^{-3}$. The element has an electronegativity of 2.2, has seven naturally occurring isotopes, $^{96}Ru$, $^{98}Ru$, $^{99}Ru$, $^{100}Ru$, $^{101}Ru$, $^{102}Ru$, and $^{104}Ru$, with relative abundance of 5.54, 1.87, 12.76, 12.60, 17.06, 31.55, and 18.62%, respectively, of which $^{99}Ru$ and $^{101}Ru$ are NMR active with a nuclear spin of 5/2. Ruthenium is one of the rare elements belonging to the platinum-group metals, and was isolated in 1844. Ruthenium is a hard, white metal with an *hcp* structure, which does not tarnish at room temperatures, as it is passivated by a coating of $RuO_2$. The bulk metal oxidizes in air at about 800 °C. The metal is inert to acids or aqua regia, but oxidizes explosively when potassium chlorate is added to the solution and it is attacked by halogens and hydroxides. Ruthenium is one of the most effective hardeners for platinum and palladium, and is often alloyed with these metals to make wear-resistant electrical contacts. The corrosion resistance of titanium is improved several fold by addition of 0.1% ruthenium. Compounds in at least eight oxidation states have been found; but, of these, the +2, +3, and +4 states are the most common. Ruthenium compounds show a marked resemblance to their osmium analogues and bear little similarity to that of iron. The higher oxidation states of +6 and +8 are more readily obtained than that for iron. Ruthenium and osmium tetroxides are highly toxic, sparingly soluble in water but soluble in $CCl_4$. Ruthenium tetroxide is thermodynamically unstable with respect to $RuO_2$ and is liable to explode. It is a very powerful oxidant, reacting violently with some organic compounds.

Ruthenium holds the prominent position as the most employed metal in catalysis and excited-state chemistry. The rich and well-studied coordination and organometallic chemistries of ruthenium result in a wide variety of compounds, featuring several oxidation states, coordination numbers, and geometries. Many ruthenium compounds have found potential use in catalysis and their relatively

*Ruthenium Complexes: Photochemical and Biomedical Applications*, First Edition.
Edited by Alvin A. Holder, Lothar Lilge, Wesley R. Browne, Mark A.W. Lawrence, and Jimmie L. Bullock Jr.

low toxicity makes them ideal for the catalytic synthesis of drugs. These facts, along with comparatively lower price than the other platinum-group metals (Pd, Pt, Rh, and Ir), have made ruthenium compounds the preferred choice for many catalytic processes. Ruthenium compounds are also being increasingly explored for use as pharmaceutical agents due to their enhanced biocompatibility relative to many metallodrugs. This feature is explored in subsequent chapters.

### 2.1.2 Stereochemistry and Common Oxidation States

The oxidation states, stereochemistries, and representative species are given in Table 2.1. The lower oxidation states of ruthenium mainly involve π-bonding ligands. The most common oxidation states of ruthenium are the +2, +3, and +4 states. These oxidations states are frequently explored since many compounds can undergo reversible conversions between them while retaining their geometry. This property makes ruthenium complexes attractive for various redox processes where reversibility is a prerequisite.

In general, ruthenium, like all the platinum group metals, has a strong tendency to form bonds with carbon, especially with alkenes and alkynes. It also forms hydrides and M—M bonds when the higher oxidation states are reduced with tertiary phosphines in the presence of alcohols or *N*,*N*′-dimethlyformamide.

Ruthenium forms anhydrous binary oxides ($RuO_2$ and $RuO_4$) when the metal is oxidized by molecular oxygen under special conditions, and hydrous oxides from the action of alkaline solutions on its binary salts. Binary compounds of the chalcogenides and phosphides are generally similar to those of other transition

**Table 2.1** Oxidation states and stereochemistry of ruthenium.

| Oxidation state | Coordination number | Geometry | Example |
|---|---|---|---|
| −2, $d^{10}$ | 4 | *tet* | $[Ru(CO)_4]^{2-}$ |
| 0, $d^8$ | 5 | *tbp* | $Ru(CO)_5$ |
| +1, $d^7$ | 6 | | $[\eta^5\text{-}C_5H_5Ru(CO)_2]_2$ |
| | 5 | *tbp* | $Ru(CO)_2(P^tBu_2Me)_2$ |
| +2, $d^6$ | 5 | *tbp* | $RuHCl(PPh_3)_3$ |
| | 6[a] | *oct* | $[Ru(bpy)_3]^{2+}$ |
| +3[b], $d^5$ | 6[a] | *oct* | $[Ru(NH_3)_6]^{2+}$ |
| +4[b], $d^4$ | 6[a] | *oct* | $K_2RuCl_6$ |
| +5[b], $d^3$ | 6 | *oct* | $KRuF_6$ |
| +6[b], $d^2$ | 4 | *tet* | $RuO_4^{2-}$ |
| | 6 | *oct* | $RuF_6$ |
| +7, $d^1$ | 4 | *tet* | $RuO_4^-$ |
| +8, $d^0$ | 4 | *tet* | $RuO_4$ |

*tet*, tetrahedral; *tbp*, trigonal bipyrimidal; *oct*, octahedral.

a) Most common oxidation states and coordination numbers.

b) Oxidation state has other coordination numbers and stereochemistry.

metals. Of the binary compounds, the halides are considered the most important, especially the trichloride, as they are used as precursors for many ruthenium species.

There are numerous examples of these oxides and binary compounds in the literature. For a review of the classical (general) chemistry of ruthenium compounds, see [1]. In the following subsections the preparations and applications of some of these ruthenium compounds are explored to highlight the common coordination numbers observed across numerous classes of compounds. Brief discussions and some applications for the common oxidation states are presented and specific examples and applications are explored in later chapters.

#### 2.1.2.1 Ruthenium in Low Oxidation States

Low-valence ruthenium species, that is, the −2, 0, and +1 (and sometimes +2) oxidation states, generally consist of π-acids and nonclassical bonds unique to d-orbitals [2]. Two of these interactions that offer this unique bonding involve metal-olefin and π-acids like CO and tertiary phosphines. The π-acid ligands consist of low-lying vacant π*-orbitals capable of accepting density from filled metal orbitals. A description consistent with the observed bonding in metal–olefin complexes is given by the Dewar–Chatt–Duncanson MO description (Figure 2.1) [3]. For a monoalkene (in the $\eta^2$ mode), the bonding is explained by an overlap that is intermediate between a σ (axial overlap) and a π (lateral overlap) interaction [4]. The doubly occupied π-orbital of the alkene can donate electron density to a metal orbital of appropriate symmetry (e.g., a hybridized $d_{z^2}$) to give a stabilizing two-electron interaction [4, 5]. A second interaction occurs with the empty $\pi^*$-orbital and another (hybridized $d_{xz}$) metal orbital which is of lower energy than the $\pi^*$-orbital. If the metal orbital is doubly

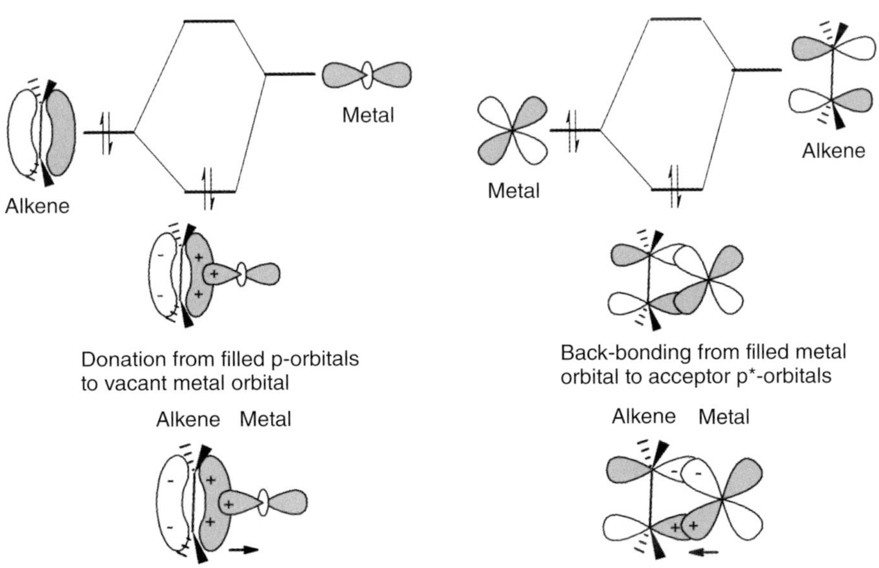


**Figure 2.1** The two stabilizing interactions that constitute the Dewar–Chatt–Duncanson model for bonding in metal–olefin interactions.

occupied (as is the case of low oxidation numbers), this interaction is stabilizing and it leads to electron density transfer from the metal to the alkene ligand (i.e., back-donation/bonding). Generally, there is a lengthening of the C—C bond resulting from this type of interaction. There is also a correlation between the bond lengthening and the electron-withdrawing power of substituents on the alkene. Both bonding components are synergically related. As one component increases, it tends to promote an increase in the other component, similar to M—CO bonding, and both components balance each other to maintain electroneutrality of the alkene. A similar explanation has also been used to explain the back-bonding observed with π-acids, such as CO; however, the symmetry of the interaction is slightly different (Figure 2.2). The extent of back-bonding in CO complexes is frequently deduced from a reduction of the C≡O stretching frequency. In the low oxidation state, M—M bonding has also been observed in polynuclear carbonyls and two such examples are $Ru_3(CO)_{12}$ and $Ru_6(CO)_{17}C$, the latter of which has a single carbon atom buried within the structure.

Complexes with two unconjugated alkene moieties such as 1,5-cyclooctadiene can be treated as having two independent olefin bonds. When there is conjugation, for example, 1,3-butadiene, the bonding is not appropriately represented by two olefin bonds. The typical "short-long-short" bond of the free diene is not observed in the complexes; but instead an averaging of the bond lengths is observed, which is consistent with a resonance stabilized system, or an electronic excited state. Of note is that the diene systems do not compete effectively with CO molecules for back-bonding. Another type of unique bonding occurs in metal-arene systems in which the metal is bonded to the π-system of an aromatic ring (Figure 2.3). These compounds are very stable toward many classical

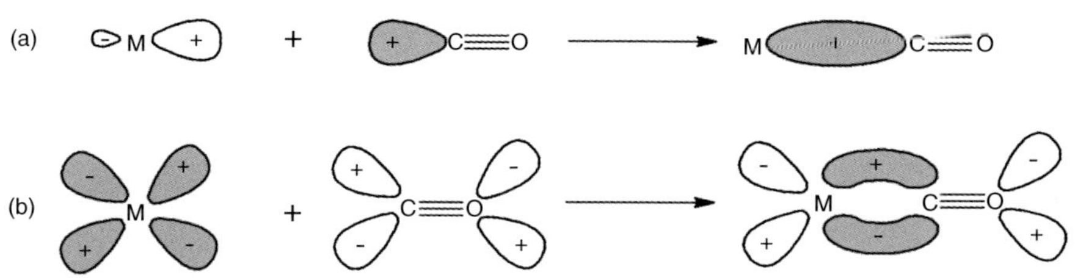


**Figure 2.2** (a) Representation of the M-C σ-bond to the unshared electron pair of the C atom. (b) Representation of the M-C π-bond (back-bonding). The other orbitals are omitted for clarity.

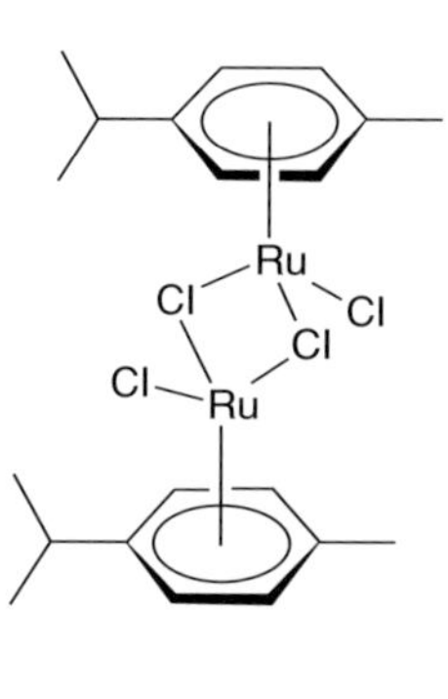


**Figure 2.3** Dichloro(*p*-cymene)ruthenium(II) dimer. An example of a ruthenium-arene.

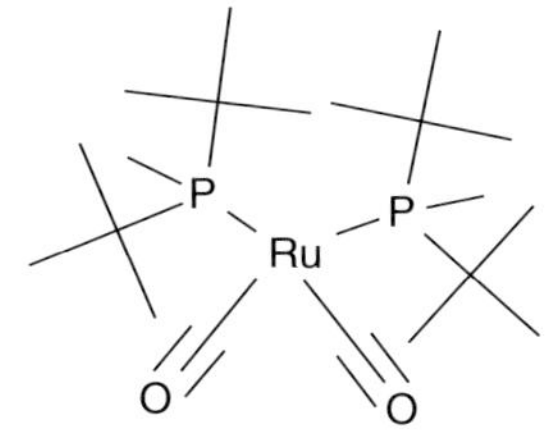


**Figure 2.4** Structure of $Ru(CO)_2(P^tBu_2Me)_2$.

reactions on the aromatic ring as well as at the metal center and presents new avenues in design for catalysts and in drug syntheses.

***Ruthenium (0) and (−2)*** Negative valent ruthenium such as $[Ru(CO)_4]^{2-}$, $d^{10}$ is very rare, very sensitive to oxygen and moisture, and have been employed in the synthesis of ammonia from dinitrogen [6]. These negative valent species are prepared by reduction of the zero-valent species, by group 1 metals (very strong reductants) in liquid ammonia in very dry solvents, under inert atmosphere. The chemistry of zero-valent ruthenium ($d^8$) [7] is primarily that of clusters (e.g., $Ru_3(CO)_{12}$ and its derivatives). There are few mononuclear species such as $Ru(CO)_5$ and its derivatives and also (arene)-$RuL_2$ species, where L = phosphine-type ligands (Figure 2.4). Zero-valent ruthenium complexes can be prepared by the reduction of the divalent species using activated magnesium.

Generally, isolated Ru(0) complexes are coordinatively unsaturated 16-electron species; they are often unstable and $Ru(CO)_5$, for example, eventually converts to $Ru_3(CO)_{12}$. With considerations of electron counting and an extension of the effective atomic number (EAN) rule for clusters, the prediction of the number of M—M bonds ($m$) and also the number of lone pairs ($n$) localized at the metal centers—and eventually used in σ-back-donation to terminal carbonyls—can be determined from Eq. (2.1) (where $V$ is the total number of available metal valence orbitals, $T$ is the total metal valence electron count of the cluster, and $L$ is the variable number of metal–carbon σ bonds).

$$
\begin{aligned}
2m + n &= V - L \\
2m + 2n &= T - 2L
\end{aligned}
\tag{2.1}
$$

For the trimeric Ru system ($Ru_3(CO)_{12}$), $V = 27$, $T = 48$, $m = 3$, $n = 9$, and hence $L = 12$ (Figure 2.5). This approach coupled to computation studies [8] shows a type of σ-aromaticity and allows for the formation of relatively strong M—M bonds despite significant electron repulsions between the nine lone pairs (Figure 2.5). The repulsion is also mitigated by back-bonding with the carbonyl ligands. Ru(0) species have been applied as catalyst to hydrogenation reactions [9], hydrogen generation [10], and electrocatalytic reduction of carbon dioxide [11], and a range of organic reactions when stabilized and dispersed as $Ru^0$-nanoparticles [12].)

***Ruthenium(I)*** Species of Fe(I), Ru(I), and Os(I) ($d^7$) are inherently unstable, and studies that extend beyond attempts to rapidly characterize them *in situ* are not readily available. Various studies involving electron paramagnetic

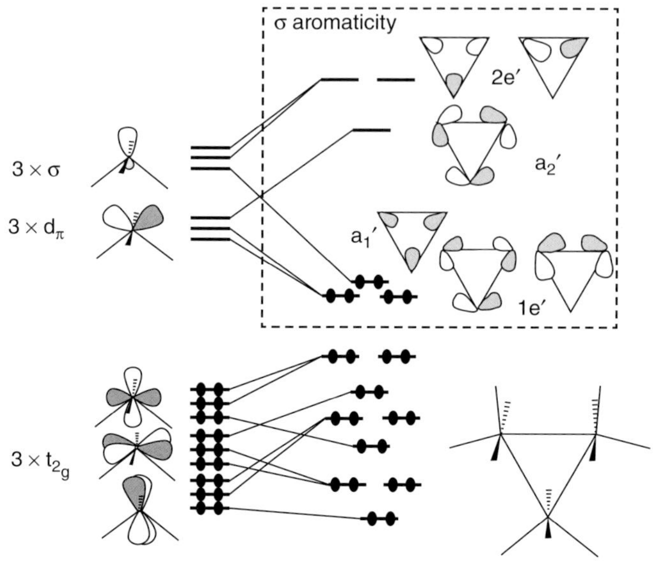


**Figure 2.5** Qualitative MO analysis of the nature of metal–metal and metal–carbonyl bonding in the $Ru_3(CO)_{12}$ trimer. (Reprinted (adapted) with permission from [8]. Copyright (1999) American Chemical Society.)

resonance (EPR) and electrochemical methods have been applied to provide evidence for the formation of, as well as to study this oxidation state [13]. Only recently Takaoka *et al.* [13a] were able to isolate and structurally characterize complexes with Ru in a formal oxidation state +1 (e.g., see Figure 2.6). The Ru(I) species were prepared by the chemical reduction of the corresponding Ru(II) species using $KC_8$ in dry diethyl ether. In these species, Ru(I) adopts a pentacoordinate (distorted *tbp*) geometry. Generally, the oxidation state +1 in iron, ruthenium, and osmium coordination chemistry is quite rare. For example, the $d^7$ Fe-group compounds are generally unstable and either disproportionate to Fe(0) and Fe(II) or to polynuclear species containing metal–metal bonds [1a, 14]. As a result, the chemistry of mononuclear Ru(I) (like Os(I)) complexes is underexplored.

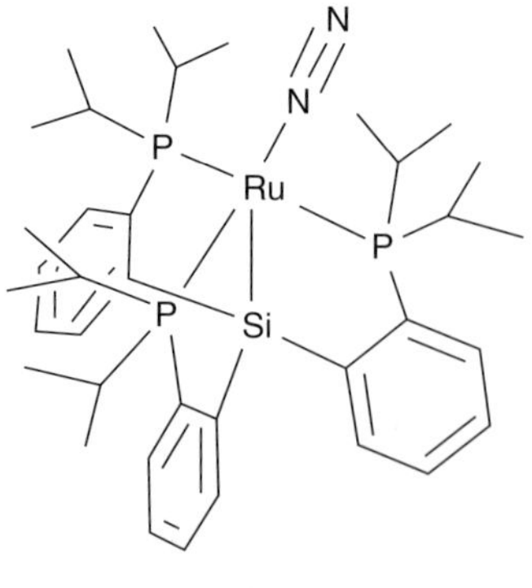


**Figure 2.6** Structure of $[(SiP^{iPr}{}_3)Ru(N_2)]$.

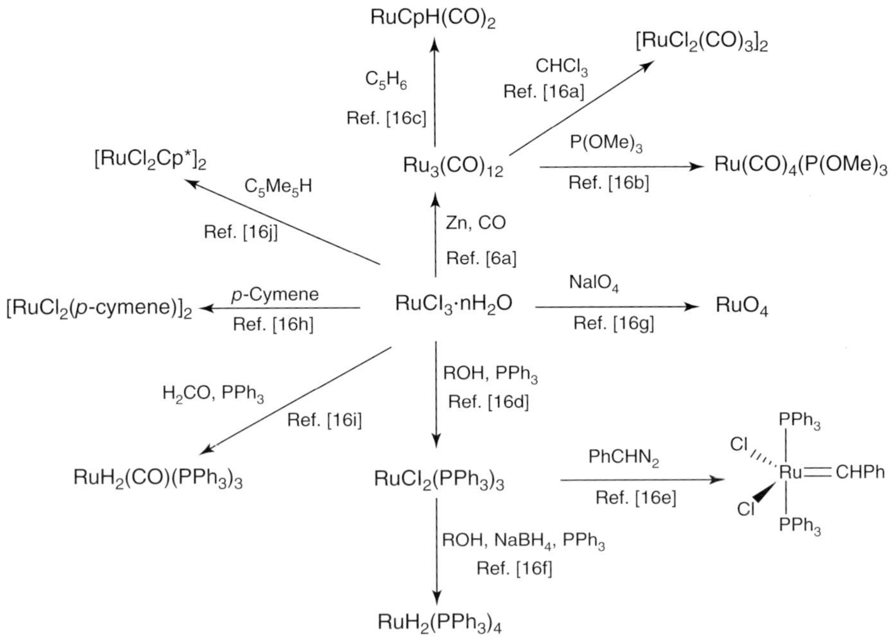


**Scheme 2.1** Examples of various approaches to prepare Ru(II) species.

#### 2.1.2.2 Chemistry of Ruthenium(II) and (III)

***Ruthenium(II)*** Ruthenium trichloride has been used as a precursor for many ruthenium(II), $d^6$ and ruthenium(III), $d^5$ species (Scheme 2.1) [15, 16]. The complexes in the +2 oxidation states are usually diamagnetic and reasonably labile, and substitution reactions often proceed with retention of configuration, suggesting an associative mechanism. Traditionally, in aqueous and alcoholic media, ruthenium(II) is generated *in situ* by a reducing agent, such as zinc powder or zinc amalgam under inert atmosphere, to give a blue solution. The exact nature of this blue species is undetermined, but it is suspected to be polymeric in nature [1]. These deep blue solutions are very air-sensitive and have been used as starting materials for the syntheses of many ruthenium(II) complexes. These solutions, in the absence of air, are slowly oxidized by water with the evolution of hydrogen. Numerous examples of complexes containing π-acid ligands such as CO and $PR_3$ are known. There are also many examples with halides and ammine ligands and one of the interests in these ligands was their ability to form complexes containing molecular nitrogen. These ruthenium(II) species generally illustrate excellent π-bonding characteristics evidenced in the reduction of the frequency of the coordinated π-acid relative to the free ligand.

The established synthetic protocols of ruthenium, especially with ammine [17], amine, and imine ligands, provide for many approaches to new metallopharmaceuticals [18]. Interestingly, Ru(II) and Ru(III) am(m)ine complexes have been shown to selectively bind to imine sites in biomolecules. The advantages of utilizing ruthenium am(m)ine complexes in drug development include (i) reliable preparations of stable complexes with predictable structures; (ii) the ability to tune the substituents toward electron transfer and substitution rates, and reduction potentials; and (iii) further understanding the biological effects of ruthenium complexes. The retention of the geometry between the +2 and +3 oxidation states lend ruthenium complexes to redox-activation and photodynamic approaches [19] to therapy. In addition, the development of radio pharmaceuticals containing one of several radionuclides of ruthenium is possible. There are a number of ruthenium compounds with anticancer activity [20]. Ruthenium complexes exhibit both nitric oxide release and scavenging functions that can affect vasodilation and synapse firing [21].

The coordination complexes of ruthenium have been successfully applied to numerous reactions. The ability of ruthenium complexes to form low-valent, coordinatively unsaturated species leads to a variety of catalytic transformations initiated by oxidative addition [22]. Terminal alkynes undergo regioselective nucleophilic addition of carboxylic acids upon heating with various ruthenium catalysts such as $RuCl_3$, $RuCl_2$(arene), and $Ru_3(CO)_{12}$ to give the corresponding enols. Low-valent ruthenium species have been applied to aromatic C—H bond activation and $sp^3$-carbons adjacent to an activating group [23] and the nucleophilic addition to alkynes [24] (Figure 2.7).

With ligands that are not sterically cumbersome, the +2 ion is readily oxidized by air to give the +3 oxidation state. To this end, approaches to produce air-stable ruthenium(II) complexes involve the use of arene-type systems [25]. These arene-type systems (e.g., $\eta^5$-$C_5R_5$, where R = H, alkyl or aryl) is assumed to occupy three coordination sites, form a pocket around the metal center, and are generally labile enough to be substituted by moderately nucleophilic systems. The use of these ruthenium arene systems in photodynamic therapy is discussed in Chapters 7 and 8. The properties of a bulky arene system in conjunction with the unique properties of ruthenium have also been incorporated in the concept of electron-fuelled molecular rotary motors [26]. In more practical applications, the use of bulky systems, such as N-heterocyclic carbenes (NHCs), to enhance the stability is also exemplified in the use of ruthenium(II) in Grubbs' first-generation, Grubbs' second-generation, and Hoveyda–Grubbs' catalysts [27]. Ruthenium(II)

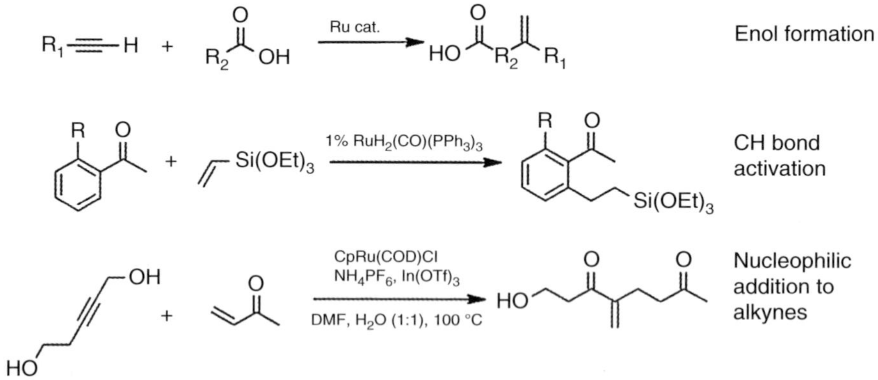


**Figure 2.7** Some selected reactions utilizing low-valent ruthenium.

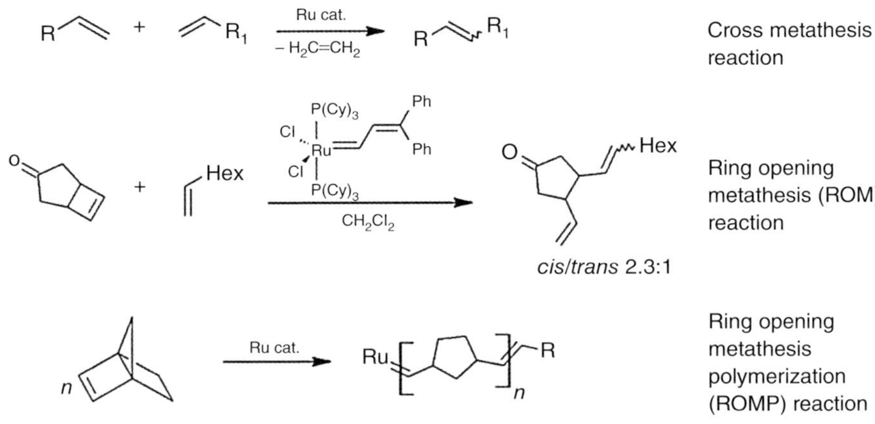


**Figure 2.8** An example of a cross-metathesis reaction.

has been pivotal in olefin metathesis. The Grubbs catalyst is typically used for cross-metathesis (also termed transalkylidenation), as shown in Figure 2.8. Various examples exist in which alkenes of different reactivity gives the cross-coupled product in excellent yield and selectivity; however, tailoring of the selectivity of the reaction is still under investigation by many researchers. Other metathesis reactions facilitated by ruthenium(II) include ring-opening metathesis (polymerization) abbreviated as ROM(P) (Figure 2.8) where strained rings are opened. For ROM [28], the product contains a terminal vinyl group and further reactions such as cross-metathesis are possible. In the absence of a partner alkene, polymerization (ROMP) occurs. Ring-closing metathesis (RCM) has been used to synthesize from 5- to 30-member cyclic alkenes. The ruthenium catalysts used in these metathesis reactions can tolerate a variety of functional groups, and the second-generation Grubbs' catalysts offer even more versatility.

Compared to other metals, for example, iridium and rhodium, ruthenium complexes are less active in catalytic hydrogenation of alkenes. This milder activity allows for their applications in chemoselective hydrogenation reactions such as the synthesis of (*S*)-Naproxen [29]. Great progress has been made in homogeneous asymmetric hydrogenation with ruthenium complexes bearing chiral phosphine ligands. A catalytic cycle for the catalyzed hydrogenation of β-keto esters using the privileged BINAP ligand (2,2′-bis(diphenylphosphino)-1,1′-binaphthyl), and other ligands has been given by Noyori and Ohkuma [30]. A general scheme has been proposed which involves the ruthenium hydride (formed in the presence of the base). The metal center and the ligands are directly involved in the bond-breaking and bond-forming reactions and influences turnover efficiency.

Ruthenium(II) has interesting luminescent properties that are well established for $[Ru(bpy)_3]^{2+}$ (bpy = 2,2′-bipyridine) and similar polypyridyl derivatives [31], and this makes it suitable for incorporation into flexible coordination polymers [32] and various photocatalysis as light antennas. Irradiation of $[Ru(bpy)_3]^{2+}$ with light below 560 nm results in the formation of a relatively long-lived (lifetime

$\tau = 0.6$ μsed in water at 25 °C) charge-transfer excited state $[^*Ru(bpy)_3]^{2+}$ potentially capable of reducing water to dihydrogen, as shown in this equation [33]:

$$[^*Ru(bpy)_3]^{2+} + H_2O \rightarrow [Ru(bpy)_3]^{3+} + \tfrac{1}{2}\ H_2 + OH^-$$

High-lying excited states generated via sequential two-photon capture by $[Ru(bpy)_3]^{2+}$ or electron capture by $[Ru(bpy)_3]^{3+}$ have been investigated by flash photolysis and pulse radiolysis techniques [34].

All the examples given *vide supra* illustrate the versatility of ruthenium and its ability to coordinate to multiple classes of ligands, as well as the two general stereochemistries of the +2 oxidation states, that is, *tbp* and *oct*.

***Ruthenium(III)*** There is an extensive chemistry of ruthenium(III) with both π-acids and σ-donor ligands, in which they predominantly adopt an *oct* geometry. The complexes are generally low spin with one unpaired electron. The commercially available "$RuCl_3{\cdot}3H_2O$" appears to contain some polynuclear Ru(IV). This $RuCl_3{\cdot}3H_2O$ species is readily soluble in water and numerous organic solvents. Generally, the $RuCl_3{\cdot}3H_2O$ starting material is activated with concentrated HCl or with Hg (primarily to reduce Ru(IV)). Ruthenium(III) forms very stable complexes with 2,2′-bipyridine and 1,10-phenanthroline and is used as a precursor for the Ru(II) congener that has shown exciting photochemistry in its electronically excited state. "Ruthenium red," which is postulated to be a linear alternating Ru(III)—O—Ru(IV)—O—Ru(III) oxo-bridged core and results from the partial (air-) oxidation of ammonia solutions of ruthenium trichloride, has been used as a stain in histology.

A different type of oxo-bridged species of the general formula $[Ru_3(\mu_3\text{-}O)(\mu_2\text{-}O_2CCH_3)_6L_3]^+$, where $L = H_2O$, ROH or pyridine derivatives (Figure 2.9), have a central 3-coordinate oxo core, are paramagnetic, readily soluble in alcoholic and aqueous media, and are structurally similar to analogous $Fe_3O$ and $Cr_3O$ carboxylates. Ruthenium oxo-centered trinuclear complexes are different from their first- and second-row congeners. They have a stronger metal–metal interaction through the $\mu_3$-oxo ligand [35]; hence, the electronic spectra of its oxo-centered trinuclear carboxylates are very different from the mononuclear and dinuclear ruthenium species [36] in that they absorb light strongly in the near-ultraviolet and visible regions, and the absorption extends

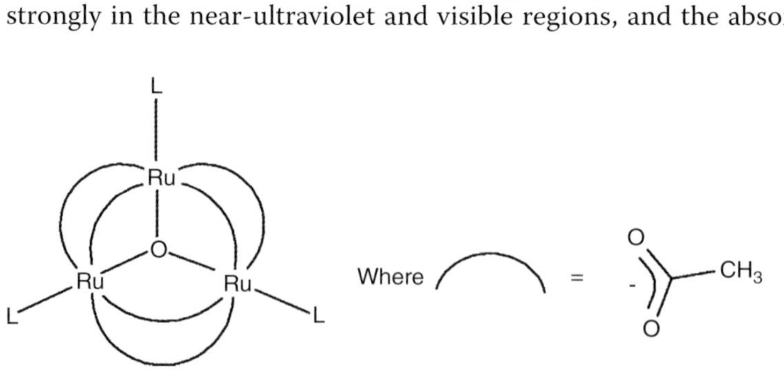


**Figure 2.9** A representation of oxo-centered trinuclear ruthenium(III) acetate. L represents coordinating solvents such as $H_2O$, ROH, py, and so on.

into the very near-infrared for the neutral species [37]. These absorption bands originate from a series of closely spaced molecular electronic transitions, composed of the metal d$\pi$-orbitals and the $\mu_3$-oxo p$\pi$-orbital. Substituting one of the ruthenium with another metal removes the strong absorption observed in the visible region (viz, disrupting the delocalization of the electrons) [35]. Unlike many other supramolecular multinuclear ruthenium systems, oxo-centered triruthenium carboxylates, $[Ru_3(\mu_3\text{-}O)(\mu_2\text{-}O_2CCH_3)_6L_3]^{n+}$ where L is a pyridine-type ligand, generally have reversible multistep and multielectron redox chemistry, which can be tuned by the nature of L and the carboxylate group [37, 38], whereas, the homonuclear and mixed-metal first-row transition-metal oxo-centered trinuclear carboxylates display irreversible redox behaviors [39]. The reversible redox properties of the $[Ru_3(\mu_3\text{-}O)(\mu_2\text{-}O_2CCH_3)_6L_3]^+$ species suggest that they are stable in a series of different redox states and they have been applied as catalysts [40–42]. Unfortunately, the reversible nature of the electron transfer as well as the stability diminishes in an aqueous environment [43], and this has been attributed to the enhanced $H_2O$ exchange rate of the oxo-centered species relative to the hexaaqua species [43, 44]. Unlike its first-row transition-metal analogues, it is reduced by $PPh_3$ to give a mixed-valence complex $[Ru_3(\mu_3\text{-}O)(\mu_2\text{-}O_2CCH_3)_6(PPh_3)_3]$. These oxo-centered trinuclear complexes continue to generate interest and ligands more tunable than carboxylates have been employed [45] The dimeric complex $[Ru_2(OOCCH_3)_4]Cl$ is obtained as a precursor or along with the oxo-centered trinuclear species [35]. The chloride anion also acts as a bridge between the ruthenium centers.

Ruthenium(III) also forms many stable complexes with nitric oxide, NO, containing various ancillary ligands of the general type $[Ru(NO)L_n]$ (where L = edta, py, bpy, etc.) and the metal is given a formal oxidation state of +2. Nitric oxide plays key roles in blood pressure regulation, in the nervous system, and in immune response to pathogens [46]. Some Ru—NO compounds have been trapped on polymeric surfaces [47] and in solution [48] have been studied as therapeutic NO-releasing agents.

Ruthenium(III) forms many complexes with a variety of sulphur- [49] and oxygen-containing ligands such as oxalates ($[Ru(ox)_3]^{3-}$) and acetylacetonate ($Ru(acac)_3$). The acetylacetonate species has been used as a precursor for other ruthenium(III) and ruthenium(II) species. It forms diolefin complexes in the presence of the olefin and a reducing agent (such as Zn). The ruthenium olefin species can be reversibly oxidized to the +3 state (a rare example) [50]. The acetylacetonate group can also be labilized by Lewis acids to open up coordination sites on the metal center.

A series of complexes that feature two Ru atoms bridged by bidentate ligands which facilitate a degree of electron transfer called Creutz–Taube ion are well studied. Named after the discoverers Carol Creutz and Henry Taube [51], these complexes illustrate the advantages that ruthenium complexes possess for examining redox reactions. Many analogues of this ion have been prepared using different bridging ligands. These ions generally possess mixed valancies, II-III (5+), but II-II (4+) and III-III (6+) salts have also been isolated. Crystal structures are consistent with symmetrical ions even in the II-III case.)

***Substitution Reactions of Ruthenium(II) and (III)*** The low-spin $Ru^{3+}$ aqua ion ($[Ru(H_2O)_6]^{3+}$, $t_{2g}{}^5$) has a water exchange rate constant of $3.5\times10^{-6}\,s^{-1}$ and is 4 orders of magnitude less labile than the $Ru^{2+}$ aqua ion ($[Ru(H_2O)_6]^{2+}$, $t_{2g}{}^6$), which has a rate constant $1.8\times10^{-2}\,s^{-1}$ [52, 53]. The exchange of $H_2O$ molecules occurs by an $I_a$ mechanism. Both exchange rate constants are relatively slow, allowing for a direct observation of the electron exchange of the $[Ru(H_2O)_6]^{3+/2+}$ redox couple in aqueous solution, which has a bimolecular rate constant of $20\,M^{-1}\,s^{-1}$ at 25 °C. Solvent exchange of acetonitrile on $Ru^{2+}$ has a rate constant of $8.9\times10^{-11}\,s^{-1}$, which is 8 orders of magnitude slower than water exchange. This has been attributed to strong back-bonding from the electron-rich $Ru^{2+}$ into the MeCN $\pi^*$ orbitals, which causes a large increase of $\Delta H^\ddagger$ (more than $50\,kJ\,mol^{-1}$). Deprotonation of the hexaaqua to give the pentaaquahydroxo species $[Ru(H_2O)_5(OH)]^{2+}$ significantly influences the water exchange rate constant by nearly 170-fold to $5.9\times10^{-4}\,s^{-1}$. The $[Ru(H_2O)_6]^{3+}$ species has a $pK_a$ calculated to be 2.40–2.90 [52, 54], which makes its acidity comparable to first-row transition metals. Replacing the water ligands with other ligand systems, whether monodentate or arene type, generally results in an increase in the water exchange rate constants for mononuclear species of both oxidation states. This is significant in ruthenium-containing pharmaceuticals.

### 2.1.2.3 Higher Oxidation States of Ruthenium

***Ruthenium(IV), (V), and (VII)*** The +4 oxidation state, $d^4$, consists of mainly neutral or anionic species, with an *oct* or distorted *oct* geometry. The information in d-d transitions of this oxidation state is sparse as these bands are masked by charge-transfer bands. The electrochemical oxidation of $[Ru(H_2O)_6]^{2+}$ in aqueous solution produces a Ru(IV) species. Its formulation as $[Ru_4O_6(H_2O)_{12}]^{4+}$ (or a protonated form depending on pH) is consistent with $^{17}O$ NMR spectroscopic data. An oxo-bridged, diamagnetic ionic species ($K_4[Ru_2OCl_{10}]$) can be prepared from the reduction of $RuO_4$ by HCl in KCl or by the oxidation of the corresponding ruthenium(III) species. The diamagnetic nature of this salt can be rationalized by considering the formation of two 3-center $\pi$-interactions involving the d-orbitals of the two low-spin Ru(IV) centers and the O atom [55]. Ruthenium(IV) alkylidene complexes have found numerous applications in olefin and enyne metathesis for the creation of unsaturated rings, functionalization of carbon–carbon double bonds, and polymerization of cyclic olefins, in a manner analogues to the Grubb's catalyst [27b, 56].

At higher oxidation states, Ru(V) species and Ru(VII), perruthenate, have been applied to oxidation reactions that take advantage of the ease of formation of ruthenium-oxo species. The Ru(V) oxidation state, $d^3$, is unfavorable and mainly the fluorides are documented. The salts of perruthenium ion $[RuO_4]^-$ with quarternary ammonium salts, show far milder oxidizing properties than $RuO_4$ and can act as efficient catalysts for selective oxidation of primary alcohols with a combined use of stoichiometric amounts of *N*-methylmorpholine *N*-oxide (NMO). When combined with large cations, the solubility of the perruthenium toward organic solvents can be improved. However, tetrapropylammonium perruthenate (TPAP) is the most common perruthenate species [57]. Water inhibits catalyst turnover; however, the catalyst has a wide tolerance of functional groups, which is typical of ruthenium.

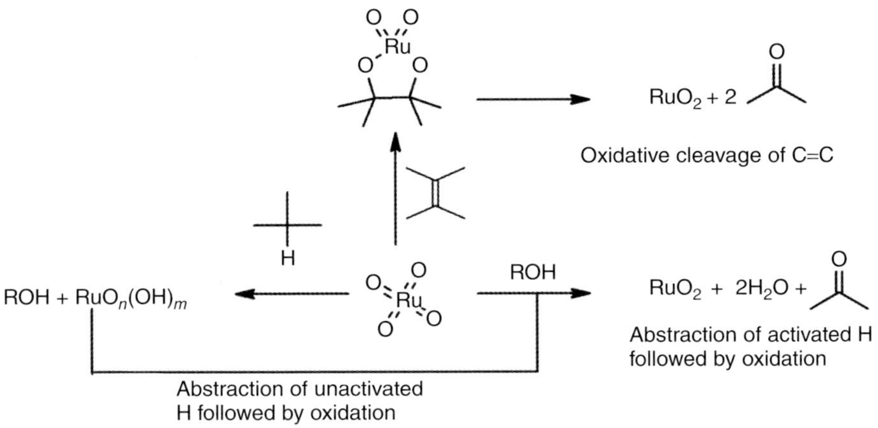


**Scheme 2.2** Typical reactions of $RuO_4$.

***Ruthenium(VIII)*** Although $RuO_4$ is volatile, toxic, and may cause explosive mixtures, it has been widely used as a powerful oxidant in the transformation of various organic compounds, which include hydrogen abstraction and oxygenation. Typical reactions of $RuO_4$ are given in Scheme 2.2. The reaction conditions are generally mild because it is an aggressive oxidant. The reactions can also be performed in biphasic conditions by generating the $RuO_4$ *in situ* from $RuCl_3$ or $RuO_2$ using strong oxidants such as $NaIO_4$. This approach, in the presence of aromatic nitrogen ligands, has been used in the epoxidation of olefins [58, 59]. These reactions are sometimes sluggish due to the deactivation of the catalyst from carboxylic acid formations and complexation giving rise to lower oxidation state of the ruthenium. This deactivation may be retarded by the addition of acetonitrile [60].

### 2.1.3 Conclusion

The chemistry of ruthenium is vast and extremely diverse, and there are still emerging fields in catalysis (homogeneous, heterogeneous, photocatalysis, etc.) and in numerous medicinal applications. In the next chapter, the photophysics and photochemistry of ruthenium complexes are discussed and in the subsequent chapters the numerous applications in energy and biology are discussed. The future of ruthenium chemistry shines bright and is beckoning all areas of modern science.

## References

1 For a general reading see (a) Cotton, F.A., Wilkinson, G., Murillo, C.A., and Bochmann, M. (2004) *Advanced Inorganic Chemistry*, 6th edn, Wiley-VCH, Singapore; (b) Cotton, S.A. (1997) *Chemistry of Precious Metals*, Blackie Academic and Professional, London; (c) Cotton, F.A. and Wilkinson, G. (1972) *Advanced Inorganic Chemistry*, 3rd edn, Wiley-VCH, New York (for more general reading on bonding and classical coordination chemistry); (d) Lide, D.R. (ed.) (2001) *CRC Handbook of Chemistry and Physics*, 82nd edn, CRC Press LLC.

2 (a) Bruce, M.I., Hambley, T.W., Snow, M.R., and Swincer, A.G. (1985) *Organometallics*, **4**, 501–508; (b) Bruce, M.I., Humphrey, M.G., Snow, M.R., and Tiekink, E.R.T. (1986) *J. Organomet. Chem.*, **314**, 213–225.
3 (a) Chatt, J. and Duncanson, L.A. (1953) *J. Chem. Soc.*, 2939–2947; (b) Dewar, M. J. S., *Bull. Soc. Chim. France*, 1951, **18**, C71–C79.
4 Jean, Y. (2005) *Molecular Orbitals of Transition Metal Complexes*, Oxford University Press, London, pp. 111–135.
5 Leigh, G.J. and Winterton, N. (eds) (2002) *Modern Coordination Chemistry: The Legacy of Joseph Chatt*, Royal Society of Chemistry, Cambridge, pp. 104–108.
6 (a) Bhattacharyya, N.K., Coffy, T.J., Quintana, W., Salupo, T.A., Bricker, J.C., Shay, T.B., Payne, M., and Shore, S.G. (1990) *Organometallics*, **9**, 2368–2374; (b) Yunusov, S.M., Kalyuzhnaya, E.S., Moroz, B.L., Ivanova, A.S., and Shur, V.B. (2011) *Russ. Chem. Bull.*, **60** (9), 1842–1847; (c) Yunusov, S.M., Kalyuzhnaya, E.S., Moroz, B.L., Ivanova, A.S., Reshetenko, T.V., Avdeeva, L.B., Likholobov, V.A., and Shur, V.B. (2004) *J. Mol. Catal. A: Chem.*, **219**, 149–153.
7 (a) Ogasawara, M., Macgregor, S.A., Streib, W.E., Folting, K., Eisenstein, O., and Caulton, K.G. (1995) *J. Am. Chem. Soc.*, **117**, 8869–8870; (b) Carrondo, M.A.A.F.d.C.T., Chaudret, B.N., Cole-Hamilton, D.J., Skapski, A.C., and Wilkinson, G. (1978) *J. Chem. Soc., Chem. Comm.*, 463–464; (c) Heyn, R.H. and Caulton, K.G. (1993) *J. Am. Chem. Soc.*, **115**, 3354–3355; (d) Ogasawara, M., Huang, D., Streib, W.E., Huffman, J.C., Gallego-Planas, N., Maseras, F., Eisenstein, O., and Caulton, K.G. (1997) *J. Am. Chem. Soc.*, **119**, 8642–8651; (e) Collman, J.P., Brothers, P.J., McElwee-White, L., and Rose, E. (1985) *J. Am. Chem. Soc.*, **107**, 6110–6111; (f) Collman, J.P., Brothers, P.J., McElwee-White, L., Rose, E., and Wright, L.J. (1985) *J. Am. Chem. Soc.*, **107**, 4570–4571.
8 Hunstock, E., Mealli, C., Calhorda, M.J., and Reinhold, J. (1999) *Inorg. Chem.*, **38**, 5053–5060.
9 Prechtl, M.H.G., Scariot, M., Scholten, J.D., Machado, G., Teixeira, S.R., and Dupont, J. (2008) *Inorg. Chem.*, **47**, 8995–9001.
10 Özkar, S. and Zahmakıran, M. (2005) *J. Alloys Compd.*, **404–406**, 728–731.
11 Collomb-Dunand-Sauthier, M.-N., Deronzier, A., and Ziessel, R. (1994) *J. Chem. Soc., Chem. Comm.*, 189–191.
12 (a) Sarmah, P.P. and Dutta, D.K. (2012) *Green Chem.*, **14**, 1086–1093; (b) Na, Y., Park, S., Han, S.B., Han, H., Ko, S., and Chang, S. (2003) *J. Am. Chem. Soc.*, **126**, 250–258; (c) Baruwati, B., Polshettiwar, V., and Varma, R.S. (2009) *Tetrahedron Lett.*, **50**, 1215–1218.
13 (a) Takaoka, A., Gerber, L.C.H., and Peters, J.C. (2010) *Angew. Chem., Int. Ed.*, **49**, 4088–4091; (b) Angelici, R.J., Zhu, B., Fedi, S., Laschi, F., and Zanello, P. (2007) *Inorg. Chem.*, **46**, 10901–10906; (c) Bianchini, C., Peuzzini, M., Ceccanti, A., Laschi, F., and Zanello, P. (1997) *Inorg. Chim. Acta*, **259**, 61–70; (d) Mulazzani, Q.G., Emmi, S., Fuochi, P.G., Hoffman, M.Z., and Venturi, M. (1978) *J. Am. Chem. Soc.*, **100**, 981–983; (e) Zotti, G., Pilloni, G., Bressan, M., and Martelli, M. (1977) *J. Electroanal. Chem. Interfacial Electrochem.*, **75**, 607–612.

14 Krusic, P.J., Cote, W.J., and Grand, A. (1984) *J. Am. Chem. Soc.*, **106**, 4642–4643.

15 Naota, T., Takaya, H., and Murahashi, S.-I. (1998) *Chem. Rev.*, **98**, 2599–2660.

16 (a) Teulon, P. and Roziere, J. (1981) *J. Organomet. Chem.*, **214**, 391–397; (b) Cobbledick, R. E., Einstein,F. W.B., Pomeroy, R. K., Spetch, E. R., *J. Organomet. Chem.*, 1980, **195**, 77–88; (c) Humphries, A.P. and Knox, S.A.R. (1975) *J. Chem. Soc., Dalton Trans.*, 1710–1714; (d) Hallman, P.S., Stephenson, T.A., and Wilkinson, G. (1970) Tetrakis(Triphenylphosphine)Dichloro-Ruthenium(II) and Tris(Triphenylphosphine)-Dichlororuthenium(II), in *Inorganic syntheses* (ed. R.W. Parry) Vol. 12, New York, McGraw-Hill Book Company, pp. 237–240; (e) Schwab, P., Grubbs, R.H., and Ziller, J.W. (1996) *J. Am. Chem. Soc.*, **118**, 100–110; (f) Yamamoto, A., Kitazume, S., and Ikeda, S. (1968) *J. Am. Chem. Soc.*, **90**, 1089–1090; (g) Nakata, H. (1963) *Tetrahedron*, **19**, 1959–1963; (h) Crisp, R.I., Hussey, C.L., and Seddon, K.R. (1995) *Polyhedron*, **14**, 2819–2825; (i) Bennett, M.A. and Smith, A.K. (1974) *J. Chem. Soc., Dalton Trans.*, 233–241; (j) Ahmad, N., Levison, J. J., Robinson, S. D., Uttley, M. F., Triphenylphosphine Complexes of Transition Metals, in *Inorganic syntheses*, G.W. Parshall (Ed.), Vol. **15**, McGraw-Hill Book Company, New York, 1974, pp 45–53; (k) Oshima, N., Suzuki, H., and Moro-Oka, Y. (1984) *Chem. Lett.*, **13**, 1161–1164.

17 Lever, F.M. and Powell, A.R. (1969) *J. Chem. Soc. A*, 1477–1482.

18 Clarke, M.J. (2002) *Coord. Chem. Rev.*, **232**, 69–93 and references therein.

19 (a) Holder, A.A., Zigler, D.F., Tarrago-Trani, M.T., Storrie, B., and Brewer, K.J. (2007) *Inorg. Chem.*, **46**, 4760–4762; (b) Holder A. A, Taylor P., Magnusen A. R., Moffett E. T, Meyer K., Hong Y., Ramsdale S. E., Gordon M., Stubbs J., Seymour L. A., Acharya D., Weber R. T., Smith P. F., Dismukes G. C., Ji P., Menocal L., Bai F., Williams J. L., Cropek D. M., Jarrett W. L., *Dalton Trans.*, 2013, 11881–11899; (c) Holder, A.A., Swavey, S., and Brewer, K.J. (2004) *Inorg. Chem.*, **43**, 303–308.

20 Schluga, P., Hartinger, C.G., Egger, A., Reisner, E., Galanski, M., Jakupec, M.A., and Keppler, B.K. (2006) *Dalton Trans.*, 1796–1802 and references therein.

21 (a) Horn, T.F.W., Wolf, G., Duffy, S., Weiss, S., Keilhoff, G., and Macvicar, B.A. (2002) *FASEB J.*, **16**, 1611–1622; (b) de Lima, R.G., Sauaia, M.G., Bonaventura, D., Tedesco, A.C., Lopez, R.F.V., Bendhack, L.M., and da Silva, R.S. (2005) *Inorg. Chim. Acta*, **358**, 2643–2650; (c) Pestana, C.R., Phelippin, D.P.S., Polizello, A.C.M., Dorta, D.J., Uyemura, S.A., Santos, A.C., Doro, F.G., Rodrigues, F.P., Tfouni, E., and Curti, C. (2009) *Nitric Oxide*, **20**, 24–30.

22 Crochet, P. and Cadierno, V. (2014) *Dalton Trans.*, **43**, 12447–12462.

23 Murai, S., Chatani, N., and Kakiuchi, F. (1997) *Pure Appl. Chem.*, **69**, 589–594.

24 Trost, B.M., Krause, L., and Portnoy, M. (1997) *J. Am. Chem. Soc.*, **119**, 11319–11320.

25 Therrien, B. (2009) *Coord. Chem. Rev.*, **253**, 493–519 and references therein.

26 Vives, G., Carella, A., Launay, J.-P., and Rapenne, G. (2008) *Coord. Chem. Rev.*, **252**, 1451–1459.

27 (a) Colacino, E., Martinez, J., and Lamaty, F. (2007) *Coord. Chem. Rev.*, **251**, 726–764 and references therein; (b) Grubbs, R.H. (ed.) (2003) *Handbook of Metathesis: Catalyst Development* Vol. 1, John Wiley & Sons, New York.

28 Randall, M.L., Tallarico, J.A., and Snapper, M.L. (1995) *J. Am. Chem. Soc.*, **117**, 9610–9611.

29 Ohta, T., Takaya, H., Kitamura, M., Nagai, K., and Noyori, R. (1987) *J. Org. Chem.*, **52**, 3174–3176.

30 Noyori, R., and Ohkuma, T., *Angew. Chem., Int. Ed.*, 2001, **40**, 40–73 (and references therein).

31 (a) Adamson, A.W. and Demas, J.N. (1971) *J. Am. Chem. Soc.*, **93**, 1800–1801; (b) Van Houten, J. and Watts, R.J. (1976) *J. Am. Chem. Soc.*, **98**, 4853–4858; (c) Caspar, J.V. and Meyer, T.J. (1983) *J. Am. Chem. Soc.*, **105**, 5583–5590; (d) Sauvage, J., Collin, J., Chambron, J., Guillerez, S., Coudret, C., Balzani, V., Barigelletti, F., De Cola, L., and Flamigni, L. (1994) *Chem. Rev.*, **94**, 993–1019;e) Meyer, T.J. (1986) *Pure Appl. Chem.*, **58**, 1193–1206.

32 Kobayashi, A., Ohba, T., Saitoh, E., Suzuki, Y., Noro, S.-I., Chang, H.-C., and Kato, M. (2014) *Inorg. Chem.*, **53**, 2910–2921.

33 (a) Navon, G. and Sutin, N. (1974) *Inorg. Chem.*, **13**, 2159–2164; (b) Demas, J.N. and Crosby, G.A. (1971) *J. Am. Chem. Soc.*, **93**, 2841–2847; (c) Harrigan, R.W., Hager, G.D., and Crosby, G.A. (1973) *Chem. Phys. Lett.*, **21**, 487–490; (d) Creutz, C. and Sutin, N. (1975) *Proc. Natl. Acad. Sci. U. S. A.*, **72**, 2858–2862.

34 Thompson, D.W., Wishart, J.F., Brunschwig, B.S., and Sutin, N. (2001) *J. Phys. Chem. A*, **105** (35), 8117–8122.

35 (a) Sasaki, Y., Yoshida, Y., Ohto, A., Tokiwa, A., Ito, T., Kobayashi, H., Uryu, N., and Mogi, I. (1993) *Chem. Lett.*, **22**, 69–72; (b) Ohto, A., Sasaki, Y., and Ito, T. (1994) *Inorg. Chem.*, **33**, 1245–1246; (c) Velayutham, M., Gopinath, C.S., and Subramanian, S. (1996) *Chem. Phys. Lett.*, **249**, 71–76.

36 (a) Sasaki, Y., Tokiwa, A., and Ito, T. (1987) *J. Am. Chem. Soc.*, **109**, 6341–6347; (b) Sasaki, Y., Suzuki, M., Nagasawa, A., Tokiwa, A., Ebihara, M., Yamaguchi, T., Kabuto, C., Ochi, T., and Ito, T. (1991) *Inorg. Chem.*, **30**, 4903–4908.

37 Baumann, J.A., Salmon, D.J., Wilson, S.T., Meyer, T.J., and Hatfield, W.E. (1978) *Inorg. Chem.*, **17**, 3342–3350.

38 Baumann, J.A., Salmon, D.J., Wilson, S.T., and Meyer, T.J. (1979) *Inorg. Chem.*, **18**, 2472–2479.

39 (a) Sowrey, F.E., MacDonald, C.J., and Cannon, R.D. (1998) *J. Chem. Soc., Faraday Trans.*, **94**, 1571–1574; (b) Manchanda, R. (1996) *Inorg. Chim. Acta*, **245**, 91–95; (c) Lawrence, M.A.W., Maragh, P.T., and Dasgupta, T.P. (2012) *Inorg. Chim. Acta*, **388**, 88–97; (d) Lawrence, M.A.W., Thomas, S.E., Maragh, P.T., and Dasgupta, T.P. (2011) *Transition Met. Chem.*, **36**, 553–563.

40 Fouda, S.A. and Rempel, G.L. (1979) *Inorg. Chem.*, **18**, 1–8.

41 Marr, S.B., Carvel, R.O., Richens, D.T., Lee, H.-J., Lane, M., and Stavropoulos, P. (2000) *Inorg. Chem.*, **39**, 4630–4638.

42 (a) Mitchell, R.W., Spencer, A., and Wilkinson, G. (1973) *J. Chem. Soc., Dalton Trans.*, 846–854; (b) Fouda, S.A., Hui, B.C.Y., and Rempel, G.L. (1978) *Inorg. Chem.*, **17**, 3213–3220; (c) Bilgrien, C., Davis, S., and Drago, R.S. (1987)

*J. Am. Chem. Soc.*, **109**, 3786–3787; (d) Davis, S. and Drago, R.S. (1988) *Inorg. Chem.*, **27**, 4759–4760.

43 Lawrence, M.A.W., Maragh, P.T., and Dasgupta, T.P. (2012) *Transition Met. Chem.*, **37**, 505–517.

44 Sasaki, Y., Nagasawa, A., Tokiwa-Yamamoto, A., and Ito, T. (1993) *Inorg. Chim. Acta*, **212**, 175–182.

45 (a) Saalfrank, R.W., Scheurer, A., Pokorny, K., Maid, H., Reimann, U., Hampel, F., Heinemann, F.W., Schunemann, V., and Trautwein, A. (2005) *Eur. J. Inorg. Chem.*, 1383–1387; (b) Stadler, C., Daub, J., Kohler, J., Saalfrank, R.W., Coropceanu, V., Schunemann, V., Ober, C., Trautwein, A.X., Parker, S.F., Poyraz, M., Inomata, T., and Cannon, R.D. (2001) *J. Chem. Soc., Dalton Trans.*, 3373–3383.

46 Ignarro, L.J. (ed.) (2000) *Nitric Oxide: Biology and Pathobiology*, Academic Press, San Diego.

47 (a) Bordini, J., Ford, P.C., and Tfouni, E. (2005) *Chem. Commun.*, 4169–4171; (b) Ferreira, K.Q., Schneider, J.F., Nascente, P.A.P., Rodrigues-Filho, U.P., and Tfouni, E. (2006) *J. Colloid Interface Sci.*, **300**, 543–552; (c) Seabra, A.B. and Durán, N. (2010) *J. Mater. Chem.*, **20**, 1624–1637.

48 Rose, M.J., Olmstead, M.M., and Mascharak, P.K. (2007) *J. Am. Chem. Soc.*, **129**, 5342–5343 and references therein.

49 (a) Chatt, J., Leigh, G.J., and Storace, A.P. (1971) *J. Chem. Soc. A*, 1380–1389;(b) Cargill Thompson, A.M.W., Bardwell, D.A., Jeffery, J.C., Rees, L.H., and Ward, M.D. (1997) *J. Chem. Soc., Dalton Trans.*, 721–726.

50 Bennett, M.A., Byrnes, M.J., and Kováčik, I. (2004) *J. Organomet. Chem.*, **689**, 4463–4474.

51 Creutz, C. and Taube, H. (1969) *J. Amer. Chem. Soc.*, **91**, 3988.

52 Rapaport, I., Helm, L., Merbach, A.E., Bernhard, P., and Ludi, A. (1988) *Inorg. Chem.*, **27**, 873–879.

53 Helm, L. and Merbach, A.E. (2005) *Chem. Rev.*, **105**, 1923–1960 (and references therein).

54 Bottcher, W., Brown, G.M., and Sutin, N. (1979) *Inorg. Chem.*, **18**, 1447–1451.

55 Housecroft, C.E. and Sharpe, A.G. (2008) *Inorganic Chemistry*, 3rd edn, Prentice Hall.

56 Bruneau, C. and Achard, M. (2012) *Coord. Chem. Rev.*, **256**, 525–536.

57 Shoair, A.-G.F. (2005) *Bull. Korean Chem. Soc.*, **26**, 1525–1528.

58 Eskénazi, C., Balavoine, G., Meunier, F., and Rivière, H. (1985) *J. Chem. Soc., Chem. Commun.*, 1111–1113.

59 Chatterjee, D. (2008) *Coord. Chem. Rev.*, **252**, 176–198.

60 Carlsen, P.H.J., Katsuki, T., Martin, V.S., and Sharpless, K.B. (1981) *J. Org. Chem.*, **46**, 3936–3938.

# Section II

# Artificial Photosynthesis

# 3

# Water Oxidation Catalysis with Ruthenium

*Andrea Sartorel*

*University of Padova, Department of Chemical Sciences, Via Marzolo 1, 35131 Padova, Italy*

## 3.1 Introduction

### 3.1.1 Energy Issue and Energy from the Sun

> *Energy is at the heart of most critical economic, environmental and developmental issues facing the world today. Clean, efficient, affordable and reliable energy services are indispensable for global prosperity.*
> Energy for a sustainable future, Report and Recommendations of United Nations Secretary-General's Advisory Group on Energy and Climate Change (AGECC), in 2010.

> *Energy is the single most important problem facing humanity today – not just the U.S., but also worldwide. The magnitude of this problem is incredible. Energy is the largest enterprise on Earth – by a large margin… While conservation efforts will help the worldwide energy situation, the problem by mid-century will be inadequate supply.*
> Richard Smalley, 1996 Nobel Prize in chemistry, Professor of chemistry and of physics, in 2004.

> *If our black and nervous civilization, based on coal, shall be followed by a quieter civilization based on the utilization of solar energy, that will not be harmful to progress and to human happiness.*
> Giacomo Ciamician, Italian photochemist, in 1912.
> And what will they burn instead of coal?

> *«Water» replied Harding*
> *«Water!» cried Pencroft,*
> *«Water as fuel for steamers and engines!»*
> *«Water to heat water! »*
> *«Yes, but water decomposed into its primitive elements» replied Cyrus*
> Jules Verne, French writer, in "The Mysterious Island," in 1874.

Energy is recognized as the most important issue of the 21st century, since other social, political, economic, and environmental problems are strictly

*Ruthenium Complexes: Photochemical and Biomedical Applications,* First Edition.
Edited by Alvin A. Holder, Lothar Lilge, Wesley R. Browne, Mark A.W. Lawrence, and Jimmie L. Bullock Jr.

entangled with the access and distribution of energy sources. The first question deals with energy supply: global consumption increased from 10 to 13.5 billion of ton of oil equivalent (toe, corresponding to an increase from 13 to 18 TW, expressed as average power) between 2000 and 2013 [1]. These values are predicted to still significantly increase along this century, as a consequence of world population growth and of the increased need from developing countries [1]. In addition, more than 80% of current consumption is provided by fossil fuels such as natural gas, coal, and oil. However, the depletion of their reserves (although there are still different opinions on fossil fuel availability in the future [2]), their nonuniform distribution on the planet (leading to political and economic instability), and the pollution associated with their combustion (associated with climate changes and health diseases) suggest that the search for alternative sources of energy – ideally renewable, safe, cheap, abundant, and equally distributed – is a mandatory task rather than an opportunity. More than 100 years ago, the Italian photochemist Giacomo Ciamician stated that sooner or later the world would face an inevitable transition from fossil to renewable fuels, recognizing the enormous potential of solar energy [3]. Sun irradiates the surface of the Earth with an amount of energy of $1.2 \times 10^5$ TW, far exceeding by 4 orders of magnitude the current human need (meaning that the energy received from the Sun in 1 h would be enough to power the planet for 1 year). However, sunlight is intermittent and is not directly exploitable; and, therefore, it has to be captured, converted, and stored into useful forms, such as electricity and fuels. Electricity from photovoltaics is a clean way to exploit this energy and, even if it contributes only as 1% of the world electricity demand, its production is fast increasing and is now 10 times higher than in 2008 [4]. Solar fuels represent an ideal way to convert and store solar energy into chemical energy, in the way that Nature has done since the first forms of photosynthetic organism appeared on the Earth.

### 3.1.2 Photosynthesis and Solar Fuels

Natural photosynthesis is the process by which plants, algae, and cyanobacteria convert solar light into chemical energy by transforming water and $CO_2$ into molecular oxygen and carbohydrates (Eq. (3.1) and Figure 3.1) [5].

$$6CO_2 + 6H_2O + h\nu \rightarrow C_6H_{12}O_6 + 6O_2 \tag{3.1}$$

This process starts with photon absorption by two protein complexes, named photosystem I (PSI) and photosystem II (PSII), promoting a series of electron transfers and ending up with a charge separation state. In this state, *high-energy electrons* are exploited to drive reduction of nicotinamide adenine dinucleotide phosphate, accumulating a reductive equivalent in the form of NADPH; this latter is then used in the light-independent part of photosynthesis, which ultimately leads to conversion of $CO_2$ into carbohydrates. The electron vacancies (usually referred to as *holes*) are collected in a $CaMn_4O_x$ complex at the heart of PSII, known as the oxygen-evolving center (OEC), and utilized to oxidize water to oxygen (Figure 3.1).

In an oversimplified view, photosynthesis consists of two functions: (i) the generation of an electric bias (viz, the charge separation state) upon light

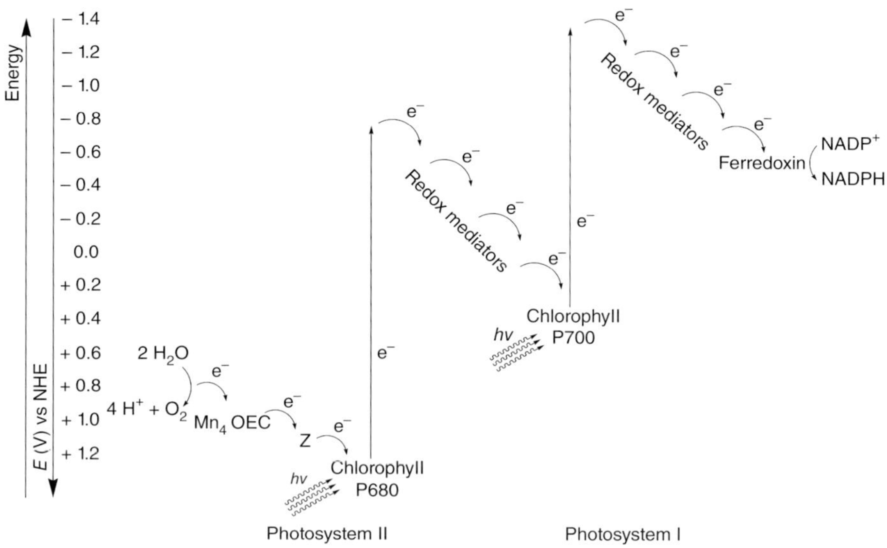


**Figure 3.1** Schematic representation of the Z-scheme of photosynthesis, within the energy diagram representing the main electron-transfer processes and the redox reactions involved.

absorption; (ii) the utilization of the electric bias to drive two redox reactions, where the reduction serves to ultimately produce the fuels, while the oxidation produces oxygen from water. Inspired by Nature, an analogous strategy could be exploited in order to perform similar transformations artificially, aimed at the production of solar fuels or of fine chemicals from abundant and cheap materials such as water and carbon dioxide [6]. Examples are reported in Eqs (3.2)–(3.6), where the valuable products are hydrogen, carbon monoxide, formic acid, methanol, and methane, respectively.

$$2H_2O + h\nu \rightarrow 2H_2 + O_2 \tag{3.2}$$

$$CO_2 + h\nu \rightarrow CO + 0.5O_2 \tag{3.3}$$

$$H_2O + CO_2 + h\nu \rightarrow HCOOH + 0.5O_2 \tag{3.4}$$

$$4H_2O + 2CO_2 + h\nu \rightarrow 2CH_3OH + 3O_2 \tag{3.5}$$

$$2H_2O + CO_2 + h\nu \rightarrow CH_4 + 2O_2 \tag{3.6}$$

Concerning light absorption and charge separation, different technologies are currently considered, such as photovoltaic modules, integrated photoelectrochemical cells, or, more ambitiously, photoactive colloids [7]. Besides this, the efficiency of a photosynthetic device is also strictly dependent on the ability of constructing the new molecules, and therefore on the efficiency, in terms of both rate and selectivity, of the redox reactions involved.

In particular, in Eqs (3.2)–(3.6), water oxidation to oxygen is a fundamental step, since through this semi-reaction water provides the electrons that feed the reductive side of the process, involving protons and/or carbon dioxide conversion.

### 3.1.3 Water Oxidation

Water oxidation [8] is represented in Eq. (3.7), while Eq. (3.8) deals with oxidation of hydroxyl anions, pertinent to alkaline environment.

$$2H_2O \rightarrow O_2 + 4H^+ + 4e^- \tag{3.7}$$

$$4OH^- \rightarrow O_2 + 2H_2O + 4e^- \tag{3.8}$$

Slow rates of the water oxidation process have long been recognized as the limiting factor that hampers the development of efficient devices for artificial photosynthesis. Therefore, this semi-reaction has recently attracted many efforts, both for the development of a new catalyst and for the identification of the reaction mechanism. The issues in water oxidation have both thermodynamic and kinetic reasons. It is indeed thermodynamically demanding, since the $E^0$ for the $O_2/H_2O$ couple is 1.23 V versus the normal hydrogen electrode (NHE); it has to be underlined that $1e^-$ oxidation of water, to hydroxyl radicals $OH^\bullet$ ($E^0$ for $OH^\bullet/H_2O = 2.31$ V), $2e^-$ oxidation of water to hydrogen peroxide $H_2O_2$ ($E^0$ for $H_2O_2/H_2O = 1.77$ V) or $3e^-$ oxidation of water to $^\bullet OOH$ ($E^0$ for the $^\bullet OOH/H_2O = 1.67$ V) are even more demanding. Moreover, water oxidation to oxygen is a four-electron process, also involving four-proton transfer and an oxygen–oxygen bond formation: given such complexity, it proceeds with high activation barriers, unless a *catalyst* is present. The mechanism of this reaction in Nature is not fully understood; however, significant progress has been achieved in the characterization of the natural OEC (Figure 3.2) [9]. This is a $CaMn_4O_x$ cluster that drives water oxidation to oxygen with a high turnover frequency (TOF, defined as the number of oxygen molecules evolved per molecule of catalyst per unit of time) in the range 100–400 $s^{-1}$, a value that for a long time has been unreachable in artificial systems. A detailed mechanistic analysis of the OEC is discussed in several recent papers and reviews [9] and is out of the scope of this chapter. Nevertheless, some of its key features, relevant in the design of synthetic catalysts, can be highlighted as follows: (i) *synergy among transition metals*: to perform the four $e^-$ oxidation of water, Nature uses a four-metal-based catalyst, where all the Mn atoms are redox active; (ii) *hole accumulation*: the OEC works as a charge pool along the Kok cycle (Figure 3.2), and in particular is

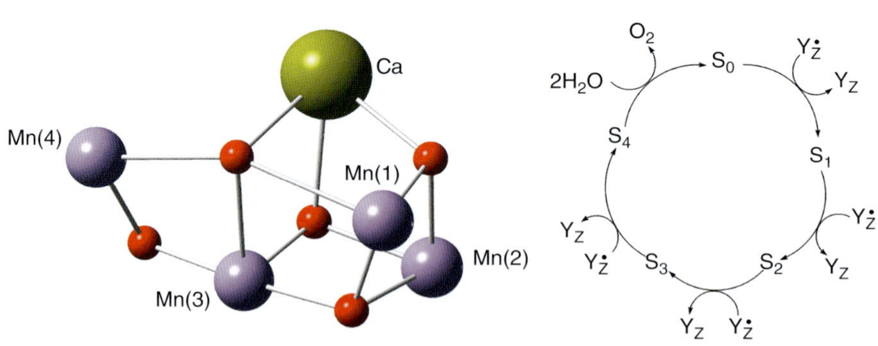


**Figure 3.2** Structural representation of the $Mn_4$ OEC (obtained from the crystallographic data in Ref [9a]), and the Kok cycle.

able to stepwise accumulate four oxidizing equivalents along the $S_i$ intermediates ($i = 0–4$), by reaction with a tyrosine radical $Y_z^\bullet$ oxidant ($E$ for $Y_z^\bullet/Y_z$ ca 1.0-1.2 V vs NHE) [10] ultimately produced upon light absorption; the four-oxidized state of the OEC is able to oxidize water to oxygen in a single step, thus avoiding generation of hydroxyl radicals, hydrogen peroxide, or peroxide radicals as partially oxidized intermediates; (iii) *proton-coupled electron transfer (PCET)*: the stepwise oxidations of the OEC are favored by the removal of one proton contemporary to the loss of one electron; this avoids charge accumulation and lowers the redox potentials of the couples (*redox potential leveling*) [11]; (iv) *high-valent Mn-O intermediate*: although the exact nature of the oxygen-evolving active state and the mechanism of $O_2$ formation are still under discussion, a key feature is the generation of a high-valent Mn—O-based intermediate, which then should react with a water molecule to form the new O—O bond; (v) *supramolecular assembly*: the OEC is embedded in a supramolecular architecture with pigments and primary electron acceptors that allow efficient light-activated water oxidation; (vi) *stability issues*: along the oxygenic cycle, the OEC undergoes damage, due to the harsh oxidative conditions required to perform water oxidation, and needs to be self-regenerated every 30-60 min.

### 3.1.4 Artificial Water Oxidation

Inspired by the astounding OEC natural machinery, efforts have been made in trying to reproduce its operating principles in synthetic catalysts; recent reviews collect the ample literature reports [12]. Given its extended redox chemistry, ruthenium was one of the most studied transition metals in this field of catalysis, either as molecular complexes or in extended phases of ruthenium oxide.

Given the complexity of inserting a catalyst in a full photosynthetic device, the catalytic performance in water oxidation is often studied in solution by taking advantage of "shunt" systems, exploiting (i) dark oxidants [13], (ii) photogenerated oxidants [14], and (iii) electrochemical techniques [15]. Examples of the most common dark oxidants are $Ce^{IV}$ salts or $NaIO_4$; $Ce^{IV}$, and, in particular, cerium ammonium nitrate $(NH_4)_2Ce(NO_3)_6$, CAN, is a one-electron oxidant with $E(Ce^{IV}/Ce^{III}) = 1.75$ V versus NHE [13]. It operates under an acidic environment and is usually employed in large excess with respect to the WOC, enabling its stepwise oxidations until the active form that evolves oxygen. Typical parameters setting catalyst performance are turnover number (TON, defined as the ratio between the moles of $O_2$ and the moles of the catalyst) and TOF (defined as the TON per unit of time). A widely used photogenerated oxidant is $Ru(bpy)_3^{3+}$ (bpy = 2,2′-bipyridine), with $E(Ru^{III}/Ru^{II}) = 1.26$ V versus NHE [13]. It is typically photogenerated from $Ru(bpy)_3^{2+}$ upon oxidative quenching, in the presence of an electron acceptor, such as $S_2O_8^{2-}$ or $Co^{III}(NH_3)_5Cl$. $Ru(bpy)_3^{3+}$ acts then as a one-electron oxidant versus the WOC, activating the oxygenic cycle. TON and TOF for the WOC can be defined also in such cycles, but may be limited by competitive routes involving $Ru(bpy)_3^{3+}$ self-bleaching. Another parameter typically reported to characterize the activity of the system is the quantum yield of oxygen production, $\phi(O_2)$, defined as the ratio between the molecules of $O_2$ produced and the number of photons absorbed by the system [14]. When dark or photogenerated oxidants are used, the possibility of activating the WOC through stepwise oxidation is determined by the redox potential of the oxidant that needs to be higher than the potentials of the redox couples of the WOC. This drawback is overcome by the use of electrochemical techniques [15], such as cyclic voltammetry: varying the potential toward anodic scan in the presence of a WOC, the onset potential for water oxidation is defined as the potential at which the catalytic process starts, as indicated by the rising of intense anodic waves. The overpotential of the catalyst can then be determined by the difference between the onset potential and the thermodynamic potential of the $O_2/H_2O$ couple, and indicates the electrochemical barrier of the catalyst to drive the reaction. Other electrochemical parameters are the current density (indicated at a given overpotential value), the faradaic efficiency (ratio of $O_2$ produced and the total charge passed at the electrode, normalized by Faraday's law), while TON and TOF values of the catalyst can be determined under specific conditions [15]. Electrochemical techniques are useful also to evaluate the stability of the catalyst under prolonged electrolysis and its behavior when linked at the electrode conducting surface.

## 3.2 Ruthenium in Water Oxidation Catalyst

### 3.2.1 Ruthenium Oxide

One of the artificial systems that exploits the synergy of neighboring redox-active transition-metal centers is a metal oxide surface. Indeed, metal oxides, ideally in nanoparticle form, have been extensively considered in water oxidation catalyst (WOC); among these, ruthenium oxide ($RuO_2$, rutile structure) is one of the most investigated catalysts. First evidences of oxygenic activity of $RuO_2$ were reported by Graetzel and coworkers in the late 1970s [16], in combination with dark oxidants such as Ce(IV) or even integrated in a light-driven system, with $Ru(bpy)_3^{2+}$ (bpy = 2,2′-bipyridine) as the photosensitizer and dimethylviologen as the primary electron acceptor [16c]. The catalytic performance is strongly influenced by the $RuO_2$ hydration grade, with the optimal value found between 12 and 14%, since more hydrated samples undergo oxidation to volatile $RuO_4$ and consequent corrosion, while conversely poorly hydrated $RuO_2$ is characterized by a reduced surface area and therefore low amount of active sites [17]. However, the most interesting property of $RuO_2$ in WOC, is its electrochemical reactivity, and in particular its low operating overpotential, as low as 0.35 V [18], that induced scientists to investigate the operating mechanism. A first mechanistic hypothesis was proposed by Trasatti *et al.* [18c, d], and is represented in Eqs (3.9)–(3.12), where {Ru} is a surface active site of $RuO_2$.

$$\{Ru\} + H_2O \rightarrow \{Ru\}—OH + H^+ + e^- \tag{3.9}$$

$$\{Ru\}—OH \rightarrow \{Ru\}—OH^* \tag{3.10}$$

$$\{Ru\}—OH^* \rightarrow \{Ru\}—O + H^+ + e^- \tag{3.11}$$

$$2\{Ru\}—O \rightarrow \{Ru\} + O_2 \tag{3.12}$$

In this mechanistic hypothesis, under applied electrochemical bias $H_2O$ adsorbs to a surface active site, with contemporary loss of one electron (to the electrode) and one proton (to the solution) and formation of a hydroxy surface site {Ru}—OH (Eq. (3.9)). After a chemical rearrangement of this hydroxy surface site, Eq. (3.10), a {Ru}-OH* intermediate undergoes a second one-electron loss concomitant to proton release, leading to the formation of an oxo surface site {Ru}-O (Eq. (3.11)); oxygen evolution involved then the reaction of two {Ru}-O units, restoring the surface metal site to its initial state (Eq. (3.12)). A change in the rate determining step (rds) was envisaged on the basis of different Tafel slopes: at low applied bias, with Tafel slopes in the range 30–60 mV/decade, the chemical rearrangement of the Ru-hydroxy surface site (Eq. (3.10)) was proposed as the rds, while at high applied bias, where Tafel slopes of 120 mV/decade were observed, $H_2O$ adsorption/one-electron oxidation on the {Ru} surface site (Eq. (3.9)) was identified as the plausible rds.

A different mechanism was however proposed by Rossmeisl and coworkers on the basis of theoretical calculations, Eqs (3.13)–(3.17) [19]. They identified water adsorption and oxidation to {Ru}-OH or {Ru}-O surface sites, similarly to the early hypothesis by Trasatti, with the presence of metal-hydroxy sites being more favored at low applied potential, while metal-oxo sites are more favored at high applied potentials ($E > 1.4$ V). The presence of {Ru}-O sites is moreover mandatory to access WOC at potentials above 1.6 V (corresponding to 0.37 V overpotential, in fair agreement with the experimental value of 0.35 V), with the most demanding step being reaction of a surface {Ru}-O with a water molecule, to form a metal-based hydroperoxide intermediate {Ru}-OOH (Eq. (3.15)). Therefore, in this mechanistic scenario, oxygen–oxygen bond formation is originated from reaction of a surface metal–oxo active site with a water molecule reacting as a nucleophile, rather than by the combination of two metal–oxo sites as originally proposed by Trasatti. More recent calculations, by adding explicit water molecules and a continuum solvent model to the $RuO_2$ surface, were performed by Fang and Liu [20], who identified a water dissociation mechanism being the most favorable at high applied bias, where the rds is the reaction of water with a surface {Ru}-O site (in line with Rossmeisl's hypothesis). The authors were also able to justify the change in Tafel slope at low applied potentials by the presence of protonated bridging oxygen at the ruthenium oxide surface, and by the consequent energy barrier for their deprotonation. Calculated Tafel slopes in the two different potential ranges were indeed consistent with the experimental values.

$$\{Ru\} + H_2O \rightarrow \{Ru\}\text{-}OH + H^+ + e^- \tag{3.13}$$

$$\{Ru\}\text{-}OH \rightarrow \{Ru\}\text{-}O + H^+ + e^- \tag{3.14}$$

$$\{Ru\}\text{-}O + H_2O \rightarrow \{Ru\}\text{-}OOH + H^+ + e^- \tag{3.15}$$

$$\{Ru\}\text{-}OOH \rightarrow \{Ru\}\text{-}OO + H^+ + e^- \tag{3.16}$$

$$\{Ru\}\text{-}OO \rightarrow \{Ru\} + O_2 \tag{3.17}$$

Such calculations were also useful for comparing the activity of different metal oxide surfaces, and to set a guideline for the development of an ideal WOC. In a plausible reaction mechanism, such as the one depicted in Eqs (3.13)–(3.17), the

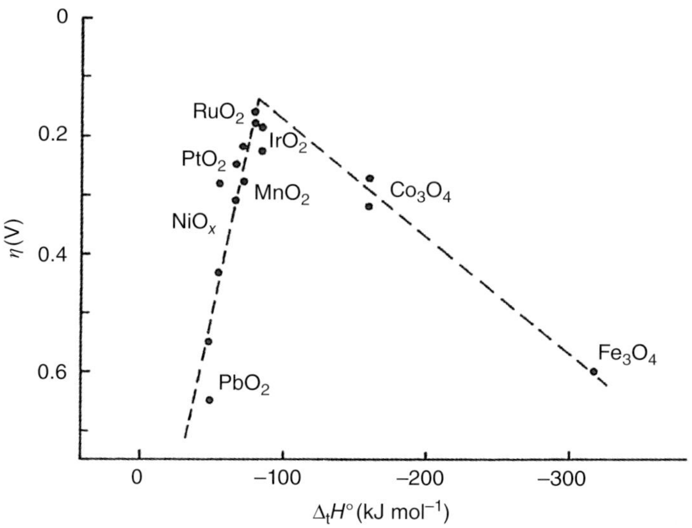


**Figure 3.3** Dependence of overpotential for water oxidation on the enthalpy of oxygen adsorption on transition metals for different heterogeneous metal oxides. (Reprinted with permission from [18a]. Copyright (2012) American Chemical Society.)

overpotential of the reaction is determined by the most energetically demanding step, referred to as the potential-determining step [21]. In an ideal catalyst, all redox steps should be characterized by the same potential of 1.23 V, in order to eliminate overpotential; for a real catalyst where oxygen–oxygen bond formation is the potential-determining step, this latter determines a finite overpotential [21]. Rossmeils and coworkers suggested that the potentials of the elementary steps are dependent on the chemisorption energies of the OH, O, and OOH intermediates, which, for a specific metal or metal-oxide surface, are linearly correlated. An ideal WOC should display a compromise of oxygen chemisorption and oxygen release abilities, and $RuO_2$ is the optimal one, even better than iridium oxide, showing the minimum overpotential and setting at the top of the volcano curve (plot of the activity versus the binding energy of surface intermediates, Figure 3.3).

### 3.2.2 Molecular Ruthenium WOC

Ruthenium was also one of the first transition metals considered to develop *molecular* WOC, namely, well-defined coordination complexes that should allow mechanistic studies through kinetic analysis and characterization of reaction intermediates, ideally exploited to optimize catalyst performance. This paragraph is not intended to be omni comprehensive of the explosion of reports in the literature of the past decade dealing with this topic, but rather comments on some selected examples, considered as major milestones in this field (Figure 3.4).

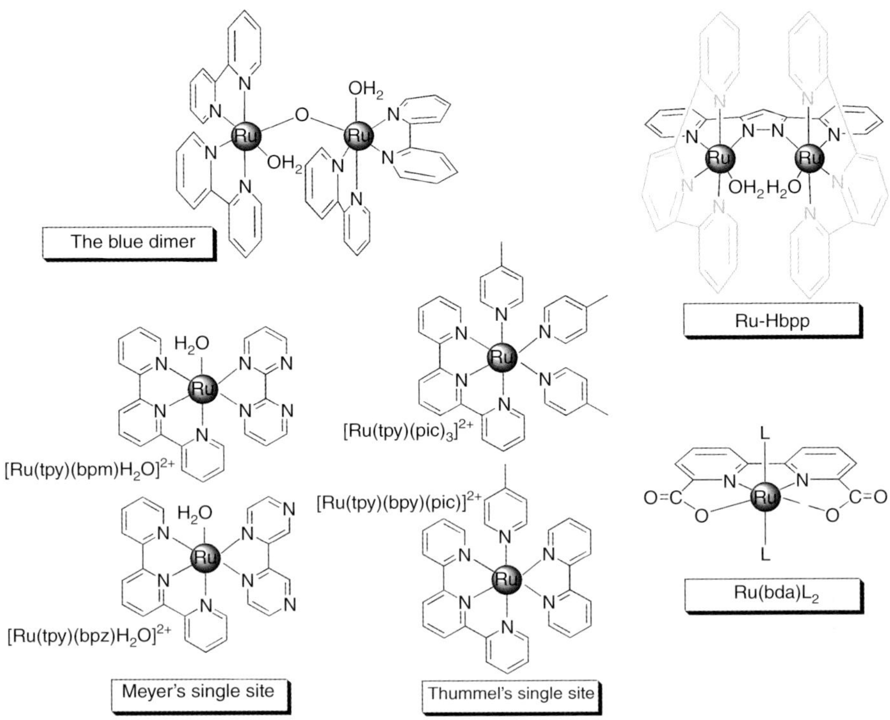


**Figure 3.4** Molecular WOC discussed in section.

### 3.2.2.1 Meyer's Blue Dimer

In 1981, pioneering work of T. J. Meyer on a $[Ru^{II}(bpy)_2(py)H_2O]^{2+}$ species (bpy = 2,2′-bipyridine; py = pyridine) [22] evidenced that the $Ru^{II}$ center in this species could undergo two sequential $1e^-$ electrochemical oxidations to $Ru^{III}$ and $Ru^{IV}$, and that in neutral aqueous solution these processes were concomitant to proton removal from the apical aquo ligand, transforming the parent species into a $Ru^{III}$–hydroxo and into a $Ru^{IV}$–oxo (Eq. (3.18)):

$$\begin{aligned}[Ru^{II}(bpy)_2(py)H_2O]^{2+} &\rightarrow [Ru^{III}(bpy)_2(py)OH]^{2+} + e^- \\ &\rightarrow [Ru^{IV}(bpy)_2(py)O]^{2+} + e^- \end{aligned} \tag{3.18}$$

The two redox potentials differ only by 110 mV, being a perfect example of redox potential leveling induced by PCET (see earlier discussion in Section 3.1.3). Although the $[Ru^{II}(bpy)_2(py)H_2O]^{2+}$ was not able to drive water oxidation, given the two-electron redox chemistry and the low potential of the $Ru^{IV}$ derivative, this study suggested that a polypyridine ligand framework could be suitable to achieve multiple PCET at a ruthenium center, while at least two metal sites would be needed to reach a four-electron process. These indications led to the report in 1982 of the first molecular WOC, $[(bpy)_2(H_2O)Ru^{III}–O–Ru^{III}(OH_2)(bpy)_2]^{4+}$, known as the blue dimer (Figure 3.4) [23]. This species is based on two ruthenium(III) centers connected through a μ-oxo bridge, with two bpy and one water molecule completing the octahedral coordination sphere of each metal, and evolved oxygen when combined with a $Ce^{IV}$ salt as the chemical oxidant, although with a modest TON of 16. Characterization of such intermediates and mechanism proposals were thus the subject of intense investigation mainly by the groups of Meyer [24] and of Hurst [25]. Although some aspects are still debated, and the occurrence of multiple reaction pathways has been envisaged, the key feature of this complex is the possibility of reaching several oxidized intermediates, through stepwise electron loss, often coupled to proton removal from the water apical ligands. It is then generally recognized that the step leading to oxygen–oxygen bond formation occurs by nucleophilic attack of a water molecule to a high-valent Ru—O moiety (usually referred to as water nucleophilic attack, WNA), leading to the formation of a RuOOH hydroperoxide. Attack of the water molecule was initially considered to occur at a $Ru^{V}$, $Ru^{V}$ blue dimer intermediate bearing oxo apical ligands (formed by $4e^-/4H^+$ removal from the pristine $Ru^{III}$, $Ru^{III}$ state) [25], while more recently reaction of water molecule with a $Ru^{IV}$, $Ru^{V}$ was postulated on the basis of spectroscopic evidence [24b]. Therefore, the mechanism of formation of the oxygen–oxygen bond in the blue dimer involves a single ruthenium center and displays striking similarities with the one proposed in heterogeneous phases of $RuO_2$. Importantly, formation of an oxygen–oxygen bond from intramolecular coupling of two high-valent Ru–O units was neglected, on the basis of the nonproximity of the two groups, while significant and energetically unfavorable distortion of the molecule would be needed to access this pathway.

#### 3.2.2.2 The Ru-Hbpp Catalyst

This reaction route is instead the one operating in the second molecular catalyst discussed here, the $\{[Ru^{II}(tpy)(H_2O)]_2(\mu\text{-bpp})\}^{3+}$, known as *Ru-Hbpp* (tpy = 2,2′:6′,2″-terpyridine; bpp = 3,5-bis-(2-pyridyl)pyrazolate) (Figure 3.4). This dinuclear ruthenium complex was reported in 2004 by Llobet and coworkers [26]; the first, clear difference with respect to the blue dimer is the nature of the bridging ligand between the two ruthenium centers, being a bpp instead of the oxo group. This confers a more rigid geometry to the molecule, arranging the two water molecules bound to the ruthenium centers in close proximity, an indispensable feature for the oxygenic reactivity of the complex; the absence of the μ-oxo moiety in this catalyst was thought also to improve stability toward reduction. The second difference between the Ru-Hbpp and the blue dimer is related to the redox states reachable upon oxidation. The isolated form of Ru-Hbpp bears two $Ru^{II}(H_2O)$ groups, connected through the μ-bpp bridge; in the presence of chemical oxidants or under anodic electrochemical scan, Ru-Hbpp undergoes a $4e^-/4H^+$ loss, to form an intermediate with two $Ru^{IV}(O)$ units responsible for $O_2$ evolution, through an intramolecular mechanism often referred to as I2M (interaction of two metal units). Isotopic labeling and kinetic, spectroscopic, and theoretical evidence [27] suggest, indeed, an internal coupling of the two oxygen atoms, evolving into a $Ru^{III}$–O–O–$Ru^{III}$ peroxide, that upon reaction with a water molecule evolves to a $Ru^{III}$–hydroperoxide (RuOOH, whereas the second Ru center at this stage bears a hydroxo group), finally evolving $O_2$ by reaction of a second water molecule, restoring the Ru-Hbpp catalyst in its initial form (Figure 3.5). It is interesting to pinpoint that the I2M mechanism

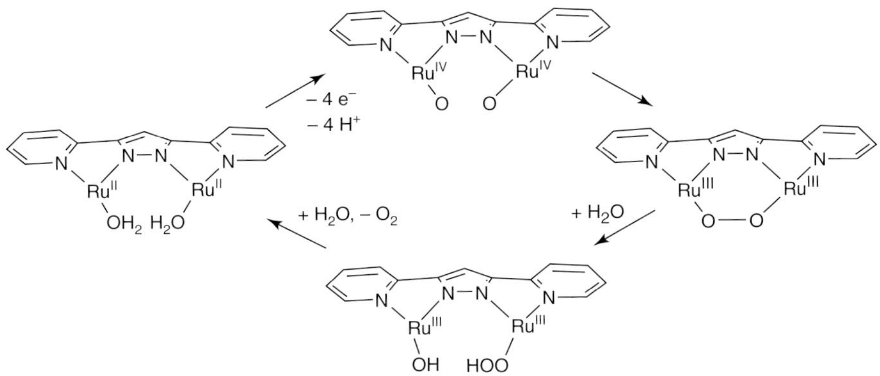


**Figure 3.5** Proposed pathway leading to oxygen evolution with Ru-Hbpp (tpy ligands of Ru-Hbpp are omitted for reasons of clarity) [27].

operating for Ru-Hbpp requires a relatively low oxidation state of $Ru^{IV}$ to build the new oxygen–oxygen bond, while the WNA described for the blue dimer and operating in other Ru-WOC (vide infra) requires at least a $Ru^{V}$ state.

#### 3.2.2.3 Single-Site Ru-WOCs

Originally, it was believed that single-metal site coordination complexes could not adequately work as WOC, since a single-metal center could hardly reach five different redox states, at the suitable potentials to be involved in a $4e^-$ cycle for water oxidation. Conversely, several single-metal site WOCs have been recently proposed, and a WNA to a metal-oxo group is the proposed mechanism for O—O bond formation in most cases. Indeed, a WNA to a high-valent M—O intermediate lowers by two units the formal oxidation state of the metal center (Eq. (3.19), the formal oxidation state of M is indicated by the apex in brackets).

$$M^{(n)} - O + H_2O \rightarrow M^{(n-2)} - OOH + H^+ \tag{3.19}$$

Therefore, if the WNA occurs at an early stage of the catalyst $4e^-$ stepwise oxidation, it is possible to reduce the number of the involved oxidation states for the metal. An example is provided by the single Ru site WOC proposed by Meyer and coworkers, $[Ru^{II}(tpy)(bpm)H_2O]^{2+}$ and $[Ru^{II}(tpy)(bpz)H_2O]^{2+}$ (bpm = 2,2′-bipyrimidine; bpz = 2,2′-bipyrazine) [28]. In the presence of $Ce^{IV}$, or under anodic electrochemical scan, the $Ru^{II}$—$OH_2$ unit undergoes a $2e^-/2H^+$ oxidation to a $Ru^{IV}$—O ($E$ ca. 1.2 V vs NHE at pH 0); a third oxidation transforms $Ru^{IV}$—O into $Ru^{V}$—O, and being pH independent this is not proton coupled and therefore occurs at a considerably high potential of 1.65 V versus NHE, concomitantly with water oxidation. It was proposed that the $Ru^{V}$—O intermediate then undergoes WNA, forming a $Ru^{III}$—OOH hydroperoxide, which is then subject to the fourth – and last – $1e^-$ oxidation, coupled to proton transfer to the solvent, forming a $Ru^{IV}$—OO hydroperoxide (Figure 3.6). This latter is the species that evolves oxygen, in an rds, restoring the catalyst in its initial form by coordination of a water molecule. At the end, the catalyst operates in the water oxidation cycle between $Ru^{II}$ and $Ru^{V}$ oxidation states. In another single-site

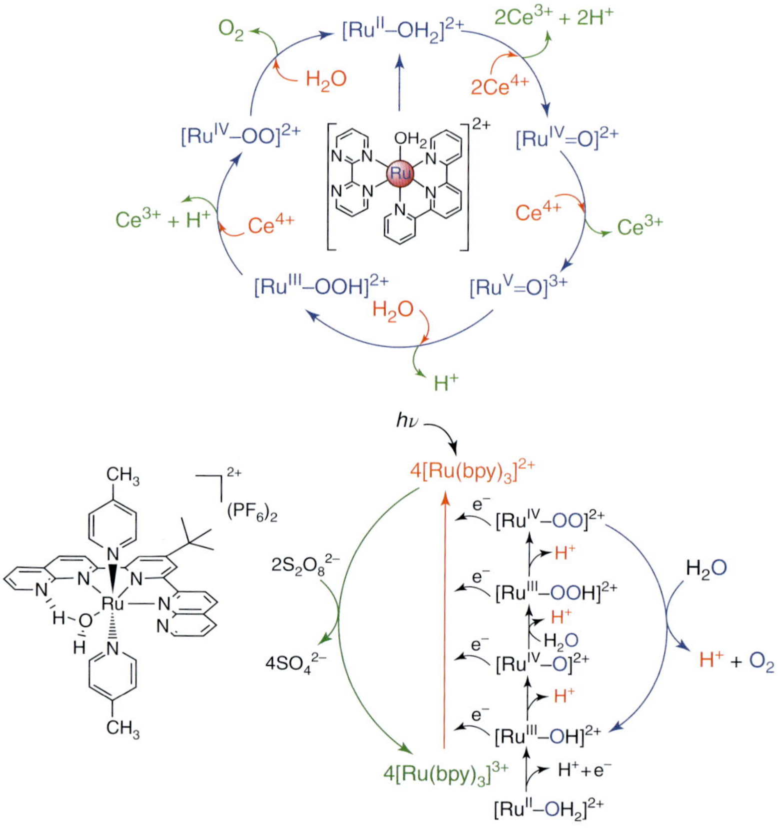


**Figure 3.6** Mechanism of water oxidation with the Meyer's. (left: Reprinted with permission from [28]. Copyright (2008) American Chemical Society.) and Thummel's (right: Reprinted with permission from [29b]. Copyright (2013) Royal Society of Chemistry.) single-site catalysts.

Ru complex with polypyridine-based ligands, $[Ru^{II}(npm)(H_2O)(pic)_2]^{2+}$ (npm = 4-*tert*-butyl-2,6-di-(1′,8′-naphthyrid-2′-yl)-pyridine, pic = 4-picoline), first reported by Thummel in 2005 [30], the WNA occurred at a $Ru^{IV}$—O site, and therefore the catalyst operates only within $Ru^{III}$ and $Ru^{IV}$ redox states in a light-activated cycle for water oxidation (Figure 3.6) [29].

#### 3.2.2.4 Heptacoordinated Ru Intermediates

In Meyer's single-site catalyst, a bidentate chelating motif of the peroxide unit to the ruthenium center in the $Ru^{IV}$—OO intermediate was envisaged on the basis of DFT calculations and of previous structural investigation of ruthenium(IV)-peroxide species [28], indicating a hepta-coordinated ruthenium. A seven-coordinated structure was considered also in another $Ru^{II}$ single-site WOC developed by Thummel, exploiting the tpy, bpy, and pic ligands: $[Ru^{II}(tpy)(pic)_3]^{2+}$ and $[Ru^{II}(tpy)(bpy)(pic)]^{2+}$ [31]. The key feature of these species is the saturated coordination sphere of ruthenium, which should, in principle, hamper their reactivity as WOCs. However, after a two-electron oxidation, $Ru^{IV}$ intermediates were envisaged to expand their coordination sphere by binding a water molecule [31]. The $Ru^{IV}—OH_2$ intermediate could then undergo a $2e^-/2H^+$ oxidation to $Ru^{VI}—O$, susceptible to WNA with hydroperoxide formation, and final oxygen release.

A hepta-coordinated intermediate, but involving a different mechanism rather than WNA, characterizes the last class of Ru WOCs discussed in this paragraph: the $Ru(bda)L_2$ family (bda = 2,2′-bipyridine-6,6′-dicarboxylic acid; L = monodentate apical ligands). The presence of negatively charged carboxylate functions at the bpy ligand lowers the redox potential of the Ru-based couples [32], while a second, key feature of such complexes is a large, O—Ru—O angle of ca 123° (where the two oxygens are those of the carboxylate functions and coordinating the Ru center). This property allows expansion of the Ru coordination sphere to hepta-coordination, by binding a water molecule. In particular, in the case of L = 4-methylpyridine (pic), second-order kinetics speak in favor of a bimolecular catalyst step involved in the mechanism, and a dimeric structure was isolated and structurally characterized, where two $[Ru^{IV}(bda)pic_2]$ moieties are connected through a $[HOHOH]^-$ bridge (Figure 3.7) [32a]. Further studies evidenced the active role of the monodentate, axial ligands on the catalytic performance [35], and, in particular, with L = isoquinoline (isoq), the catalyst reached astounding activity in terms of TOF, up to $300\ s^{-1}$, close to the natural OEC benchmark, although in the presence of $Ce^{IV}$ dark oxidant [33]; the enhanced reactivity was ascribed to favorable intermolecular interactions between the aromatic isoq ligands, favoring the bimolecular O—O bond formation through radical coupling of two $Ru^V—O$ species, at low overpotential of ca 0.2 V (Figure 3.7). Catalysts with similar activity and TOF up to $1000\ s^{-1}$ were indeed obtained by introducing methoxy substituents at the isoq [36a] or halogen substituents at the isoq or phthalazine (ptz) monodentate axial ligands [36b]. Tuning of the apical ligand and use of 1,4-bis(pyrid-3-yl)benzene (bpb) as linkers were recently exploited by Wurthner to build a supramolecular architecture composed of three Ru(bda)-based units (Figure 3.7) [34]. Interestingly, in this system, kinetic and isotopic labeling experiments ruled out a bimolecular (I2M) pathway and suggested a WNA to a $Ru^V—O$ unit as the operating mechanism. In addition, observed TOF values $> 100\ s^{-1}$ indicate an unusual, very low activation barrier for a WNA mechanism; the authors hypothesized a suitable organization of water molecules inside the cavity of the catalyst triad as responsible for the low barrier for O—O bond formation.

### 3.2.3 Polyoxometalates: The Bridge Between Metal Oxides and Coordination Complexes

Given the harsh oxidizing conditions required to access water oxidation, one of the key properties required for a catalyst in order to achieve long-lasting performance is of course robustness toward oxidation. This is why *polyoxometalates* (POMs) have been considered in this field of catalysis. These are molecular, metal–oxo clusters of early transition metals in their highest oxidation state ($V^V$, $Nb^V$, $Ta^V$, $Mo^{VI}$, $W^{VI}$; in particular, $W^{VI}$ forms polyoxotungstates, that are the most studied POMs), and therefore inert to oxidation (although POMs may

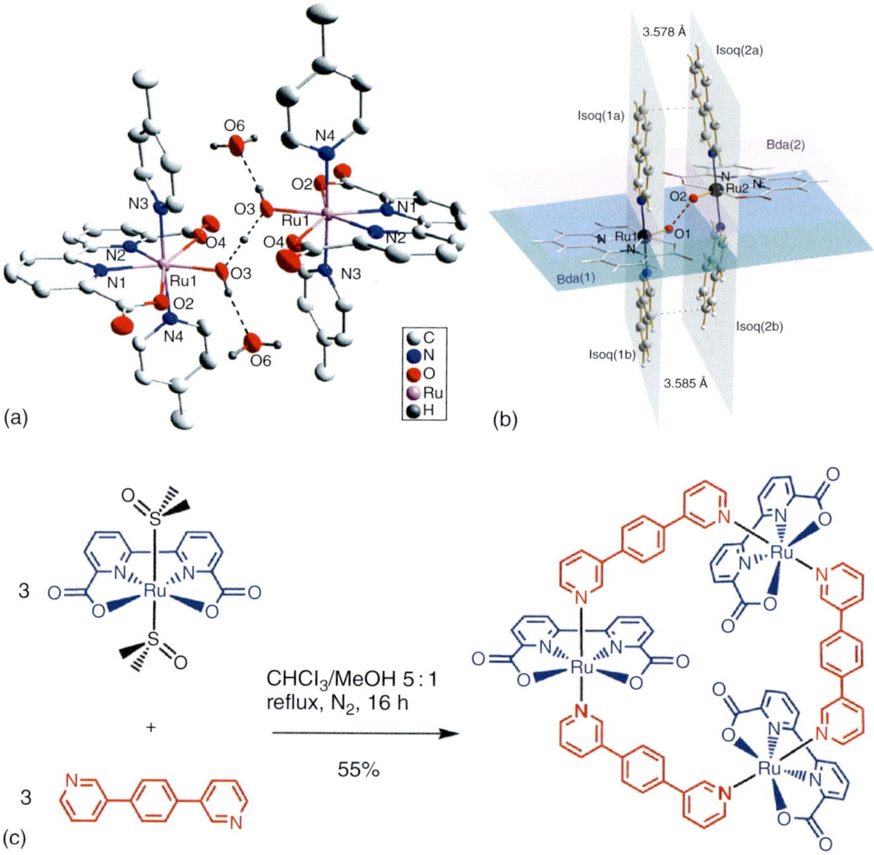


**Figure 3.7** (a) Crystal structure of the Ru(bda)pic$_2$ dimer showing the [HOHOH]$^-$ bridge. (Reprinted with permission from [32a]. Copyright (2009) American Chemical Society.) (b) Calculated bimolecular complex of the Ru(bda)isoq$_2$ catalyst. (Reprinted with permission from [33]. Copyright (2012) Nature Publishing Group.) (c) Synthesis of the macrocycle trimer [Ru(bda)bpb]$_3$. (Reprinted with permission from [34]. Copyright (2016) Nature Publishing Group.)

suffer from hydrolytic stability, especially in alkaline environment) [37]. Other key features of POMs are that (i) they can act as ligands for catalytically active transition-metal centers, and extensive studies have been considered in the field of oxidation catalysis; (ii) a polyanionic charge, allowing solubility in a wide range of solvents, depending on the choice of the countercation; (iii) dimensions in the range of a few nanometers. The combination of ruthenium and polyoxometalates showed striking similarities with Ru-porphyrin chemistry [38a], with the POM scaffold that provides an electron-withdrawing environment, suitable to access high-valent Ru states [38a], as observed in Ru-perfluorinated porphyrins [38b, c].

Electrocatalytic oxidation of water by a sandwich-type ruthenium-based polyoxotungstate $[Ru^{III}_2Zn_2(H_2O)_2(ZnW_9O_{34})_2]^{14-}$ was reported in 2004 by Shannon and coworkers [39], although this species displayed some compositional variation and the mechanism was not deeply investigated. In 2008, our group and that of

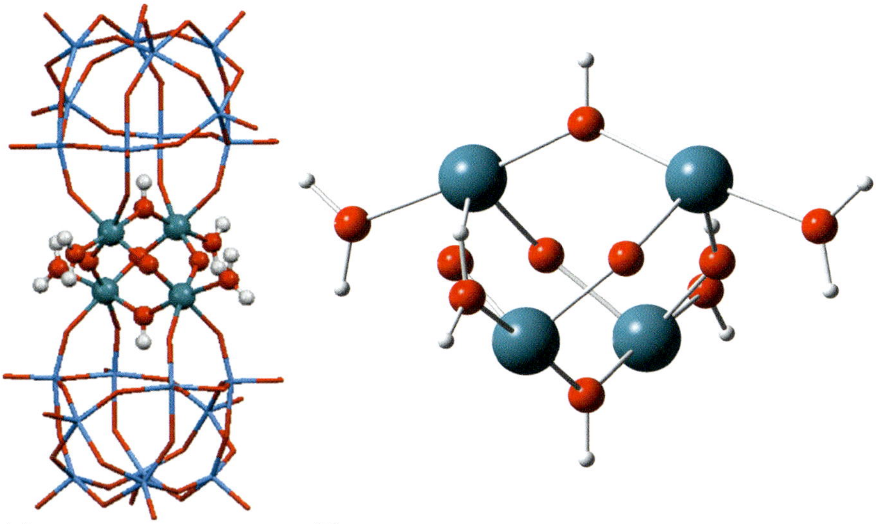

(a) (b)

**Figure 3.8** Structure of $[Ru^{IV}{}_4(\mu\text{-}O)_4(\mu\text{-}OH)_2(H_2O)_4(\gamma\text{-}SiW_{10}O_{36})_2]^{10-}$ (a) and of the adamantane-like $[Ru^{IV}{}_4(\mu\text{-}O)_4(\mu\text{-}OH)_2(H_2O)_4]^{6+}$ tetraruthenium-oxo core (b).

Prof. Hill reported a structurally characterized tetraruthenium polyoxometalate, $[Ru^{IV}{}_4(\mu\text{-}O)_4(\mu\text{-}OH)_2(H_2O)_4(\gamma\text{-}SiW_{10}O_{36})_2]^{10-}$ (Figure 3.8), hereafter $Ru_4$POM, able to evolve oxygen in aqueous solution in the presence of dark oxidants such as $Ce^{IV}$ or $Ru(bpy)_3{}^{3+}$, with turnovers up to 500 and TOF up to $0.125\,s^{-1}$, and low operating overpotential of 0.20-0.35 V, depending on the pH [40].

The key feature of this catalyst is the active, adamantane-like tetraruthenium-oxo core, $[Ru^{IV}{}_4(\mu\text{-}O)_4(\mu\text{-}OH)_2(H_2O)_4]^{6+}$. This can be considered as an optimized fragment of $RuO_2$, embedded within the two polyoxometalate units ("a fragment of a metal oxide embedded in a molecular metal oxide") [41], although in a more irreverent interpretation it can be viewed as a totally inorganic model of the natural OEC ("an inorganic synzime" [14, 42]), being constituted by four redox-active metals, connected through μ-oxo and μ-hydroxo bridges. However, besides structural features, similarities of $Ru_4$POM with both $RuO_2$ and the $Mn_4$-OEC are observed in both the reactivity and on the mechanism of oxygen evolution. Although there was not an unambiguous attribution of the processes, electrochemical characterization of $Ru_4$POM under anodic scan shows the presence of several, equally spaced redox waves in a narrow potential range; these shift depending on the pH [43] and suggest the involvement of PCETs in $Ru^{IV}\text{–}OH_2 \rightarrow Ru^{V}\text{—}OH$ transitions [44]. Concerning the formation of the oxygen–oxygen bond, a WNA to a formal $Ru^{VI}\text{—}O$ (although this should be better described as a $Ru^{V}$—oxyl, with radical character on the oxygen atom) was suggested on the basis of DFT calculations [41]; intramolecular coupling involving two Ru—O moieties is hampered by a nonfavorable orientation of the units, while intermolecular couplings were discarded on the basis of first-order kinetics. Besides its mechanism, there are two further aspects of WOC by $Ru_4$POM that deserve consideration: (i) exploiting its polyanionic charge,

$Ru_4POM$ can be integrated in positively charged conducting surfaces to access heterogeneous catalysis. In particular, carbon nanostructured materials such as multiwall nanotubes [45] and graphene layers [46] have been considered, where the catalyst maintains solution performance in terms of overpotential and TOF. (ii) $Ru_4POM$ displays high activity in light-activated systems, in combination with ruthenium polypyridine photosensitizers and sodium persulfate as the sacrificial electron acceptor (see Introduction) [47]. When the photosensitizer is $Ru(bpy)_3^{2+}$, one of the key features for successful oxygenic performance are multiple electron transfers from $Ru_4POM$ to $Ru(bpy)_3^{3+}$, occurring in a short timescale of tens of milliseconds [48]. When the photosensitizer is $[Ru\{(\mu\text{-dpp})Ru(bpy)_2\}_3]^{8+}$ (dpp = 2,3-bis(2′-pyridyl)pyrazine) strong ionic association with $Ru_4POM$ is observed [49], and the photocatalytic mechanism was proposed to be different from the one involved with $Ru(bpy)_3^{2+}$ photosensitizer, since direct, reductive quenching by $Ru_4POM$ of the triplet excited state $^{3*}[Ru\{(\mu\text{-dpp})Ru(bpy)_2\}_3]^{8+}$ was observed and characterized with transient absorption spectroscopy [49a].

## 3.3 Conclusions and Perspectives

The scenario in the field of WOC has drastically changed in the past decade. In 2008, after the publication of a novel class of iridium-based WOCs, Prof. Meyer commented that "*catalysts for water oxidation are so rare that the discovery of a new family is cause for celebration*" [50]. The first objective of researchers in this field was to develop new, highly active WOC, and since then, hundreds of papers have been published on this topic [12] with impressive improvements reached in terms of catalyst performance. For instance, focusing on the TOF parameter, a vertical jump of 3 to 4 orders of magnitude has been reached from earlier catalysts to current benchmarks; nowadays some Ru-based catalysts, discussed in this chapter, overtake the natural OEC.

A second target, as Prof. Meyer was envisaging in 2008, was that "*barring serendipitous discoveries, further progress in designing catalysts for water oxidation will require detailed knowledge of the mechanism by which these reactions occur*" [50]. A convergent opinion was envisioned 6 years later by Llobet and coworkers [51]. Again, studies on Ruthenium complexes provided several insights also in mechanistic analysis; the reaction routes involved in O—O bond formation identified for ruthenium complexes, in particular the WNA, have been then recognized in other classes of catalysts, based on different transition metals, including earth-abundant ones.

A third, and most important issue for the development of devices for solar energy conversion, is to consider the interface of WOCs with light-activated systems. In this direction, efforts have been directed toward the study and comprehension of "sacrificial systems," by combining the WOC with a photosensitizer and an electron acceptor, often the $Ru(bpy)_3^{2+}/S_2O_8^{2-}$ couple, although other photosensitizers have been recently considered [52]. These studies are pivotal for the comprehension of photoinduced events, in particular the electron transfers involving the photosensitizer/WOC couple [14a, 48, 49], and are introductory to the development of photoanodes, where the photosensitizer (P) and the WOC are anchored to the surface of a semiconductor (SC) [53]. In such devices, the semiconductor acts as an electron acceptor from the excited P (*electron injection*, Eqs (3.20)–(3.21)), ultimately transferring the electrons to a cathode for protons or carbon dioxide reduction; the oxidized photosensitizer $P^+$ oxidizes then the WOC (Eq. (3.22)); processes in Eqs. (3.20)–(3.22) are then repeated until the WOC reaches its form capable of oxidizing water. However, the heterogenization of the photosensitizer/WOC system is not always straightforward, since the rates and the dynamics of photoinduced electron transfers are drastically different from those observed in solution in a sacrificial system. Injection of the PS into the SC, and oxidation of the WOC by $P^+$ may indeed occur in timescales ranging from hundreds of femtoseconds to tens of nanoseconds [14a].

$$\mathrm{SC{-}P{-}WOC} + h\nu \rightarrow \mathrm{SC{-}{}^{*}P{-}WOC} \tag{3.20}$$

$$\mathrm{SC{-}{}^{*}P{-}WOC} \rightarrow \mathrm{SC(e^-){-}P^+{-}WOC} \tag{3.21}$$

$$\mathrm{SC(e^-){-}P^+{-}WOC} \rightarrow \mathrm{SC(e^-){-}P{-}WOC^+} \tag{3.22}$$

However, the proximity of the SC/P/WOC components implies also very fast unproductive charge recombination processes, detrimental to the efficiency of the photoelectrode; in addition, the stability of the material is often limited by photosensitizer desorption or bleaching from the surface [53]. Deeper comprehension and knowledge of the photoactive surface, aimed at the development of novel technological solutions, are definitely needed to overcome the current limits.

## References


1 2015 World Energy Statistics from International Energy Agency (IEA), www.iea.org (accessed 24 July 2016).

2 (a) Mitka, M. (2012) *J. Am. Med. Ass.*, **307**, 2135; (b) Vidic, R.D., Brantley, S.L., Vandenbossche, J.M., Yoxtheimer, D., and Abad, J.D. (2013) *Science*, **340**, article number 1235009; (c) Armaroli, N. and Balzani, V. (2007) *Angew. Chem., Int. Ed.*, **46**, 52–66.

3 Ciamician, G. (1912) *Science*, **36**, 385–394.

4 IEA (2014) *A Snapshot of Global PV (1992–2014)*, IEA – International Energy Agency.

5 Senge, M.O., Ryan, A.A., Letchford, K.A., MacGowan, S.A., and Mielke, T. (2014) *Symmetry*, **6**, 781–843.

6 Berardi, S., Drouet, S., Francàs, L., Gimbert-Surinach, C., Guttentag, M., Richmond, C., Stoll, T., and Llobet, A. (2014) *Chem. Soc. Rev.*, **43**, 7501–7519.

7 McKone, J.R., Lewis, N.S., and Gray, H.B. (2014) *Chem. Mater.*, **26**, 407–414.

8 Hetterscheid, D.G.H. and Sun, L. (2014) *Eur. J. Inorg. Chem.*, 568–790.

9 (a) Suga, M., Akita, F., Hirata, K., Ueno, G., Murakami, H., Nakajima, Y., Shimizu, T., Yamashita, K., Yamamoto, M., Ago, H., and Shen, J.-R. (2015) *Nature*, **517**, 99–103; (b) Yano, J. and Yachandra, V. (2014) *Chem. Rev.*, **114**,

4175–4205; (c) Cox, N., Pantazis, D.A., Neese, F., and Lubitz, W. (2013) *Acc. Chem. Res.*, **46**, 1588–1596; (d) Cox, N., Retegan, M., Neese, F., Pantazis, D.A., Boussac, A., and Lubitz, W. (2014) *Science*, **345**, 804–808.

10 McEvoy, J.P. and Brudvig, G.W. (2006) *Chem. Rev.*, **106**, 4455–4483.

11 (a) Weinberg, D.R., Gagliardi, C.J., Hull, J.F., Fecenko Murphy, C., Kent, C.A., Westlake, B.C., Paul, A., Ess, D.H., Granville McCafferty, D., and Meyer, T.J. (2012) *Chem. Rev.*, **112**, 4016–4093; (b) Liu, F., Concepcion, J.J., Jurss, J.W., Cardolaccia, T., Templeton, J.L., and Meyer, T.J. (2008) *Inorg. Chem.*, **47**, 1727–1752; (c) Gagliardi, C.J., Vannucci, A.K., Concepcion, J.J., Chen, Z., and Meyer, T.J. (2012) *Energy Environ. Sci.*, **5**, 7704–7717.

12 (a) Kärkäs, M.D., Verho, O., Johnston, E.V., and Åkermark, B. (2014) *Chem. Rev.*, **114**, 11863–12001; (b) Blakemore, J.D., Crabtree, R.H., and Brudvig, G.W. (2015) *Chem. Rev.*, **115**, 12974–13005.

13 Parent, A.R., Crabtree, R.H., and Brudvig, G.W. (2013) *Chem. Soc. Rev.*, **42**, 2247–2252.

14 (a) Sartorel, A., Bonchio, M., Campagna, S., and Scandola, F. (2013) *Chem. Soc. Rev.*, **42**, 2262–2280; (b) Puntoriero, F., Sartorel, A., Orlandi, M., La Ganga, G., Serroni, S., Bonchio, M., Scandola, F., and Campagna, S. (2011) *Coord. Chem. Rev.*, **255**, 2594–2601.

15 Costentin, C., Drouet, S., Robert, M., and Savéant, J.-M. (2012) *J. Am. Chem. Soc.*, **134**, 11235–11242.

16 (a) Kiwi, J. and Graetzel, M. (1979) *Chimia*, **33**, 289–291; (b) Kiwi, J. and Graetzel, M. (1979) *Angew. Chem., Int. Ed. Engl.*, **18**, 624–626; (c) Kalayanasundaram, K. and Graetzel, M. (1979) *Angew. Chem., Int. Ed. Engl.*, **18**, 701–702.

17 Mills, A. (1989) *Chem. Soc. Rev.*, **18**, 285–316.

18 (a) Over, H. (2012) *Chem. Rev.*, **112**, 3356–3426; (b) Chang, C.C. and Wen, T.C. (1997) *J. Appl. Electrochem.*, **27**, 355; (c) Lodi, G., Sivieri, E., De Battisti, A., and Trasatti, S. (1978) *J. Appl. Electrochem.*, **8**, 135; (d) Castelli, P., Trasatti, S., Pollak, F.H., and O'Grady, W.E. (2003) *J. Electroanal. Chem.*, **545**, 89.

19 (a) Rossmeils, J., Logadottir, A., and Nørskov, J.K. (2005) *Chem. Phys.*, **319**, 178–184; (b) Rossmeils, J., Qu, Z.-W., Zhu, H., Kroes, G.-J., and Nørskov, J.K. (2007) *J. Electroanal. Chem.*, **607**, 83–89; (c) Man, I.C., Su, H.-Y., Calle-Vallejo, F., Hansen, H.A., Martinez, J.I., Inoglu, N.G., Kitchin, J., Jaramillo, T.F., Nørskov, J.K., and Rossmeils, J. (2011) *ChemCatChem*, **3**, 1159–1165.

20 Fang, Y.-H. and Liu, Z.-P. (2010) *J. Am. Chem. Soc.*, **132**, 18214–18222.

21 Dau, H., Limberg, C., Reier, T., Risch, M., Roggan, S., and Strasser, P. (2010) *ChemCatChem*, **2**, 724–761.

22 Moyer, B.A. and Meyer, T.J. (1981) *Inorg. Chem.*, **20**, 436.

23 Gersten, S.W., Samuels, G.J., and Meyer, T.J. (1982) *J. Am. Chem. Soc.*, **104**, 4030–4032.

24 (a) Concepcion, J.J., Jurss, J.W., Brennaman, M.K., Hoertz, P.G., Patrocinio, A.O.T., Murakami Iha, N.Y., Templeton, J.L., and Meyer, T.J. (2009) *Acc. Chem. Res.*, **42**, 1954–1965; (b) Moonshiram, D., Jurss, J.W., Concepcion, J.J.,

Zakharova, T., Alperovich, I., Meyer, T.J., and Pushkar, Y. (2012) *J. Am. Chem. Soc.*, **134**, 4625–4636.
25 Clark, A.E. and Hurst, J.K. (2012) *Prog. Inorg. Chem.*, **57**, 1–54.
26 Sens, C., Romero, I., Rodriguez, M., Llobet, A., Parella, T., and Benet-Buchholz, J. (2004) *J. Am. Chem. Soc.*, **126**, 7798–7799.
27 (a) Romain, S., Bozoglian, F., Sala, X., and Llobet, A. (2009) *J. Am. Chem. Soc.*, **131**, 2768–2769; (b) Bozoglian, F., Romain, S., Ertem, M.Z., Todorova, T.K., Sens, C., Mola, J., Rodriguez, M., Romero, I., Benet-Buchholz, J., Fontrodona, X., Cramer, C.J., Gagliardi, L., and Llobet, A. (2009) *J. Am. Chem. Soc.*, **131**, 15176–15187; (c) Romain, S., Vigara, L., and Llobet, A. (2009) *Acc. Chem. Res.*, **42**, 1944–1953.
28 Concepcion, J.J., Jurss, J.W., Templeton, J.L., and Meyer, T.J. (2008) *J. Am. Chem. Soc.*, **130**, 16462–16463.
29 (a) Polyansky, D.E., Muckerman, J.T., Rochford, J., Zong, R., Thummel, R.P., and Fujita, E. (2011) *J. Am. Chem. Soc.*, **133**, 14649–14655; (b) Lewandowska-Andralojc, A., Polyansky, D.E., Zong, R., Thummel, R.P., and Fujita, E. (2013) *Phys. Chem. Chem. Phys.*, **15**, 14058–14068.
30 Zong, R. and Thummel, R.P. (2005) *J. Am. Chem. Soc.*, **127**, 12802–12803.
31 Tseng, H.-W., Zong, R., Muckerman, J.T., and Thummel, R. (2008) *Inorg. Chem.*, **47**, 11763–11773.
32 (a) Duan, L., Fischer, A., Xu, Y., and Sun, L. (2009) *J. Am. Chem. Soc.*, **131**, 10397–10399; (b) Duan, L., Xu, Y., Zhang, P., Wang, M., and Sun, L. (2010) *Inorg. Chem.*, **49**, 209–215.
33 Duan, L., Bozoglian, F., Mandal, S., Stewart, B., Privalov, T., Llobet, A., and Sun, L. (2012) *Nat. Chem.*, **4**, 418–423.
34 Schulze, M., Kunz, V., Frischmann, P.D., and Würthner, F. (2016) *Nat. Chem.*, **8**, 576–583.
35 (a) Wang, L., Duan, L., Stewart, B., Pu, M., Liu, J., Privalov, T., and Sun, L. (2012) *J. Am. Chem. Soc.*, **134**, 18868–18880; (b) Jiang, Y., Li, F., Zhang, B., Li, X., Wang, X., Huang, F., and Sun, L. (2013) *Angew. Chem., Int. Ed.*, **52**, 3398–3401; (c) Duan, L., Araujo, C.M., Ahlquist, M.S.G., and Sun, L. (2012) *Proc. Natl. Acad. Sci. U. S. A.*, **109**, 15584–15887.
36 (a) Richmond, C.J., Matheu, R., Poater, A., Falivene, L., Benet-Bucholz, J., Sala, X., Cavallo, L., and Llobet, A. (2014) *Chem. Eur. J.*, **20**, 17282–17286; (b) Wang, L., Duan, L., Wang, Y., Ahlquist, M.S.G., and Sun, L. (2014) *Chem. Commun.*, **50**, 12947–12950.
37 Wang, S.-S. and Yang, G.-Y. (2015) *Chem. Rev.*, **115**, 4893–4962.
38 (a) Sartorel, A., Miró, P., Carraro, M., Berardi, S., Bortolini, O., Bagno, A., Bo, C., and Bonchio, M. (2014) *Chem.–Eur. J.*, **20**, 10932–10943; (b) Groves, J.T., Bonchio, M., and Carofiglio, T. (1996) *J. Am. Chem. Soc.*, **118**, 8961–8962; (c) Wang, C., Shalyaev, K.V., Bonchio, M., Carofiglio, T., and Groves, J.T. (2006) *Inorg. Chem.*, **45**, 4769–4782.
39 Howells, A.R., Sankarraj, A., and Shannon, C. (2004) *J. Am. Chem. Soc.*, **126**, 12258–12259.
40 (a) Sartorel, A., Carraro, M., Scorrano, G., De Zorzi, R., Geremia, S., McDaniel, N.D., Bernhard, S., and Bonchio, M. (2008) *J. Am. Chem. Soc.*,

**130**, 5006–5007; (b) Geletii, Y.V., Botar, B., Köegerler, P., Hillesheim, D.A., Musaev, D.G., and Hill, C.L. (2008) *Angew. Chem., Int. Ed.*, **47**, 3896–3899.
41 Piccinin, S., Sartorel, A., Aquilanti, G., Goldoni, A., Bonchio, M., and Fabris, S. (2013) *Proc. Natl. Acad. Sci. U. S. A.*, **110**, 4917–4922.
42 Bonchio, M., Carraro, M., Sartorel, A., Scorrano, G., and Kortz, U. (2006) *J. Mol. Cat. A: Chem.*, **251**, 93–99.
43 Liu, Y., Guo, S.-X., Bond, A.M., Zhang, J., Geletii, Y.V., and Hill, C.L. (2013) *Inorg. Chem.*, **52**, 11986–11996.
44 (a) Sartorel, A., Mirò, P., Salvadori, E., Romain, S., Carraro, M., Scorrano, G., Di Valentin, M., Llobet, A., Bo, C., and Bonchio, M. (2009) *J. Am. Chem. Soc.*, **131**, 16051–16053; (b) Geletii, Y.V., Besson, C., Hou, Y., Yin, Q., Musaev, D.G., Quiñonero, D., Cao, R., Hardcastle, K.I., Proust, A., Kögerler, P., and Hill, C.L. (2009) *J. Am. Chem. Soc.*, **131**, 17360–17370.
45 Toma, F.M., Sartorel, A., Iurlo, M., Carraro, M., Parisse, P., Maccato, C., Rapino, S., Gonzalez, B.R., Amenitsch, H., Da Ros, T., Casalis, L., Goldoni, A., Marcaccio, M., Scorrano, G., Scoles, G., Paolucci, F., Prato, M., and Bonchio, M. (2010) *Nat. Chem.*, **2**, 826–831.
46 (a) Quintana, M., Lopez, A.M., Rapino, S., Toma, F.M., Iurlo, M., Carraro, M., Sartorel, A., Maccato, C., Ke, X.X., Bittencourt, C., Da Ros, T., Van Tendeloo, G., Marcaccio, M., Paolucci, F., Prato, M., and Bonchio, M. (2013) *ACS Nano*, **7**, 811–817; (b) Guo, S.X., Liu, Y.P., Lee, C.Y., Bond, A.M., Zhang, J., Geletii, Y.V., and Hill, C.L. (2013) *Energy Envron. Sci.*, **6**, 2654–2663.
47 (a) Geletii, Y.V., Huang, Z., Hou, Y., Musaev, D.G., Lian, T., and Hill, C.L. (2009) *J. Am. Chem. Soc.*, **131**, 7522–7523; (b) Puntoriero, F., La Ganga, G., Sartorel, A., Carraro, M., Scorrano, G., Bonchio, M., and Campagna, S. (2010) *Chem. Commun.*, **46**, 4725–4727.
48 (a) Orlandi, M., Argazzi, R., Sartorel, A., Carraro, M., Scorrano, G., Bonchio, M., and Scandola, F. (2010) *Chem. Commun.*, **46**, 3152–3154;(b) Natali, M., Orlandi, M., Berardi, S., Campagna, S., Bonchio, M., Sartorel, A., and Scandola, F. (2012) *Inorg. Chem.*, **51**, 7324–7331.
49 (a) Natali, M., Puntoriero, F., Chiorboli, C., La Ganga, G., Sartorel, A., Bonchio, M., Campagna, S., and Scandola, F. (2015) *J. Phys. Chem. C*, **119**, 2371–2379; (b) Burian, M., Syrgiannis, Z., La Ganga, G., Puntoriero, F., Natali, M., Scandola, F., Campagna, S., Prato, M., Bonchio, M., Amenitsch, H., and Sartorel, A. (2017) *Inorg. Chim. Acta.*, **454**, 171–175.
50 Meyer, T.J. (2008) *Nature*, **451**, 778–779.
51 Sala, X., Maji, S., Bofill, R., Garcia-Antón, J., Escriche, L., and Llobet, A. (2014) *Acc. Chem. Res.*, **47**, 504–516.
52 (a) Natali, M., Deponti, E., Vilona, D., Sartorel, A., Bonchio, M., and Scandola, F. (2015) *Eur. J. Inorg. Chem.*, 3467–3477; (b) Chen, H.C., Hetterscheid, D.G.H., Williams, R.M., van der Vlugt, J.I., Reek, J.N.H., and Brouwer, A.M. (2015) *Energy Environ. Sci.*, **8**, 975–982.
53 (a) Youngblood, W.J., Lee, S.-H.A., Kobayashi, Y., Hernandez-Pagan, E.A., Hoertz, P.G., Moore, T.A., Moore, A.L., Gust, D., and Mallouk, T.E. (2009) *J. Am. Chem. Soc.*, **131**, 926–927; (b) Ronconi, F., Syrgiannis, Z., Bonasera, A., Prato, M., Argazzi, R., Caramori, S., Cristino, V., and Bignozzi, C.A. (2015) *J. Am. Chem. Soc.*, **137**, 4630–4633; (c) Fielden, J., Sumliner, J.M.,

Han, N., Geletii, Y.V., Xiang, X., Musaev, D.G., Lian, T., and Hill, C.L. (2015) *Chem. Sci.*, **6**, 5531–5543; (d) Gao, Y., Ding, X., Liu, J., Wang, L., Lu, Z., Li, L., and Sun, L. (2013) *J. Am. Chem. Soc.*, **135**, 4219–4222; (e) Swierk, J.R., Mendez-Hernandez, D.D., McCool, N.S., Liddell, P., Terazono, Y., Pahk, I., Tomlin, J.J., Oster, N.V., Moore, T.A., Moore, A.L., Gust, D., and Mallouk, T.E. (2015) *Proc. Natl. Acad. Sci. U. S. A.*, **112**, 1681–1686; (f) Poddutoori, P.K., Thomsen, J.M., Milot, R.L., Sheenan, S.W., Negre, C.F.A., Garapati, V.K.R., Schmuttenmaer, C.A., Batista, V.S., Brudvig, G.W., and van der Est, A. (2015) *J. Mater. Chem. A*, **3**, 3868–3879.

# 4

# Ruthenium- and Cobalt-Containing Complexes and Hydrogenases for Hydrogen Production

*Michael J. Celestine, Raj K. Gurung, and Alvin A. Holder*

*Old Dominion University, Department of Chemistry and Biochemistry, 4541 Hampton Boulevard, Norfolk, VA 23529-0126, USA*

## 4.1 Introduction

It is a known fact that the amount of fossil fuels in global reserves is limited. Countries such as Venezuela and Saudi Arabia make up two of the eight countries that hold the majority of the crude oil in reserves (81%), while Russia and the United States are a part of the six countries that hold the majority of natural gas reserves (70%). The United States is one of eight countries that hold most of the world's coal reserves (89%) [1].

Greenhouse gases, such as carbon dioxide, chlorofluorocarbons, nitrous oxide, carbon monoxide, methane, and many others, have been observed to have an impact on global weather while producing numerous negative effects on the environment [2]. Many of these gases can be released into the environment through the burning of fossil fuels. The burning of these fossil fuels has been noted to have adverse effects on the environment, one of which is worldwide climate change. It has been reported that the United States alone makes a significant contribution to the world's average carbon dioxide emission, with 78% coming from motor vehicles [1]. As such, there is a need to reduce the use of fossil fuels in order to circumvent such negative effects and focus on the use of hydrogen as a possible alternative.

The efficient storage of solar energy in chemical fuels, such as hydrogen, is essential for the large-scale utilization of solar energy systems. Hydrogen, a potential alternative to fossil fuels, is seen to have many advantages as a fuel source [3]. The sun is a large and long-lasting source of energy [4]. Only a small amount of solar energy is being used; thus, harnessing solar energy can become one of the prominent sources of cleaner energy on Earth. Today, many researchers are carrying out studies to make this a reality, and, as reported, hydrogenases and photocatalysts (from various metal centers) can produce molecular hydrogen from several proton sources [4, 5]. Some of the advantages of using hydrogen are as follows: (i) it is a stable element; (ii) it is also noncorrosive, (iii) it is three times lighter than traditional fuels such as gasoline and diesel, but has a high energy density, and (iv) upon burning produces water as the only

*Ruthenium Complexes: Photochemical and Biomedical Applications*, First Edition.
Edited by Alvin A. Holder, Lothar Lilge, Wesley R. Browne, Mark A.W. Lawrence, and Jimmie L. Bullock Jr.

product and heat [3]. Just as there are advantages to using hydrogen, there are also disadvantages, and they are as follows: (i) ways of safe transportation and its storage as a fuel; (ii) its production requires a significant energy input; (iii) during its production via electrocatalysis, the electrodes can be poisoned by the hydrogen, thus affecting the efficiency of the process; and (iv) most of the catalysts used in its production are very expensive [3, 6].

In this chapter we provide an insight into the production of hydrogen through the use of cobalt-containing complexes and hydrogenases, all in the presence of ruthenium-containing complexes. The chapter is divided into two sections.

## 4.2 (A) Ruthenium- and Cobalt-Containing Complexes for Hydrogen Production

### 4.2.1 Nonbridged Systems

The photoexcited states of various polypyridyl ruthenium(II) complexes have been studied for many years [7]. Many ruthenium polypyridyl complexes have been observed to possess strong metal-to-ligand charge-transfer (MLCT) bands that occur when an electron from the ruthenium's $t_{2g}$ orbital is transferred to a $\pi^*$ orbital on the ligand [8]. The resulting MLCT is known to produce a powerful reductant, which then transfers an electron to an electron acceptor [9]. In the early 1980s, there were reports on photoinduced electron transfer processes involving $[Ru(N—N)_3]^{2+}$ (where N—N = 2,2′-bipyridne (bpy) or 1,10-phenanthroline (phen)) [10]. In one study, the excited-state redox potentials that were calculated from the ground state, as well as the excited state, proved that the photoexcited species, $[Ru^*(bpy)_3]^{2+}$ is not only a strong reductant but can also be a good oxidant [11]. Such ruthenium(II) species are known as photosensitizers, and, over the years, photochemical hydrogen evolution studies aimed at evaluating new molecular catalysts have usually exploited $[Ru(bpy)_3]^{2+}$ as the reference photosensitizer, thanks to its suitable optical and redox properties. In principle, an additional improvement of the photocatalytic performances can be achieved also by a careful adjustment of the photophysical and/or electrochemical characteristics of the ruthenium-based sensitizer.

Recently Deponti and Natali [12] reported homogeneous molecular systems for photocatalytic hydrogen evolution composed of a series of ruthenium polypyridine complexes as the photosensitizers, a cobaloxime catalyst, $[Co(dmg)_2Cl(py)]$ (where dmg = dimethylglyoximato), and L-ascorbic acid as the sacrificial electron donor. Suitable functionalizations of the 4 and 4′ positions of the 2,2′-bipyridine ligand were utilized in order to modify the redox properties of the chromophores rather than their optical ones. A careful and detailed kinetic characterization of the relevant processes at the basis of hydrogen-evolving photocatalysis was addressed to rationalize the observed behavior. It was reported that a good balance among (i) the excited-state redox properties of the sensitizer, affecting the reductive quenching rate by the sacrificial donor, (ii) reducing ability of the photogenerated reduced species of the chromophore, determining the rate of electron transfer to the catalyst, and (iii) steric hindrance of the complex, affecting the electronic coupling within the encounter complex in both electron transfer processes, was deemed as the winning strategy used to identify a successful sensitizer in the study as reported by Deponti and Natali [12]. The results showed that the ruthenium complex involving two 2,2′-bipyridine ligands and one 4,4′-dimethyl-2,2′-bipyridine ligand may outperform the standard $[Ru(bpy)_3]^{2+}$, combining the right balance of structural and redox properties, thus posing as an alternative benchmark photosensitizer for the study of new hydrogen-evolving catalysts [12].

Recent advances in the photocatalytic production of $H_2$ were highlighted in the literature by Teets and Nocera [10a]. Future directions and challenges in photocatalytic $H_2$ generation were highlighted by Teets and Nocera [10a]. In this highlight, two general approaches for the photocatalytic hydrogen generation by homogeneous catalysts were considered: HX (X = Cl, Br) splitting involving both proton reduction and halide oxidation via an inner-sphere mechanism with a single-component catalyst; and sensitized $H_2$ production, employing sacrificial electron donors to regenerate the active catalyst [10a].

Homogeneous catalysis of the photoreduction of water to produce hydrogen by visible light can also be mediated by a tris(2,2′-bipyridine)ruthenium(II)-cobalt(II) macrocycle system and cobalt(II)-polypyridyl catalysts [9a, 13]. For the latter, two cobalt(II)-polypyridyl catalysts, $[Co(pdt)_3]^{2+}$(where pdt = 3,5,6-triphenyl-1,2,4-triazine) and $[Co(tpp)_2]^{2+}$ (where tpp = tetra(pyridin-2-yl)pyrazine) were used as homogeneous catalysts for the production of hydrogen in the presence of $[Ru(bpy)_3]^{2+}$ as a photosensitizer and L-ascorbic acid as a sacrificial electron donor [13]. From these studies, it was observed that the initial quantum yields were up to 20% for $[Co(pdt)_3]^{2+}$ [13].

Numerous systems with cobalt-containing catalysts have been reported to have the ability to produce hydrogen in the presence of a ruthenium(II) polypyridium photosensitizer. Some examples of cobalt-containing catalysts are shown in Figure 4.1.

There have been investigations into the mechanisms and kinetics of the reduction of protons by cobaloximes, which resulted in the postulation of three different pathways that proceed through the same intermediate, namely, a cobalt(III)-hydride, as illustrated in Figure 4.2 [14]. The cobalt(III)-hydride is either protonated and releases $H_2$ in a heterolytic pathway or it can be reduced to a cobalt(II)-hydride, which can also produce hydrogen via a heterolytic pathway [14c]. The third pathway is homolytic and energetically more favorable. Calculations have been utilized to prove that two cobalt(III)-H species can produce hydrogen after a reductive elimination step which could result in the formation of a cobalt(II) metal center [14b, c]. The likelihood of any of the pathways depends on the relative concentrations of protons and cobalt(I). There was a report of theoretical studies of the standard reduction potentials of the $[Co^{II}(dmgBF_2)_2]^0$ (where $dmgBF_2$ = difluoroboryldimethylglyoximato) in acetonitrile solution. Such studies were used to shed light on the complex's electrocatalytic mechanism for hydrogen production. Three such mechanisms were proposed, all proceeding through the formation of cobalt(III)-H. The report indicated that the mechanism involving a cobalt(III)-H intermediate is the most likely pathway [15].

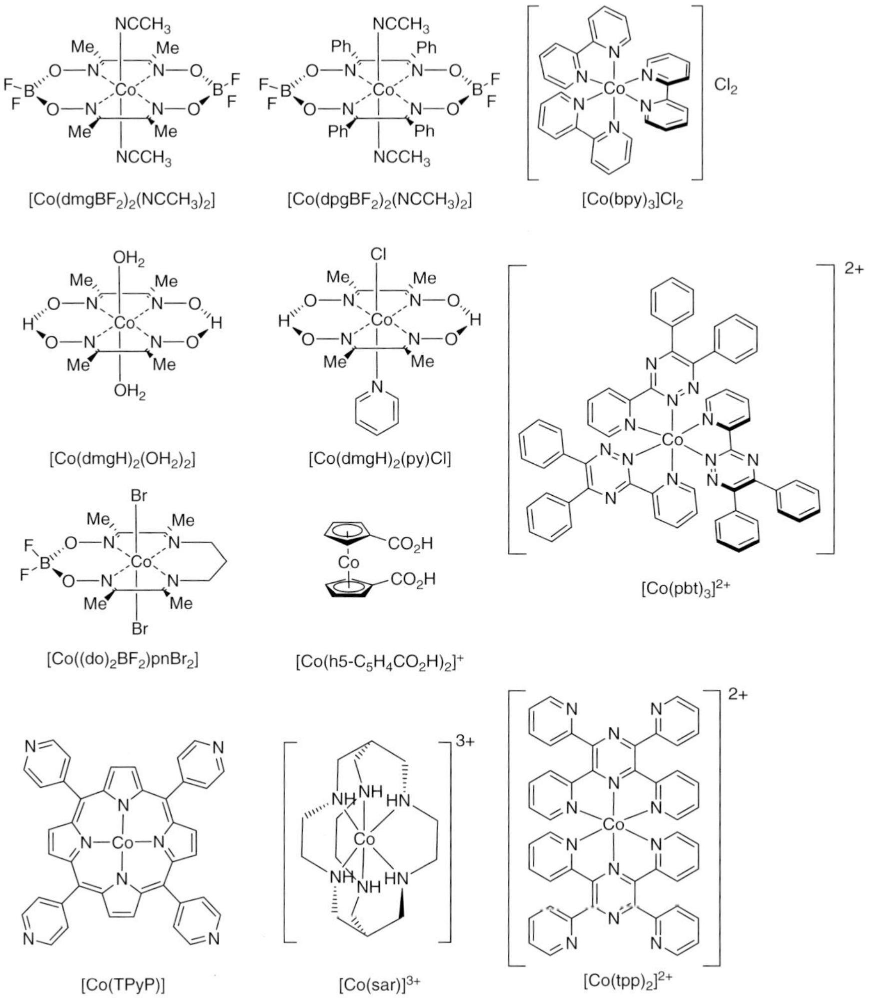


**Figure 4.1** Examples of cobalt-containing catalysts that are able to produce hydrogen in the presence of a ruthenium photosensitizer.

### 4.2.2 Bridged Systems

In the catalytic production of hydrogen, two types of systems are usually utilized, namely, (i) multicomponent systems and (ii) bridged systems [16]. In the case of multicomponent systems, a bridge between a ruthenium(II) photocatalyst and a cobalt-containing catalyst is normally absent, and the electron can be transferred via an outer-sphere mechanism, whereas in the bridged systems, the electron transfer is not slowed by the random chance of the excited photosensitizer and oxidized form of the activated catalyst approaching each other [17]. Thus, the bridged systems are more efficient when compared to the multicomponent systems. The two types of electron transfer processes that may occur are illustrated in Figure 4.3. Figure 4.3a shows a multicomponent system where

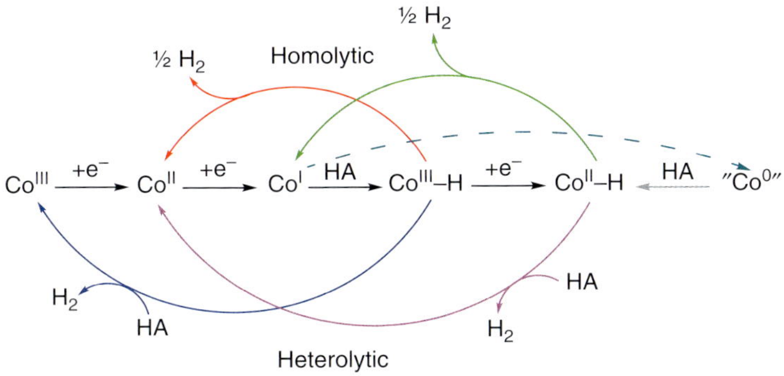


**Figure 4.2** Proposed mechanisms for the production of hydrogen in acidic media [14c].

the electron transfer occurred via an outer-sphere mechanism [17], whereas Figure 4.3b shows a binuclear mixed-metal complex where the two metal centers are connected via a bridging ligand, and the electron transfer occurred via an inner-sphere electron transfer mechanism [18]. For these bridged systems, an additional pathway involving an outer-sphere mechanism is also believed to be possible.

It was observed that with the cobaloximes, when the two H-bonded are substituted with two $BF_2$ caps, the overall efficiency of such complexes increases during a catalytic process. This substitution of the two H with $BF_2$ caps also resulted in the complex being relatively easier to reduce and simultaneously less susceptible to side reactions, such as acidic hydrolysis and hydrogenation [18, 19]. There have been many cases dating back to the early 2000s where a ruthenium(II) photosensitizer is bridged to a cobalt(II) metal center, for example, as cobaloximes, as a means of increasing the overall efficiency of the catalyst [18, 19]. The photosensitizer is typically bridged to the catalyst via a pendent pyridine, which coordinates by the substitution of an axial ligand on the cobalt(II) metal center. The bridging ligand has been noted as an electron reservoir, whereby it transfers the electron from the ruthenium(II) photoactive center to the catalyst through an inner-sphere electron transfer process [18, 20]. Irradiation of the complexes in the presence of a proton source, such as $[Et_3NH]BF_4$, has been observed to be more efficient in the production of hydrogen in contrast to the multicomponent system of the photosensitizer and catalyst under the same conditions.

Studies have proved that linking of cobalt hydrogen evolution catalysts (HECs) such as cobaloximes to ruthenium(II)-containing photoactive catalytic complexes can produce binuclear mixed-metal complexes such as $[Ru(pbt)_2(\text{L-pyr})Co(dmgBF_2)_2(OH_2)]^{2+}$(where pbt = 2-(2′-pyridyl)benzothiazole and L-pyr = (4-pyridine)oxazolo[4,5]phenanthroline) (complex **1**, Figure 4.4) [21]. Upon irradiation, the binuclear mixed-metal complexes in acidic acetonitrile undergo an intramolecular electron transfer from a photoexcited ruthenium(II) photosensitizer to a cobalt(II) metal center. This then leads to the production of $H_2$ [14c], but an outer-sphere electron transfer cannot be excluded. These Ru(II)—Co(II) mixed-metal binuclear complexes were reported to have

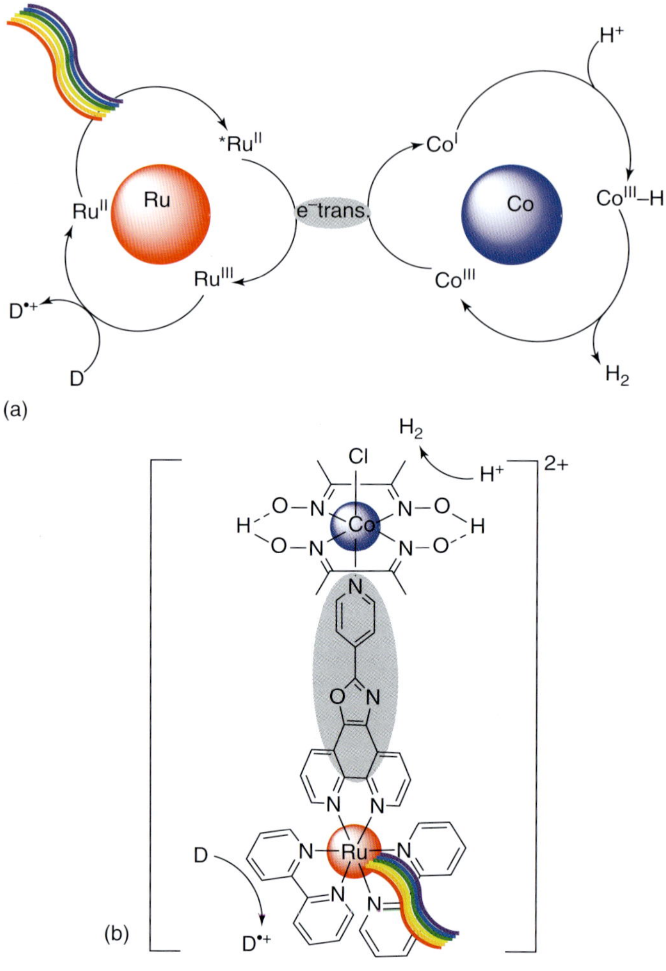


**Figure 4.3** (a and b) Homogeneous systems for photocatalytic $H_2$ production with a sacrificial electron donor D. (a) Multicomponent system consisting of $[Ru(bpy)_3]Cl_2$ and a cobaloxime, $[Co(dmgH)_2(H_2O)_2]$ [17]. (b) Binuclear mixed-metal ruthenium(II)–cobalt(II) complex, where the ruthenium(II) photosensitizer is covalently linked to the cobaloxime [18].

efficiencies up to 8.5 times more than analogous systems under similar conditions, such as $[Ru(phen)_2(\text{L-pyr})Co(dmgBF_2)_2(OH_2)]^{2+}$ (complex **2**, Figure 4.4) [14c]. Studies carried out on $[Ru(bpy)_2(\text{L-pyr})Co(dmgH)_2(Cl)]^{2+}$ (complex **3**, Figure 4.4), $[Ru(bpy)_2(\text{L-pyr})Co(dmgBF_2)_2(OH_2)]^{2+}$ (complex **4**, Figure 4.4), and $[Ru(bpy)_2(\text{L-pyr})Co(dpgBF_2)_2(OH_2)]^{2+}$ (complex **5**, Figure 4.4) showed that these complexes were able to produce hydrogen over a 4-h period with a turnover number (TON) of 17, 56, and 12, respectively [18]. From the study, two important points were made about the efficiency of complex **4** over complex **3**, which was in accordance with what was observed with the mononuclear cobaloximes. They are (i) the $BF_2$-capped ligand increased the stability of the cobalt catalyst to hydrolysis and hydrogenation and (ii) the $BF_2$-capped ligand

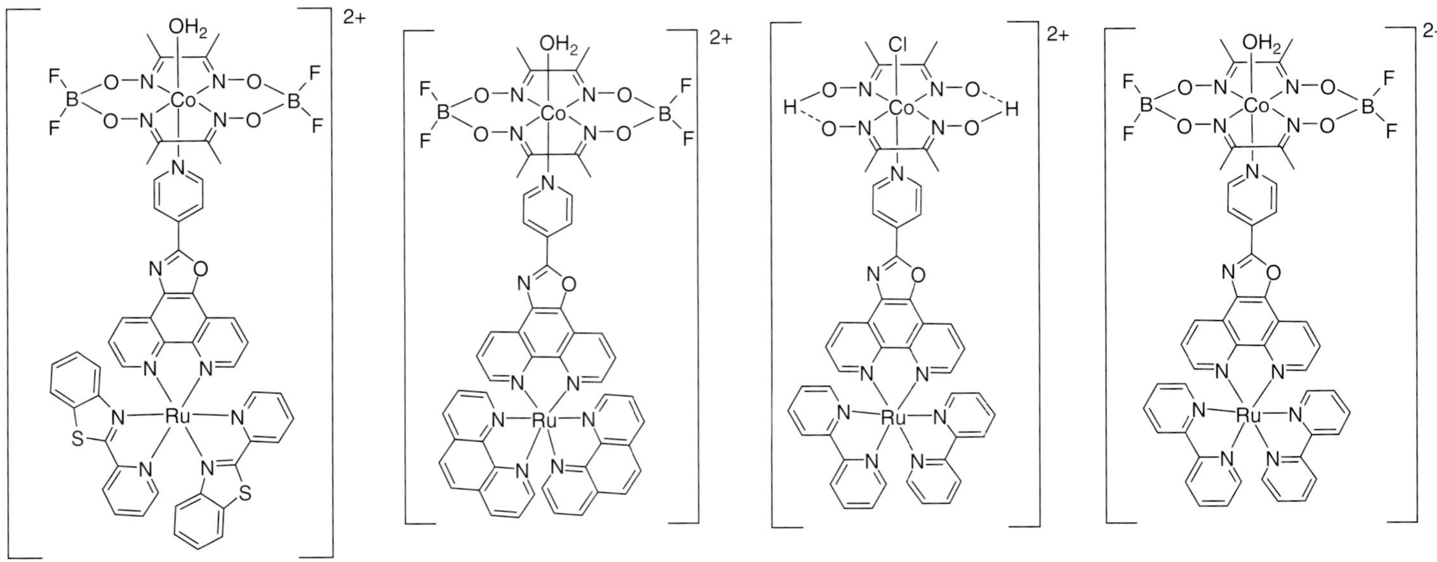

1 2 3 4

Figure 4.4 Binuclear mixed-metal complexes used for the production of hydrogen in acidic media [14c, 21, 22].

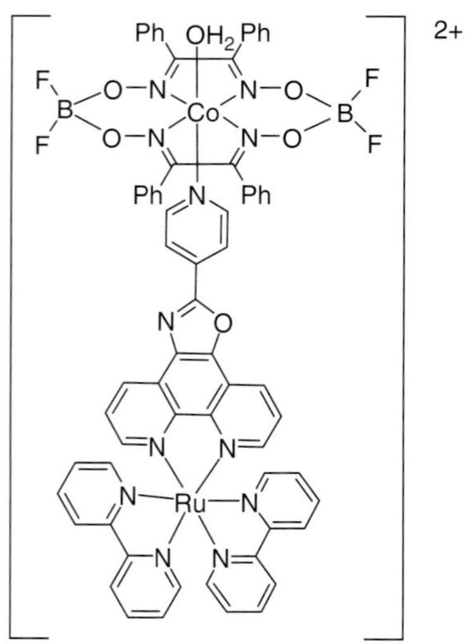

**5**

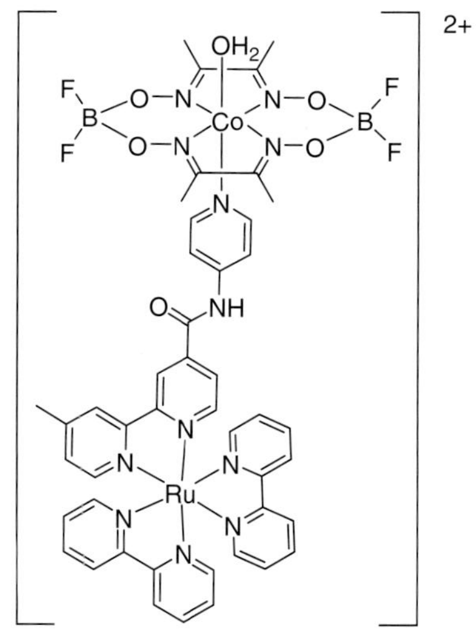

**6**

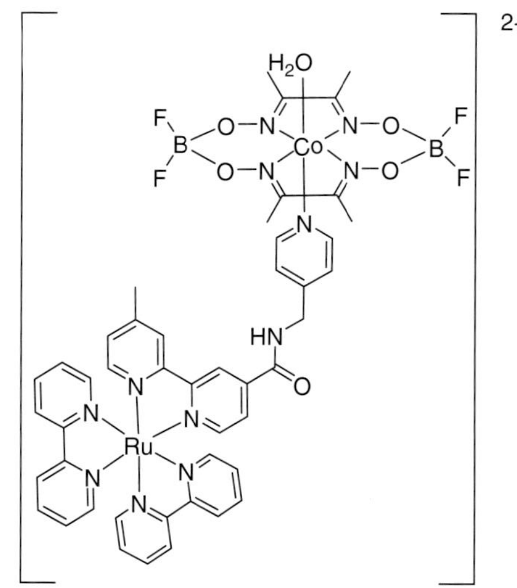

**7**

**Figure 4.4** *(Continued)*

made the cobalt(II) metal center much easier to be reduced [18]. Studies on $[(bpy)_2Ru(bpy\text{-}4\text{-}CH_3,4'\text{-}CONH(4\text{-}py)Co(dmgBF_2)_2(OH_2)](PF_6)_2$ (complex **6**, Figure 4.4) and $[(bpy)_2Ru(bpy\text{-}4\text{-}CH_3,4'\text{-}CONHCH_2(4\text{-}py)Co(dmgBF_2)_2(OH_2)](PF_6)_2$ **7** showed that the latter had a higher TON when compared to the former complex. This is surprising since the former is conjugated, whereas the latter is not [14c, 22].

Such a study was conducted in acetone with $Et_3N$ as the sacrificial electron donor and $[Et_3NH]BF_4$ as the proton source, where the maximum TON for hydrogen production was 38 for the conjugated complex and 48 for the nonconjugated complex [22]. In another study, complex **1** in the presence of $Et_3N$ was irradiated with light in acidified acetonitrile, where the system was observed to produce hydrogen uninterrupted over a 42-h period [21]. When the photocatalytic efficiency of complex **1** was tested using $(EtOH)_3N$, it was observed that the turnover frequency (TOF) increased from ~1 to $\sim 2\,h^{-1}$ when compared to $Et_3N$ [21]. When complex **1** was compared to $[Ru(pbt)_2(\text{L-pyr})]^{2+}$ photocatalyst in solution with the cobaloximes, the amount of hydrogen produced was very negligible in the latter, thus concluding that the effectiveness of the former was attributed to that of the bridging ligand expediting the electron transfer process [21].

The mechanism proposed for the mixed-metal binuclear complexes involved the following processes: (i) the ruthenium(II) photosensitizer being photoexcited at first, and the excited electron is then transferred to the cobalt(II) metal center through the bridging ligand via an inner-sphere mechanism, which resulted in the formation of a reactive cobalt(I) species [21]; (ii) the cobalt(I) metal center then reacts and undergoes a two-electron transfer to a proton to form a cobalt(III)-hydride species, which further reacts with another proton, and (iii) hydrogen is formed when the cobalt–hydrogen bond is cleaved through heterolytic fission. Finally, $Et_3N$ is utilized as a sacrificial electron donor to reduce the oxidized species in the catalytic process, thus allowing the cycle to be repeated [21].

Time-resolved spectroscopic studies carried out on complex **1** (Figure 4.5) showed that an intramolecular electron transfer from the excited ruthenium(II) metal center to the cobalt(II) metal center did occur via the bridging L-pyr ligand [21]. This resulted in the formation of a cobalt(I)-containing species that

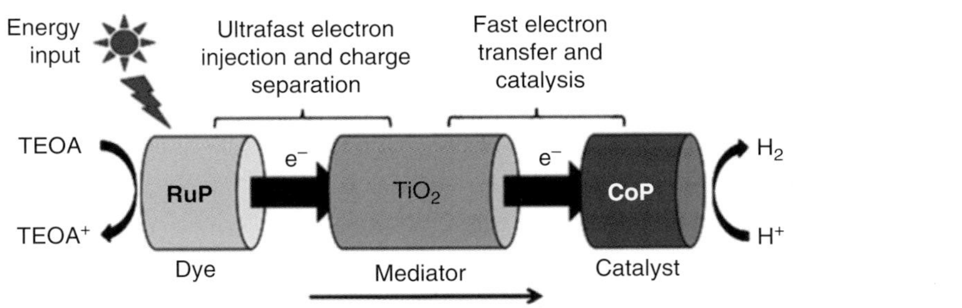


**Figure 4.5** Schematic representation of photo-$H_2$ evolution with RuP/CoP-modified $TiO_2$.

is essential for the production of $H_2$ gas in the presence of $H^+$ ions [21]. There was a report on a self-assembled system comprising a molecular $H_2$ production cobalt catalyst attached on a ruthenium(II) dye-sensitized $TiO_2$ nanoparticle. Visible-light irradiation of the dispersed nanoparticles in the presence of the sacrificial electron donor triethanolamine produced $H_2$ photocatalytically in pH neutral water and at room temperature [23].

A visible-light-driven $H_2$ evolution system comprising a ruthenium(II) dye (**RuP**) and cobalt(III) proton reduction catalysts (**CoP**) immobilized on $TiO_2$ nanoparticles and mesoporous films was reported (Figure 4.5) [24]. The heterogeneous system evolved $H_2$ efficiently during visible-light irradiation in a pH-neutral aqueous solution at 25 °C in the presence of a hole scavenger [24]. Photodegradation of the self-assembled system occurred at the ligand framework of **CoP**, which was reported to be readily repaired by addition of fresh ligand, resulting in TONs above 300 mol $H_2$ (mol **CoP**)$^{-1}$ and above 200 000 mol $H_2$ (mol $TiO_2$ nanoparticles)$^{-1}$ in water [24]. The reported studies supported that a molecular Co species, rather than metallic Co or a Co-oxide precipitate, was responsible for $H_2$ formation on $TiO_2$ [24]. Electron transfer in the system was studied by transient absorption spectroscopy and time-correlated single-photon counting techniques. Essentially quantitative electron injection took place from **RuP** into $TiO_2$ in approximately 180 ps. Upon dye regeneration by the sacrificial electron donor, a long-lived $TiO_2$ conduction band electron was thereby formed with a half-lifetime of approximately 0.8 s. Electron transfer from the $TiO_2$ conduction band to the **CoP** catalysts occurred quantitatively on a 10-μs timescale and was about a hundred times faster than charge recombination with the oxidized **RuP** [24]. The study provided a benchmark for future investigations in photocatalytic fuel generation with molecular catalysts integrated in semiconductors [24].

More recently, a series of Ru–protein–Co biohybrids were prepared using the electron transfer proteins ferredoxin (Fd) and flavodoxin (Fld) as scaffolds for photocatalytic hydrogen production [25]. The light-generated charge separation within these hybrids were monitored by transient optical and ESR spectroscopies. Two distinct electron transfer pathways were observed. The Ru—Fd—Co biohybrid produced up to 650 turnovers of $H_2$ utilizing an oxidative quenching mechanism for Ru(II)* and a sequential electron transfer pathway via the native [2Fe—2S] cluster to generate a Ru(III)—Fd—Co(I) charge separated state that lasts for ~6 ms. In contrast, a direct electron transfer pathway occurred for the Ru—ApoFld—Co biohybrid, which lacked an internal electron relay, generating Ru(I)—ApoFld—Co(I) charge-separated state that persisted for ~800 μs and produced 85 turnovers of $H_2$ by a reductive quenching mechanism for Ru(II)* [25]. These two mechanisms are highlighted in Figure 4.6. The authors demonstrated the utility of protein architectures for linking donor and catalytic function via direct or sequential electron transfer pathways to enable stabilized charge separation which facilitated photocatalysis for solar fuel production [25].

It is a known fact that $NaBH_4$ is one of the most safe and attractive hydrogen-storage materials for $H_2$ production. Recently, it was reported that a zero-valent ruthenium–cobalt (Ru—Co)-based nanocluster incorporated conducting poly-3,4-ethylenedioxythiophene/poly-styrenesulfonate (PEDOT/PSS)

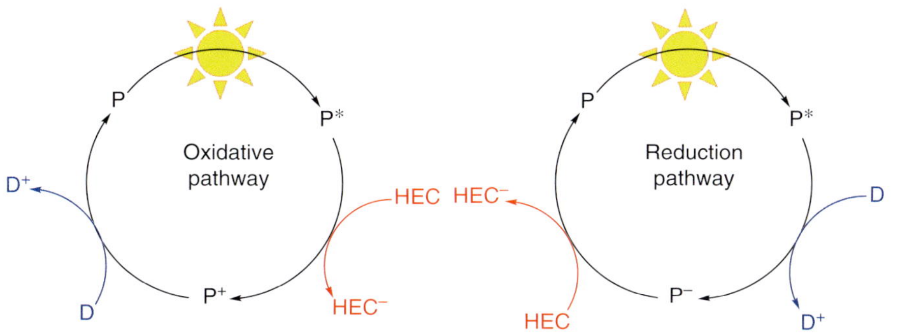


**Figure 4.6** Possible photochemical mechanisms for catalyst reduction in a homogeneous system for hydrogen production involving a hydrogen evolution catalyst (HEC) [10a, 25, 26].

nanocomposite as the catalyst for rapid hydrogen production from $NaBH_4$ [27]. Initially, the Ru—Co nanocluster was synthesized by reduction of a mixture of ruthenium(III) chloride and cobalt(II) chloride using an aqueous sodium borohydride solution. The Ru—Co cluster itself showed good synergistic catalytic effect when compared to "free" Ru and Co nanoparticles, but their performance was found to be exceptionally good when incorporated into the conducting polymer (PEDOT/PSS). The catalyst was characterized by transmission electron microscopy (TEM), energy-dispersive X-ray analysis (EDX), X-ray powder diffraction (XRD), four-probe conductivity measurements, and so on [27]. The hydrolysis kinetics showed that the 85 wt% $NaBH_4$ + 15 wt% Ru—Co PEDOT/PSS nanocomposite sample in 1 M NaOH yielded the best result among all other combinations. A hydrogen generation rate of $40.1\,l\,min^{-1}\,g^{-1}$ at 25 °C was achieved [27].

## 4.3 (B) Ruthenium(II)-Containing Complexes and Hydrogenases for Hydrogen Generation in Aqueous Solution

### 4.3.1 Hydrogenases

The harnessing of solar energy is one of the prominent precursors to the generation of clean energy. As such, biomimetic enzymes like hydrogenase, when used alongside different catalytic systems, can produce molecular hydrogen using solar energy.

$$H_2 \xrightleftharpoons{\text{Hydrogenase}} 2H^+ + 2e^- \tag{4.1}$$

Hydrogenases are used in tandem as powerful catalysts for light-driven hydrogen ($H_2$) production in combination with photosensitizers [28]. However, except oxygen-tolerant hydrogenases, they are immediately deactivated under aerobic conditions [28]. Hydrogenases have been reported to utilize hydrogen as a substrate, or produce hydrogen ($H_2$) by the reduction of protons [29].

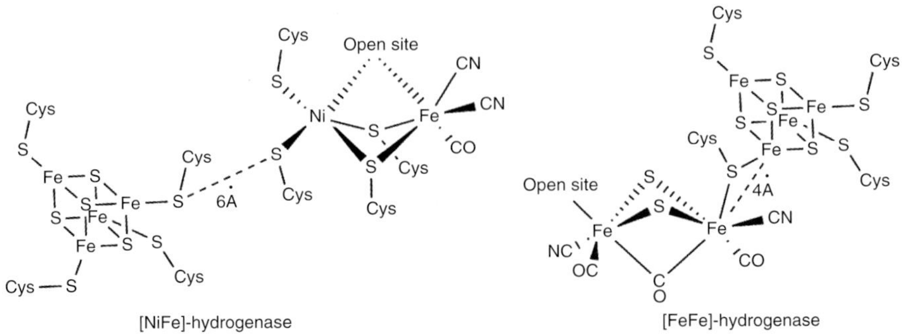


**Figure 4.7** Two different types of [FeFe]- and [NiFe]-hydrogenases [30].

Proton-reducing hydrogenase enzymes are classified into two subsets (Figure 4.7): (i) [FeFe]-hydrogenases, which have highest rate of hydrogen evolution but are sensitive to aerobic conditions, and (ii) [NiFe]-hydrogenases which have lower hydrogen evolution activity but are active in an oxygen-containing atmosphere, or the inhibition by oxygen is at least reversible [4]. The iron atoms in both the [FeFe]- and [NiFe]-hydrogenases are coordinated with small inorganic ligands like CO and $CN^-$. The enzymes have an open coordination site on one of the metal centers and there occurs the sulfur bridge in between two metal centers of the enzymes. Similarly, the subgroup of the [NiFe]-hydrogenase are also known in the form of [NiFeSe]-hydrogenase, in which selenocysteine replaces one of the nickel's cysteine ligands. Also, another class of such enzymes has a single iron atom (i.e., [Fe] hydrogenase or iron–sulfur cluster-free hydrogenases) [31].

[FeFe]-hydrogenases are widely distributed in fermentative anaerobic microorganisms and likely evolved under selective pressure to couple hydrogen production to the recycling of electron carriers that accumulate during anaerobic metabolism. In contrast, many [NiFe]-hydrogenases catalyze hydrogen oxidation as part of energy metabolism, and are likely key enzymes in early life and arguably represent the predecessors of modern respiratory metabolism [29]. When combined with ruthenium-based photosensitizers, one can create efficient and environmentally friendly photoactive model systems for photosystem II and hydrogenase [32]. A long lifetime of the charge-separated state is the key characteristic of efficient sensitizers as the strong reductive force disappears with charge recombination [4].

### 4.3.2 Hydrogenases with Ruthenium(II) Complexes

Photoinduced hydrogen evolution from water has been studied extensively using four component systems, namely, an electron donor (D), a photosensitizer (P), an electron carrier (C), and a catalyst, as shown in Figure 4.8 [33]. One such system was featured in a report where a cytochrome $c_3$-viologen-ruthenium(II) triad complex, Ru-V-cyt.$c_3$, was prepared and characterized using spectroscopic techniques [33]. Effective quenching of the photoexcited state of ruthenium complex moiety by the bound viologen was observed in Ru-V-cyt.$c_3$. When the system containing Ru-V-cyt.$c_3$ and hydrogenase was irradiated by visible light, photoinduced hydrogen evolution was observed, showing the effective two-step electron

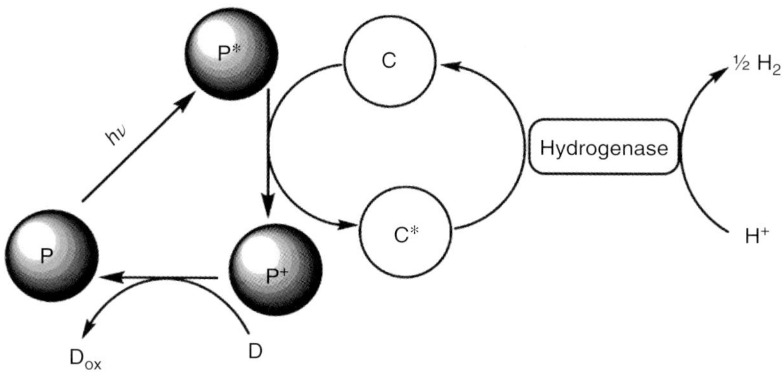


**Figure 4.8** Scheme of photoinduced hydrogen evolution system [33].

transfer from the photoexcited state of ruthenium complex moiety to cytochrome $c_3$ via bound viologen [33].

There was a report where a dinuclear iron complex, related to the active site of Fe hydrogenases, was covalently linked to a redox-active tris-2,2′-bipyridineruthenium(II)-type photosensitizer [34]. Photophysical and electrochemical studies of this system were carried out in solution. The photoexcited ruthenium(II) complex, when oxidatively quenched by the dinuclear iron site, generated a reduced iron species [34].

After regeneration of the photosensitizer (by an external electron donor), this process was repeated to accumulate two electrons on the diiron unit [34]. It was suggested that electron transfer from the excited state of ruthenium complex to the iron binuclear moiety was not favored because the reduction potential of the excited state of ruthenium complex was only ca. −1.1 V when compared to the −1.5 V for the iron complex [34]. Alternatively, electron injection to the diiron site was believed to be feasible from a ruthenium(I) complex, formed through reductive quenching of the ruthenium(II) excited state. By having acquired sufficient reductive power, the diiron complex was believed to be capable of reducing protons and generating hydrogen [34].

In an article published by a group of Swedish researchers, a [FeFe]-hydrogenase-active site was synthesized, but covalently linked with a $[Ru(tpy)_2]^{2+}$ (where tpy = 2,2′:6′,2″-terpyridine) photosensitizer [35]. A trinuclear complex (complex **8**, Figure 4.9) was useful for the solar-driven photo-production of hydrogen. In the synthesized complex (complex **8**, Figure 4.9), phenylaacetatylene, a rigid linker, was used to incorporate the [FeFe]-hydrogenase into the ruthenium(II) complex. One major purpose of this incorporation was to prolong the lifetime of the short-lived excited state of the ruthenium(II) photosensitizer [35]. Also, this arrangement had an important role in a natural system, providing a kinetically and thermodynamically favored pathway for hydrogen production. The comparison of the photophysical properties of complex demonstrated the success of this strategy [35].

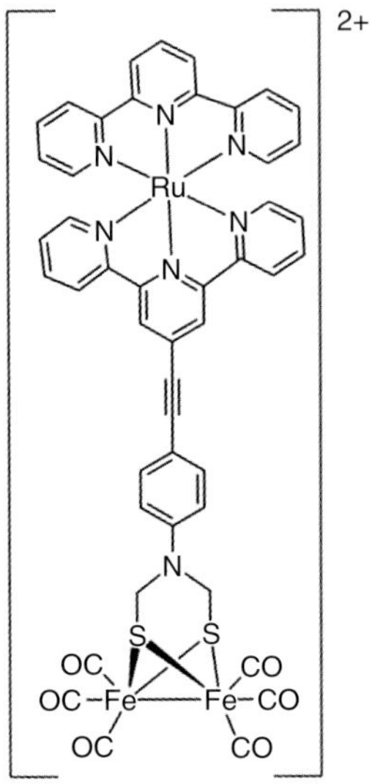

8

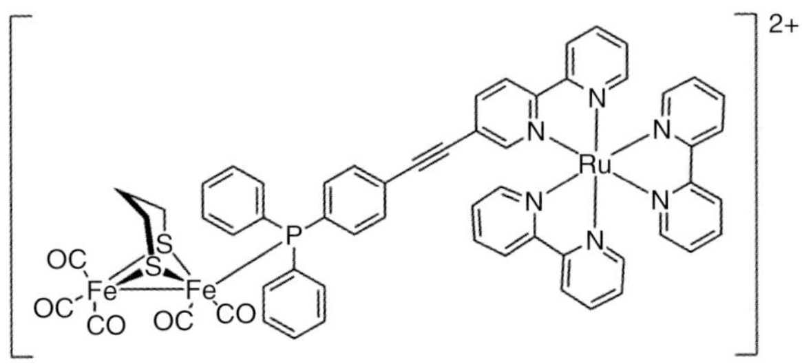

9

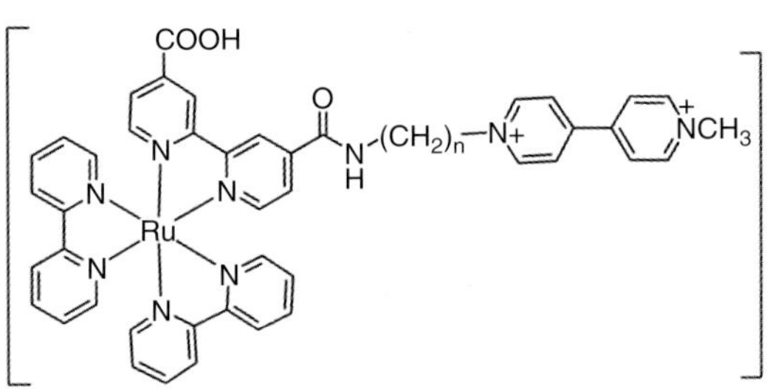

10

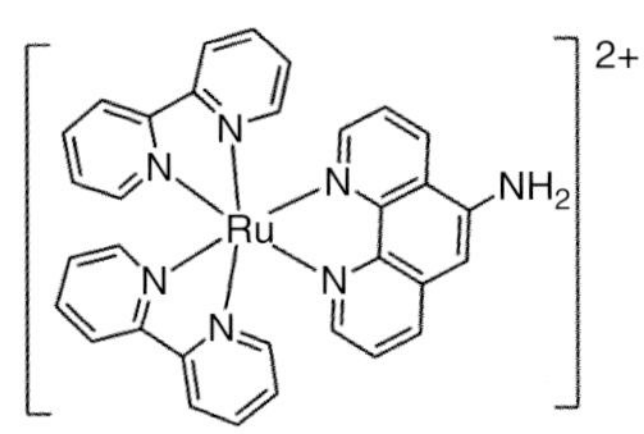

11

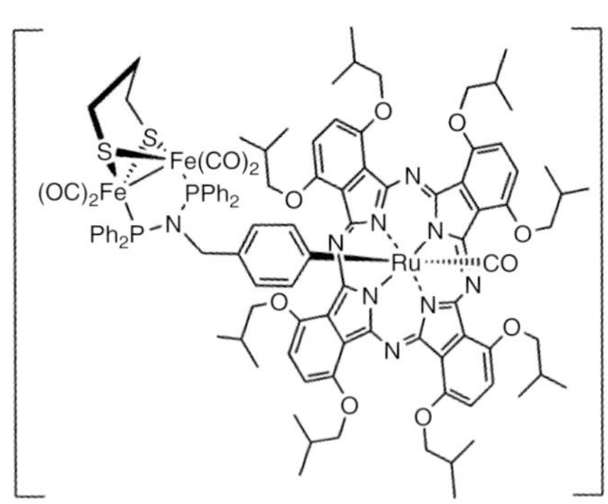

12

**Figure 4.9** Various ruthenium(II) photosensitizers used with hydrogenase for hydrogen production.

$[(\mu\text{-pdt})Fe_2(CO)_5(PPh_2(C_6H_4CCbpy))Ru(bpy)_2]^{2+}$ (where pdt = propyldithiolate) was the very first reported photoactive tris-2,2′-bipyridineruthenium(II) unit linked to a iron hydrogenase mimic active site by a ligand directly attached to one of the iron centers [36]. The light-induced MLCT excited state of the ruthenium(II) photosensitizer in complex **9** (Figure 4.9) was localized toward the potential diiron acceptor unit. Complex **9** had a relatively mild reduction potential for the acetylenic 2,2′-bipyridine analogue and led to reductive quenching of the excited state, forming a transiently oxidized diiron entity [36].

It was reported that ruthenium(II) complexes were coupled with viologen by a covalent bond, all of which were characterized by spectroscopic techniques, which confirmed the successful synthesis of the viologen(V)-linked ruthenium(II) complexes [37]. By differing the number of methyl groups attached to various viologen-linked ruthenium(II), $[Ru(bpy)_2(dcbpy)C_nVCH_3]$ (where dcbpy = 4,4′-dicarboxy-2,2′-bipyridine and $n = 2$ or 3) complexes (Figure 4.9, complex **10**) were synthesized. The system with nicotinamide adenine dinucleotide phosphate, a reduced form of NADPH, $[Ru(bpy)_2(dcbpy)C_nVCH_3]$ and hydrogenase, when irradiated with visible light, was efficient in photoinduced hydrogen production [37]. A ruthenium(II) metal center was found to act as both a photosensitizer and an electron carrier when linked with viologen. An intramolecular electron transfer occurred when viologen binds to oxidatively quench photoexcited state of $[Ru(bpy)_2(dcbpy)]$ [37].

There was a report of a light-driven $H_2$ evolution system that worked stably even under aerobic conditions [28]. A [NiFe]-hydrogenase from *Desulfovibrio vulgaris* Miyazaki F was immobilized inside nanoporous glass plates (PGPs) with a pore diameter of 50 nm together with a ruthenium(II) complex and methyl viologen as a photosensitizer and an electron mediator, respectively. After immersion of PGP into the medium containing the catalytic components, an anaerobic environment was automatically established inside the nanopores even under aerobic external conditions upon irradiation with solar-simulated light. The system constantly evolved $H_2$ with an efficiency of 3.7 μmol $H_2$ $m^{-2}$ $s^{-1}$. The PGP system proposed in this work represented a promising first step toward the development of an oxygen-tolerant solar energy conversion system [28].

It was reported that when $[Ru(bpy)_2(phen\text{-}NH_2)]^{2+}$ (where phen-$NH_2$ = 5-amino-1,10-phenanthroline)] (Complex **11**, Figure 4.9) was covalently attached to the *Thiocapsa roseopersicina* hydrogenase, photoinduced hydrogen production was observed in aqueous solution. In the presence of the redox mediator and the ruthenium(II) photosensitizer in aqueous solution, the efficiency of photocatalytic hydrogen generation was increased [38]. When the excited state undergoes quenching by a redox mediator, such methyl viologen electron transfer occurs, which can subsequently deliver the reducing equivalents to the resting state of the enzyme. Oxidation of a sacrificial terminal reductant such as EDTA completed the catalytic cycle (Figure 4.10) by regenerating a ruthenium(II) species. It was also reported that sustained hydrogen production occurred even in the presence of oxygen by presumably creating a local anoxic environment through the reduction of oxygen, similar to what is proposed for oxygen-tolerant hydrogenases. [38] These results provided a strong proof of

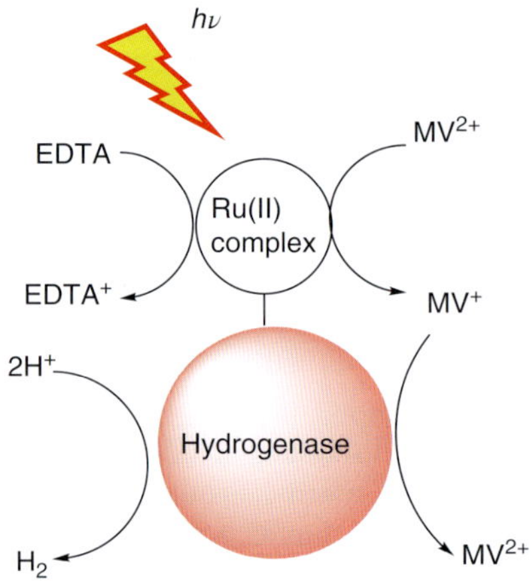


**Figure 4.10** Hydrogenase covalently linked to a Ru(II) complex with EDTA as a sacrificial reductant and methyl viologen (MV) as a redox mediator for hydrogen production. (Reprinted from Ref. [38]. Copyright (2012), with permission from Elsevier.)

concept for engineering photocatalytic hydrogen production in the presence of oxygen using biohybrid mimetic systems [38].

The effects of surfactants, lipids, and amphiphilic viologen mediators on $H_2$ production from dithionite as well as on a $[Ru(bpy)_3]^{2+}$-cation-sensitized $H_2$ photoproduction by hydrogenase from *T. roseopersicina* was studied [39]. Three systems which differed as to the nature of the hydrophobic matrix around the hydrogenase were tested. An enhanced hydrogenase activity was observed in the presence of surfactants, in the 1- to 6-mM concentration range [39]. Hydrogenase showed a selectivity for the amphiphilic viologens, and 2C7-diCl was the most efficient electron mediator in both reactions. Hydrogen photoproduction seemed not to be feasible in the detergent-hydrogenase system because of intensive foaming. Hydrogenase incorporated into liposomes catalyzed $H_2$ photoevolution efficiently, but the rate was decreasing in time, although reversibly [39]. Using intact bacterial cells instead of purified hydrogenase yielded stable $H_2$ photoevolution for at least 12 h. The system is believed to offer several advantages for potential practical applications [39].

Several novel ruthenium(II) phythalocyanine macrocycles-containing hydrogenase model complexes, especially complex **12** (Figure 4.9), were synthesized and characterized using various techniques [40]. Also, the structure of the complex was solved by X-ray crystallography. Generation of molecular hydrogen was the motive behind construction of complex 12 [40]. One of the novel ruthenium phtyalocyanine macrocycles (complex **12**, Figure 4.9) was identified as a catalyst for photoinduced hydrogen generation [40]. In order to test whether the complex could be used as a photoactive catalyst to produce $H_2$ production, the researchers carried out a study on photoinduced $H_2$ production using a three-component catalytic system that contained light-driven model complex **12**, electron donor $Et_3N$, and water as a proton source. It was found that $H_2$ was indeed generated when a tetrahydrofuran (THF) solution of the three-component system was irradiated by a 50 W Xe lamp with a UV cutoff filter ($\lambda > 400$ nm) [40]. However, when the same experiment was carried out in the absence of any one of the three components or without light irradiation, no $H_2$ evolution was observed. It was noted that the presence of electron donor $Et_3N$, water as a proton source, light-driven complex **12**, and light irradiation were all essential for photoinduced $H_2$ production [40]. In the reported study, the $H_2$ production increased linearly during 140 min of irradiation; then became very slow. A total of 180 min of irradiation produced $0.13 \times 10^{-3}$ mmol of $H_2$. Such a low catalytic efficiency was mainly due to serious decomposition of complex **12** as evidenced by a color change of the dark green solution to blue. It was interesting to note that complex **12** was the first phthalocyanine-macrocycle-containing [FeFe]-Hase model found to be a catalyst for photoinduced $H_2$ production, although the catalytic efficiency was low [40].

There was a report of the photosensitized production of hydrogen by hydrogenase in reversed micelles, where hydrogenase (hydrogen:ferricytochrome $c_3$ oxidoreductase, EC 1.12.2. 1) from *D. vulgaris* was encapsulated in reversed micelles with cetyltrimethylammonium bromide as surfactant and a chloroform/octane mixture as solvent [41]. Reducing equivalents for hydrogenase-catalyzed hydrogen production were provided by vectorial photosensitized electron transfer from a donor (thiophenol) in the organic phase through a surfactant-$[Ru(bpy)_3]^{2+}$, photosensitizer located in the interphase to methyl viologen concentrated in the aqueous core of the reversed micelle [41]. The maximum rate of hydrogen obtained was 0.7 ml $min^{-1}$ per mg of hydrogenase, with a quantum yield of 2%. However, no long-term stabilization was achieved. The results showed that reversed micelles provided a microenvironment that (i) stabilizes hydrogenase against inactivation and (ii) allowed an efficient vectorial photosensitized electron and proton flow from the organic phase to hydrogenase in the aqueous phase [41].

As a model of the active site of [NiFe]-hydrogenases, a dinuclear nickel–ruthenium complex $[Ni(xbsms)Ru(CO)_2Cl_2]$ **13** (where $H_2$xbsms = 1,2-bis (4-mercapto-3,3-dimethyl-2-thiabutyl)benzene) was synthesized and fully characterized [42]. The three-dimensional structure revealed a nickel center in a square-planar dithioether–dithiolate environment connected to a ruthenium moiety via a $[Ni(\mu\text{-}SR)_2Ru]$ bridge [42]. Complex **13** catalyzed hydrogen evolution by electroreduction of the weakly acidic $[Et_3NH]^+$ ions in DMF, and was reported as the first functional bioinspired model of [NiFe]-hydrogenase [42].

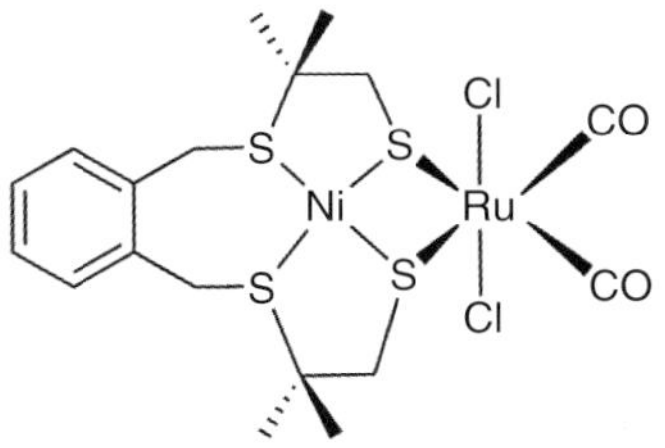


**13**

There was a report where the self-assembled vesicles with membrane-embedded or adsorbed ruthenium(II) polypyridine complexes were functionalized by the adsorption of an [FeFe]-hydrogenase subunit, which then mimicked the membrane interface enhancing the photocatalytic hydrogen production in water under acidic conditions [43]. The resulting two-dimensional membrane assembly placed the photosensitizer and hydrogen-evolving diiron complex in close proximity, resulting in a 6- to 12-fold increase in the TON as compared to the same system in the absence of lipid membranes. The interface assembly then enabled the combining of hydrophilic and hydrophobic catalytic entities for light-driven proton reduction in acidic water and provided a flexible method for membrane functionalization [43].

There was a book chapter as written by Reisner and Armstrong [44] where a hybrid system comprising a hydrogenase and a photosensitizer coattached to a nanoparticle served as a rational model for fast $H_2$ production using visible light. The chapter described a stepwise procedure for preparing $TiO_2$ nanoparticles functionalized with a hydrogenase from *Desulfomicrobium baculatum* (*Db* [NiFeSe]-H) and a tris(bipyridyl)ruthenium photosensitizer (RuP) [44]. Upon irradiation with visible light, such particles produced $H_2$ from neutral water at room temperature in the presence of a sacrificial electron donor, which was a test system for the cathodic half reaction of water splitting. In particular, Reisner and Armstrong [44] described how a hydrogenase and a photosensitizer with desired properties, including strong adsorption on $TiO_2$, can be selected by electrochemical methods. The catalyst *Db* [NiFeSe]-H was selected for its high $H_2$ production activity even when $H_2$ and traces of $O_2$ were presented. Adsorption of *Db* [NiFeSe]-H and RuP on $TiO_2$ electrodes resulted in high electrochemical and photocatalytic activities that translate into nanoparticles exhibiting efficient light harvesting, charge separation, and sacrificial $H_2$ generation [44].

Combined with a simple water-soluble [FeFe]-hydrogenase mimic, $[Ru(bpy)_3]^{2+}$ and L-ascorbic acid enabled hydrogen production photocatalytically. More than 88 equivalents of $H_2$ were achieved in water, which was much better than that obtained in an organic solvent or a mixture of organic solvent and water [45].

## 4.4 Conclusions

The path to producing very active and stable systems for the production of hydrogen as a viable alternative to fossil fuels is becoming a straight road to travel; however, there are still many hurdles to overcome. Ruthenium(II) photosensitizers with either cobalt-containing complexes or hydrogenases, or one of their mimics can offer much promise for the production of hydrogen as an alternative fuel. As such, ruthenium(II)-containing species can act as a light antenna and also transfer electrons to a metal-containing catalytic site which then produces hydrogen through the reduction of protons. Supramolecular systems containing mixed-metal centers can offer many advantages over the multicomponent system and present an avenue for continued growth in this area of research. In concluding, the readers are encouraged to read the reviews about systems that may contain cobalt, ruthenium, and other metal centers that can be used to produce hydrogen [14c, 46, 47].

## References


1 Stambouli, A.B. and Traversa, E. (2002) *Renew. Sust. Energy Rev.*, **6**, 295–304.

2 (a) Hansen, J., Sato, M., Ruedy, R., Lacis, A., and Oinas, V. (2000) *Proc. Natl. Acad. Sci. U. S. A.*, **97**, 9875–9880; (b) Rubin, E.S., Cooper, R.N., Frosch, R.A., Lee, T.H., Marland, G., Rosenfeld, A.H., and Stine, D.D. (1992) *Science*, **257**, 148–149.

3 Sobrino, F.H., Monroy, C.R., and Pérez, J.L.H. (2010) *Renew. Sust. Energy Rev.*, **14**, 772–780.

4 Krassen, H., Ott, S., and Heberle, J. (2011) *Phys. Chem. Chem. Phys.*, **13**, 47–57.

5 (a) Fukuzumi, S. (2015) *Curr. Opin. Chem. Biol.*, **25**, 18–26; (b) Lomoth, R. and Ott, S. (2009) *Dalton Trans.*, 9952–9959; (c) Ott, S., Styring, S., Hammarstroem, L., and Johansson, O. (2010) *Encyclopedia of Inorganic Chemistry*, John Wiley & Sons Ltd., pp. 199–227; (d) Sekar, N., Ramasamy, R.P., and Photochem, J. (2015) *Photobiol. C*, **22**, 19–33; (e) Xu, Y. and Xu, R. (2015) *Appl. Surf. Sci.*, **351**, 779–793.

6 Gordon, R.B., Bertram, M., and Graedel, T.E. (2006) *Proc. Natl. Acad. Sci. U. S. A.*, **103**, 1209–1214.

7 (a) Bellelli, A., Brunori, M., Brzezinski, P., and Wilson, M.T. (2001) *Methods*, **24**, 139–152; (b) Durham, B. and Millett, F. (2012) *Biochim. Biophys. Acta, Bioenerg.*, **1817**, 567–574; (c) Horvath, R., Huff, G.S., Gordon, K.C., and George, M.W. (2016) *Coord. Chem. Rev.*, **325**, 41–58; (d) Kaneko, M. (2001) *Prog. Polym. Sci.*, **26**, 1101–1137; (e) Sun, L., Hammarstrom, L., Akermark, B., and Styring, S. (2001) *Chem. Soc. Rev.*, **30**, 36–49; (f) Vlcek, A. Jr., (1998) *Chemtracts*, **11**, 621–625; (g) Wenger, O.S. (2013) *Acc. Chem. Res.*, **46**, 1517–1526.

8 Cotton, F.A., Wilkinson, G., Murillo, C.A., Bochmann, M., and Grimes, R. (1999) *Advanced Inorganic Chemistry* vol. 5, Wiley, New York.

9 (a) Brown, G.M., Brunschwig, B.S., Creutz, C., Endicott, J.F., and Sutin, N. (1979) *J. Am. Chem. Soc.*, **101**, 1298–1300; (b) Ziessel, R., Hawecker, J., and Lehn, J.-M. (1986) *Helv. Chim. Acta*, **69**, 1065–1084; (c) Ihara, M., Nishihara, H., Yoon, K.-S., Lenz, O., Friedrich, B., Nakamoto, H., Kojima, K., Honma, D., Kamachi, T., and Okura, I. (2006) *Photochem. Photobiol.*, **82**, 676–682.

10 (a) Teets, T.S. and Nocera, D.G. (2011) *Chem. Commun.*, **47**, 9268–9274; (b) Balzani, V. and Juris, A. (2001) *Coord. Chem. Rev.*, **211**, 97–115.

11 Kalyanasundaram, K. (1982) *Coord. Chem. Rev.*, **46**, 159–244.

12 Deponti, E. and Natali, M. (2016) *Dalton Trans.*, **45**, 9136–9147.

13 Konigstein, C. and Bauer, R. (1997) *Int. J. Hydrogen Energy*, **22**, 471–474.

14 (a) Connolly, P. and Espenson, J.H. (1986) *Inorg. Chem.*, **25**, 2684–2688; (b) Razavet, M., Artero, V., and Fontecave, M. (2005) *Inorg. Chem.*, **44**, 4786–4795; (c) Dempsey, J.L., Brunschwig, B.S., Winkler, J.R., and Gray, H.B. (2009) *Acc. Chem. Res.*, **42**, 1995–2004; (d) Dempsey, J.L., Esswein, A.J., Manke, D.R., Rosenthal, J., Soper, J.D., and Nocera, D.G. (2005) *Inorg. Chem.*, **44**, 6879–6892.

15 Muckerman, J.T. and Fujita, E. (2011) *Chem. Commun.*, **47**, 12456–12458.

16 Artero, V., Chavarot-Kerlidou, M., and Fontecave, M. (2011) *Angew. Chem., Int. Ed.*, **50**, 7238–7266.
17 Hawecker, J., Lehn, J.M., and Ziessel, R. (1983) *Nouv. J. Chim.*, **7**, 271–277.
18 Fihri, A., Artero, V., Razavet, M., Baffert, C., Leibl, W., and Fontecave, M. (2008) *Angew. Chem., Int. Ed.*, **47**, 564–567.
19 Fihri, A., Artero, V., Pereira, A., and Fontecave, M. (2008) *Dalton Trans.*, 5567–5569.
20 Tschierlei, S., Presselt, M., Kuhnt, C., Yartsev, A., Pascher, T., Sundström, V., Karnahl, M., Schwalbe, M., Schäfer, B., Rau, S., Schmitt, M., Dietzek, B., and Popp, J. (2009) *Chem.–Eur. J.*, **15**, 7678–7688.
21 Cropek, D.M., Metz, A., Muller, A.M., Gray, H.B., Horne, T., Horton, D.C., Poluektov, O., Tiede, D.M., Weber, R.T., Jarrett, W.L., Phillips, J.D., and Holder, A.A. (2012) *Dalton Trans.*, **41**, 13060–13073.
22 Li, C., Wang, M., Pan, J., Zhang, P., Zhang, R., and Sun, L. (2009) *J. Organomet. Chem.*, **694**, 2814–2819.
23 Lakadamyali, F. and Reisner, E. (2011) *Chem. Commun.*, **47**, 1695–1697.
24 Lakadamyali, F., Reynal, A., Kato, M., Durrant, J.R., and Reisner, E. (2012) *Chem.–Eur. J.*, **18**, 15464–15475.
25 Soltau, S.R., Dahlberg, P.D., Niklas, J., Poluektov, O.G., Mulfort, K.L., and Utschig, L.M. (2016) *Chem. Sci.*, **7**, 7068–7078.
26 Natali, M., Badetti, E., Deponti, E., Gamberoni, M., Scaramuzzo, F.A., Sartorel, A., and Zonta, C. (2016) *Dalton Trans.*, **45**, 14764–14773.
27 Chowdhury, A.D., Agnihotri, N., and De, A. (2015) *Chem. Eng. J.*, **264**, 531–537.
28 Noji, T., Kondo, M., Jin, T., Yazawa, T., Osuka, H., Higuchi, Y., Nango, M., Itoh, S., and Dewa, T. (2014) *J. Phys. Chem. Lett.*, **5**, 2402–2407.
29 Peters, J.W., Schut, G.J., Boyd, E.S., Mulder, D.W., Shepard, E.M., Broderick, J.B., King, P.W., and Adams, M.W.W. (2015) Biochim. *Biophys. Acta Mol. Cell Res.*, **1853**, 1350–1369.
30 Thauer, R.K., Kaster, A.-K., Goenrich, M., Schick, M., Hiromoto, T., and Shima, S. (2010) *Annu. Rev. Biochem.*, **79**, 507–536.
31 Liu, T.-M., Lin, K.-T., Li, F.-J., Lee, G.-H., Chen, M.-C., and Lai, C.K. (2015) *Tetrahedron*, **71**, 8649–8660.
32 D. G. Giarikos, in *Natural and Artificial Photosynthesis: Solar Power as an Energy Source* (Ed.: R. Razeghifard), John Wiley & Sons, Inc., 2013, pp. 143–171.
33 Asakura, N., Hiraishi, T., Kamachi, T., and Okura, I. (2001) *J. Mol. Catal. A: Chem.*, **172**, 1–7.
34 Wolpher, H., Borgstrom, M., Hammarstrom, L., Bergquist, J., Sundstrom, V., Styring, S., Sun, L., and Akermark, B. (2003) *Inorg. Chem. Commun.*, **6**, 989–991.
35 Ott, S., Borgstroem, M., Kritikos, M., Lomoth, R., Bergquist, J., Aakermark, B., Hammarstroem, L., and Sun, L. (2004) *Inorg. Chem.*, **43**, 4683–4692.
36 Ekstroem, J., Abrahamsson, M., Olson, C., Bergquist, J., Kaynak, F.B., Eriksson, L., Sun, L., Becker, H.-C., Aakermark, B., Hammarstroem, L., and Ott, S. (2006) *Dalton Trans.*, 4599–4606.

37 Hiraishi, T., Kamachi, T., and Okura, I. (1999) *J. Mol. Catal. A: Chem.*, **138**, 107–113.

38 Zadvornyy, O.A., Lucon, J.E., Gerlach, R., Zorin, N.A., Douglas, T., Elgren, T.E., and Peters, J.W. (2012) *J. Inorg. Biochem.*, **106**, 151–155.

39 Kovács, K.L. and Dér, A. (1986) *Biochimie*, **68**, 211–215.

40 Song, L.-C., Luo, F.-X., Liu, B.-B., Gu, Z.-C., and Tan, H. (2016) *Organometallics*, **35**, 1399–1408.

41 Hilhorst, R., Laane, C., and Veeger, C. (1982) *Proc. Natl. Acad. Sci. U. S. A.*, **79**, 3927–3930.

42 Oudart, Y., Artero, V., Pécaut, J., and Fontecave, M. (2006) *Inorg. Chem.*, **45**, 4334–4336.

43 Troppmann, S., Brandes, E., Motschmann, H., Li, F., Wang, M., Sun, L., and König, B. (2016) *Eur. J. Inorg. Chem.*, **2016**, 554–560.

44 E. Reisner, F. A. Armstrong, in *Nanoscale Biocatalysis: Methods and Protocols* (Ed.: P. Wang), Humana Press, Totowa, NJ, 2011, pp. 107–117.

45 Cao, W.-N., Wang, F., Wang, H.-Y., Chen, B., Feng, K., Tung, C.-H., and Wu, L.-Z. (2012) *Chem. Commun.*, **48**, 8081–8083.

46 Artero, V., Chavarot-Kerlidou, M., and Fontecave, M. (2011) *Angew. Chem., Int. Ed.*, **50**, 7238–7266.

47 (a) Losse, S., Vos, J.G., and Rau, S. (2010) *Coord. Chem. Rev.*, **254**, 2492–2504; (b) Claeys, M. (2016) *Nature*, **538**, 44–45; (c) Creutz, C. and Sutin, N. (1985) *Coord. Chem. Rev.*, **64**, 321–341; (d) Demirci, U.B. and Miele, P. (2010) *Phys. Chem. Chem. Phys.*, **12**, 14651–14665; (e) Eckenhoff, W.T., McNamara, W.R., Du, P., and Eisenberg, R. (2013) *Biochim. Biophys. Acta, Bioenerg.*, **1827**, 958–973; (f) Han, Z. and Eisenberg, R. (2014) *Acc. Chem. Res.*, **47**, 2537–2544; (g) Jin, S. (2016) *Prepr. - Am. Chem. Soc. Div. Energy Fuels*, **61**, 268; (h) Michalak, W.D. and Somorjai, G.A. (2013) *Top. Catal.*, **56**, 1611–1622; (i) Mulfort, K.L. (2017) *C. R. Chim.*, **20**, 221–229; (j) Singh, A. and Spiccia, L. (2013) *Coord. Chem. Rev.*, **257**, 2607–2622; (k) Sutin, N., Creutz, C., and Fujita, E. (1997) *Comments Inorg. Chem.*, **19**, 67–92; (l) Thoi, V.S., Sun, Y., Long, J.R., and Chang, C.J. (2013) *Chem. Soc. Rev.*, **42**, 2388–2400; (m) Zee, D.Z., Chantarojsiri, T., Long, J.R., and Chang, C.J. (2015) *Acc. Chem. Res.*, **48**, 2027–2036.

# Section III

# Applications in Medicine

# 5

# Ligand Photosubstitution Reactions with Ruthenium Compounds: Applications in Chemical Biology and Medicinal Chemistry

*Samantha L. Hopkins and Sylvestre Bonnet*

*Leiden University, Leiden Institute of Chemistry, Einsteinweg 55, Leiden, 2300 RA, The Netherlands*

## 5.1 Introduction

Other chapters in this book show how the rich photochemistry of ruthenium compounds has been exploited in the fields of supramolecular chemistry, photocatalysis, and photoactive materials. However, the limitations of platinum-based anticancer drugs, such as their toxicity toward patients, and the inherent as well as acquired resistances observed in tumors treated with cisplatin, resulted in the flourishing development of ruthenium-based anticancer compounds. Two of them have reached phase II of clinical trials, and many others are currently being developed [1].

Of course, making ruthenium-based dyes for photodynamic therapy (PDT) is appealing as intersystem crossing to the triplet metal-to-ligand charge-transfer ($^3$MLCT) excited state is very efficient and is also effective for sensitizing reactive oxygen species (ROS) [2]. However, many compounds without ruthenium are also considered for PDT, and ruthenium itself does not show perspectives that are qualitatively different from other transition metals. There is one type of reactivity that clearly differentiates Ru(II) complexes from other metal-containing compounds: their ability to photosubstitute ligands. When the energy difference between the $^3$MLCT and the metal-centered triplet state ($^3$MC) of a Ru(II) complex is low enough, thermal population of the $^3$MC state from the photochemically generated $^3$MLCT state can occur (see Chapter 1). In other words, an electron from a $t_{2g}$ orbital is promoted into a metal-ligand antibonding $e_g$ orbital, which weakens one or two of the coordination bonds and facilitates the substitution of a ligand by solvent molecules. This reactivity, initially discovered as a detrimental decomposition pathway for phosphor such as $[Ru(bpy)_3]^{2+}$ (bpy = 2,2′-bipyridine) [3], was later used in supramolecular applications such as light-controlled molecular machines [4]. It is now rapidly developing in a biological setting [5]. Indeed, ligand photosubstitution reactions can be seen as the light-induced cleavage of a chemical bond, notably when the ruthenium complex and/or the photosubstituted ligand are biologically active (Figure 5.1). The ability to photochemically "uncage" bioactive compounds is an important

*Ruthenium Complexes: Photochemical and Biomedical Applications*, First Edition.
Edited by Alvin A. Holder, Lothar Lilge, Wesley R. Browne, Mark A.W. Lawrence, and Jimmie L. Bullock Jr.

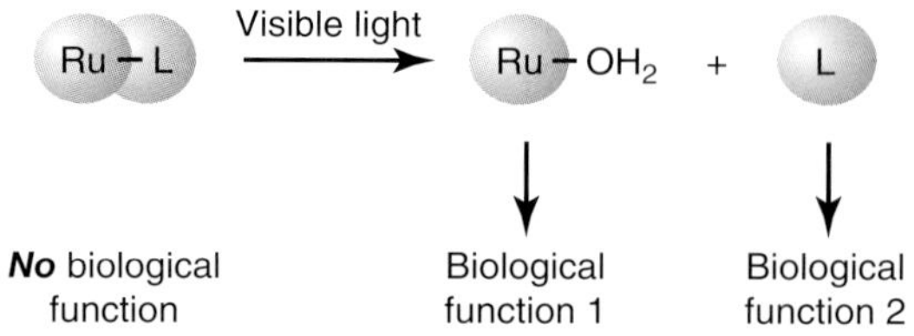


**Figure 5.1** Photosubstitution of a ligand L bound to a Ru(II) or Ru(III) center in chemical biology. Either the ruthenium complex, or the ligand, or both, can have a biological function.

part of organic photochemistry, but many organic caging groups require ultraviolet (UV) light to be cleaved, which is problematic for biological applications [6]. Alternatively, photosubstitutionally active ruthenium compounds display $^1$MLCT absorption in the visible part of the spectrum, which has played a critical role in the development of such compounds in photopharmacology and chemical biology. Visible light is generally less toxic to cells and penetrates further into biological tissues, compared to UV light.

In a ruthenium(II) or ruthenium(III) compound containing a photolabile Ru–L coordination bond, either the ruthenium-containing part, or the photosubstituted ligand L, or both, can have a biological function (Figure 5.1). In the first part of this chapter, the use of nontoxic ruthenium complexes to cage biologically active organic ligands L is introduced (Section 5.2). In the second part, the photosubstituted organic ligand, L, conceals the toxicity of a ruthenium complex (Section 5.3). The third part focuses on strategies developed to move the wavelength of activation of photosubstitutionally active ruthenium compounds to lower energy (Section 5.4).

## 5.2 Caging and Uncaging Biologically Active Ligands with a Nontoxic Ruthenium Complex

When the aqua ruthenium complex Ru—$OH_2$ is poorly toxic, it can be used to "cage" a biologically active organic ligand L. A "caged" ligand is, in this context, a Ru–L complex that can be considered as biologically inactive because it does not interact (or interacts much less) with, for example, proteins or DNA, compared to the uncaged ligand L. For example, a neurotransmitter such as GABA can be caged by preparing the complex $[Ru(bpy)_2(PMe_3)(\kappa N\text{-}GABA)]^{2+}$ ($[\mathbf{1}]^{2+}$, see Table 5.1) [7]. This complex has a very different charge and shape compared to GABA, which will strongly reduce the interaction of the caged GABA complex $[\mathbf{1}]^{2+}$ with the GABA receptors. Visible-light irradiation in the MLCT band of the ruthenium complex $[\mathbf{1}]^{2+}$ leads to the selective cleavage of the Ru–L amine bond, that is, GABA is photosubstituted by a water molecule. This photoreaction releases the biologically active ligand GABA and the ruthenium complex $[Ru(bpy)_2(PMe_3)(OH_2)]^{2+}$, which in cells is essentially biologically inactive at the concentrations used.

Table 5.1 shows an overview of biologically active organic molecules L that have been caged using nontoxic or poorly toxic ruthenium(II) complexes. Introduced by the Etchenique group for neurotransmitters [7, 8], the ruthenium caging concept has extended to amino acids [8d, e, 12], peptides [12a], protein inhibitors

**Table 5.1** Overview of biologically active ligands L caged with a nontoxic ruthenium complex.

| Cpd. # | Formula of Ru-L | Structure of L | Biological function of L | References |
|---|---|---|---|---|
| [**1**]$^{2+}$ | $[Ru(bpy)_2(PPh_3)(L)]^{2+}$ | 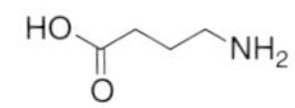 GABA | Neurotransmitter | [7, 8c, d, f] |
| [**2**]$^{2+}$ | $[Ru(bpy)_2(PMe_3)(L)]^{2+}$ | | | |
| [**3**]$^{2+}$ | $[Ru(bpy)_2(L)_2]^{2+}$ | 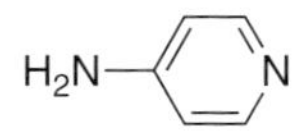 4-Aminopyridine | Glial $K^+$ channel blocker [9] | [8a, 10] |
| [**4**]$^{2+}$ | $[Ru(bpy)_2(PMe_3)(L)]^{2+}$ | 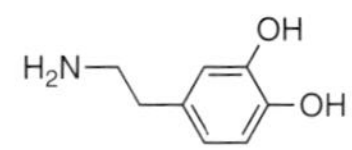 DOPA | Neurotransmitter | [11] |
| [**5**]$^{2+}$ | $[Ru(bpy)_2(L)_2]^{2+}$ | 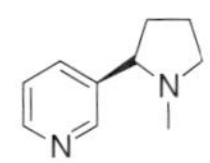 Nicotine | Neurotransmitter | [8b] |
| [**6**]$^{2+}$ | $[Ru(bpy)_2(PMe_3)(L)]^{2+}$ | 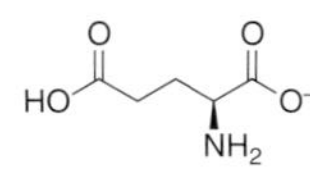 Glutamate | Neurotransmitter, amino acid | [8e, 12c, d] |
| [**7**]$^{2+}$ | $[Ru(tmp)(bpy)(L)]^{2+}$ | | | |
| [**8**]$^{2+}$ | $[Ru(tpy)(bpy)(L)]^{2+}$ | 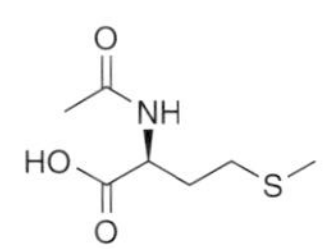 *N*-Acetylmethionine | Amino acid | [12b] |

*(Continued)*

**Table 5.1** (Continued)

| Cpd. # | Formula of Ru-L | Structure of L | Biological function of L | References |
|---|---|---|---|---|
| [**9**]$^{2+}$ | $[Ru(bpy)_2(PPh_3)(L)]^{2+}$ | 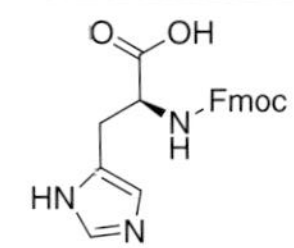 | Amino acid | [12a] |
| [**10**]$^{2+}$ | $[Ru(bpy)_2(PPh_3)(L)]^{2+}$ | 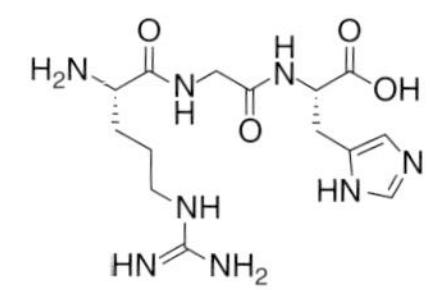 Arg-Gly-His | Ni-dependent peptide nuclease | [12a] |
| [**11**]$^{2+}$ | $[Ru(bpy)_2(L)_2]^{2+}$ | 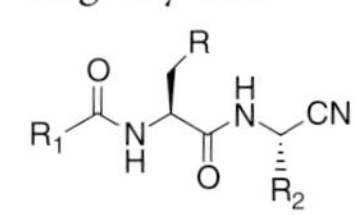 R = Ph or i-Pr, $R_1$ = Me or $OCH_2Ph$, $R_2$ = H or $CH_2OCH_2Ph$ | Cathepsin K inhibitors | [13b, c] |
| [**12**]$^{2+}$ | $[Ru(bpy)_2(PMe_3)(L)]^{2+}$ | 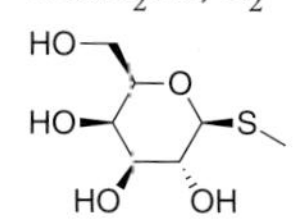 Methyl-β-D-thiogalactoside (MTG) | Inducer of Lac operon [14] | [8g] |
| [**13**]$^{2+}$ | $[Ru(tpy)(bpy)(L)]^{2+}$ | 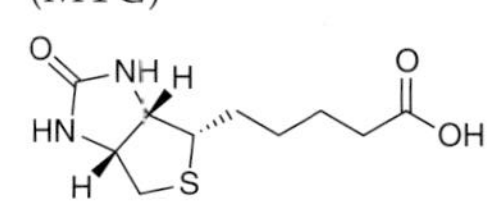 D-(+)-Biotin | Vitamin (B7) | [12b] |

| | | | | |
|---|---|---|---|---|
| [**14**]$^{2+}$ | $[Ru(bpy)_2(L)_2]^{2+}$ | 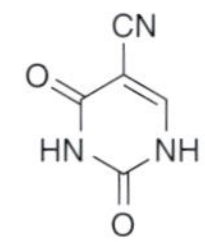<br>5-Cyanouracyl | Rat liver pyridmidine reductase [15] | [13a, 16] |
| [**15**]$^{2+}$ | $[Ru(tpy)(L)_3]^{2+}$ | | | |
| [**16**]$^{2+}$ | $[Ru(bpy)_2(Cl)(L)]^{+}$ | 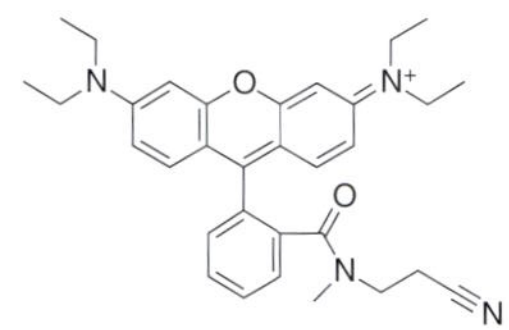<br>RhodB-MAPN | Fluorophore | [17] |
| [**17**]$^{2+}$ | $[Ru(bpy)_2(PMe_3)(L)]^{2+}$ | *n*-BuNH$_2$ | Brönsted base | [18] |
| [**18**]$^{2+}$ | $[Ru(bpy)_2(L)_2]^{2+}$ | 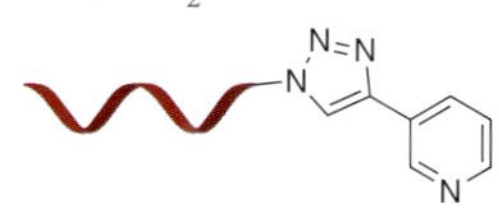 | 25-mer antisense morpholino | [19] |
| [**19**]$^{2+}$ | $[Ru(bpy)_2(MeCN)(L)]^{2+}$ | 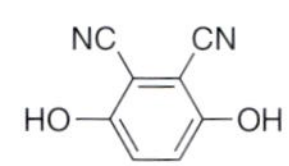<br>Dicyanohydroquinone | Fluorophore | [20] |

[13], vitamins [12b], thiosaccharides [8g], and, recently, even Brönsted bases such as n-butylamine [18]. The principle of caging biologically active ligands has been and is still being thoroughly investigated by organic chemists [6, 21]; compounds such as caged ATP [22] or caged phosphate [23] have been around since the late 1970s. However, most organic caging groups require near-UV light to be removed, which is suboptimal for biological applications and particularly *in vivo* as UV photons penetrate poorly through tissues and damage cells significantly [24]. The use of ruthenium polypyridyl complexes instead of nitrophenyl caging groups (organic caging) allows for uncaging of the biological ligand L using visible light, usually in the blue or green region of the spectrum. The uncaging quantum efficiency or quantum yield of photorelease ($\Phi_{PR}$) is high with caged amine compounds based on the $[Ru(bpy)_2(PMe_3)(H_2NR)]^{2+}$ building block: typical values of 0.08 have been found at room temperature, which, combined with continuous green lasers available today, leads to half reaction times of tens of microseconds [7]. Green light is not yet in the "phototherapeutic window" (600–900 nm) where photons optimally penetrate living tissues, but it is perfect for *in vitro* studies in cell cultures and chemical biology as it is nontoxic to cells even at relatively high intensities (up to 20 mW cm$^{-2}$ at 520 nm with doses of ~40 J cm$^{-2}$) [25]. A recent study even discussed the uncaging of GABA in the cortex of anesthetized female mice, opening the application of caging ruthenium complexes *in vivo* [8e, f, 12d]. Finally, two-photon uncaging [8e, 12d, 26] and new upconverting strategies [27] are being developed for achieving ligand photosubstitution reactions using red or near-infrared (NIR) light (Section 5.4).

## 5.3 Caging Cytotoxic Ruthenium Complexes with Organic Ligands

Although not all heavy-metal-based compounds are toxic, many show significant anticancer properties; and, in particular, platinum-based compounds such as cisplatin, carboplatin, or oxaliplatin have become very successful in the clinics. Ruthenium-based anticancer compounds, and among them light-activated metallodrugs, are being developed to solve the main issues of platinum-based antineoplastic compounds: (i) their lack of selectivity toward cancer cells, which leads to side effects in cancer patients and (ii) the acquired or inherent resistances of certain cancer types, which limits treatment efficacy.

Photosubstitutionally labile metallodrugs are different from dyes used in conventional photodynamic therapy in that the triggering of toxicity is based on a mechanism of the type shown in Figure 5.1, where it is the aqua ruthenium photoproduct Ru—$OH_2$ that is the cytotoxic species, instead of the ligand. Importantly, such a light activation mechanism does not depend on the presence of dioxygen, whereas in conventional PDT light activation requires $O_2$ to be present at the place of irradiation to lead to ROS production, oxidative stress, and ultimately to cell death. The basic assumption in photoactivated chemotherapy (PACT) is that in the metal compound Ru–L the ligand L acts as a "protecting" group that prevents the metal complex from binding to biomolecules because its coordination sphere is saturated. Once irradiated with light, photosubstitution of L opens one or more coordination sites at the metal center, thus allowing the aqua photoproduct Ru—$OH_2$ to bind to biomolecules such as DNA, proteins, or lipids. Such binding may lead, like for cisplatin, to significant cytotoxicity, and the aqua species is usually seen as the "active species."

Rhodium-based photosubstitutionally labile compounds were discovered first [28], followed by platinum-based compounds where a combination of photoredox reactions and substitution ultimately led to the active species [29]. More recently, the concept has demonstrated its full power with ruthenium-based photosensitive compounds because of their excellent light absorption properties in the visible region. Ruthenium compounds have a long history for light-induced DNA binding [30], DNA cleavage [31], and PDT action [5b, 32], but the use of photosubstitution reactions as a means to switch on cell toxicity is relatively new [12b, 33]. In 2011 the Turro group introduced the complex $[Ru(bpy)_2(5CNU)_2]^{2+}$ ($[\mathbf{14}]^{2+}$, see Table 5.1), which bears 5-cyanouracyl (5CNU), an analogue of the clinically used chemotherapeutic agent 5-fluorouracyl. Upon irradiation, $[\mathbf{14}]^{2+}$ releases 5CNU and the bis-aqua complex $[Ru(bpy)_2(OH_2)_2]^{2+}$. This bis-aqua complex can, in turn, interact with DNA (linearized pUC18 plasmid), whereas no metal-DNA interaction was observed in the dark. At that stage Turro *et al.* realized that "the transition metal portion of the complex itself may be biologically active." [16] The demonstration that the uncaging of 5CNU would lead to increased cellular toxicity after visible-light irradiation came in a 2013 paper comparing the dark and light toxicity on HeLa cancer cells of complex $[Ru(tpy)(5CNU)_3]^{2+}$ ($[\mathbf{15}]^{2+}$, Figure 5.2a) to that of $[Ru(tpy)(MeCN)_3]^{2+}$ ($[\mathbf{20}]^{2+}$) (tpy = 2,2′,6′,2"-terpyridine) [13a]. Cytotoxicity was assayed by confocal microscopy using SYTOX Green, a cell-permeable dye that can only translocate cells when the cell membrane has been compromised—for example, by cell death. Fluorescence was found in the cells only when complex $[\mathbf{15}]^{2+}$ was irradiated in the cells using white light ($\lambda_{irr} > 400$ nm). $[Ru(tpy)(MeCN)_3]^{2+}$ did not show any membrane permeabilization both in the dark and after irradiation. The concentration of $[\mathbf{15}]^{2+}$ needed to kill 50% of the cells ($LC_{50}$) was almost identical ($156 \pm 18$ μM) to that of the toxic compound 5CNU ($151 \pm 33$ μM), which was interpreted as a sign that one equivalent of 5CNU was released per ruthenium ion under their experimental conditions ($\lambda_{irr} > 400$ nm, 1 h, UV–vis photoreactor). Finally, binding to double-stranded DNA took place only when $[\mathbf{15}]^{2+}$ was irradiated with visible light. However, the authors did not know whether the compound would penetrate into the nucleus, and they did not conclude on whether DNA binding of $[Ru(tpy)]^{2+}$-containing fragments would contribute to the phototoxicity observed *in vitro*, or whether phototoxicity was solely due to the photochemical uncaging of the 5CNU ligand.

In 2012, Glazer's group published two seminal papers on the strained ruthenium compounds $[Ru(bpy)_2(dmbpy)]^{2+}$ ($[\mathbf{21}]^{2+}$, dmbpy = 6,6′-dimethyl-2,2′-bipyridine, Figure 5.2a), $[Ru(bpy)_2(dmdpq)]^{2+}$ ($[\mathbf{22}]^{2+}$, dmdpq = 2,9-dimethyl-dipyrido[3,2-*f*:2′,3′-*h*]-quinoxaline), $[Ru(phen)_2(biq)]^{2+}$ ($[\mathbf{23}]^{2+}$, biq = 2,2′-biquinoline), and $[Ru(phen)(biq)_2]^{2+}$ ($[\mathbf{24}]^{2+}$) (see Figure 5.2b) [34]. These complexes are also photosubstitutionally active, but it is the hindered bidentate ligand that is photosubstituted, that is, dmdpq, dmbpy or biq, instead of a monodentate ligand. Like for $[\mathbf{14}]^{2+}$ or $[\mathbf{15}]^{2+}$ DNA binding takes place

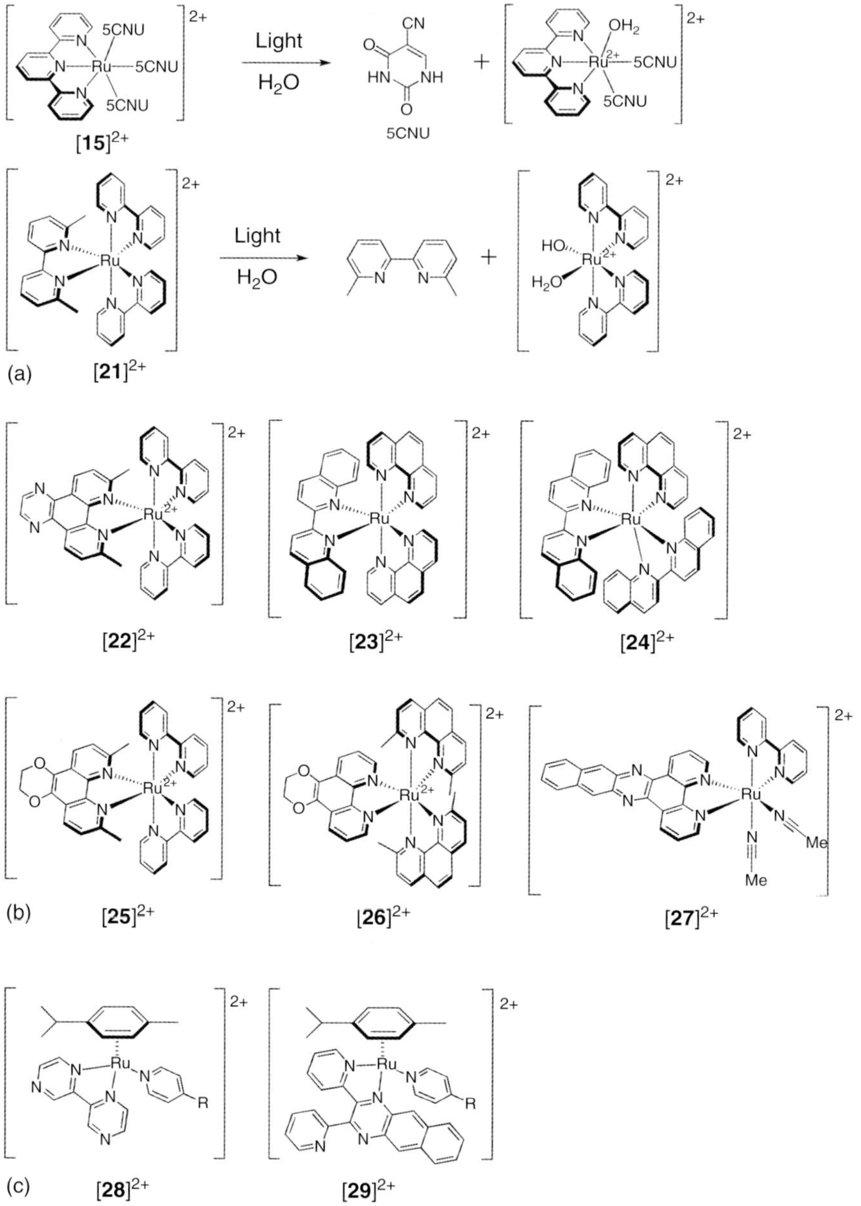


**Figure 5.2** (a) Photosubstitution reactions of ruthenium polypyridyl complexes leading to phototoxicity *in vitro*, (b) photosubstitutionally active ruthenium polypyridyl complexes with phototoxicity *in vitro*, and (c) photosubstitutionally active ruthenium arene complexes.

specifically after visible-light irradiation for [**21**]$^{2+}$, [**23**]$^{2+}$ and [**24**]$^{2+}$, whereas for [**22**]$^{2+}$ both photobinding and photocleavage occur. The cytotoxicity in the dark and after light irradiation was determined for these four complexes in HL-60 leukemia cells, and for [**21**]$^{2+}$ and [**22**]$^{2+}$ in A549 cancer cell lines as well as in A549 tumor spheroids. In 2D cell cultures, dark cytotoxicity $LC_{50}$ values >100 μM were found for these compounds, whereas following visible-light irradiation (>450 nm cutoff filter, 3 min, 410 W projector) the $LC_{50}$ decreased to 1.1-3 μM, resulting in phototherapeutic indices as large as 200 (Figure 5.3). These results were confirmed by another group using slightly different compounds with hydroxyl substituents instead of methyl groups [35]. The latest development of these type of compounds showed that compound [**26**]$^{2+}$, for example, has a record phototoxic index of >1880, with an $IC_{50}$ above 300 μM in the dark versus 0.16 μM after white light irradiation [36].

The last family of ruthenium-based cytotoxic compounds, which display photosubstitution reactions, is the ruthenium-arene anticancer compounds [1, 1g, 37]. When an aza-bipyridyl ligand is also bound to the ruthenium center of the piano stool complex, the resulting Ru-arene compound may show some photosubstitution properties. Complexes [**28**]$^{2+}$ and [**29**]$^{2+}$ (Figure 5.2c) are typical examples of such molecules [38]. Sadler and Marchán demonstrated that by linking the pyridine monodentate ligand in [**28**]$^{2+}$ to a cancer-targeting peptide such as octreotide, it was possible to prepare a receptor-targeted, light-activated anticancer prodrug [39]. The photosubstitution quantum yield of this type of compound is, however, rather low (typically 0.001) [38c]. Compound [**29**]$^{2+}$ is an interesting case of a phototoxic prodrug with a double mode of action: in parallel to ligand photosubstitution occurring via the $^3$MLCT excited state of the complex, $^1O_2$ is also produced from a $^3\pi\pi^*$ excited state based on the polyazapyridyl ligand, which contributes to the observed phototoxicity. Similar effects were thoroughly studied the same year for the polypyridyl compound [**27**]$^{2+}$, which shows a phototoxic index of 1110 on HeLa cancer cells after 20-min irradiation at 466 nm [40].

Developments in the field of photoactivated chemotherapy are aimed at improving the phototoxic index of this type of compounds and increasing the

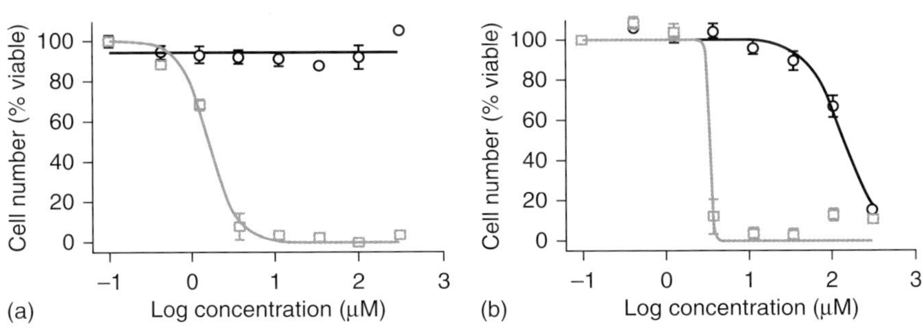


**Figure 5.3** Dose–response curves for compound [**21**]$^{2+}$ (a) and [**22**]$^{2+}$ (b) in HL60 leukemia cells in the dark (circles) and after white light irradiation (squares, >450 nm light, 410 W, 3 min). (Adapted with permission from reference [34]. Copyright (2012) American Chemical Society.)

wavelength at which activation by light irradiation may occur (see Section 5.4). For most substitutionally active ruthenium compounds, DNA studies in reconstituted chemical models of the intracellular medium demonstrate that photosubstitution occurs, followed by (or in parallel with) DNA binding and/or DNA cleavage. However, few experiments have been fulfilled on the mode of action of these compounds *in vitro*, for example, to determine their molecular target after light irradiation. In addition, when both the photo-uncaged organic ligand L and the aqua-metal-containing photoproduct Ru—$OH_2$ (see Figure 5.1) are biologically active, it is difficult to distinguish their respective contributions to the phototoxicity of the caged compound *in vitro*. New chemical biology methods for determining the cellular target(s) of a drug were recently applied to organometallic ruthenium-arene compounds [41]. The field of photosubstitutionally labile ruthenium prodrugs would certainly profit from the development of these methods to light-activated compounds.

## 5.4 Low-Energy Photosubstitution

### 5.4.1 Introduction

Photoactive Ru-based caged complexes provide space- and time-resolved release of the bioactive moiety, and a shift of the light activation wavelength from the UV region to the visible. However, the wavelength of the light needed for activation is often in the high-energy blue domain, which poorly transmits through organs and, in certain cases, could cause undesired side effects in both normal and cancerous cells [25, 42]. Thus, it would be most beneficial to shift the wavelength of light activation toward the red or even NIR region of the spectrum. In this region, also known as the phototherapeutic window (600–900 nm), light transmits best through epidermal layers due to reduced light absorption by endogenous chromophores and water, as well as decreased light scattering [43]. Two strategies have been considered to shift the wavelength of activation for Ru-based complexes: either the modulation of the ruthenium complex photophysical properties via ligand modifications, or the generation of blue photons *in situ* using upconversion. These strategies are explained in more detail below.

### 5.4.2 Modulating Ru Photophysics by Ligand Modulation

The general ligand design principles for tuning the equilibrium between the $^3$MLCT and $^3$MC excited states of ruthenium(II) polypyridyl complexes have been extensively studied and reviewed elsewhere [5b, 8g, 44]. Here, we more specifically discuss the ligand modifications aimed at shifting the activation window toward the phototherapeutic window.

The most straightforward method for extending the light absorption properties of Ru complexes toward the phototherapeutic window is to extend the conjugation of the π-system of some of the ligands. The most prominent example of increased conjugation of the polypyridyl ligands is the phen (1,10-phenanthroline), dpq (dipyrido[3,2-*f*-2′,3′-*h*] quinoxaline), dppz (dipyrido[3,2-*a*-2′,3′-*c*] phenazine), and dppn (benzo[*i*]dipyrido-[3,2-*a*:2′,3′-*c*] quinoxaline) series [32c, 45]. For example, in the $[Ru(bpy)(NN)(CH_3CN)_2]^{2+}$ complexes, substitution of NN = bpy ($[\mathbf{30}]^{2+}$) by dppn ($[\mathbf{27}]^{2+}$) shifts the tail of the lowest energy absorption band of the ruthenium complex toward the red [40]. However, the most extended ligand of the series, dppn, does not offer high extinction coefficients in the red region to its ruthenium polypyridyl complexes. Indeed, it is only the ligand-based $^1\pi\pi^*$ absorption band of the complex, which usually lies in the UV region with the smallest ligand of the series (NN = bpy or phen), that move "toward the red." Thus, dppn ruthenium complexes such as $[Ru(bpy)(dppn)(CH_3CN)_2]^{2+}$ ($[\mathbf{27}]^{2+}$) $[Ru(bpy)_2(dppn)]^{2+}$ ($[\mathbf{31}]^{2+}$) or $[Ru(tpy)(dppn)(pyridine)]^{2+}$ ($[\mathbf{32}]^{2+}$) have their $^1$MLCT absorption band typically between 450 and 500 nm, while the $^1\pi\pi^*$ absorption attributed to dppn usually occurs near 400 nm. These complexes, when irradiated with visible light, populate not only the $^3$MLCT and $^3$MC states but also the long-lived, ligand-based $^3\pi\pi^*$ excited states that lead to significant production of $^1O_2$ instead of photosubstitution. Steric hindrance can be introduced in the coordination sphere of the metal complex by methylation of the position ortho to the coordinating nitrogen atoms, which reestablishes higher photosubstitution quantum yields [46]. In the $[Ru(tpy)(NN)(pyridine)]^{2+}$ system, for example, where NN = dppn ($[\mathbf{32}]^{2+}$) or NN = $Me_2$dppn ($[\mathbf{33}]^{2+}$, $Me_2$dppn = 6,6-dimethylbenzo[*i*]dipyrido-[3,2-*a*:2′,3′-*c*] quinoxaline), the photosubstitution quantum yield ($\Phi_{PR}$) increased from <0.0001 for $[\mathbf{32}]^{2+}$ to 0.053 for $[\mathbf{33}]^{2+}$ at 500 nm [47].

Alternatively, Ru cyclometalated complexes can be of interest for shifting the wavelength of the absorption maximum of ruthenium compounds to the red [38b, 48]. Generally, introducing a carbon–metal bond in the coordination sphere enhances the contribution of the ligand to the metal-based $t_{2g}$ orbitals of the ruthenium complex, which in turn increases the energy of the highest occupied molecular orbital (HOMO) and shifts the $^1$MLCT absorption band of the complex toward the red. However, this effect is detrimental for the photosubstitution reactivity of the complex, as the $^3$MLCT-$^3$MC energy gap increase significantly when the 3MC is destabilized, which prevents thermal population of the $^3$MC state from the photochemically generated $^3$MLCT state. Turro's group studied the replacement of bipyridyl ligands by $phpy^-$ ligands (Hphpy = 2-phenylpyridine) in two series of ruthenium compounds [49]. For the $[Ru(phpy)(NN)(CH_3CN)]^+$ complexes, where NN = bpy or phen, the electronic absorption spectrum indeed showed a tail extending into the red region of the spectrum, and photosubstitution of the acetonitrile ligand by chlorides in dichloromethane occurred upon 430- to 450-nm irradiation in the presence of tetrabutylamonium chloride. Appreciable phototoxicity was observed for $[Ru(phpy)(phen)(CH_3CN)]^+$ ($[\mathbf{34}]^+$) in OVCAR-5 cancer cells, with a high toxicity ($LC_{50}$ = 70 nM) after 100 s irradiation at 690 nm (5 J cm$^{-2}$, 50 mW cm$^{-2}$), compared to an $LC_{50}$ of 1 μM in the dark [49a]. However, in the sterically hindered heteroleptic series $[Ru(biq)_2(NX)]^{2+/1+}$, where NX = bpy ($[\mathbf{35}]^{2+}$), phen ($[\mathbf{36}]^{2+}$), or $phpy^-$($[\mathbf{37}]^+$), the biq photosubstitution quantum yield at 600 nm was negligible for NX = $phpy^-$, whereas it was 0.05 and 0.02 for NX = bpy and phen, respectively [49b]. These examples clearly show that modifying the ligands around the ruthenium center to increase the absorption maximum of the complex in the red region must be done carefully, and that a compromise must be found between low-energy light absorption and efficiency of the photosubstitution reaction.

Another strategy to obtain red-light-absorbing ruthenium complexes is to hybridize one of the polypyridyl ligands with an organic-based chromophore such as coumarin, pyrene, thiophene, or rhodamine [17, 50]. Hybrid metal complex-chromophore dyads can indeed combine low-energy light absorption with ultrafast energy transfer processes (Table 5.2). In donor–acceptor molecular compounds nonradiative energy transfer between an energy donor (D) and an energy acceptor (A) usually occurs via Förster resonance energy transfer (FRET) or Dexter energy transfer (DET) (Figure 5.4a) [51]. Normal FRET occurs when the donor is excited at higher energy wavelengths compared to the donor (Figure 5.4b). However, reverse FRET can also occur if the acceptor has a long-lived singlet lifetime, if the donor emission overlaps with the acceptor absorption significantly, and if the donor–acceptor distance is short enough (Figure 5.4c). As explained subsequently, energy transfer can be used to obtain photosubstitution reactions on ruthenium complexes by excitation of the low-energy-absorbing organic dye attached to it [27a].

In 2010, $[Ru(bpy)_2(RhodB\text{-}MAPN)Cl]^+$ (compound [**16**]$^+$) (Table 5.1) was reported to undergo photoactivated uncaging of the rhodamine-based fluorescent probe [17b]. Further studies of [**16**]$^+$ indicated that the rhodamine-B was acting as an antenna and funneling its excitation energy into the $^1$MLCT

**Table 5.2** Summary of the ruthenium-chromophore dyad photochemical properties.

| Compound | $\lambda^{abs}_{max}$ (nm) | $\varepsilon_{\lambda abs}$ ($M^{-1}$ $cm^{-1}$) | $\lambda^{PR}_{exc}$ (nm) | $\varepsilon_{\lambda_{exc}}$ ($M^{-1}$ $cm^{-1}$) | $\Phi_{PR}$ | $\xi$ ($\Phi_{PR} \times \varepsilon_{\lambda exc}$) | References |
|---|---|---|---|---|---|---|---|
| $[16]^+$ | | | 473 | | 0.12$^{a)}$ | | [17b] |
| $[38]^+$ | 470 | 6300 | 532 | 1700 | 0.14$^{b)}$ | 240 | |
| $[16]^+$ | | | 532 | 84500 | 0.070$^{a)}$ | 5915 | [17a] |
| $[39]^{2+}$ | | | 532 | 101 | 0.026$^{c)}$ | 3 | [50b] |
| $[40]^+$ | | | 532 | 727 | 0.0098$^{c)}$ | 7 | [50b] |
| $[41]^{2+}$ | | | 532 | | 0.03$^{c)}$ | | [50b] |
| $[42]^{2+}$ | | | 532 | | 0.03$^{c)}$ | | [50b] |
| $[43]^{3+}$ | 570 | 44000 | 452 | 4800 | 0.0092$^{d)}$ | 44 | [50c] |
| $[43]^{3+}$ | | | 570 | 44000 | 0.0085$^{d)}$ | 370 | [50c] |
| [45] | 395 | 5300 | 500 | | 0.0008$^{e)}$ | | [50d] |
| [46] | 500 | 11900 | 500 | 11900 | 0.052$^{e)}$ | 620 | [50d] |
| [47] | 495 | 24000 | 500 | 24000 | 0.31$^{e)}$ | 7440 | [50d, e] |
| $[48]^+$ | 475 | 24700 | 500 | | 0.041$^{e)}$ | | [50e] |

a) Release of RhB-MAPN.
b) Release of MAPN.
c) Release of $Cl^-$.
d) Release of Hmte.
e) Release of NO.

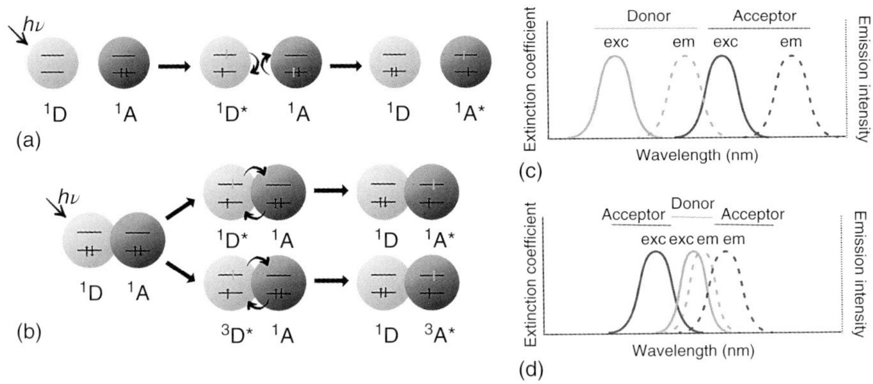


**Figure 5.4** Förster resonance and Dexter energy transfer mechanisms, (a) forward FRET (b) showing donor and acceptor excitation (solid) and emission (dashed), and (c) reverse FRET (d) showing donor and acceptor excitation (solid) and emission (dashed).

state of the $Ru(bpy)_2$ subunit [17a]. Such a reverse-FRET mechanism resulted in photosubstitution at low-energy excitation (532 nm instead of 450 nm), as well as a 25-fold increase in photorelease activity $\xi = \Phi_{PR} \times \varepsilon_{\lambda_{exc}}$, compared to $[Ru(bpy)_2(MAPN)Cl]^+$ (**[38]**$^+$, see Figure 5.5) at the same wavelength (where $\Phi_{PR}$ is the photosubstitution quantum yield and $\varepsilon_{\lambda_{exc}}$ is the extinction coefficient of the ruthenium-rhodamine dyad at the excitation wavelength). More recently, work with the same structural motif, but using rhodamine 6G and varied ligand-linker combinations (picolylamine **[39]**$^{2+}$/**[40]**$^+$, 1,2-diaminoethane **[41]**$^{2+}$, and 1,3-diaminopropane **[42]**$^{2+}$, see Figure 5.5), were reported [50b]. However, in these complexes green light irradiation led to the photosubstitution of the chloride ligand rather than that of the fluorophore. Another example of photosubstitution triggered by reverse FRET is the complex $[Ru(bpy)(RhodB\text{-}tpy)(Hmte)]^{3+}$ (**[43]**$^{3+}$, Figure 5.5), where rhodamine B was coupled to the spectator terpyridine ligand rather than to the photosubstituted ligand (here the thioether Hmte = 2-(methylthio)ethanol) [50c]. Irradiation of **[43]**$^{3+}$ or of its rhodamine-deprived analogue complex $[Ru(bpy)(tpy)(Hmte)]^{2+}$ (**[44]**$^{2+}$) using blue light (452 nm) resulted in similar $\xi$ values (44 and 100, respectively). However, irradiation with yellow light (570 nm) resulted in $\xi$ values of 370 and 4.8, respectively, which showed the excellent photon-collecting properties of the rhodamine dye at the appropriate wavelength. In this complex, the reverse-FRET mechanism allows the Hmte ligand to be photosubstituted quickly utilizing yellow light instead of blue light, with a photosubstitution activity increased by a fourfold factor compared to when irradiation was performed with blue light.

Mascharak's group reported the first Ru-organic fluorophore hybrid antenna for applications in lower energy light-activated NO photosubstitution [50d, e, 52]. In the NO-releasing Ru(III) compound $[Ru(Me_2bpb)(NO)(L)]$, where $H_2Me_2bpb$ is the tetradentate chelate 1,2-bis(pyridine-2-carboxamido)5-dimethylbenzene and L = $Cl^-$ (**[45]**), resorufin (**[46]**) or the fluorescein anion (**[47]**), the organic chromophore is coordinated directly or via a pyridinyloxy moiety (**[48]**$^+$) to the Ru center. Replacing the chloride ligand with resorufin or fluorescein results in

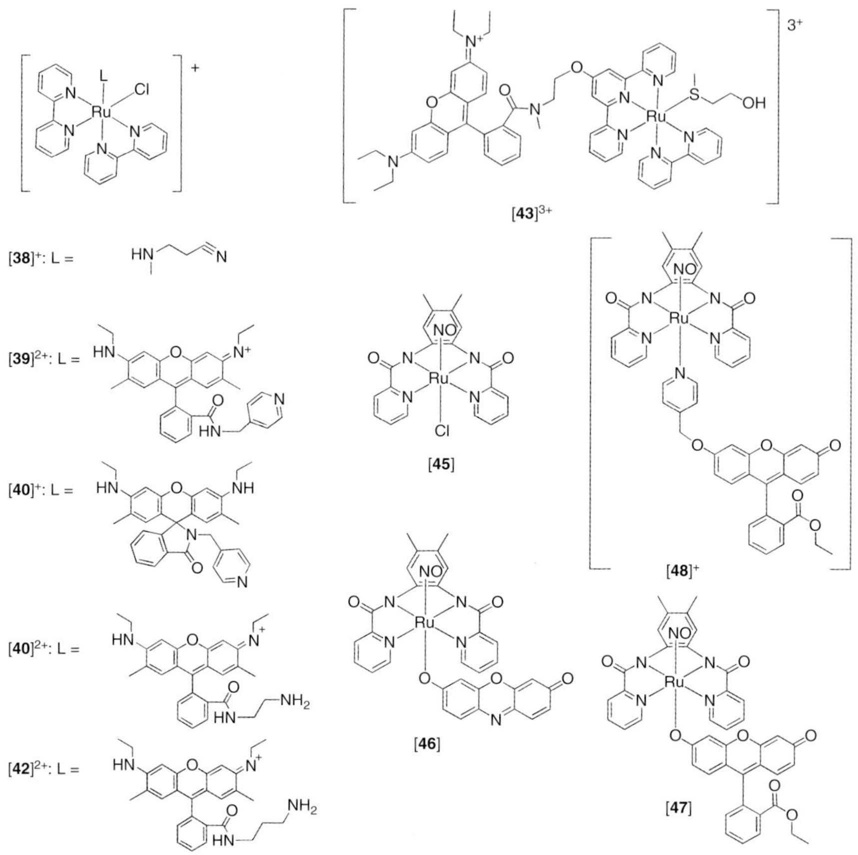


**Figure 5.5** Ruthenium-chromophore molecular dyads for low-energy photosubstitution reactions via chromophore-to-ruthenium nonradiative energy transfer.

a shift of the absorption maximum from 395 nm ($\varepsilon = 5300\,M^{-1}\,cm^{-1}$) to 500 nm ($\varepsilon = 11\,900\,M^{-1}\,cm^{-1}$) or 495 nm ($\varepsilon = 24\,000\,M^{-1}\,cm^{-1}$), respectively. Irradiation of these complexes with blue-green light (500 nm) resulted in increasing NO photosubstitution quantum yields from 0.0008 for **[45]** to 0.05 for **[46]** and even 0.3 for **[47]**. By changing the linking unit to a pyridinyloxy moiety, the absorption maximum of complex $[\mathbf{48}]^+$ was blueshifted to 475 nm ($\varepsilon = 24\,700\,M^{-1}\,cm^{-1}$), which is lower than the directly coordinated complex [**47**]. In addition, the NO photosubstitution quantum yields for $[\mathbf{48}]^+$ ($\Phi_{PR} = 0.04$) decreased by a factor 8 compared to [**47**]. The authors hypothesized that direct coordination of the fluorophore favored DET, shifted the light activation toward the red, and facilitated photosubstitution of NO, whereas without the organic fluorophore absorption in the green was negligible, and with the pyridinyloxy moiety the Dexter mechanism from the fluorescein antenna to the Ru center was disrupted, resulting in significant reduction of the NO photosubstitution efficiency.

As summarized, the antenna effect coupled with energy transfer can lead to increased ligand photosubstitution activity $\xi = \Phi_{PR} \times \varepsilon_{\lambda_{exc}}$ due to the excellent photon collection properties (high $\varepsilon_{\lambda_{exc}}$) of an attached organic chromophore. However, the energy levels, spectral overlap, molecular geometry, and donor–acceptor distance must be carefully controlled to allow for efficient energy transfer from the low-energy-absorbing chromophores to the high-energy-absorbing ruthenium complex, and to insure that $\Phi_{PR}$ does not become too low. Finally, one should note that the toxicity of complexes [**38**]$^+$-[**48**]$^+$ has not been tested *in vitro*, and that the role that photosubstitution could play on their biological properties in a cell remains, in most circumstances, highly hypothetical.

### 5.4.3 Upconversion (UC)

#### 5.4.3.1 Triplet–Triplet Annihilation Upconversion

Among the different upconversion mechanisms used for realizing photosubstitution on ruthenium complexes, triplet–triplet annihilation upconversion (TTA-UC) is the oldest one, as it was first observed in the 1960s by Parker and Hatchard [53]. However, it took 50 years to realize that this mechanism could be used for biological applications [54]. A scheme for TTA-UC is shown in Figure 5.6. In short, a molecular photosensitizer $^1$PS absorbs the low-energy light to generate a triplet excited state, $^3$PS* that is transferred via DET to a second molecular dye called the annihilator, $^1$A, forming a triplet excited-state annihilator, $^3$A*. When two triplet annihilators $^3$A* collide, a fraction of the encounter complex will produce one ground-state annihilator $^1$A and one singlet excited-state annihilator $^1$A*, which emits a high-energy (upconverted) photon.

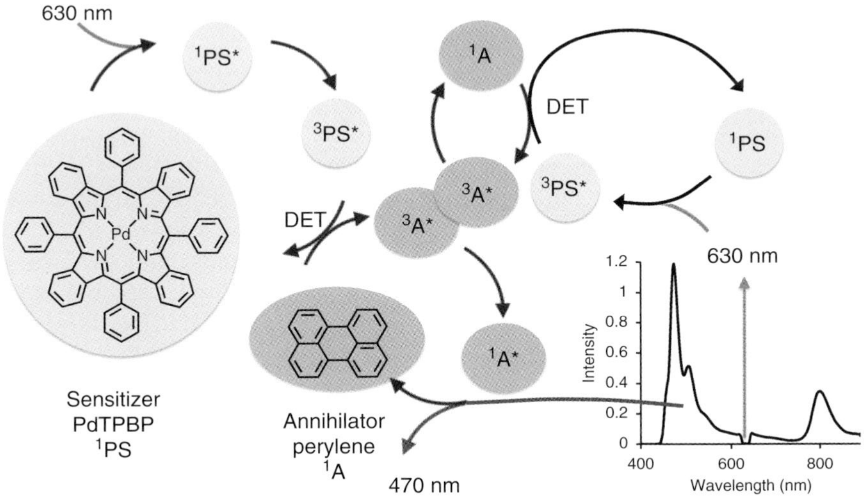


**Figure 5.6** Triplet–triplet annihilation upconversion (TTA-UC) scheme for PdTPBP (sensitizer PS) and perylene (annihilator A). Irradiation (630 nm) of the TTA-UC couple results in excitation of $^1$PS to $^3$PS*, which undergoes Dexter energy transfer (DET) to the annihilator. Two annihilators in the triplet state $^3$A* undergo TTA, which finally results in the emission of higher energy light (473 nm). The luminescence spectrum of the sensitizer and annihilator in toluene (298 K) is shown with excitation and emission wavelengths. (Adapted from references [54a, c].)

There are several TTA-UC sensitizer-annihilator couples reported in the literature. However, the most interesting one in the context of this chapter is the red-to-blue upconversion couple made of palladium tetraphenyltetrabenzoporphyrin [PdTPBP] ($^1$PS) and perylene ($^1$A) [55]. The major drawback of these dyes for biological applications is their negligible solubility in aqueous solutions. To use them in a biological context, supramolecular systems that can, on the one hand, deliver the light-activated metal prodrug to its biological target (e.g., a tumor) and, on the other hand, upconvert red light *in situ* into blue light capable of exciting the nearby ruthenium complex are necessary. Recently, upconverting liposomes containing the TTA-UC pairs described that can upconvert the red light of a commercial PDT laser (630 nm) into blue light (473 nm) were prepared. Such blue light overlaps optimally with the typical absorption spectrum of ruthenium polypyridyl compounds (Figure 5.7) [54c]. Upon physically mixing the red-to-blue upconverting liposomes to liposomes functionalized with $[Ru(tpy)(bpy)(S\text{-}chol)]^{2+}$ ($[\mathbf{49}]^{2+}$, Figure 5.7), where S-chol was a thioether-cholesterol conjugate anchoring the Ru-prodrug to the liposome, the system was able to radiatively transfer the upconverted blue light to the ruthenium complexes, which triggered the photosubstitution of the thioether ligand by water and released the ruthenium complex from the liposome surface. In this initial system, bubbling with argon was necessary to avoid quenching of the triplet states involved in TTA-UC by molecular oxygen. Recently, it was shown that antioxidants can solve the problem of oxygen sensitivity of TTA-UC, leading to red-to-blue upconversion in air [54d]. Further optimization is currently under way to obtain ruthenium prodrug activation in the PDT window in biologically relevant conditions.

#### 5.4.3.2 Upconverting Nanoparticles (UCNPs)

Upconverting nanoparticles (UCNPs) are inorganic phosphors consisting of a crystalline lattice, that is, $NaYF_4$, doped with a low concentration of lanthanide ions, for example, $Yb^{3+}$ and $Er^{3+}$ and/or $Tm^{3+}$. Depending on the dopant, these nanoparticles upconvert NIR light (980 nm) to near-UV, blue, green, or red light [27a, 56]. The development of UCNP for applications in biological fields has exploded in the past years, especially for the photoactivated release of Pt(IV) prodrugs, organic compounds, and PDT [57]. However, for Ru complexes this area has been less exploited, with only two examples utilizing UCNP for ruthenium prodrug activation [27b, 58]. Ruggiero *et al.* showed that a two-component system consisting of $NaYF_4{:}Yb^{3+}/Er^{3+}$ UCNP and $[Ru(bpy)_2(py)_2]^{2+}$ was able to upconvert NIR light (980 nm, 25 W cm$^{-2}$) into green light with enough efficiency to observe photosubstitution of a pyridine to form the aquated $[Ru(bpy)_2(py)(H_2O)]^{2+}$ complex. [27b] Blue-light-emitting UCNPs had a better spectral overlap with the absorption of the ruthenium complex $[Ru(bpy)_2(py)_2]^{2+}$ but were also much less bright, which prevented any activation occurring even under 35 W cm$^{-2}$ power. More recently, $[Ru(bpy)_2(PMe_3)(L)]^{2+}$ ($L = NH_2(CH_2)_3Si(OC_2H_5)_3$) was used as a "stopper" to hold doxorubicin in mesoporous Si channels, which covered $NaYF_4{:}Yb^{3+}/Tm^{3+}$ UCNP with a $NaYF_4$ shell (Figure 5.8) [58a]. Irradiation of the delivery system at (974 nm, 0.35 W cm$^{-2}$) resulted in 20% doxorubicin release over 1 h. A 30–40%

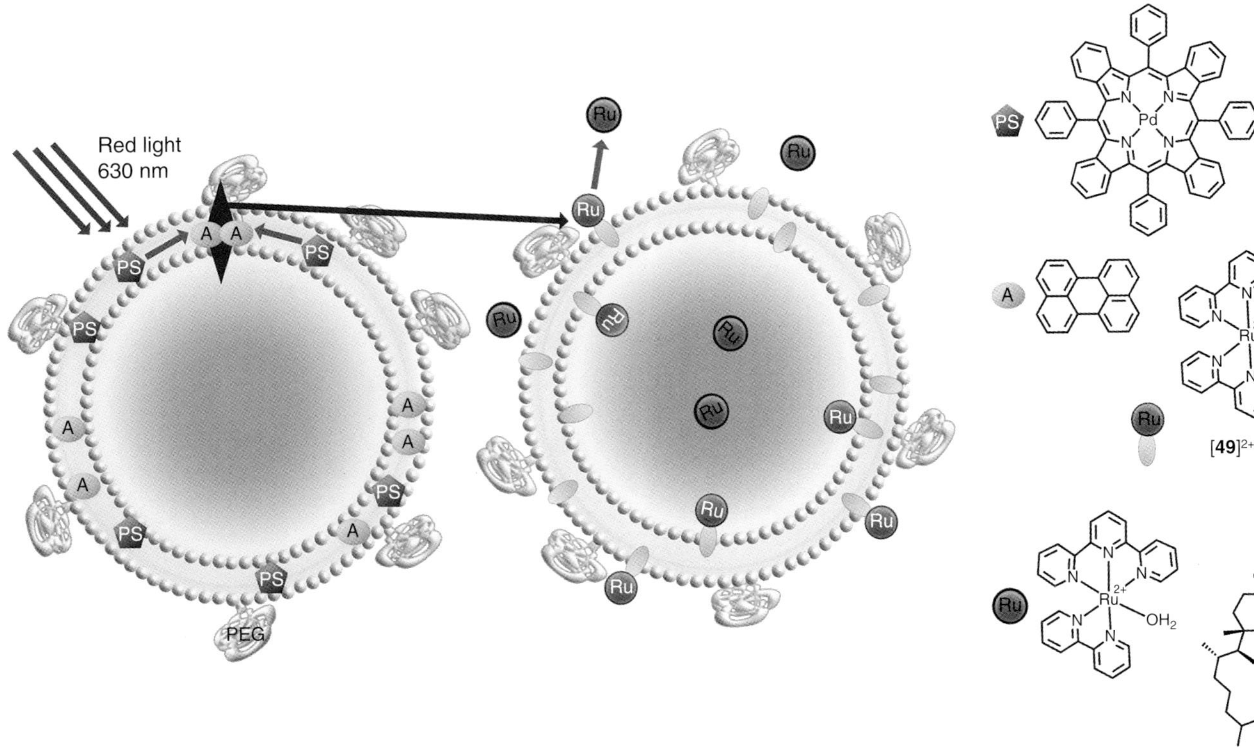


**Figure 5.7** Two-liposome system for red light activation of ruthenium polypyridyl compound, [**49**]$^{2+}$ using triplet–triplet annihilation upconversion (TTA-UC). TTA-UC liposomes consist of DMPC liposomes with 4 mol% DSPE-mPEG-2000 loaded with PdTPBP as the photosensitizer, and perylene as the annihilator. They were mixed in a 1 : 1 volume ratio with Ru-loaded liposomes made of a 96 : 4 : 4 DMPC:DSPE-mPEG-2000:[**49**]$^{2+}$ lipid mixture. (Adapted from reference [54c].)

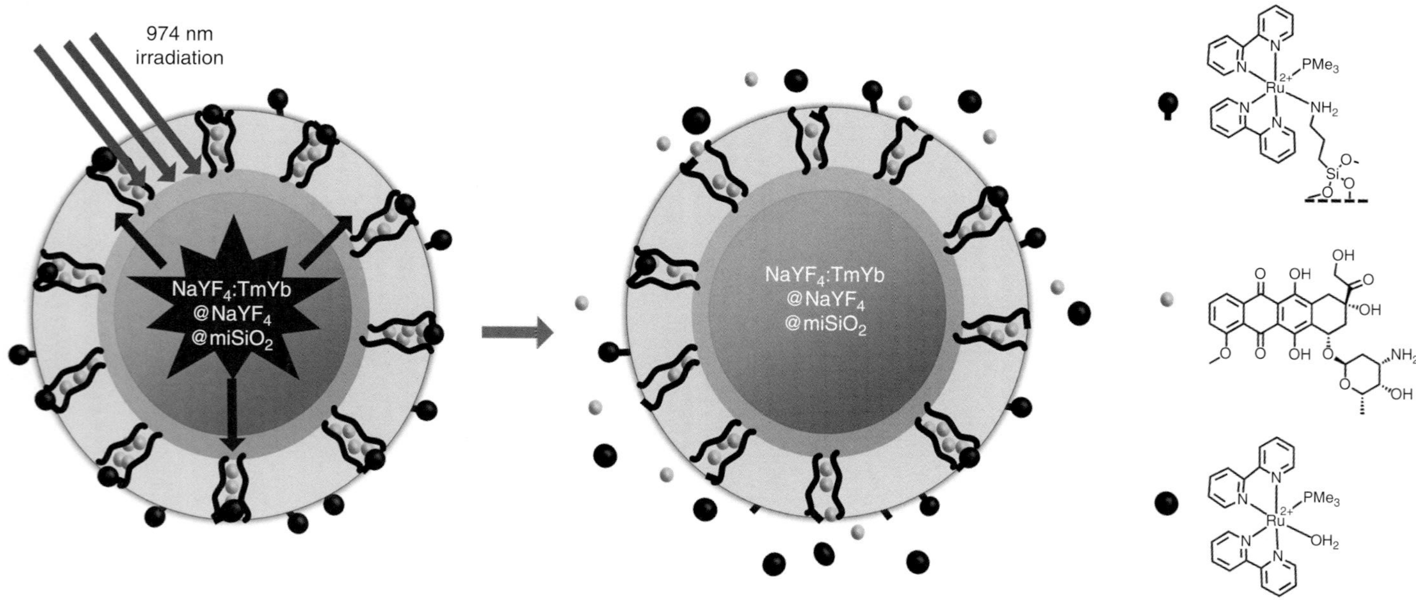


**Figure 5.8** Lanthanide-doped $NaYF_4$:$Yb^{3+}$/$Tm^{3+}$ upconverting nanoparticle for photorelease of $[Ru(bpy)_2(PMe_3)(L)]^{2+}$ ($L = NH_2(CH_2)_3Si(OC_2H_5)_3$] to deliver doxorubicin. (Adapted from reference [58a].)

HeLa cell viability reduction was observed for the UCNP system following 10- to 30- min NIR irradiation (974 nm, 0.35 W $cm^{-2}$). The toxicity was attributed to the delivery of the doxorubicin.

#### 5.4.3.3 Two-Photon Absorption (TPA) Photosubstitution

In 1930, Maria Göppert-Mayer predicted that an atom could simultaneously absorb two photons in one event. However, it took several decades until Kaiser and Garret were able to prove it [59]. In fact, high-intensity monochromatic light sources (lasers) were necessary to experimentally observe two-photon absorption upconversion (TPA-UC). Due to the low probability of three-particle collisions, it is necessary to have an extremely high instantaneous light intensity to obtain the (nearly) simultaneous absorption of two photons by one molecule. Expensive femtosecond pulsed lasers in the kW power regime are used for this purpose. By the mid-1990s Denk and coworkers developed two-photon laser scanning fluorescence microscopy, which provided visualization of the phenomena in living pig kidney (LLC-PK-1) cells. They proposed using the two-photon photolysis combined with microscopy for localized, fast, and spatially controlled release of bioactive chemicals (i.e., nucleotides or neurotransmitters) from organic-based compounds such as 4-methoxy-7-nitroindolinyl-amino (MNI) glutamate [60]. This step resulted in renewed interest in organic and inorganic TPA systems. However, not all molecules are of interest for TPA photochemistry due to their small two-photon excitation (TPE) cross section ($\sigma_{TPE} < 0.1$ GM, Göppert-Mayer units, 1 GM $= 10^{-50}$ $cm^4$ $s^{-1}$ $photon^{-1}$) [57b, 61]. The TPE cross section of several organic molecular probes have been analyzed, which have since been used to spatially resolve fluorophore localization in cells and as calibrated standards for new photochemically active TPA systems (Table 5.3) [62].

In 1997, Castellano *et al.* observed TPA-UC emission for the first time in a ruthenium complex using $[Ru(bpy)_3]^{2+}$ [63]. As expected, the one-photon excitation (OPE, 440 nm) and TPE (880 nm) emission spectra displayed similar overlapping $^3$MLCT-based emissions. The TPE cross section, $\sigma_{TPE}$, of $[Ru(bpy)_3]^{2+}$ was determined to be 4.3 GM in comparison to the standard

**Table 5.3** Summary of two-photon absorption properties of selected compounds.

| Compound | Solvent | $\lambda_{exc}$ (nm) | $\sigma_{TPE}$ [a] (GM) | $\Phi_{PR}$ [b] | References |
|---|---|---|---|---|---|
| Rhodamine B | Methanol | 840 | $210 \pm 55$ | | [62a] |
| $[Ru(bpy)_3]^{2+}$ | $H_2O$ | 880 | 4.3 | - | [63] |
| $[Ru(phen)_3]^{2+}$ | $CH_3CN$ | 850 | 10 | - | [64] |
| MDNI-Glu (caged glutamate) | Physiological buffer | 730 | 0.06 | 0.47 | [61c] |
| $[3]^{2+}$ | $H_2O$ | 740 | <0.1 | 0.05 | [26] |
| $[4]^{2+}$ | $H_2O$ | 800 | $0.24 \pm 0.04$ | 0.12 | [11] |
| $[6]^{2+}$ | $H_2O$ | 800 | 0.14 | 0.13 | [12d] |

a) $\sigma_{TPE}$ is the two-photon excitation cross section.
b) $\Phi_{PR}$ is the two-photon excitation substitution quantum yield.

rhodamine B at 880 nm. More recently, TPA ruthenium-based systems have been developed as cellular probes [20, 65], PDT agents [65d, 66], as well as photouncaging agents based on photosubstitution reactions [8e, 11, 12d, 26, 50f, 67]. Etchenique and coworkers have dominated the field of two-photon uncaging of Ru-based complexes, using the complexes $[Ru(bpy)_2(PMe_3)(L)]^{2+}$, where L = GABA ([**2**]$^{2+}$), glutamate ([**6**]$^{2+}$), or dopamine ([**4**]$^{2+}$); and $[Ru(bpy)_2(L)_2]^{2+}$ ([**3**]$^{2+}$), where L = 4-aminopyridine (4AP) [8e, 11, 12d, 26]. According to NMR and electronic absorption spectroscopy, light irradiation of these complexes in water under two-photon conditions resulted in clean photosubstitution reactions, with the caged neurotransmitter ligand being released together with $[Ru(bpy)_2(PMe_3)(H_2O)]^{2+}$ or $[Ru(bpy)_2(4AP)(H_2O)]^{2+}$. A typical example of conditions required to observe TPA photosubstitution is a capillary filled with 5 μL of a 500-mM $D_2O$ solution of the ruthenium complex. This capillary was irradiated for 1 h at 800 nm using a pulsed Ti–sapphire laser focused with a 10× microscope objective. The pulses were ultrashort (100 fs) and the repetition rate of 80 MHz led to an average power of 300 mW [26]. A major difference between one-photon and two-photon photosubstitution is that TPA reactions require highly concentrated solutions to increase the probability of absorption at lower energy wavelengths of irradiation. The TPE cross section ($\sigma_{TPE}$) of most ruthenium-based caged compounds are similar, and their photosubstitution quantum yield ($\Phi_{PR}$) is close to that of the corresponding one-photon uncaging quantum yield, indicating that the same $^1$MLCT excited state is probably populated under OPA and TPA. All of these complexes have been readily used *in vitro* and show promise as a time- and spatially controlled delivery system triggered by NIR irradiation.

## 5.5 Conclusions

Ruthenium complexes capable of photosubstituting one of their ligands have demonstrated great potential for applications in anticancer phototherapy and more generally for the delivery of a biologically active compound only at the place and time of irradiation. The use of Ru(II)-based compounds offers major advantages compared to purely organic, Rh(III)-based, and Pt(IV)-based caged compounds. Indeed, Ru-based complexes absorb light in the visible region of the spectrum, that is, in a biologically safe zone where light damage remains most of the time negligible. The main limitations of photoactivated ligand substitution reactions are the diminishing extinction coefficients when the excitation wavelength reaches the phototherapeutic window. Extending the wavelength of activation of this family of compounds into the phototherapeutic window (red or IR light) is possible, using one of the methods summarized in Section 4. Finally, Ru-based compounds in the photouncaging field are usually nontoxic, whereas those used in photoactivated chemotherapy can be highly toxic. No generality about the toxicity of ruthenium compounds after light irradiation can be made, as it results from a delicate balance between lipophilicity, coordination properties of the ruthenium-based photoproduct, and singlet oxygen generation properties. Detailed biochemical studies are necessary in each case to separate the effect of the ruthenium-containing (aqua) photoproduct versus that of the photoreleased ligand, and to determine whether the phototoxicity comes from singlet oxygen generation (PDT) or from the specific interactions of one of the photoproducts with biological target(s).

## References


1 (a) Antonarakis, E.S. and Emadi, A. (2010) *Cancer Chemother. Pharmacol.*, **66**, 1; (b) Groessl, M., Zava, O., and Dyson, P.J. (2011) *Metallomics*, **3**, 591; (c) Smith, G.S. and Therrien, B. (2011) *Dalton Trans.*, **40**, 10793; (d) Suess-Fink, G. (2010) *Dalton Trans.*, **39**, 1673; (e) Jakupec, M.A., Galanski, M., Arion, V.B., Hartinger, C.G., and Keppler, B.K. (2008) *Dalton Trans.*, 183; (f) Levina, A., Mitra, A., and Lay, P.A. (2009) *Metallomics*, **1**, 458; (g) Peacock, A.F.A. and Sadler, P.J. (2008) *Chem–Asian J*, **3**, 1890.

2 Ruggi, A., van Leeuwen, F.W.B., and Velders, A.H. (2011) *Coord. Chem. Rev.*, **255**, 2542.

3 Van Houten, J. and Watts, R. (1976) *J. Am. Chem. Soc.*, **98**, 4853.

4 (a) Bonnet, S., Collin, J., and Koizumi, M. (2006) *Adv. Mater.*, **18**, 1239; (b) Bonnet, S. and Collin, J.-P. (2008) *Chem. Soc. Rev.*, **37**, 1207.

5 (a) Salassa, L. (2011) *Eur. J. Inorg. Chem.*, **2011**, 4931; (b) Mari, C., Pierroz, V., Ferrari, S., and Gasser, G. (2015) *Chem. Sci.*, **6**, 2660.

6 Klán, P., Šolomek, T., Bochet, C.G., Blanc, A., Givens, R., Rubina, M., Popik, V., Kostikov, A., and Wirz, J. (2013) *Chem. Rev.*, **113**, 119.

7 Filevich, O. and Etchenique, R. (2013) *Photochem. Photobiol. Sci.*, **12**, 1565.

8 (a) Zayat, L., Calero, C., Albores, P., Baraldo, L., and Etchenique, R. (2003) *J. Am. Chem. Soc.*, **125**, 882; (b) Filevich, O., Salierno, M., and Etchenique, R. (2010) *J. Inorg. Biochem.*, **104**, 1248; (c) Zayat, L., Noval, M.G., Campi, J., Calero, C.I., Calvo, D.J., and Etchenique, R. (2007) *ChemBioChem*, **8**, 2035; (d) Verde, E.M.R., Zayat, L., Etchenique, R., and Yuste, R. (2008) *Front. Neur. Circ.*, **2**, 2; (e) Fino, E., Araya, R., Peterka, D.S., Salierno, M., Etchenique, R., and Yuste, R. (2009) *Front. Neur. Circ.*, **3**, 2; (f) Lopes-dos-Santos, V., Campi, J., Filevich, O., Ribeiro, S., and Etchenique, R. (2011) *Br. J. Med. Biol. Res.*, **44**, 688; (g) Zayat, L., Filevich, O., Baraldo, L.M., and Etchenique, R. (2013) *Phil. Trans. Royal. Soc. A Math. Phys. Engin. Sci*, **371**, 20120330.

9 Muller, M., Dierkes, P.W., and Schlue, W.R. (1999) *Brain Res.*, **826**, 63.

10 Zayat, L., Salierno, M., and Etchenique, R. (2006) *Inorg. Chem.*, **45**, 1728.

11 Araya, R., Andino-Pavlovsky, V., Yuste, R., and Etchenique, R. (2013) *ACS Chem. Neurosci.*, **4**, 1163.

12 (a) Mosquera, J., Sánchez, M.I., Mascareñas, J.L., and Eugenio Vázquez, M. (2015) *Chem. Commun.*, **51**, 5501; (b) Goldbach, R.E., Rodriguez-Garcia, I., Lenthe, J.H.v., Siegler, M.A., and Bonnet, S. (2011) *Chem.–Eur. J.*, **17**, 9924; (c) Salierno, M., Fameli, C., and Etchenique, R. (2008) *Eur. J. Inorg. Chem.*, 1125; (d) Salierno, M., Marceca, E., Peterka, D.S., Yuste, R., and Etchenique, R. (2010) *J. Inorg. Biochem.*, **104**, 418.

13 (a) Sgambellone, M.A., David, A., Garner, R.N., Dunbar, K.R., and Turro, C. (2013) *J. Am. Chem. Soc.*, **135**, 11274; (b) Respondek, T., Garner, R.N.,

Herroon, M.K., Podgorski, I., Turro, C., and Kodanko, J.J. (2011) *J. Am. Chem. Soc.*, **133**, 17164; (c) Respondek, T., Sharma, R., Herroon, M.K., Garner, R.N., Knoll, J.D., Cueny, E., Turro, C., Podgorski, I., and Kodanko, J.J. (2014) *ChemMedChem*, **9**, 1306.
14 Jacob, F., Perrin, D., Sanchez, C., and Monod, J. (1960) *C. R. Acad. Sci.*, **250**, 1727.
15 (a) Porter, D.J., Chestnut, W.G., Merrill, B.M., and Spector, T. (1992) *J. Biol. Chem.*, **267**, 5236; (b) Gentry, G.A., Morse, P.A., and Dorsett, M.T. (1971) *Cancer Res.*, **31**, 909.
16 Garner, R.N., Gallucci, J.C., Dunbar, K.R., and Turro, C. (2011) *Inorg. Chem.*, **50**, 9213.
17 (a) Filevich, O., García-Acosta, B., and Etchenique, R. (2012) *Photochem. Photobiol. Sci.*, **11**, 843; (b) del Mármol, J., Filevich, O., and Etchenique, R. (2010) *Anal. Chem.*, **82**, 6259.
18 Filevich, O., Carrone, G., Andino Pavlovsky, V., and Etchenique, R. (2012) *Anal. Chem.*, **84**, 5618.
19 Griepenburg, J.C., Rapp, T.L., Carroll, P.J., Eberwine, J., and Dmochowski, I.J. (2015) *Chem. Sci.*, **6**, 2342.
20 Carrone, G. and Etchenique, R. (2015) *Anal. Chem.*, **87**, 4363.
21 (a) Mayer, G. and Heckel, A. (2006) *Angew. Chem., Int. Ed.*, **45**, 4900; (b) Pelliccioli, A.P. and Wirz, J. (2002) *Photochem. Photobiol. Sci.*, **1**, 441.
22 Kaplan, J.H., Forbush, B., and Hoffman, J.F. (1978) *Biochemistry*, **17**, 1929.
23 Walker, J.W., Reid, G.P., McCray, J.A., and Trentham, D.R. (1987) *J. Am. Chem. Soc.*, **67**, 583.
24 McMillan, T.J., Leatherman, E., Ridley, A., Shorrocks, J., Tobi, S.E., and Whiteside, J.R. (2008) *J. Pharm. Pharmacol.*, **60**, 969.
25 Hopkins, S.L., Siewert, B., Askes, S., Veldhuizen, P., Zwier, R., and Bonnet, S. (2016) *Photochem. Photobiol. Sci.*, **15**, 644.
26 Nikolenko, V., Yuste, R., Zayat, L., Baraldo, L., and Etchenique, R. (2005) *Chem. Commun.*, 1752.
27 (a) Bonnet, S. (2015) *Comments Inorg. Chem.*, **35**, 179; (b) Ruggiero, E., Habtemariam, A., Yate, L., Mareque Rivas, J., and Salassa, L. (2014) *Chem. Commun.*, **50**, 1715.
28 (a) Mahnken, R.E., Billadeau, M.A., Nikonowicz, E.P., and Morrison, H. (1992) *J. Am. Chem. Soc.*, **114**, 9253; (b)Morrison, H., Menon, E., Loganathan, D. (2004) in *WO/2004/012667* (Ed.: N. I. o. Health), USA, pp. 1; (c) Lutterman, D.A., Fu, P.K.L., and Turro, C. (2006) *J. Am. Chem. Soc.*, **128**, 738.
29 (a) Kratochwil, N.A., Zabel, M., Range, K.-J., and Bednarski, P.J. (1996) *J. Med. Chem.*, **39**, 2499; (b) Kratochwil, N.A., Parkinson, J.A., Bednarski, P.J., and Sadler, P.J. (1999) *Angew. Chem., Int. Ed.*, **38**, 1460; (c) Farrer, N.J., Woods, J.A., Salassa, L., Zhao, Y., Robinson, K.S., Clarkson, G., Mackay, F.S., and Sadler, P.J. (2010) *Angew. Chem., Int. Ed.*, **49**, 8905.
30 (a) Barton, J.K., Basile, L.A., Danishefsky, A., and Alexandrescu, A. (1984) *Proc. Natl. Acad. Sci. U. S. A.*, **81**, 1961; (b) Yamagishi, A. (1983) *J. Chem. Soc., Chem. Commun.*, 572.
31 (a) Sun, Y., Joyce, L.E., Dickson, N.M., and Turro, C. (2010) *Chem. Commun.*, **46**, 2426; (b) Zhou, Q.-X., Lei, W.-H., Li, C., Hou, Y.-J., Wang, X.-S., and

Zhang, B.-W. (2010) *New J. Chem.*, **34**, 137; (c) Goulle, V., Lehn, J.M., and Schoentjes, B. (1991) *Helv. Chim. Acta*, **74**, 1471.

32 (a) Gianferrara, T., Bratsos, I., Iengo, E., Milani, B., Oštrić, A., Spagnul, C., Zangrando, E., and Alessio, E. (2009) *Dalton Trans.*, 10742; (b) Gianferrara, T., Bergamo, A., Bratsos, I., Milani, B., Spagnul, C., Sava, G., and Alessio, E. (2010) *J. Med. Chem.*, **53**, 4678; (c) Yin, H., Stephenson, M., Gibson, J., Sampson, E., Shi, G., Sainuddin, T., Monro, S., and McFarland, S.A. (2014) *Inorg. Chem.*, **53**, 4548.

33 (a) Crespy, D., Landfester, K., Schubert, U.S., and Schiller, A. (2010) *Chem. Commun.*, **46**, 6651; (b) Schatzschneider, U. (2010) *Eur. J. Inorg. Chem.*, 1451; (c) Schatzschneider, U. (2011) *Inorg. Chim. Acta*, **374**, 19; (d) Betanzos-Lara, S., Salassa, L., Habtemariam, A., and Sadler, P.J. (2009) *Chem. Commun.*, 6622; (e) Farrer, N.J., Salassa, L., and Sadler, P.J. (2009) *Dalton Trans.*, 10690.

34 Wachter, E., Heidary, D.K., Howerton, B.S., Parkin, S., and Glazer, E.C. (2012) *Chem. Commun.*, **48**, 9649.

35 Hufziger, K.T., Thowfeik, F.S., Charboneau, D.J., Nieto, I., Dougherty, W.G., Kassel, W.S., Dudley, T.J., Merino, E.J., Papish, E.T., and Paul, J.J. (2014) *J. Inorg. Biochem.*, **130**, 103.

36 Hidayatullah, A.N., Wachter, E., Heidary, D.K., Parkin, S., and Glazer, E.C. (2014) *Inorg. Chem.*, **53**, 10030.

37 Ang, W.H., Casini, A., Sava, G., and Dyson, P.J. (2011) *J. Organomet. Chem.*, **696**, 989.

38 (a) Barragán, F., López-Senín, P., Salassa, L., Betanzos-Lara, S., Habtemariam, A., Moreno, V., Sadler, P.J., and Marchán, V. (2011) *J. Am. Chem. Soc.*, **133**, 14098; (b) Huang, H., Zhang, P., Yu, B., Chen, Y., Wang, J., Ji, L., and Chao, H. (2014) *J. Med. Chem.*, **57**, 8971; (c) Habtemariam, A., Garino, C., Ruggiero, E., Alonso-de Castro, S., Mareque Rivas, J., and Salassa, L. (2015) *Molecules*, **20**, 7276; (d) Betanzos-Lara, S., Salassa, L., Habtemariam, A., Novakova, O., Pizarro, A.M., Clarkson, G.J., Liskova, B., Brabec, V., and Sadler, P.J. (2012) *Organometallics*, **31**, 3466.

39 Barragán, F., López-Senín, P., Salassa, L., Betanzos-Lara, S., Habtemariam, A., Moreno, V., Sadler, P.J., and Marchán, V. (2011) *J. Am. Chem. Soc.*, **133**, 14098.

40 Albani, B.A., Peña, B., Leed, N.A., de Paula, N.A.B.G., Pavani, C., Baptista, M.S., Dunbar, K.R., and Turro, C. (2014) *J. Am. Chem. Soc.*, **136**, 17095.

41 Babak, M.V., Meier, S.M., Huber, K.V.M., Reynisson, J., Legin, A.A., Jakupec, M.A., Roller, A., Stukalov, A., Gridling, M., Bennett, K.L., Colinge, J., Berger, W., Dyson, P.J., Superti-Furga, G., Keppler, B.K., and Hartinger, C.G. (2015) *Chem. Sci.*, **6**, 2449.

42 (a) Liebel, F., Kaur, S., Ruvolo, E., Kollias, N., and Southall, M.D. (2012) *J. Invest. Dermatol.*, **132**, 1901; (b) Liebmann, J., Born, M., and Kolb-Bachofen, V. (2010) *J. Invest. Dermatol.*, **130**, 259; (c) Matsumoto, N., Yoshikawa, K., Shimada, M., Kurita, N., Sato, H., Iwata, T., Higashijima, J., Chikakiyo, M., Nishi, M., Kashihara, H., Takasu, C., Eto, S., Takahashi, A., Akutagawa, M., and Emoto, T. (2014) *Anticancer Res.*, **34**, 4709.

43 Anderson, R.R. and Parrish, J.A. (1981) *J. Invest. Dermatol.*, **77**, 13.

44 (a) Knoll, J.D. and Turro, C. (2015) *Coord. Chem. Rev.*, **282–283**, 110; (b) Smith, N.A. and Sadler, P.J. (2013) *Phil. Trans. Royal. Soc. A Math. Phys. Engin. Sci*, **371**, 20120519; (c) Wagenknecht, P.S. and Ford, P.C. (2011) *Coord. Chem. Rev.*, **255**, 591;(d) Pal, A.K. and Hanan, G.S. (2014) *Chem. Soc. Rev.*, **43**, 6184.

45 Zhou, Q., Lei, W., Chen, Y., Li, C., Hou, Y., Zhang, B., and Wang, X. (2012) *Chem.–Eur. J.*, **18**, 8617.

46 Bonnet, S., Collin, J.P., Sauvage, J.P., and Schofield, E. (2004) *Inorg. Chem.*, **43**, 8346.

47 Knoll, J.D., Albani, B.A., and Turro, C. (2015) *Chem. Commun.*, **51**, 8777.

48 (a) Finck, S., Issenhuth, J.-T., Despax, S., Sirlin, C., Pfeffer, M., Poidevin, C., Gourlaouen, C., Boeglin, A., and Daniel, C. (2014) *J. Organomet. Chem.*, **760**, 248; (b) Huang, H., Zhang, P., Chen, H., Ji, L., and Chao, H. (2014) *Chem.–Eur. J.*, **21**, 715; (c) Jäger, M., Smeigh, A., Lombeck, F., Görls, H., Collin, J.-P., Sauvage, J.-P., Hammarström, L., and Johansson, O. (2009) *Inorg. Chem.*, **49**, 374; (d) Peña, B., David, A., Pavani, C., Baptista, M.S., Pellois, J.-P., Turro, C., and Dunbar, K.R. (2014) *Organometallics*, **33**, 1100; (e) Wadman, S.H., Lutz, M., Tooke, D.M., Spek, A.L., Hartl, F., Havenith, R.W.A., van Klink, G.P.M., and van Koten, G. (2009) *Inorg. Chem.*, **48**, 1887; (f) Klajner, M., Licona, C., Fetzer, L., Hebraud, P., Mellitzer, G., Pfeffer, M., Harlepp, S., and Gaiddon, C. (2014) *Inorg. Chem.*, **53**, 5150; (g) Frik, M., Martínez, A., Elie, B.T., Gonzalo, O., de Mingo, D.R., Sanaú, M., Sánchez-Delgado, R., Sadhukha, T., Prabha, S., Ramos, J.W., Marzo, I., and Contel, M. (2014) *J. Med. Chem.*, **57**, 9995; (h) Gaiddon, C., Jeannequin, P., Bischoff, P., Pfeffer, M., Sirlin, C., and Loeffler, J.P. (2005) *J. Pharmacol. Exp. Ther.*, **315**, 1403; (i) Fetzer, L., Boff, B., Ali, M., Xiangjun, M., Collin, J.-P., Sirlin, C., Gaiddon, C., and Pfeffer, M. (2011) *Dalton Trans.*, **40**, 8869.

49 (a) Palmer, A.M., Pena, B., Sears, R.B., Chen, O., El Ojaimi, M., Thummel, R.P., Dunbar, K.R., and Turro, C. (2013) *Phil. Trans. R. Soc. A*, **371**, 20120135; (b) Albani, B.A., Peña, B., Dunbar, K.R., and Turro, C. (2014) *Photochem. Photobiol. Sci.*, **13**, 272.

50 (a) Castellano, F.N. (2015) *Acc. Chem. Res.*, **48**, 828; (b) Carrone, G., Gantov, F., Slep, L.D., and Etchenique, R. (2014) *J. Phys. Chem. A*, **118**, 10416; (c) Bahreman, A., Cuello-Garibo, J.-A., and Bonnet, S. (2014) *Dalton Trans.*, **43**, 4494; (d) Fry, N.L., Wei, J., and Mascharak, P.K. (2011) *Inorg. Chem.*, **50**, 9045; (e) deBoer-Maggard, T.R., Fry, N.L., and Mascharak, P.K. (2013) *Inorg. Chim. Acta*, **406**, 190; (f) Filevich, O., Zayat, L., Baraldo, L.M., and Etchenique, R. (2014) *Structure and Bonding*, Springer, Berlin, Heidelberg, p. 1; (g) Lincoln, R., Kohler, L., Monro, S., Yin, H., Stephenson, M., Zong, R., Chouai, A., Dorsey, C., Hennigar, R., Thummel, R.P., and McFarland, S.A. (2013) *J. Am. Chem. Soc.*, **135**, 17161; (h) Monro, S., Scott, J., Chouai, A., Lincoln, R., Zong, R., Thummel, R.P., and McFarland, S.A. (2010) *Inorg. Chem.*, **49**, 2889; (i) Shi, G., Monro, S., Hennigar, R., Colpitts, J., Fong, J., Kasimova, K., Yin, H., DeCoste, R., Spencer, C., Chamberlain, L., Mandel, A., Lilge, L., and McFarland, S.A. (2014) *Coord. Chem. Rev.*, **282–283**, 127–138; (j) Stephenson, M., Reichardt, C., Pinto, M., Wächtler, M., Sainuddin, T., Shi,

G., Yin, H., Monro, S., Sampson, E., Dietzek, B., and McFarland, S.A. (2014) *J. Phys. Chem. A*, **118**, 10507; (k) Ford, W.E. and Rodgers, M.A.J. (1992) *J. Phys. Chem.*, **96**, 2917.

51 (a) Olaya-Castro, A. and Scholes, G.D. (2011) *Int. Rev. Phys. Chem.*, **30**, 49; (b) Clegg, R.M., Sener, M., and Govindjee (2010) *BiOS*, **7561**, 75610C; (c) Sahoo, H. (2011) *J. Photochem. Photobiol., C*, **12**, 20.

52 deBoer, T.R. and Mascharak, P.K. (2015) *Adv. Inorg. Chem.*, **67**, 145.

53 Parker, C.A. and Hatchard, C.G. (1962) *Proc. Chem. Soc.*, 386.

54 (a) Schmidt, T.W. and Castellano, F.N. (2014) *J. Phys. Chem. Lett.*, **5**, 4062; (b) Wohnhaas, C., Turshatov, A., Mailänder, V., Lorenz, S., Baluschev, S., Miteva, T., and Landfester, K. (2011) *Macromol. Biosci.*, **11**, 772; (c) Askes, S.H., Bahreman, A., and Bonnet, S. (2014) *Angew. Chem., Int. Ed.*, **53**, 1029; (d) Askes, S.H.C., Mora, N.L., Harkes, R., Koning, R.I., Koster, B., Schmidt, T., Kros, A., and Bonnet, S. (2015) *Chem. Commun.*, **51**, 9137; (e) Gray, V., Dzebo, D., Abrahamsson, M., Albinsson, B., and Moth-Poulsen, K. (2014) *Phys. Chem. Chem. Phys.*, **16**, 10345.

55 (a) Singh-Rachford, T.N. and Castellano, F.N. (2009) *J. Phys. Chem. Lett.*, **1**, 195; (b) Turshatov, A., Busko, D., Avlasevich, Y., Miteva, T., Landfester, K., and Baluschev, S. (2012) *ChemPhysChem*, **13**, 3112; (c) Jiang, Z., Xu, M., Li, F., and Yu, Y. (2013) *J. Am. Chem. Soc.*, **135**, 16446.

56 Li, X., Zhang, F., and Zhao, D. (2015) *Chem. Soc. Rev.*, **44**, 1346.

57 (a) Feng, W., Zhu, X., and Li, F. (2013) *NPG Asia Mater.*, **5**, e75; (b) Bao, C., Zhu, L., Lin, Q., and Tian, H. (2015) *Adv. Mater.*, **27**, 1647; (c) Carling, C.-J., Nourmohammadian, F., Boyer, J.-C., and Branda, N.R. (2010) *Angew. Chem., Int. Ed.*, **49**, 3782; (d) Punjabi, A., Wu, X., Tokatli-Apollon, A., El-Rifai, M., Lee, H., Zhang, Y., Wang, C., Liu, Z., Chan, E.M., Duan, C., and Han, G. (2014) *ACS Nano*, **8**, 10621; (e) Min, Y., Li, J., Liu, F., Yeow, E.K.L., and Xing, B. (2014) *Angew. Chem., Int. Ed.*, **53**, 1012;(f) Dai, Y., Xiao, H., Liu, J., Yuan, Q., Ma, P.a., Yang, D., Li, C., Cheng, Z., Hou, Z., Yang, P., and Lin, J. (2013) *J. Am. Chem. Soc.*, **135**, 18920.

58 (a) He, S., Krippes, K., Ritz, S., Chen, Z., Best, A., Butt, H.-J., Mailänder, V., and Wu, S. (2015) *Chem. Commun.*, **51**, 431; (b) Li, X., Wu, Y., Liu, Y., Zou, X., Yao, L., Li, F., and Feng, W. (2014) *Nanoscale*, **6**, 1020.

59 (a) Göppert Mayer, M. (1931) *Annalen der Physik*, **401**, 273; (b) Kaiser, W. and Garrett, C.G.B. (1961) *Phys. Rev. Lett.*, **7**, 229.

60 Denk, W., Strickler, J.H., and Webb, W.W. (1990) *Science*, **248**, 73.

61 (a) Ellis-Davies, G.C.R. (2013) *Beilstein J. Org. Chem.*, **9**, 64; (b) Furuta, T., Wang, S.S., Dantzker, J.L., Dore, T.M., Bybee, W.J., Callaway, E.M., Denk, W., and Tsien, R.Y. (1999) *Proc. Natl. Acad. Sci. U. S. A.*, **96**, 1193; (c) Fedoryak, O.D., Sul, J.-Y., Haydon, P.G., and Ellis-Davies, G.C.R. (2005) *Chem. Commun.*, 3664.

62 (a) Xu, C. and Webb, W.W. (1996) *J. Opt. Soc. Am. B*, **13**, 481; (b) Albota, M.A., Xu, C., and Webb, W.W. (1998) *Appl. Opt.*, **37**, 7352.

63 Castellano, F.N., Malak, H., Gryczynski, I.L., and Joseph, R. (1997) *Inorg. Chem.*, **36**, 5548.

64 Girardot, C., Lemercier, G., Mulatier, J.C., Chauvin, J., Baldeck, P.L., and Andraud, C. (2007) *Dalton Trans.*, 3421.

**65** (a) Baggaley, E., Gill, M.R., Green, N.H., Turton, D., Sazanovich, I.V., Botchway, S.W., Smythe, C., Haycock, J.W., Weinstein, J.A., and Thomas, J.A. (2014) *Angew. Chem., Int. Ed.*, **126**, 3435; (b) Qiu, K., Yu, B., Huang, H., Zhang, P., Ji, L., and Chao, H. (2015) *Dalton Trans.*, **44**, 7058; (c) Xu, W., Zuo, J., Wang, L., Ji, L., and Chao, H. (2014) *Chem. Commun.*, **50**, 2123; (d) Liu, J., Chen, Y., Li, G., Zhang, P., Jin, C., Zeng, L., Ji, L., and Chao, H. (2015) *Biomaterials*, **56**, 140.

**66** Boca, S.C., Four, M., Bonne, A., van der Sanden, B., Astilean, S., Baldeck, P.L., and Lemercier, G. (2009) *Chem. Commun.*, 4590.

**67** Peterka, D.S., Nikolenkoa, V., Finoa, E., and Arayaa, R. (2010) *Proc. SPIE*, **7548**, 1.

# 6

# Use of Ruthenium Complexes as Photosensitizers in Photodynamic Therapy

*Lothar Lilge*

*University Health Network and University of Toronto, Princess Margaret Cancer Centre 101 College Street, Rm 15-310, Toronto, ON, Canada M5G1L7*

## 6.1 Introduction

The inception of photodynamic therapy (PDT) dates back over 100 years, and while metallocomplex photosensitizers (PSs) are well established, interest in using ruthenium (Ru) as a central metal atom started only in the mid-1990s with the replacement of the central metal ions in established second-generation PSs such as phthalocyanines [1] and naphtalocyanines [2] with Ru. The initial aim of replacing the central ion, typically aluminum, was to utilize the well-known long-wavelength absorption provided by this transitional metal in order to reduce light attenuation in tissue, improving its penetration depth in tissue and facilitating PDT in large volumes. The interest in Ru-based PSs was also fueled by observations, made two decades earlier, showing that multiligand Ru(II) complexes, such as tris(2,2′-bipyridyl)ruthenium(II) dichloride [3], provide for high triplet–triplet energy and electron transfer reaction yields [4] opening the door to oxygen- and oxygen-independent-mediated photochemical reactions. The fact that some of the early Ru(II) complexes exhibited singlet oxygen quantum yields, $\varphi_\Delta$, approaching unity [5] further strengthened the interest of chemists. It was suggested [3] that the complexes could combine all the desired attributes for an ideal PS, including long-wavelength absorption, localization at sensitive intracellular sites, and high, $\varphi_\Delta$. One initial drawback for Ru complexes was rapid photobleaching, with some Ru-PSs losing $1/e$ activity after only 45–50 J cm$^{-2}$ radiant exposure, **H**, (J cm$^{-2}$) in solution [2, 4]. While ligands such as triphenylphosphine-m-disulfonic acid suggested a reduced photobleaching rate, if plotted as a function of **H** they also resulted in a reduced molar absorptivity, $\varepsilon$, and thus, bleaching as a function of absorbed photon dose remained high. More recent Ru(II) complexes provide improved photostability, which is comparable to or better than those of first-generation PSs, notably Rose Bengal and phthalocyanines [6]. It should be noted that photobleaching is a double-edged sword when it comes to PDT dosimetry, as rapid bleaching can facilitate simplified PDT dosimetry, as discussed subsequently.

*Ruthenium Complexes: Photochemical and Biomedical Applications*, First Edition.
Edited by Alvin A. Holder, Lothar Lilge, Wesley R. Browne, Mark A.W. Lawrence, and Jimmie L. Bullock Jr.

Recent developments in coordination chemistry have dramatically increased the generation of novel and particular Ru(II)-based PSs as mediators for PDT, as made evident by over 50 indexed peer-reviewed articles in 2014. However, two decades after their first introduction, Ru(II) PSs have yet to make it to the clinic, either in antimicrobial activities, or for oncological applications.

In this chapter we briefly review the principle of PDT, the definition of the PDT dose and dosimetry models, and the state of Ru(II) PSs *in vitro* and *in vivo* in preclinical animal models.

In the final paragraphs of this chapter, methods for drug delivery are discussed, aiming for improved enhanced selective PS accumulation in malignant cells.

## 6.2 The Basics of Photodynamic Therapy

The observation of a photomediated effect was first reported over 100 years ago [7] by von Trappeiner and Betz, yet it remained largely a low-priority curiosity within the scientific community until, beginning in the 1970s, the work by Thomas Dougherty at Roswell Park ushered in the new era of PDT [8]. Currently, the thrust of PDT research is aimed equally against premalignant and malignant growth *in vivo* and the inactivation of fungi and bacteria, as originally performed by von Trappeiner and Betz.

Selective, *in situ*, and spatially controlled destruction of aberrant tissue, achieved utilizing the PS's properties and light confinement, are the basis for PDT's significant potential to influence medicine in the coming decade. PDT's high spatial selectivity is underpinned first by utilizing the PS's low dark toxicity and pharmacokinetics, which provide a high uptake and/or retention within the target cells or tissues versus the surrounding normal host tissue; and, second, through exploiting the spatial confinement of the activating photon density by the light propagation properties of the target and host tissues. For the latter, the light scattering properties of the biological tissues are predominantly determining by the light dose field. Further improving the therapeutic efficacy of PDT will require building on the biochemistry and coordination chemistry knowledge in the development of new PSs and their delivery vehicles, exploiting the improved understanding of light transport in tissue and focusing on patient safety.

PDT comprises a sequence of energy transfers involving photophysical, photochemical, and photobiological steps, as depicted in Figure 6.1. The process commences with photon absorption and ends with the destruction of the biological target. This process is followed by the healing of the surrounding normal tissue, which receives a lower PDT dose, defined subsequently.

In the case of oncology, one needs to remember that treating only a primary tumor may not result in a cure. As discussed subsequently, it is becoming increasingly apparent that the activation of the immune system is also required for long term survival [8].

Cytotoxic compounds generated by the photophysical process are principally $^1\Delta_gO_2$ or $^1O_2$ and other radical oxygen species such as $OH^*$ and $H_2O_2$; see

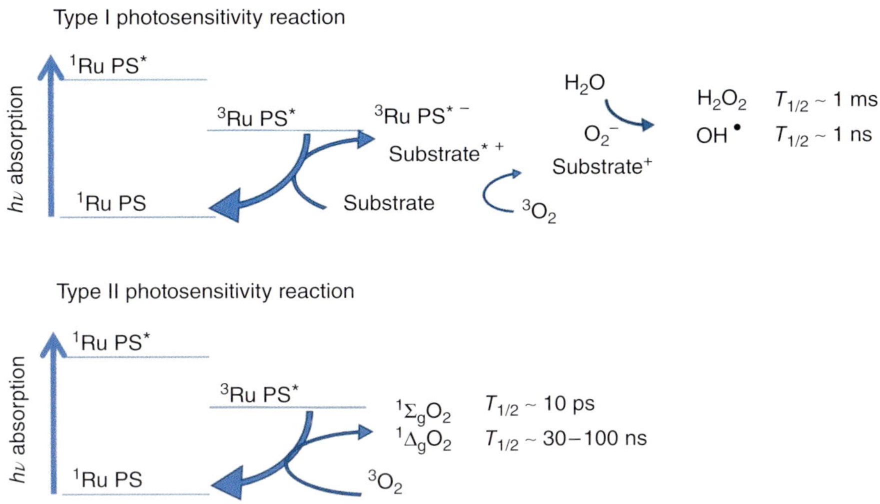


**Figure 6.1** Type I/II photochemical reactions for PS, here indicated as Ru-PS, resulting in the generation of short- and long-lived reactive oxygen species (ROS) with various lifetimes as indicated.

Mroz *et al.* or Wilson and Patterson [9] for a detailed coverage of possible reactive oxygen species (ROS) generation. In the classical type II photodynamic reaction, which results in the generation of $^1O_2$, a photon absorption event raises a PS from its singlet ground state, here Ru(II) complex ($^1$Ru(II)), to the first singlet excited state $^1$Ru(II)$^*$. The excited-state lifetime of $^1$Ru(II)$^*$ can range from the ns [10] to tenth of microseconds [11], decaying via one of four photophysical and photochemical deactivation pathways: (i) nonradiative relaxation or (ii) fluorescence emission into its ground-state $^1$Ru(II); (iii) direct photochemistry initiated from the singlet excited state for direct DNA cleavage, as originally reported by Lecomte *et al.* and since then reported by various other groups [12], or (iv) intersystem crossing into the triplet excited state resulting in $^3$Ru(II)*, which enables energy and spin exchange with ground-state oxygen to arrive at different forms of $^1O_2$, whereby the dominant species is $^1\Delta_gO_2$ [13]. The lifetime of $^1O_2$ has been determined to be in the range of 30 ns to several hundred nanoseconds [3, 14], depending on the microenvironment. In either case, the ROS diffusion distance in biological media is less than a cell diameter, with the exception of $H_2O_2$; Subcellular localization is an important parameter for the efficacy of Ru complex and other PSs. While some Ru complexes with, $\varphi_\Delta$, approaching unit have been reported, as noted earlier, theoretically providing a very effective PS, as shown in Figure 6.1 localization *in vitro* and *in vivo* away from critical cellular structures can reduce or abolish a PS's efficacy, negating some or all benefits from a high $\varphi_\Delta$. The ability of the various published Ru complexes to act as type I or type II PSs and their resulting effects in cellular environments are further discussed.

Based on the known relationships of the established PDT efficacy-determining parameters, it is paramount that the future development of Ru complexes as PSs pays close attention to the desirable attributes of a PS, not only based on its pharmacokinetics, leading to a high specific uptake ratio [15], but also to its subcellular localization, as they will both impact the therapeutic efficacy.

As shown in Figure 6.1, photochemical destruction of biomolecules, such as proteins and lipids, is either due to energy-transfer reactions (ETR) or charge-transfer reactions (CTR) both occurring within type I and type II reactions. For type I reactions the substrate can be proteins, lipids, DNA or a solvent, commonly water. All reactions are initiated from the $^3$Ru-PS$^*$ state, so a protonated excited state is also possible, resulting in a substrate or solvent molecule accepting a proton.

### 6.2.1 Singlet Oxygen Production

The quantum energy of the dominant ROS in type II photosensitization, $^1\Delta_gO_2$, lies at 0.98 eV [16] above its triplet ground state. Hence, the energy difference between the $^1$Ru(II) and its excited triplet state needs to be sufficiently above this value to unidirectional drive the generation of $^1O_2$ with high overall quantum efficiency.

For currently employed PSs, the $\varphi_\Delta$ in various solutions ranges from 0.56 for protoporphyrin IX (PpIX) to 0.84 for benzoporphyrin derivative monoacid ring-A [17].

### 6.2.2 Other Radical Production

Other non-oxygen-related photosensitization was demonstrated for phthalocyanines excited at $\lambda = 670$ nm in $D_2O$, where the $\varphi_\Delta$, dropped below 0.01, or for nitrogen-saturated methanol, where the excited-State lifetime is almost halved, both leading to the suggestion of a non-singlet-oxygen-mediated mechanism, possibly involving nitric oxide (NO) radicals [1].

Using NO production to control or modulate physiological processes was also demonstrated for some Ru-NO nitrosyls upon ultraviolet (UV) irradiation [18]. Employing production of the radical NO in tissues could be achieved with thiolate ligands and visible-light activation ($\lambda = 455$ nm). Translation *in vivo* has not been demonstrated for this or any other system at longer wavelength and hence is not further addressed here.

### 6.2.3 PDT Dose Definition

PS concentration and photon density provide the practical basis to define a PDT dose, whereby the PDT dose-determining parameter varies according to the experimental or clinical indications. Following the reactions depicted in Figure 6.1, the PDT dose is ideally given by the density of the cytotoxic species produced. While using the singlet oxygen sensor green [19] for detection of $^1O_2$ or 3′-p-(hydroxyphenyl) fluorescein [20] for hydroxyl radicals, the cytotoxic dose is quantifiable *in vitro*: however, this is not applicable *in vivo*.

For all indications, the PS dose, [PS], can be sufficiently defined in terms of its molarity, concentration (mg kg$^{-1}$), or variations thereof. One of the keys to appreciating dosimetry concepts is the difference between irradiation of a surface, as applicable for *in vitro* experiments and irradiation of the skin, versus interstitial therapy of a tissue volume as in orthotopic grown tumor and clinical situations. For irradiation of a body cavity, for example, with bladder or oral malignancies, the concepts of interstitial or surface illumination can be employed. For *in vitro* or trans-surface PDT, the photon density is given by the irradiance, ($E$, mW $cm^{-2}$) or its time-integrated version, which yields the photon dose, by radiant exposure ($L$, J $cm^{-2}$). For surface irradiation, only photons arriving from one hemisphere, or $2\pi$ [sr], are considered, as diffuse reflected photons exiting the tissue are lost for further PS activation. Irradiance can be measured with standard photometers, which provide high accuracy in the photon dose. In these situations, contributions of back-reflected photons from the bottom of the plastic tissue culture plate or subcutaneous or subsurface tissues are not considered for the definition of the photon density or PDT dose, and it is assumed that for the respective *in vitro* or *in vivo* application this additional photon dose is reproducible.

For interstitial studies of orthotopic tumors *in vivo* or clinically. PDT therapy of large tissue volumes, the diffuse scattering of the photon density by the tissue can no longer be ignored. Hence, the omnidirectional photon density is given by the fluence rate, ($\phi$, W $cm^{-2}$); and its time-integrated photon dose the fluence ($H$, J $cm^{-2}$). They differ from irradiance and radiant exposure such that the photon density and dose are integrated over $4\pi$ [sr]. By integrating the photon flux over the surface of a sphere, the fluence rate takes into account collimated transmitted photons and diffuse photons according to the diffusion equation as solved by the Boltzman transport equation [21].

Partial oxygen pressure ($pO_2$), the third often-required parameter for ROS generation, is difficult to quantify *in vitro* and *in vivo*. For *in vitro* experiments, the shallow depth of the media above the cells enables direct diffusion of $^3O_2$ to the monocellular cell layer, typically satisfying the need for metabolic and PDT consumption. To a certain extent, this applies also for PDT of *in situ* lesions at a tissue's surface, where for an irradiance of <200 mW $cm^{-2}$ and molecular attenuation coefficient $\varepsilon < 10^5$ $cm^{-1}$, a $pO_2$ above 2% is sufficient for $pO_2$ to not limit PDT efficacy. For example, using 1 mg $kg^{-1}$ Photofrin and 30 mW $cm^{-2}$ oxygen availability did not limit PDT outcome.

As the radiant exposure, $L$, and fluence, $H$, given in SI units (J $cm^{-2}$), are under the direct control of the experimenter or the surgeon, they are often selected as dose descriptors for a PS's efficacy; however, one needs to realize that since PDT is a quantum effect, a PS's efficacy needs to be determined for a target outcome as a function of absorbed photon density. The biological outcome can be given by DNA cleavage, cell survival, tumor reduction, growth delay, or survival times. A photon's quantum energy, $E_p$ (eV) is given by Eq. (1) based on the PDT activation wavelength. The resultant photon density, $P_n$, ($h\upsilon$ $cm^{-2}$), given by Eq. (2), can be combined with the concentration of the PS and its molar absorption coefficient at the activation wavelength producing the absorbed photon density, $P_a$ ($h\upsilon$ $cm^{-3}$), according to Eq. (3).

$$E_p = h\frac{c}{\lambda} \tag{6.1}$$

whereby $h$ is Planck's constant, $c$ is the speed of light, and $\lambda$ is the wavelength in nanometer.

$$P_{\mathrm{n}} = \frac{H \times 6.24 \times 10^{18}}{E_{\mathrm{p}}} \tag{6.2}$$

whereby $6.24 \times 10^{18}$ is the conversion factor from Joule to electronvolt as units of energy. It follows that the absorbed photon density $P_{\mathrm{a}}$ is given by

$$P_{\mathrm{a}} = 2.3\varepsilon[\mathrm{PS}]\frac{H \times 6.24 \times 10^{18}}{h\frac{c}{\lambda}} \tag{6.3}$$

whereby [PS] is the molar concentration of the PS in the tissue.

The biological outcome for a given photon density is, however, largely determined by the quantum efficacy of the ROS generated and by the tissue or subcellular localization of the PS. *In vivo,* for short time intervals between PS administration and light activation relative to the sensitizers plasma half-life of the PDT, results in vascular-mediated effects through the local destruction of capillary and feeder vessels, thus not targeting and destroying the malignant cells. This is the case for BPD-MA, m-THPC, and, to a limited extent, Photofrin [22]. For PSs which are accumulated in particularly sensitive subcellular organelles, such as the mitochondria for aminoluvelinic acid (ALA)-induced-PpIX [23], liposome for PC4 [24], or the nucleus for Ru-PS [25], to be discussed subsequently, the importance of that organelle's cellular function and overall biologically relevance, is essential for a PS's efficacy, as they reduce the concentration of ROS required to cause cellular destruction. For example, during *in vitro* experiments, Niedre *et al.* and Jarvi *et al.* [26] determined that for ALA-induced PpIXmediated PDT $10^{7}\,{}^{1}O_2$ molecules are required per cell to cause its death. Hence, dosimetry comprising all PDTefficacy-determining parameters – fluence, PS concentration, and $pO_2$ – is ideally required to evaluate the efficacy of a particular photosensitizing compound when associated with a particular cell or tissue response model. Comparing [PS], $H$, or $\phi$ and $pO_2$ between PSs is very difficult, particularly when different wavelengths and time delays are utilized. These limitations are further complicated as no PDT treatment model integrating these three major PDT-efficacy-determining parameters has been validated, and an effective ROS dose has never been determined.

In the following sections, the principal published dosimetry approaches are reviewed, followed by PDT tissue response models and the impact of PS localization.

#### 6.2.3.1 PDT Dosimetry *In Vitro*

When determining the PDT dose *in vitro,* a basic assumption is that there is an equilibrium between the PS concentration in solution and in the cells, whereby the actual intracellular concentration depends on passive or active uptake and drug export mechanisms. Intracellular PS quantification approaches are typically based on fluorescence microscopy, although the high triplet quantum yield, $\varphi_{\mathrm{T}}$, of some PS or Ru complexes reduces their luminescence emission significantly, complicating its detection by this method [27]. This is further complicated by the fact that many Ru complexes' luminescence quantum yields, $\varphi_{\mathrm{L}}$, are sensitive to environmental effects such as pH [28], $pO_2$ [29] or even $^1O_2$ [30]. The latter, in conjunction with the high $\varphi_T$, also suggests that the luminescence is phosphorescence emanating from the triplet state, rather than fluorescence from the singlet state. Hence, knowledge of the mircoenvironmental factors affecting $\varepsilon$ and $\varphi_F$ is paramount to optical quantification of the PS. Other techniques, such as inductively coupled plasma-mass spectrometry (ICP-MS) [31], can provide quantitative information pertaining to the Ru(II) concentration. However, due to the minimum mass required, a cell pellet containing $10^6$–$10^7$ cells is needed. Variability in uptake and subcellular localization cannot be assessed. A report of Ru-PS quantification by ICP-MS was not found in literature databases at the time of writing this chapter.

If Ru(II) complex uptake and retention is linear with its concentration in the media, there will be reciprocity between the irradiance and the Ru-PS concentration; doubling one of the parameters and halving the other will yield the same cell toxicity. Reciprocity in the sensitizer and photon concentration is commonly assumed for *in vitro* conditions [32], often requiring a limit to the maximum permissible photon density rate. $^3O_2$ depletion, self-shielding, and PS bleaching are the primary causes for drug–light reciprocity breakdown. The effect of $^3O_2$ depletion is of less concern for Ru(II) complexes, as for the majority of these complexes, $\varepsilon$, is low at the clinically relevant wavelength, the compounds can be remarkably photostable. Thus, for low-fluence-rate $^3O_2$ consumption, the dose-limiting effect and self-shielding of the Ru-PS are of no concern. Reciprocity between PS concentration and absorbed photon density is anticipated to hold, but it should still be validated as it was shown to fail in particular situations for some Ru-PS [33].

As irradiance in Ru-PS-mediated PDT research is commonly provided by either white light sources [10, 11, 34], collimated lasers, or light-emitting diode (LED) arrays [35], the power density needs to be quantified only over $2\pi$ [sr] and a standard NIST calibration traceable power meter is sufficient [36]. Calculating the absorbed dose for LEDs and white light sources can become complicated, as the emission profile of the source needs to be convoluted with the absorption spectrum of the PS and normalized to the power density to determine the effective irradiance in comparison to other PSs. When using white light sources and photovoltaic-based power meters, an additional issue pertains to the spectral responsivity of the latter, which needs to be taken into consideration when determining the power density. In any case, white light sources need to be spectrally confined to the spectral range that is both detectable by the power meter and clinically relevant. The efficacy of Ru-PSs with strong absorption in their Soret band, covering the UVA and blue spectral bands, can be overstated, as their translation into clinical indications will be hampered by the short penetration depth of these wavelengths. In these situations, the use of UVA blocking and heat rejection filters should always be considered.

To compare the efficacy between Ru-PS and approved PSs, the *in vitro* dose should be given based on the absorbed photon dose. It is important to realize that the microenvironment does not only modify the fluorescence quantum yield but also the absorption coefficient. Figure 6.2 illustrates the difficulty in obtaining a correct $\varepsilon$ for [Ru(4,4′-dimethyl-2,2′-bipyridine)$_2$(2-(2′,2″:5″,2‴-terthiophene)

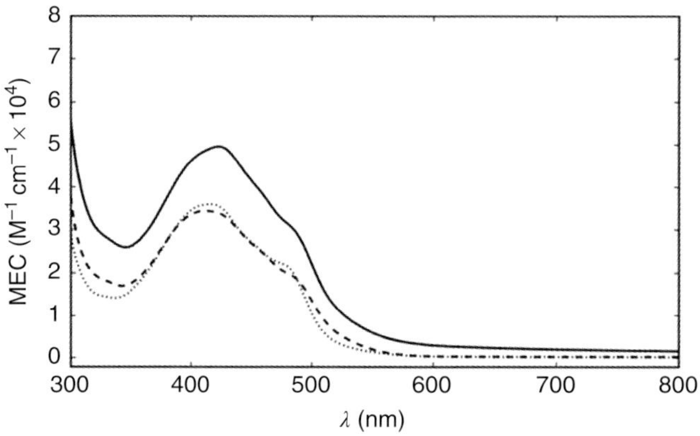


**Figure 6.2** Molar absorptivity (extinction) coefficient (MEC) of a 10 μM TLD1433 solution in DI-water (dotted), in incomplete DMEM (dashed black) and in complete DMEM (solid black) as function of wavelength.

-imidazo[4,91][1,10]phenanthroline)]$^{2+}$, termed TLD1433 [34, 37], due to its modification by an aqueous solution, versus incomplete media and complete media. It becomes apparent that the determination of the absorbed photons will be different depending on the growth media used for a particular experiment and also the *in vivo* situation.

#### 6.2.3.2 PDT Dosimetry *In Vivo*

The complications of establishing an effective dose *in vitro* are further exacerbated during *in vivo* testing. The additional complication derives from the gradient in the photon density due to the tissue optical properties and the continuously varying microenvironment.

It is well established that accurate light delivery during PDT increases efficacy and safety of the therapy, whereby the local time-integrated photondensity [38], governed by the absorption, $\mu_a$, and light scattering, $\mu_s$, properties of the tissue, provide the confinement of the photon density. However, during intracavity irradiation, such as in the esophagus, oral cavity, and, particularly, the bladder, the photondensity provided by the photon source emission is augmented by the integrating sphere effects provided by even moderate backscattering power, or albedo, of these tissues [38, 39].

In cavities as well as in bulk tissues, predicting the fluence rate throughout the clinical target volume necessitates either knowledge of the tissue's absorption or the scattering coefficient to be used in a validated method of calculating the photon density throughout this volume. Methods use either finite element models based on diffusion theory [40] or Monte Carlo methods [21, 39, 41].

Conversely, methods to measure the fluence rate intracavity or interstitially [42] and the concentrations of [PS] [43] and [$^1O_2$], [14, 26, 44] have been demonstrated *in vivo*, and so are not necessarily translatable into clinical practice. Translation was achieved for surface-based probes or imaging systems, as these are considered easier to implement clinically, but they are limited to established PDT centers [45] and ultimately require approval by the regulatory bodies.

For PSs with low photo stability, as with ALA-induced PpIX versus Ru-PS, the need to quantify the [PS] in real time is paramount, and various, mostly spectroscopy-based, solutions have been presented [46] for preclinical and clinical studies; their validity needs to be verified against chemical extraction of the PS from biopsies [47] demonstrated also for a few Ru(II) PSs [48]. For PDT response prediction, one needs to be aware of limitations in the techniques used to determine the [PS] from bulk tissue. Only the tissue average [PS] is provided, and no information about the extracellular versus cytoplasmic or vascular PS distribution is provided. For example, in early brain PDT, the PS Photofrin remained at high concentration in the blood–brain barrier, causing a predominantly vascular effect [49]. Integrating the [PS] quantification over a large tissue volume makes the interpretation of a subject's tissue optical properties essential, which in turn requires small optical contact probes [50] or imaging approaches such as "spatial frequency domain imaging" [51].

Photobleaching can be a powerful PDT dose measure if the rate of bleaching correlates strongly with the rate of ROS generation. Particularly for ALA-induced PpIX, the strong photobleaching provides a good correlation [52] with [ROS]. However, due to the high stability of recently presented Ru-PS, multiexponential bleaching rates and the low $\varphi_L$, photobleaching-based dosimetry will be difficult to apply for Ru(II)-complexes [53]. An alternative is the use of fluorescent ROS-sensitive compounds to monitor the deposited dose [54]. For Ru compounds designed to exert their cytotoxic effect via DNA cleavage [12, 55], the fluorescent reporter molecule will also need to localize at the DNA.

#### 6.2.3.3 Oxygen Consumption Model

While the photodynamic and metabolic consumption of oxygen has value in determining a PS's efficacy [43, 56] when the dominant radicals generated are ROS, in particular $^1O_2$, their utility is not well established for Ru-PS. For Ru-PS, the situation is more complicated under anoxic conditions when a PDT effect may not be abolished [34, 57], and additionally for the case of direct DNA damage with the Ru-PS becoming the radical, PDT activity becomes $O_2$ consumption independent, rendering this PDT dose metric ineffective.

Breathing hyperbaric oxygen or carboxygen (95% $O_2$ and 5% $CO_2$) has been shown to have benefit *in vivo* for Photofrin-mediated PDT [58], but may have limited impact for Ru-PS-mediated PDT.

Conversely, the potential oxygen consumption shortfall in PDT may necessitate limiting the irradiance or fluence rate to retain PDT efficacy but this is not necessarily true for Ru-PS complexes relying on oxygen-independent pathways. As the photodynamic consumption of oxygen will not limit the therapeutic efficacy [59], irradiances up to the maximum permissible exposure of skin can be used [60].

#### 6.2.3.4 *In Vivo* Tissue Response Models

There is a significant lag in translating the various Ru-PSs to *in vivo* preclinical or even clinical experiments. As indicated, there are several ability-determining parameters which need to be considered to quantify a PS's efficacy to achieve a particular biological endpoint. General tissue destruction by necrosis, as required for tumor mass debulking or eradication is an obvious endpoint, as are the induction of apoptosis or vascular stasis. For necrosis, there is strong evidence of a threshold effect for PDT [33, 61]. For PDT doses below this threshold value, tissue damage is reversible. This threshold value depends on the PS's tissue concentration, time after injection, and the target tissue [33]. In general, the threshold model states that once the number of photons absorbed by the PS per unit volume exceeds this critical value, T, tissue necrosis occurs; as stated also by Eq. (6.4).

$$T \leq P_{\mathrm{a}}(d) \tag{6.4}$$

whereby $P_{\mathrm{a}}(d)$ reflects the diminishing fluence as a function of distance, $d$, from the photon emitter. A simpler version of this equation was proposed by the group of Selman and Jankun, and considers only the photon density $P_{\mathrm{n}}(d)$, assuming homogeneous PS uptake for a given tissue and species [62]. For either case, the density of the effective activating photons at the histologically established boundary needs to be determined, using experimental or numerical methods outlined earlier. Simplified analytical solutions for the fluence rate as a function of distance from the photon emitter have been published.

### 6.2.4 PDT and Immunology

The topic of PDT and immunology was recently reviewed by Castano *et al.* [63]. One of the key findings is that PDT differs from standard oncological therapies such as surgery, radiotherapy, and chemotherapy in respect to immune modulation. Standard therapies are generally immunosuppressive, whereas PDT, if provided without surgery, causes acute inflammation made evident by the expression of heat-shock proteins, followed by the invasion and infiltration of the damaged malignant tissue by leukocytes. Attracted by the increasing availability of antigens, T cells derived from the malignant tissue are reprogrammed and ultimately yields memory cells capable of destroying viable tumor cells beyond the PDT treatment volume. This is evidence that an immunogenic cell death requires changes in the cell surface composition, as achieved, for example, by membrane-bound lipophilic PSs. Cell surface damage changing the release of soluble factors, including heat-shock proteins [64] in specific sequences, can operate on dendritic cells' receptors to stimulate the presentation of tumor antigens for T cells [65]. Henderson and colleagues demonstrated that Photofrin-mediated PDT-generated tumor cell lysates were able to activate dendritic cells to express IL-12, critical for the development of an immune response [66]. While these effects are observable *in vitro* only for a narrow PDT-dose range, one needs to recall that *in vivo* the PDT dose covers a wide dynamic range, and by default always includes the immune-stimulating dose, when overtreatment of the target volume is not a planned or realized outcome.

## 6.3 Status of Ru Photosensitizing Complexes

The introduction mentions that only a few published reports are available for Ru complexes currently *in vivo* tested. Despite the impressive gains made for Ru-PS complexes from a biochemical or coordination chemistry perspective, translation is rather limited, or outright disappointing, considering that more than 20 years have passed since their initial introduction. Only a few reports have been presented on the *in vivo* efficacy of Ru-PS constructs. Using the chicken chorioallantoic membrane assay, Nowak-Sliwinska *et al.* [67] demonstrated that for a group of [Ru($\eta^6$-*p*-cymene)$Cl_2$(1,3,5-triaza-7-phosphaadamantane)] and analogue [Ru($\eta^6$-*p*-cymene)$Cl_2$(3,7-diacetyl-1,3,7-triaza-5-phosphabicyclo[3.3.1]nonane)] complexes, vascular-mediated PDT at sub millimolar concentrations was successful and, interestingly, it also prevented revascularization. Pernot *et al.* demonstrated a PDT-mediated growth delay in a subcutaneous oral carcinoma model, mediated by enhanced uptake in the tumor as determined spectroscopically. However, neither of the two PS complexes tested, [Ru($\eta^6$-*p*-$Pr^iC_6H_4Me$)(5-(3-pyridyl)-10,15,20-triphenylporphyrin)$Cl_2$] and [$Ru_4$($\eta^6$-*p*-$Pr^iC_6$-$H_4Me$)$_4$(5,10,15,20-tetra(3-pyridyl)porphyrin)$Cl_8$], were effective at sub nanomolar concentrations, and no complete responses were obtained [68]. Microscopic studies executed on tumor samples did demonstrate cytoplasmic but not nuclear localization, as anticipated on the basis of *in vitro* studies.

PDT garners its efficacy from both the biochemical properties of the PS and its photophysical attributes, which influence the light delivery concepts. Looking at the former as they pertain to optimizing Ru complexes, various factors are to be considered. A biochemist's goal when designing effective Ru-PSs for PDT is to achieve high singlet oxygen yield, long wavelength absorption, and possible nuclear targeting for DNA cleavage to facilitate translation *in vivo* and into the clinic. For example, a low ROS quantum yield or unfavorable tissue and subcellular localization can be compensated for by an increased exposure of PDT-activating photons when the complex is highly photostable. Photons are generated cheaply, and depending on the wavelength and location, $10^{20}$–$10^{23}$ $h\upsilon\,cm^{-1}$ for cylindrical diffusers or $10^{21}$ to $10^{24}$ $h\upsilon\,cm^{-2}$ in surface- or intracavity delivery, can be provided during a clinically acceptable time period, even when intracavity delivery needs to be limited to <20 min. The rate of photon delivery should be limited to avoid thermal effects when nonspecific absorption by tissue components exceeds the convective cooling capacity of the blood [69]. The ANSI standards for the exposure of skin is a practical guideline for maximizing surface and intracavity exposures while avoiding a thermal effect during PDT [60]. However, the majority of currently developed Ru(II)-PS complexes require a high quantum energy for electronic excitation and short excitation wavelength, <600 nm, restricting the maximum permissible irradiance due to the absorption by oxy- and deoxy-hemoglobin. This nonspecific absorption also restricts the fraction of photons to be absorbed by the PSs, reducing the efficacy of PDT *in vivo*.

Solubility is another issue requiring attention for drug development prior to translation. Delivery vehicles are required to transport the Ru-PS as active pharmaceutical ingredient through the blood and/or barriers, such as the stratum corneum of the skin or the blood–brain barrier in the central nervous system. This complicates drug development in a way similar to that of non-photosensitizing pharmaceutical Ru complexes; some of the lessons learned for these complexes should be directly applicable to Ru-PSs.

### 6.3.1 Photostability for Ru-PS Complexes

As indicated, the photostability of the PS during light activation has a direct impact on the drug dose required to surpass the minimum PDT damage threshold value. High photostability allows the replacement of the expensive drug, which might be carrying dark toxicity effects, with inexpensive and safe photons preferably in the visible or near-infrared (NIR) spectral range.

Photostability can be improved by ligand substitution, as for Ru(II)-2,3-naphthalocyanine-bis-benzonitrile, where the particular axial position confers solubility and stability [2]. However, high photostability can result in prolonged photosensitivity if the Ru-PS has either a long plasma half-life, is not exported from cells, or is not metabolized. As Ru-PSs can be excited by single-photon or multiphoton processes, an intelligent use of photobleaching was proposed by Ishii *et al.* [70]. Photoactivation of ruthenium phthalocyanines by continuous wave (CW) light proximal to the light source or tissue surface resulted in ROS production, and was followed by photobleaching of the PS in this layer by a nanosecond-pulsed laser. This resulted in selective photo-decarbonylation of the sensitizer, diminishing the Q-band by a two-photon process. Thus, the effective attenuation of the surface proximal tissue layer is reduced, enabling deeper sensitizer activation with a subsequent CW treatment interval.

### 6.3.2 Long Wavelength Activation of Ru(II)-PS Complexes

PDT would be the ideal treatment modality for malignancies or carcinogenic events, as it is a localized therapy and it provides excellent healing properties. Due to the superficial nature of early malignancies that originate in the epithelial layers, it is possible to achieve an adequate PDT dose using even short wavelengths for the required depth of several 100 μm. Hence, most Ru(II)-PSs synthesized to date could be ideal mediators. However, their translation into the clinic for these indications will be difficult due to the competition with other effective therapies, such as radiofrequency ablation (RFA) and other photosensitizing drugs. The short activation wavelengths of Ru(II)-PSs limit use in larger tumors, due to insufficient light penetration and volumetric coverage. Developing long wavelength activation for the PSs via their Q-absorption band, similar to other PSs, is a promising approach. Ruthenium(II)-2,3-naphthalocyanines can have their Q-band excited in the NIR (710–760 nm), while maintaining a high $\varepsilon$ of $\sim 10^5\ M^{-1}\ cm^{-1}$, the dose at which 50% of the cells are dead, $LD_{50}$, remaining in the micromolar range during *in vitro* testing [2], and the favorable photophysical properties were not translated in a biological system. Chemists, realizing the need for longer wavelength absorption properties, are developing various Ru-PSs that are excitable in the red and NIR range.

An alternative approach is the use of two-photon absorption events for photoactivation of PSs demonstrated for vascular-acting PSs *in vivo* [71]. Several Ru(II)-PSs, able to initiate cell death following NIR two-photon absorption events, have been presented [72]. While a topic of interest, photon delivery for two-photon-induced PDT remains a challenge outside of free space delivery to the skin or the oral cavity the treatment of small lesions. The high electric fields generated by sub-pico and picosecond pulses are difficult to transmit through optical fibers, so the recent development of femtosecond fiber lasers [73] will assist in overcoming this limitation, possibly opening up the treatment of interstitial indications. Nevertheless, two photophysical concerns remain. First, the high photon densities required for multiphoton absorption which necessitates a tight focus, and focusing on tissue targets deeper than 1–2 mm will be challenging, as photon losses are not suffered only by spatial but also by temporal dilution of the laser pulse. Second, the required photon densities, particularly for activation via a virtual excited state, can approximate those sufficient to initiate other nonlinear absorption effects, particularly ablative effects [74].

As already indicated in Figure 6.2, association of Ru complexes with proteins can modify their absorption properties. Kaspler *et al.* [75] recently demonstrated a significant augmentation of the NIR absorption of [Ru(II)(4,4′-dimethyl-2,2′-bipyridine(dmb))$_2$(2-(2′,2′′:5′′,2′′′-terthiophene)-imidazo[4,91] [1,10]phenanthroline)]$^{2+}$, called TLD1433 and originally synthesized by McFarland's team at Arcadia, following binding to transferrin. Premixing the Ru-PS with apo-transferrin resulted in reduced dark toxicity *in vitro* and *in vivo* and gains in PDT efficacy given by the ratio of light over dark sensitivity. Transferrin binding augmented in particular the long wavelength absorption properties. While similar transferrin binding was reported for non-photosensitive Ru complexes [4, 76], transferrin binding was not reported for other Ru-PSs, despite it being likely. Extensive spectral studies investigating the effects of Ru-PS binding to biological macromolecules have not been reported.

## 6.4 Issues to Be Considered to Further Develop Ru-Based Photosensitizers

In order to compare the efficacy of novel PSs versus approved ones, the ratio of maximum tolerated dose over effective drug concentration needs to be provided. In addition, the product of $\varepsilon$ and ROS quantum yield must be high enough to enable photoactivation within a clinically reasonable time, which may vary from tenths of a minute for cutaneous application to maximum 40 min needed for some intracavity or interstitial indications.

*In vitro* effects always need to be demonstrated in multiple cell lines, including nonmalignant ones, with the dark toxicity, $ED_{50}$, and the light toxicity, $LD_{50}$, quantified on the basis of photon dose absorbed by the PSs. The ratio of $LD_{50}$ over $ED_{50}$ needs to exceed at least 10 for all cell lines to achieve an effective and safe treatment during translation into preclinical and clinical studies as this provides a minimum of two effective attenuation depth of PDT. In the skin using an effective attenuation coefficient of ~1 mm$^{-1}$ for green wavelength, this would equal only 2 mm selectivity in depth over dark toxicity. Next to improving the biochemical ROS generation and photophysical properties of the Ru-PS, their subcellular localization, particularly *in vivo*, is of highest interest, as indicated.

### 6.4.1 Subcellular Localization

As already alluded to, a PS's efficacy is in large part determined by its association with critical subcellular structures and organelles, due to the short diffusion distances of the cytotoxic radicals. The fluorescent imaging approaches [77] commonly utilized to validate the subcellular localization have limited use for Ru-PSs whose $\varphi_\Delta$ have been maximized at the expense of the luminescence quantum yield. Nevertheless, mitochondria, lysosomes, and the endoplasmic reticulum [78] are formidable targets.

The desire to exploit the DNA targeting and cleaving properties of Ru(II) complexes is driven by the need to reduce the deliverable physical energies, which are currently unfavorable for PDT when compared to ionizing radiation. For the latter, a treatment course comprises the delivery of 50 Gy or $0.05\,J\,g^{-1}$ of tissue to be destroyed, whereas optical modifications including PDT require delivery of $1–1000\,J\,g^{-1}$. However, aiming to exploit DNA damage mechanisms is a double-edged sword. For one, nuclear localizing PSs can achieve cell kill at nanomolar concentrations [57, 79] and lower energy delivery by targeting the most sensitive organelles. Also, nanomolar drug concentration can facilitate easier regulatory approval processes. For ionizing radiation, very steep dose gradients covering several orders of magnitude can be obtained, particularly for intensity-modulated radiation therapy. Dose gradients in PDT are determined by the tissue's effective attenuation coefficient described earlier if targeting toward the malignant cells is limited. Hence, regions with sublethal DNA damage induced by Ru-PS-mediated PDT will be invariably present in the target and the host tissue; genetic modifications including the development of possible PDT resistance are likely. To date, PDT or PS resistance is not reported in literature, as targeted subcellular organelles for current approved PSs do not include the nucleus. Singh and coworkers attempted to generate PDT-resistant cell lines, but with limited success [80]. Lack of development of PDT resistance is a considerable advantage as it permits repeat therapies for field cancerization or in palliative settings [81]. Through targeting, the DNA, as an extremely sensitive subcellular structure, may lose this advantage for Ru-PS complexes.

Nevertheless, the focus on nuclear localization remains. Zhao *et al.* [79] demonstrated that for $[Ru([2,2';6',2']\text{-terpyridine})(3\text{-(pyrid-2'-yl)-4,5,9,16-tetraaza-dibenzo[a,c]naphthacene})]^{2+}$ in particular the ([2,2′;6′,2′]-terpyridine) ligand promotes cellular uptake, whereas the other ligand is predominantly responsible for the cross-linking of p53 and inhibiting DNA replication.

The DNA cleavage capabilities of the vast majority of Ru-PSs published to date have been commonly demonstrated only using plasmids or CT-DNA in solution. Imaging studies *in vitro* showing nuclear localization are rare and no *in vitro* DNA cleavage was demonstrated using, for example, the comet assay [82], showing only sublethal damage to DNA for PDT [83]. While the group of Gasser [57] presents nuclear localization for some $[Ru(II)(2,2'\text{-bipyridine})_2$ $\text{dipyrido[3,2-a:2',3'-c]phenazine}]^{2+}$ derivatives, the co-localization between the DAPI stain within the nucleus and the weak PS luminescence is not convincing enough to suggest equally strong binding or association with the DNA as required for electron transfer reaction or energy transfers. Non-PDT Ru-complex interactions include DNA stabilization and telomerase inhibition are demonstrated *in vivo* [68], so none of these mechanisms have been described for Ru-PS-mediated PDT.

### 6.4.2 Ruthenium Complex Photosensitizers and the Immune Response

It is known that transition metal complexes can induce an immune response [84] or modulate inflammatory and immune pathways in cells, potentially augmenting the direct tumor burden reduction by the acute PDT effects. However, as Clark [4] pointed out, basic ruthenium complexes inhibit T-cell proliferation, thus suppressing the immune response. In particular, a Ru(II) complex with nitrogen ligands is a potent immunosuppressant for antigen-induced T-cell proliferation, even at nanomolar concentrations. However, immunosuppression is not desirable following PDT. Nevertheless, it was noted that simple ruthenium complexes are unusually effective in suppressing the immune response by inhibiting T-cell proliferation and hence memory cell formation [4]. No publications pertaining to the immune-modulating effects of Ru-metal-based PSs are available at the time of writing this chapter. The effect of Ru-PSs on T-cell proliferation and the organism's ability to generate memory cells must be investigated prior to initiating clinical trials.

## 6.5 Future Directions for Ru-PS Research

The use of low fluence rates and extended light exposures has been proposed by several groups to increase the efficacy of PDT, particularly for cellular acting PSs, such as ALA. These concepts are following the approaches for metronomic chemotherapy first proposed by Kerbel *et al.* [85] and others [86], and are shown to achieve equal cytotoxicity toward tumors while reducing the toxic load to the normal tissues. The continuous toxic load impedes recovery of malignant cells more than normal tissue and also allows a deeper drug diffusion from the blood vessels [87]. This approach was applied for ALA-induced PpIX-mediated PDT by the Wilson group in Toronto [88] and validated in various models [89], albeit not used to date for Ru-PSs. The reasons for the latter are not that Ru complexes are not suitable for this approach; indeed, their low photobleaching may make them ideal, but it is probably a reflection of the currently restricted translation of these complexes outside of chemistry laboratories.

Indeed, establishing the pharmacokinetics, plasma half-life, clearance rates and, ultimately, selectivity of Ru-PS *in vivo* is paramount to further their development as mediators for PDT. While currently approximately 80 active clinical PDT studies are registered at www.clinicaltrials.gov, a prerequisite for later publishing of results from these clinical trials, only one drug containing ruthenium is being tested, but it is not a PS.

## 6.6 Conclusion

Since their introduction, the initial photophysical and biochemical limitations of Ru-PSs hindering translation into *in vivo,* short-wavelength activation and rapid photobleaching, have been addressed. The translation of Ru-PSs into light-activated drugs or PSs with ultimate clinical promise may be hindered by the same reasons which make them so promising for medical indications in the first place: their broad absorption range that approximately matches solar irradiance and their ability to act as electron donors or acceptors. From a medical point of view, the former raises the danger of prolonged skin photosensitivity and the latter results in a large variety of cellular damage mechanisms. From a commercial point of view, broad absorption properties and a range of energy transfer mechanisms makes Ru-PSs desirable in the energy sector, which also holds the promise of far higher and faster return on investment.

Nevertheless, chemists must partner with experts in appropriate animal models for oncological and nononcological indications to drive the next steps in the translation process. Novel *in silico* simulation tools for PDT light delivery [90] and pharmacokinetics [91] will assist in the design of the most promising approaches for preclinical and, ultimately, also clinical studies.

Completing toxicology and pharmacology studies on these drugs will become the next challenge. While initial studies do not require good manufacturing practice (GMP), large Ru-PS quantities are required, posing new challenges to the chemists who will be scaling up these reactions. The promise of Ru-complex-based PSs for clinical use, in particular for oncology, needs to be harvested by initiating these translational steps.

## References


1 (a) Bossard, G.E., Abrams, M.J., Darkes, M.C., Vollano, J.F., and Brooks, R.C. (1995) *Inorg. Chem.*, **34**, 1524–1527; (b) Charlesworth, P., Truscott, T.G., Brooks, R.C. *et al.* (1994) *J. Photochem. Photobiol. B*, **26**, 277–282.

2 Vollano, J.F., Bossard, G.E., Martellucci, S.A. *et al.* (1997) *J. Photochem. Photobiol. B*, **37**, 230–235.

3 Spiller, W., Kliesch, H., Wohrle, D. *et al.* (1998) *J. Porphyrins Phthalocyanines*, **2**, 145–158.

4 (a) Clarke, M.J. (2002) *Coord. Chem. Rev.*, **232**, 69–93; (b) Garner, R.N., Joyce, L.E., and Turro, C. (2011) *Inorg. Chem.*, **50**, 4384–4391.

5 Pefkianakis, E.K., Christodouleas, D., Giokas, D.L. *et al.* (2013) *Eur. J. Inorg. Chem.*, **2013**, 4628–4635.

6 Suzuki, M., Bartels, O., Gerdes, R. *et al.* (2000) *Phys. Chem. Chem. Phys.*, **2**, 109–114.

7 (a) Meyer-Betz, F. (1913) *Dtsch. Arch. Klin. Med.*, **112**, 476–503; (b) von Tappeiner, H.A. and Betz, J.A. (1903) *Munch. Med. Wochenschr.*, **47**, 2042–2044; (c) von Tappeiner, H.A. and Betz, J.A. (1907) *Die Sensibilisierende*

*Wirkung Fluorescierender Substanzen: Gesammelte Untersuchungen über die Photodynamische Erscheinung*, Germany, Leipzig.

8 (a) Dougherty, T.J., Kaufman, J.E., and Goldfarb, A. (1978) *Cancer Res.*, **38**, 2628–2635; (b) Dougherty, T.J., Lawrence, G., Kaufman, J.H. *et al.* (1979) *J. Natl. Cancer Inst.*, **62**, 231–237.

9 (a) Mroz, P., Hashmi, J.T., Huang, Y.-Y. *et al.* (2011) *Expert Rev. Clin. Immunol.*, **7**, 75–91; (b) Wilson, B.C. and Patterson, M.S. (2008) *Phys. Med. Biol.*, **53**, R61–R109.

10 Shi, G., Monro, S., Hennigar, R. *et al.* (2015) *Coord. Chem. Rev.*, **282**, 127–138.

11 Lincoln, R., Kohler, L., Monro, S. *et al.* (2013) *J. Am. Chem. Soc.*, **135**, 17161–17175.

12 (a) Lecomte, J.P., Kirsch-De Mesmaeker, A., Feeney, M.M. *et al.* (1995) *Inorg. Chem.*, **34**, 6481–6491; (b) Zhang, Q.L., Liu, J.G., Chao, H. *et al.* (2001) *J. Inorg. Biochem.*, **83**, 49–55; (c) Vicendo, P., Mouysset, S., and Paillous, N. (1997) *Photochem. Photobiol.*, **65**, 647–655; (d) Nagababu, P., Latha, J.N.L., and Satyanarayana, S. (2006) *Chem. Biodivers.*, **3**, 1219–1229; (e) Musetti, C., Spagnul, C., Mion, G. *et al.* (2015) *ChemPlusChem*, **80**, 158–168.

13 Weishaupt, K.R., Gomer, C.J., and Dougherty, T.J. (1976) *Cancer Res.*, **36**, 2326–2329.

14 (a) Niedre, M., Patterson, M.S., and Wilson, B.C. (2002) *Photochem. Photobiol.*, **75**, 382–391; (b) Skovsen, E., Snyder, J.W., Lambert, J.D.C. *et al.* (2005) *J. Phys. Chem. B*, **109**, 8570–8573.

15 (a) Verma, S., Watt, G.M., Mal, Z. *et al.* (2007) *Photochem. Photobiol.*, **83**, 996–1005; (b) Jori, G. (1996) *J. Photochem. Photobiol. B*, **36**, 87–93; (c) Vrouenraets, M.B., Visser, G.W.M., Snow, G.B. *et al.* (2003) *Anticancer Res.*, **23**, 505–522.

16 Juris, A., Balzani, V., Barigelletti, F. *et al.* (1988) *Coord. Chem. Rev.*, **84**, 85–277.

17 Fernandez, J.M., Bilgin, M.D., and Grossweiner, L.I. (1997) *J. Photochem. Photobiol. B*, **37**, 131–140.

18 (a) Sauaia, M.G., de Lima, R.G., Tedesco, A.C. *et al.* (2005) *Inorg. Chem.*, **44**, 9946–9951; (b) Rose, M.J. and Mascharak, P.K. (2008) *Coord. Chem. Rev.*, **252**, 2093–2114.

19 Kim, S., Fujitsuka, M., and Majima, T. (2013) *J. Phys. Chem. B*, **117**, 13985–13992.

20 Price, M., Reiners, J.J., Santiago, A.M. *et al.* (2009) *Photochem. Photobiol.*, **85**, 1177–1181.

21 (a) Wang, L.H., Jacques, S.L., and Zheng, L.Q. (1995) *Comput Methods Programs Biomed*, **47**, 131–146; (b) Jacques, S.L. (1998) *Photochem. Photobiol.*, **67**, 23–32.

22 (a) Wang, W., Moriyama, L.T., and Bagnato, V.S. (2013) *Laser Phys. Lett.*, **10**, 023001; (b) Casas, A., Di Venosa, G., Hasan, T. *et al.* (2011) *Curr. Med. Chem.*, **18**, 2486–2515; (c) Chen, B., Pogue, B.W., Hoopes, P.J. *et al.* (2006) *Crit. Rev. Eukaryot. Gene Expr.*, **16**, 279–305.

23 (a) Ogura, S.-I., Hagiya, Y., Tabata, K. *et al.* (2012) *Curr. Top. Med. Chem.*, **12**, 176–184; (b) Milla Sanabria, L., Rodriguez, M.E., Cogno, I.S. *et al.* (2013) *Biochim. Biophys. Acta*, **1835**, 36–45.
24 (a) Rodal, G.H., Rodal, S.K., Moan, J. *et al.* (1998) *J. Photochem. Photobiol. B*, **45**, 150–159; (b) Mroz, P., Sharma, S.K., Zhiyentayev, T. *et al.* (2012) in *Drug Delivery in Oncology: From Basic Research to Cancer Therapy*, vols 1–3 (eds F. Kratz, P. Senter, and H. Steinhagen), Wiley, pp. 1569–1603.
25 Hsieh, Y.J., Wu, C.C., Chang, C.J. *et al.* (2003) *J. Cell. Physiol.*, **194**, 363–375.
26 (a) Niedre, M.J., Yu, C.S., Patterson, M.S. *et al.* (2005) *Br. J. Cancer*, **92**, 298–304; (b) Jarvi, M.T., Niedre, M.J., Patterson, M.S. *et al.* (2006) *Photochem. Photobiol.*, **82**, 1198–1210.
27 Zava, O., Zakeeruddin, S.M., Danelon, C. *et al.* (2009) *Chembiochem*, **10**, 1796–1800.
28 (a) Mazuryk, O., Magiera, K., Rys, B. *et al.* (2014) *J. Biol. Inorg. Chem.*, **19**, 1305–1316; (b) Zheng, Z.B., Wu, Y.-Q., Wang, K.-Z. *et al.* (2014) *Dalton Trans.*, **43**, 3273–3284; (c)Zhao, X.-L., Y.-Z. Ma, and K.-Z. Wang, *J. Inorg. Biochem.*, 2012, **113**, 66–76.
29 Paxian, M., Keller, S.A., Cross, B. *et al.* (2003) *FASEB J.*, **17**, A548.
30 Manjon, F., Santana-Magana, M., Garcia-Fresnadillo, D. *et al.* (2010) *Photochem. Photobiol. Sci.*, **9**, 838–845.
31 Polec-Pawlak, K., Abramski, J.K., Semenova, O. *et al.* (2006) *Electrophoresis*, **27**, 1128–1135.
32 (a) Bown, S.G. (1986) *Br. J. Cancer*, **54**, 43–52; (b) Marijnissen, J.P., Versteeg, J.A.C., Star, W.M. *et al.* (1992) *Int. J. Radiat. Oncol. Biol. Phys.*, **22**, 963–972; (c) Lilge, L. and Wilson, B.C. (1998) *J. Clin. Laser Med. Surg.*, **16**, 81–91.
33 Moesta, K.T., Greco, W.R., Nursefinlay, S.O. *et al.* (1995) *Cancer Res.*, **55**, 3078–3084.
34 (a) Arenas, Y., Monro, S., Shi, G. *et al.* (2013) *Photodiag Photodyn. Ther.*, **10**, 615–625; (b) Stephenson, M., Reichardt, C., Pinto, M. *et al.* (2014) *J. Phys. Chem. A*, **118**, 10507–10521.
35 (a) Chen, D., Zheng, H., Huang, Z. *et al.* (2012) *Int. J. Photoenergy*, 920671; (b) Schmidt, M.H., Reichert, K.W., Ozker, K. *et al.* (1999) *Pediatr. Neurosurg.*, **30**, 225–231.
36 Zhu, T.C., Bonnerup, C., Colussi, V.C. *et al.* (2013) *Med. Phys.*, **40**, 081501.
37 Fong, J., Kasimova, K., Arenas, Y. *et al.* (2015) *Photochem. Photobiol. Sci.*, **14**, 2014–2023.
38 (a) Wilson, B.C., Muller, P.J., and Yanch, J.C. (1986) *Phys. Med. Biol.*, **31**, 125–133; (b) van Staveren, H.J., Beek, J.F., Keijzer, M. *et al.* (1995) *Phys. Med. Biol.*, **40**, 1307–1315; (c) Star, W.M., Marijnissen, J.P., and van Gemert, M.J. (1988) *Phys. Med. Biol.*, **33**, 437–454; (d) Marijnissen, J.P., Baas, P., Beek, J.F. *et al.* (1993) *Photochem. Photobiol.*, **58**, 92–99.
39 (a) Wilson, B.C. and Adam, G. (1983) *Med. Phys.*, **10**, 824–830; (b) Cassidy, J., Betz, V., and Lilge, L. (2015) *Frontiers in Physics*, **3**, 6.
40 (a) Davidson, S.R., Weersink, R.A., Haider, M.A. *et al.* (2009) *Phys. Med. Biol.*, **54**, 2293–2313; (b) Zhu, T.C. and Finlay, J.C. (2006) *Photodiag. Photodyn. Ther.*, **3**, 234–246; (c) Dehghani, H., Eames, M.E., Yalavarthy, P.K. *et al.* (2009) *Commun. Numer. Methods Eng.*, **25**, 711–732.

41 Lo, W.C.Y., Redmond, K., Luu, J. *et al.* (2009) *J. Biomed. Opt.*, **14**, 029801.

42 Grossweiner, L.I. (1986) *Lasers Surg. Med.*, **6**, 462–466.

43 (a) Jarvi, M.T., Patterson, M.S., and Wilson, B.C. (2012) *Biophys. J.*, **102**, 661–671; (b) Liu, B., Farrell, T.J., and Patterson, M.S. (2012) *Phys. Med. Biol.*, **57**, 825.

44 (a) Niedre, M.J., Secord, A.J., Patterson, M.S. *et al.* (2003) *Cancer Res.*, **63**, 7986–7994; (b) Georgakoudi, I. and Foster, T.H. (1998) *Photochem. Photobiol.*, **67**, 612–625.

45 (a) Valdes, P.A., Leblond, F., Kim, A. *et al.* (2012) *Opt. Lett.*, **37**, 1817–1819; (b) Stringer, M.R., Hudson, E.J., and Smith, M.A. (1995) *Lasers Med. Sci.*, **10**, 19–24; (c) Vulcan, T.G., Zhu, T.C., Rodriguez, C.E. *et al.* (2000) *Lasers Surg. Med.*, **26**, 292–301.

46 (a) Johansson, A., Faber, F., Kniebuehler, G. *et al.* (2013) *Lasers Surg. Med.*, **45**, 225–234; (b) Thompson, M.S., Johansson, A., Johansson, T. *et al.* (2005) *Appl. Opt.*, **44**, 4023–4031.

47 (a) Lilge, L., O'Carroll, C., and Wilson, B.C. (1997) *J. Photochem. Photobiol. B*, **39**, 229–235; (b) Bourdon, O., Laville, I., Carrez, D. *et al.* (2002) *Photochem. Photobiol. Sci.*, **1**, 709–714.

48 Pernot, M., Bastogne, T., Barry, N.P.E. *et al.* (2012) *J. Photochem. Photobiol. B*, **117**, 80–89.

49 (a) Yoshida, Y., Dereski, M.O., Garcia, J.H. *et al.* (1992) *J. Neuropathol. Exp. Neurol.*, **51**, 91–100; (b) O'Connor, A.E., Gallagher, W.M., and Byrne, A.T. (2009) *Photochem. Photobiol.*, **85**, 1053–1074.

50 (a) Lilge, L., Pomerleau-Dalcourt, N., Douplik, A. *et al.* (2004) *Phys. Med. Biol.*, **49**, 3209–3225; (b) Pomerleau-Dalcourt, N. and Lilge, L. (2006) *Phys. Med. Biol.*, **51**, 1929–1940; (c) Lilge, L., Haw, T., and Wilson, B.C. (1993) *Phys. Med. Biol.*, **38**, 215–230; (d) Pogue, B.W. and Burke, G. (1998) *Appl. Opt.*, **37**, 7429–7436.

51 (a) Dognitz, N. and Wagnieres, G. (1998) *Lasers Med. Sci.*, **13**, 55–65; (b) Saager, R.B., Cuccia, D.J., Saggese, S. *et al.* (2013) *Lasers Surg. Med.*, **45**, 207–215; (c) Sunar, U., Rohrbach, D.J., Morgan, J. *et al.* (2013) *Biomed. Opt. Exp.*, **4**, 531–537.

52 Robinson, D.J., Rohrbach, D.J., Morgan, J. *et al.* (1998) *Photochem. Photobiol.*, **67**, 140–149.

53 Wilson, B.C., Patterson, M.S., and Lilge, L. (1997) *Lasers Med. Sci.*, **12**, 182–199.

54 Gudgin, D.E.F. and Pottier, R.H. (1995) *J. Photochem. Photobiol. B*, **29**, 91–93.

55 (a) Dhar, S., Nethaji, M., and Chakravarty, A.R. (2005) *Inorg. Chem.*, **44**, 8876–8883; (b) Hergueta-Bravo, A., Jimenez-Hernandez, M.E., Montero, F. *et al.* (2002) *J. Phys. Chem. B*, **106**, 4010–4017; (c) Mei, W.J., Liu, J., Chao, H. *et al.* (2003) *Trans. Metal Chem.*, **28**, 852–857.

56 (a) Weston, M.A. and Patterson, M.S. (2011) *Photochem. Photobiol.*, **87**, 1129–1137; (b) Weston, M.A. and Patterson, M.S. (2014) *Photochem. Photobiol.*, **90**, 1359–1367.

57 (a) Mari, C. and Gasser, G. (2015) *Chimia*, **69**, 176–181; (b) Mari, C., Pierroz, V., Rubbiani, R. *et al.* (2014) *Chem. Eur. J.*, **20**, 14421–14436; (c) Frei, A., Rubbiani, R., Tubafard, S. *et al.* (2014) *J. Med. Chem.*, **57**, 7280–7292.

58 Chen, Q., Huang, Z., Chen, H. *et al.* (2002) *Photochem. Photobiol.*, **76**, 197–203.

59 Henderson, B.W., Busch, T.M., Vaughan, L.A. *et al.* (2000) *Cancer Res.*, **60**, 525–529.

60 American National Standard (2007) *American National Standard for Safe Use of Laser*, Laser Institute of America, Orlando, FL.

61 (a) Patterson, M.S., Wilson, B.C., and Graff, R. (1990) *Photochem. Photobiol.*, **51**, 343–349; (b) Georgakoudi, I., Nichols, M.G., and Foster, T.H. (1997) *Photochem. Photobiol.*, **65**, 135–144; (c) Farrell, T.J., Wilson, B.C., Patterson, M.S. *et al.* (1998) *Photochem. Photobiol.*, **68**, 394–399.

62 (a) Jankun, J., Keck, R.W., Skrzypczak-Jankun, E. *et al.* (2005) *Br. J. Urol. Int.*, **95**, 1237–1244; (b) Jankun, J., Lilge, L., Douplik, A. *et al.* (2004) *J. Urol.*, **172**, 739–743.

63 Castano, A.P., Mroz, P., and Hamblin, M.R. (2006) *Nat. Rev. Cancer*, **6**, 535–545.

64 Korbelik, M. and Dougherty, G.J. (1999) *Cancer Res.*, **59**, 1941–1946.

65 (a) Kroemer, G., Galluzzi, L., Kepp, O. *et al.* (2013) *Ann. Rev. Immunol.*, **31**, 51–72; (b) Panzarini, E., Inguscio, V., and Dini, L. (2013) *Biomed. Res. Int.*, **2013**, 482160.

66 Gollnick, S.O., Vaughan, L., and Henderson, B.W. (2002) *Cancer Res.*, **62**, 1604–1608.

67 Nowak-Sliwinska, P., van Beijnum, J.R., Casini, A. *et al.* (2011) *J. Med. Chem.*, **54**, 3895–3902.

68 Chen, Z.-F., Qin, Q.-P., Qin, J.-L. *et al.* (2015) *J. Med. Chem.*, **58**, 4771–4789.

69 Hertzman, A.B. (1959) *Physiol. Rev.*, **39**, 280–306.

70 Ishii, K., Shiine, M., Shimizu, Y. *et al.* (2008) *J. Phys. Chem. B*, **112**, 3138–3143.

71 (a) Collins, H.A., Khurana, M., Moriyama, E.H. *et al.* (2008) *Nat. Photon.*, **2**, 420–424; (b) Drobizhev, M., Karotki, A., Kruk, M. *et al.* (2003) *Chem. Phys. Lett.*, **370**, 690–699.

72 (a) Girardot, C., Lemercier, G., Mulatier, J.-C. *et al.* (2007) *Dalton Trans*, **31**, 3421–3426; (b) Poon, C.-T., Chan, P.-S., Man, C. *et al.* (2010) *J. Inorg. Biochem.*, **104**, 62–70; (c) Natrajan, L.S., Toulmin, A., Chew, A. *et al.* (2010) *Dalton Trans.*, **39**, 10837–10846; (d) Ke, H., Wang, H., Wong, W.-K. *et al.* (2010) *Chem. Commun.*, **46**, 6678–6680; (e) Hanczyc, P., Norden, B., and Samoc, M. (2012) *Dalton Trans.*, **41**, 3123–3125.

73 Kobtsev, S., Kukarin, S., Smirnov, S. *et al.* (2009) *Opt. Express.*, **17**, 20707–20713.

74 Vogel, A. and Venugopalan, V. (2003) *Chem. Rev.*, **103**, 577–644.

75 Kaspler, P., Lazic, S., Forward, S. *et al.* (2016) *Photochem. Photobiol. Sci.*, **15**, 481–495.

76 (a) Levina, A., Mitra, A., and Lay, P.A. (2009) *Metallomics*, **1**, 458–470; (b) Ang, W.H. and Dyson, P.J. (2006) *Eur. J. Inorg. Chem.*, **20**, 4003–4018.

77 Ali, S.M. and Olivo, M. (2002) *Int. J. Oncol.*, **21**, 531–540.

78 Teiten, M.H., Bezdetnaya, L., Morliere, P. *et al.* (2003) *Br. J. Cancer*, **88**, 146–152.

79 (a) Zhao, R., Hammitt, R., Thummel, R.P. *et al.* (2009) *Dalton Trans.*, **48**, 10926–10931; (b) Joshi, T. and Gasser, G. (2015) *Synlett*, **26**, 275–284.

80 Adams, K., Rainbow, A.J., Wilson, B.C. *et al.* (1999) *J. Photochem. Photobiol. B*, **49**, 136–141.

81 Zvi, M. (2015) *Photomed. Laser Med.*, **4**, 19–26.

82 Kumaravel, T.S. and Jha, A.N. (2006) *Mutat. Res*, **605**, 7–16.

83 Pouget, J.P., Ravanat, J.L., Douki, T. *et al.* (1999) *J. Chim. Phys.*, **96**, 143–146.

84 Leung, C.-H., Lin, S., Zhong, H.-J. *et al.* (2015) *Chem. Sci.*, **6**, 871–884.

85 (a) Kerbel, R.S., Emmenegger, U., Munoz, R. *et al.* (2008) *Cancer Drug Discov. Dev.*, 277–296; (b) Kerbel, R.S. and Kamen, B.A. (2004) *Nat. Rev. Cancer*, **4**, 423–436.

86 Gasparini, G. (2001) *Lancet Oncol.*, **2**, 733–740.

87 (a) Saggar, J.K. and Tannock, I.F. (2015) *Clin. Cancer Res.*, **21**, 2107–2114; (b) Primeau, A.J., Rendon, A., Hedley, D. *et al.* (2005) *Clin. Cancer Res.*, **11**, 8782–8788.

88 (a) Bisland, S., Lilge, L., Lin, A. *et al.* (2003) *Proc. SPIE*: **5260**, 342–351; (b) Bisland S.K., Austin J., Wilson B.C., *et. al.* (2003) *Proc. SPIE.*, **5260**, 352–360.

89 van Zaane, F., Subbaiyan, D., van der Ploeg-van den Heuvel, A. *et al.* (2010) *J. Biophotonics*, **3**, 347–355.

90 (a) Alerstam, E., Lo, W.C.Y., Han, T.D. *et al.* (2010) *Biomed. Opt. Express*, **1**, 658–675; (b) Barajas, O., Ballangrud, A.M., Miller, G.G. *et al.* (1997) *Phys. Med. Biol.*, **42**, 1675–1687; (c) Baran, T.M., Wilson, J.D., Mitra, S. *et al.* (2012) *J. Biomed. Opt.*, **17**, 098002; (d) Cassidy, J., Betz, V., and Lilge, L. (2015) *Proc. SPIE*, **9308**, 930800; (e) Oakley, E., Wrazen, B., Bellnier, D.A. *et al.* (2015) *Lasers Surg. Med.*, **47**, 60–67; (f) Tromberg, B.J., Svaasand, L.O., Fehr, M.K. *et al.* (1996) *Phys. Med. Biol.*, **41**, 223–237.

91 (a) Liu, B., Farrell, T.J., and Patterson, M.S. (2012) *J. Biomed. Opt.*, **17**, 088001; (b) Valentine, R.M., Brown, C.T.A., Moseley, H. *et al.* (2011) *J. Biomed. Opt.*, **16**, 078901.

# 7

# Photodynamic Therapy in Medicine with Mixed-Metal/Supramolecular Complexes

*Jimmie L. Bullock and Alvin A. Holder*

*Old Dominion University, Department of Chemistry and Biochemistry, 4541 Hampton Boulevard, Norfolk, VA 23529-0126, USA*

## 7.1 Introduction

Over the past few decades, coordination complexes containing $d^6$ metal centers and polypyridyl ligand architectures have been developed as structure- and site-specific DNA-binding agents [1]. Due to the attractive photophysical properties, much research has been focused on complexes containing ruthenium metal centers and, more recently, attention has shifted to the use of these complexes in biological systems. As the rules that govern the cellular uptake and localization of these systems are determined, the number of applications ranging from cellular imaging to therapeutics is rising [2]. As the search for adequate photodynamic therapeutic agents continues to evolve, the complexes/systems being utilized as photodynamic therapy (PDT) agents are also constantly evolving. While DNA interactions with mononuclear ruthenium complexes have flooded literature recently [3], similar interaction are possible with dinuclear ruthenium-containing species [4]; however, their applications as potential DNA-binding agents remain largely unexplored. Dinuclear complexes offer increased variation in shape and size and have the potential to show greater specificity. For instance, $[\{Ru(phen)_2\}_2(l\text{-}HAT)]^{4+}$ (where phen = 1,10-phenanthroline and HAT = 1,4,5,8,9,12-hexaazatriphenylene) has been shown to strongly bind partially denatured DNA [4e], $[\{Ru(Me_2bpy)_2\}_2(bpm)]_4)$ (where $Me_2bpy$ = 4,4′-dimethyl-2,2′-bipyridine and bpm = 2,2′-bipyrimidine) specifically binds DNA bulge sequences [4f], $[\{Ru(dpq)_2\}_2(phen\text{-}x\text{-}SOS\text{-}phen)]^{4+}$ ($x$ = 3, 4, 5) (where dpq = dipyrido[3,2-f:2′,3′-h]-quinoxaline, SOS = 2-mercaptoethyl ether) exhibit high DNA affinity at 100-mM NaCl concentration [4g]. Similarly, Chao *et al.* [5] reported on $[(phen)_2Ru(mbpibH_2)Ru(phen)_2]^{4+}$ (where $mbpibH_2$ = 1,3-bis([1,10]phenanthroline[5,6-*d*]imidazol-2-yl)benzene), which binds to DNA by inserting the bridging moiety into adjacent base pairs via an intercalative mechanism. Mixed-metal dinuclear complexes offer many advantages over mononuclear and ruthenium–ruthenium dinuclear complexes, as the second metal center utilized generally possesses its unique biological application, making for dual-application complex or enhanced biological function. In this

*Ruthenium Complexes: Photochemical and Biomedical Applications*, First Edition.
Edited by Alvin A. Holder, Lothar Lilge, Wesley R. Browne, Mark A.W. Lawrence, and Jimmie L. Bullock Jr.

chapter, we discuss first common bioactive metal centers, followed by a survey of mixed-metal and supramolecular complexes which incorporate ruthenium metal centers.

## 7.2 Platinum and Rhodium Centers as Bioactive Sites

The study of smaller complexes that exhibit DNA binding is important to understand the interactions of larger complexes that combine light-activated PDT agents with DNA targeting bioactive sites (BASs). A variety of Pt and Rh complexes have been reported to interact covalently with DNA, and a select few examples are provided herein.

### 7.2.1 Platinum(II)-Based Chemotherapeutics

Of all Pt(II) chemotherapeutic agents, the most widely recognized is cisplatin, *cis*-[$PtCl_2(NH_3)_2$]. The reactivity of cisplatin with biomolecules was first reported in 1965 by Rosenberg *et al.* [6]. While studying the effect of electric fields on the bacteria *Escherichia coli* using a Pt electrode, a compound released from the electrode was found to inhibit replication of the bacteria. The active compound, later determined to be cisplatin, was also shown to inhibit the replication of mammalian cells and cause apoptosis (programmed cell death) in cancerous cells [7].

Following its U.S. Food and Drug Administration (FDA) approval in 1978, it has remained one of the most widely utilized anticancer drugs, particularly in the treatment of testicular cancer and ovarian carcinoma [8].

The mechanism of cisplatin activation has also been well studied [8, 9]. It is known that the complex remains intact in the blood stream where the $Cl^-$ concentration is high (≈100 mM) and undergoes rapid sequential aquation of the chloride ligands following diffusion in cells ([$Cl^-$] ≈ 4 mM), as shown in Figure 7.1 [10]. The mono- and bis-aquated complexes are the active species

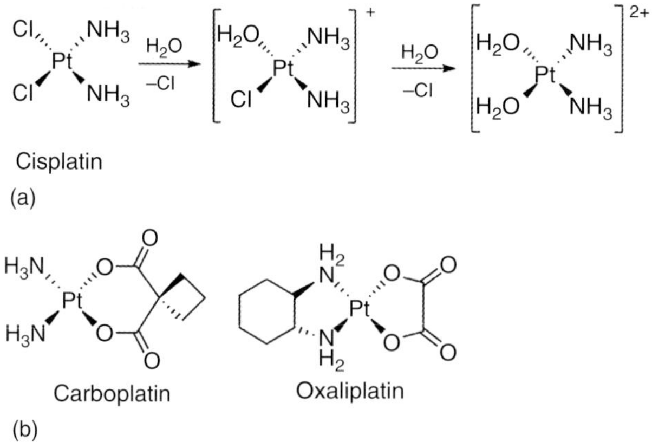


Figure 7.1 (a) Aquation of cisplatin. (b) Second-generation derivatives of cisplatin carboplatin and oxaliplatin.

that diffuse across the nuclear membrane and bind to DNA, primarily forming 1,2-interstrand cross-links. These species primarily access DNA via the major groove, covalently binding to the $N^7$ of guanine and adenine bases due to the accessibility and high nucleophilicity of these sites [11]. While $Pt^{II}$ drugs are effective in initiating apoptosis, many limitations have arisen from the utilization of cisplatin including nephrotoxicity (kidney poisoning), ototoxicity (loss of high-frequency hearing), peripheral neuropathy, and the emergence of cisplatin-resistant cancers [8d, 12]. Second- and third-generation $Pt^{II}$ drugs have been developed in an effort to improve upon the successes and overcome the limitations of cisplatin [13]. For instance, both carboplatin and oxaliplatin (Figure 7.1) have been FDA approved, in 1989 and 2002, respectively. However, much research has focused on developing more effective anticancer agents.

### 7.2.2 Rhodium(III) as a Bioactive Site

Octahedral Rh(III) complexes such as those shown in Figure 7.2, $[Rh^{III}(NN)_2X_2]^{2+}$ (where NN = bidentate polyazine ligands and X = halide) are also known to bind to DNA [14]. $[Rh(phen)_2Cl_2]^+$ was designed with a *cis*-dihalide moiety to mimic cisplatin, providing labile ligands to undergo dissociation followed by covalent binding of the metal center to DNA [15]. The major advantage of using these complexes for DNA modification is their inherent photoactivity.

This complex absorbs ultraviolet (UV) and near-UV light with transitions centered at 334 and 351 nm assigned to intraligand transition and a transition centered at 380 nm assigned to a ligand field (LF) transition [16]. Excitation of an aqueous solution of $[Rh(phen)_2Cl_2]^+$ resulted in photolablization of a $Cl^-$, which is substituted by water molecules to form the aqua chloro $[Rh(phen)_2(OH_2)Cl]^{2+}$ species [15]. In addition, the positive charge on the $[Rh(phen)_2Cl_2]^+$ allows the

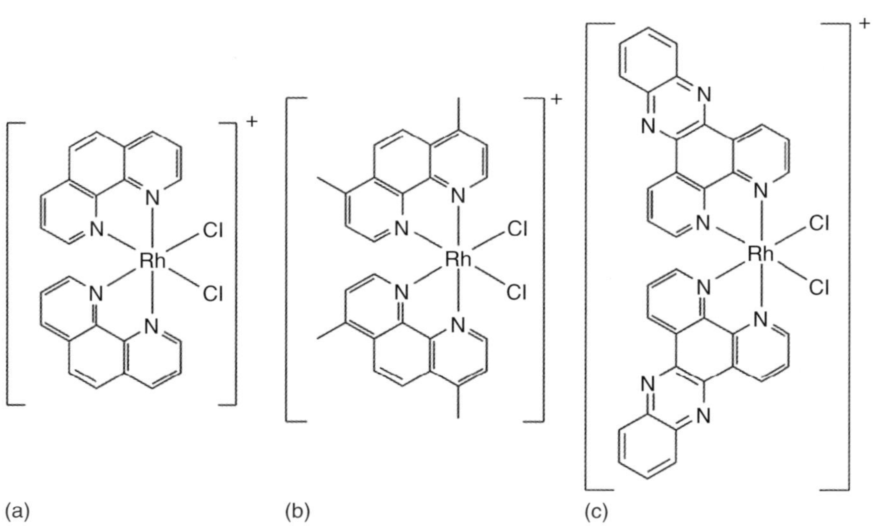

Figure 7.2 Structural representations of (a) $[Rh(phen)_2Cl_2]^+$, (b) $[Rh(Me_4phen)_2Cl_2]^+$, and (c) $[Rh(phen)(dppz)Cl_2]^+$.

complex to interact with DNA via an electrostatic interaction [14]. Photolysis of the complex in the presence of calf thymus DNA with $\lambda_{irr} = 360$ nm results in the formation of covalent adducts with moderate efficiency, most notably with the purine base guanine (similar to cisplatin) [14, 17]. The complex does not exhibit the same binding event under dark conditions or when the *bis*-aquated complex is added directly to DNA. Binding of the complex is initiated by reductive quenching of the excited state where an electron is transferred from the base to the complex in the $^3$LF excited state. This process is followed by $Cl^-$ loss and the subsequent binding to DNA [17b].

The photochemistry and photobiological effect of these complexes can be impacted by variation of the bidentate ligands coordinated. Replacing the phen ligands in $[Rh(phen)_2Cl_2]^+$ with $Me_4$phen (where Me4phen = 3,4,7,8-tetramethyl-1,10-phenanthroline) to give $[Rh(Me_4phen)_2Cl_2]^+$ results in higher hydrophobicity to enable the complex to more readily pass through the cell membrane and undergo more efficient ground-state association with DNA [18]. Methylation results in a minor shift in the electronic absorption and emission spectrum of this complex compared to traditional phen; however, photoaquation to form aqua chloro species $[Rh(Me_4phen)_2(OH_2)Cl]^{2+}$ is greatly enhanced with 347-nm excitation ($\Phi = 0.63$ as opposed to 0.03 for phen). This difference is attributed to the $\sigma$-donating ability of $Me_4$phen, resulting in enhanced stabilization of the pentacoordinate complex formed upon chloride dissociation when the complex is directly excited into an LF excited state. A similar effect can be observed for $[Rh(phen)(dppz)Cl_2]^+$ (where dppz = dipyrido[3,2-a:2′,3′-c]phenazine). In addition, this complex has been shown to photocleave DNA into smaller fragments. Binding and cleavage of DNA by Rh(III)-polyazine mononuclear complexes provided a means toward efficient type II PDT agents.

## 7.3 Supramolecular Complexes as DNA Photomodification Agents

A supramolecular complex is, as defined by Verhoeven [19], an assembly of components that individually perform a specific role, but act together to execute a complex function. When the function is initiated by light, the assembly is known more commonly as a photochemical molecular device (PMD). Supramolecular chemistry has played an important role in many new pharmaceutical therapies, especially in those involved in mapping of drug interaction/binding sites [20]. The extensive properties of supramolecular complexes have thus naturally found applications in PDT, as they allow for the development of multifunctional agents that provide combination therapy in a single molecule [21]. This is typically accomplished by coupling photosensitizers (PSs) to promote light-activated DNA damage with a BAS to prelocalize the PS in close proximity to the DNA double helix through covalent binding [22]. The PS with two bidentate polyazine terminal ligands (TLs) or one tridentate TL and one monodentate TL is coupled to the BAS through a bidentate or tridentate polyazine bridging ligand (BL). The modulation of redox and spectroscopic properties, as well as control of the size, stereoisomerization, lipophilicity, and water solubility of the supramolecule, can be achieved by component variation, providing the possibility for a wide array of PMDs to assay the effects of varied structures and orbital energetics. This ability to easily modify structures can be exploited to avoid many of the inherent problems associated with platinum-based, anticancer agents. Discussed in Sections 4.1 and 4.2 are PMDs that utilize Ru(II)–polyzine PSs coupled to a Pt(II) or Rh(III) BAS.

## 7.4 Mixed-Metal Complexes as Photodynamic Therapeutic Agents

### 7.4.1 Photosensitizers with a Ru(II) Metal Center Coupled to Pt(II) Bioactive Sites

Many of these complexes are designed to possess a *cis*-$PtCl_2$ BAS to afford binding to DNA and achieve efficient cleavage of the DNA backbone through $^1O_2$ production. This provides for more effective DNA photocleavage even at reduced oxygen concentrations.

#### 7.4.1.1 Binuclear Complexes with Ru(II) and Pt(II) Metal Centers with Bidentate Ligands

Bidentate polyazine BLs allow for $[Ru(bpy)_3]^{2+}$ (where bpy = 2,2′-bipyridine)-type complexes to be coupled into larger supramolecular systems through replacement of a bpy TL with a BL (e.g., dpp, dpq, dpb, or bpm) (where dpp = 2,3-Bis(2-pyridyl)pyrazine, dpq = 2,3-bis(2-pyridyl) benzoquinoxaline). These BLs possess two bidentate sites for remote metal coordination, providing for the construction of complex systems [23]. Importantly, $(TL)_2Ru^{II}(BL)$ motifs undergo metal-to-ligand charge-transfer (MLCT) excitation to move an excited electron to the BL, toward the coupled BAS.

The first binuclear complexes featuring Ru(II)-PS units bridged to a *cis*-$PtCl_2$ unit that were reported to interact with DNA were of the supramolecular architecture $[(bpy)_2Ru(dpb)PtCl_2]^{2+}$ (Figure 7.3) [24]. This complex was designed to feature two structural motifs: the planar benzoquinoxaline portion of the BL that provided strong groove binding or intercalation to DNA and the Pt BAS for covalent DNA binding.

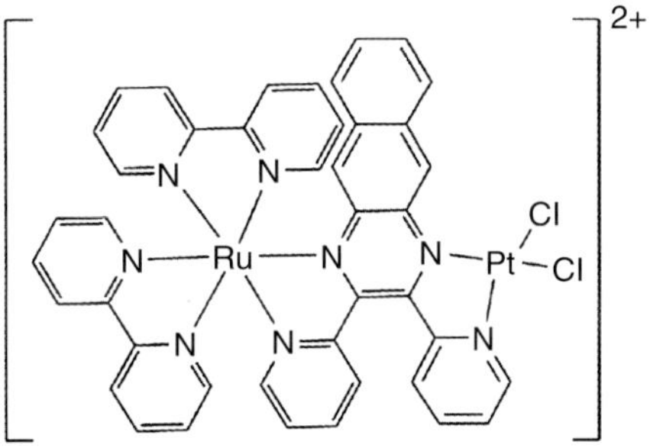


**Figure 7.3** Structural representation of $[(bpy)_2(Ru(dpb)PtCl_2]^{2+}$.

Concentration- and time-dependent thermal-binding studies of the $[(bpy)_2Ru(dpb)PtCl_2]Cl_2$ with pBluescript KS+ linear plasmid DNA were performed using gel electrophoresis and where a known DNA binder, cisplatin, was utilized as a control [24b]. Incubation of $[(bpy)_2Ru(dpb)PtCl_2]Cl_2$ with DNA at 37 °C resulted in retardation in the migration of DNA compared to cisplatin. This result, along with a comparison to the lack of retarded migration of the plasmid that was incubated with $[(bpy)_2Ru(dpb)]Cl_2$, suggests the primary binding mode of $[(bpy)_2Ru(dpb)PtCl_2]Cl_2$ is facilitated through covalent binding through the Pt site. The more dramatically retarded migration using the binuclear complex compared to cisplatin is a result of the overall decreased negative charge of DNA upon binding of a more positively charged cation and the increased size or change in the DNA three-dimensional structure upon metal complex binding. Denaturing gel electrophoresis was utilized to study whether the $[(bpy)_2Ru(dpb)PtCl_2]Cl_2$ complexes bind through intrastrand or interstrand cross-linking [22]. The primarily intrastrand cross-linker cisplatin and the primarily interstrand cross-linker $[\{t\text{-}PtCl(NH_3)_2\}_2\{\mu\text{-}H_2N(CH_2)_4HN_2\}]Cl_2$ were studied as controls. Results indicated that $[(bpy)_2Ru(dpb)PtCl_2]Cl_2$ exhibits approximately 90% intrastrand and 10% interstrand cross-linking. The light-absorbing properties in the low-energy visible region make this complex potentially a PDT agent; however, no photolysis studies were reported to date.

Two Ru(II)—Pt(II) PMDs were designed to couple the enhanced $^1O_2$ generation of $[Ru\text{-}(Ph_2phen)_3]^{2+}$ (where $Ph_2phen$ = 4,7-diphenyl-1,10-phenanthroline) with the DNA-binding ability of cisplatin [25]. These complexes with the architecture $[(Ph_2phen)_2Ru(BL)PtCl_2]^{2+}$ (BL = dpp or dpq), possess redox, spectroscopic, and photophysical properties that make them potential PDT agents. The thermal-binding abilities of the $[(Ph_2phen)_2Ru(BL)PtCl_2]Cl_2$ complexes were probed with pUC18 plasmid DNA and assayed by gel electrophoresis. When incubated at 37 °C in the dark, both complexes retard migration through the gel with base pair/metal complex (BP/MC) ratios as high as 20 : 1. This result is similar to the activity of cisplatin under the same conditions, suggesting that the binuclear complexes are efficient DNA-binding agents [25].

Interestingly, these new complexes possess a previously unprecedented reactivity for Ru,Pt complexes: DNA photobinding. The DNA photocleavage of circular pUC18 DNA imparted by $[(Ph_2phen)_2Ru(BL)PtCl_2]Cl_2$ was assayed by gel electrophoresis studies in which a solution of 20 : 1 BP/MC was photolyzed at 455 nm. In the presence of $O_2$, the supercoiled (SC) DNA (Form I) was converted to open circular (OC, Form II) for BL = dpp, evidenced by slowed migration. When BL = dpq, a band corresponding to double-strand cleavage to produce the linear form is observed.

The addition of $^1O_2$ quencher, sodium azide ($NaN_3$), resulted in most of the plasmid remaining in the SC form, suggesting that $^1O_2$ caused the DNA cleavage converting SC DNA to relaxed, or cleaved DNA. In addition, when the solutions were degassed with $N_2$ prior to addition of the compounds, the photocleavage was inhibited [25].

The Ru(II)—Pt(II) binuclear complex reported by Sakai and coworkers was designed to couple a PS to a *cis*-$PtCl_2$ BAS through aliphatic amines and afforded the complex: $[Ru(bpy)_2\{\mu\text{-}bpy(CONH(CH_2)_3NH_2)_2\}PtCl_2]^{2+}$ [26]. The

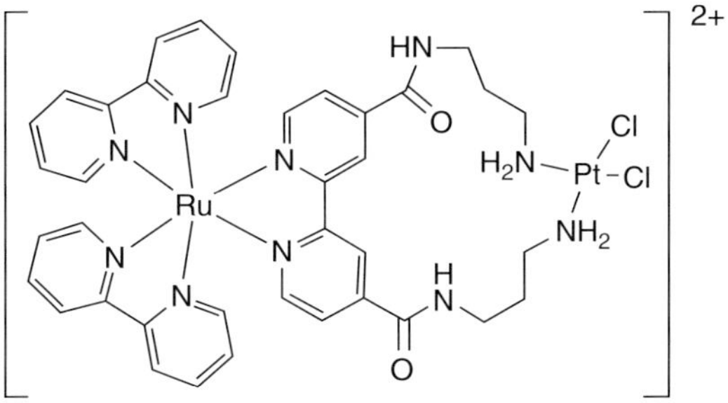


**Figure 7.4** Structural representation of $[Ru(bpy)_2\{\mu\text{-}bpy(CONH(CH_2)_3NH_2)_2\}PtCl_2]^{2+}$. (Reprinted with permission [26a].)

structural representation of the complex is shown in Figure 7.4. The utilization of aliphatic amines rather than α-diimines more closely mimics the coordination environments of cisplatin. In addition, N heterocycles typically have an impact on the spectroscopic properties of the PS once the Pt(II) moiety is coordinated, so aliphatic amies were utilized to maintain the photophysical properties of the PS Ru(II) moiety. The electronic absorption spectroscopy of the mononuclear $[(bpy)_2Ru\{\mu\text{-}bpy(CONH(CH_2)_3NH_2)_2\}]^{2+}$ and the platinated binuclear are quite similar as a result of the intervening aliphatic linkers between the chromophore and the Pt unit [26b]. The enhanced emission at ≈660 nm and the longer lived excited state observed upon Pt(II) coordination (244 ns for the mononuclear and 518 ns for the binuclear) is the result of the formation of a rigid metallcycle that decreases the nonradiative decay processes that are promated when the aliphatic ligands are fixable in the mononuclear complex. The complex showed efficient cleavage of pBR322 DNA to nicked and linearized forms only following irradiation ($\lambda_{irr} = 470 \pm 10$ nm) under atmospheric conditions [26a]. Dark incubation at 5 : 1, 10 : 1, 20 : 1, and 50 : 1 BP/MC ratios without photolysis resulted in no cleavage, indicating the necessity of the light source. Photoinduced electron transfer to cleave DNA is suggested as the mechanism; however, no $O_2$ free studies have been reported to support this process.

The tetranuclear supramolecule, $[\{(bpy)_2Ru(dpp)\}_2Ru(dpp)PtCl_2]^{6+}$, features three ruthenium-centered PS coupled to a *cis*-$PtCl_2$ unit through a dpp BL on the central Ru (Figure 7.5) [27]. The movement of the $Ru^{II}$ PSs, which possess the lowest lying $^3$MLCT state, away from the $Pt^{II}$ BAS provides for much longer lived excited states to enhance photoreactivity. DNA binding and photocleavage experiments using $[\{(bpy)_2Ru(dpp)\}_2Ru(dpp)PtCl_2]^{6+}$ were performed in comparison to the known DNA binder cisplatin, as well as to the trinuclear $[\{(bpy)_2Ru(dpp)\}_2Ru(dpp)]^{6+}$, to observe the impact of the *cis*-$PtCl_2$ unit. The trinuclear complex $[\{(bpy)_2Ru(dpp)\}_2Ru(dpp)]^{6+}$ does not exhibit thermal binding, consistent with the lack of BAS [27].

In the presence of $O_2$ and visible light, a significant band corresponding to cleaved DNA is observed, indicating that the complex is able to produce $^1O_2$; however, SC (Form I) DNA persists, potentially due to the inability of the PS to remain in close enough proximity for efficient photocleavage. Thermal binding is observed for the tetrauclear complex $[\{(bpy)_2Ru(dpp)\}_2Ru(dpp)PtCl_2]^{6+}$, and the retardation of migration compared to cisplatin is consistent with the larger size

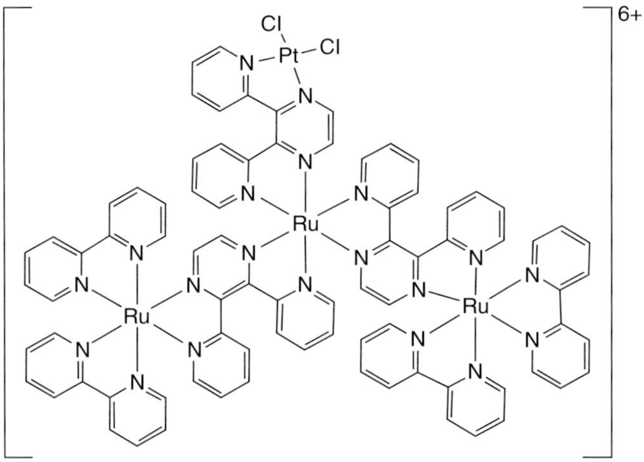


**Figure 7.5** Structural representation of $[\{(bpy)_2Ru(dpp)\}_2Ru(dpp)PtCl_2]^{6+}$.

and higher positive charge. Exposure to $O_2$ upon photolysis results in efficient conversion of SC (Form I) DNA to OC (Form II), a result of the $Pt^{II}$ BAS covalently bound to prelocalize the PS close to the DNA for efficient photocleavage. In the absence of molecular oxygen, photocleavage is not observed, suggesting that an $O_2$-dependent mechanism is operative. This multifunctional DNA modification agent is the first of its architecture to be reported [27].

#### 7.4.1.2 Binuclear and Trinuclear Complexes with Ru, Pt with Tridentate Ligands

Tridentate TLs and BLs are of interest in supramolecular complexes for DNA modification as they limit the number of optical and/or geometric isomers ($\Lambda$ and $\Delta$ stereoisomers), providing the potential for additional methods of analysis. Isolating one isomer of a complex can also give more detailed information about how the complex interacts with DNA.

Two Ru(II), Pt(II) binuclear complexes featuring a tridentate tpy (where tpy = 2,2′:6′,2″-terpyridine) TL, a monodentate $PEt_2Ph$ TL, and a bidentate dpp or bpm BL,were designed to limit the number of stereoisomers with the tpy and bpm ligands and provide a $^{31}P$ NMR tag for structure and reactivity studies by incorporation of the phosphine ligand [28]. Photocleavage studies were not reported for $[(tpy)Ru(PEt_2Ph)(BL)PtCl_2]^{2+}$; however, thermal-binding studies were reported and compared to cisplatin [28b]. A concentration-dependent study was performed in which solutions of metal complex and pBluescript DNA (5 : 1, 10 : 1, 20 : 1, 100 : 1, 200 : 1, and 300 : 1 BP/MC ratios) were incubated at 37 °C in the dark for 4 h. Binding with cisplatin becomes apparent at a lowest relative metal complex concentration of 20 : 1 BP/MC, while the $[(tpy)Ru(PEt_2Ph)(BL)PtCl_2]^{2+}$ complexes appear to be more avid DNA binders with binding apparent at a low concentration of 100 : 1 BP/MC. Slightly enhanced thermal binding is observed with BL = dpp compared to the bpm analogue [28b]. The efficient binding through the Pt(II) BAS to bring the PS in close proximity to the DNA target makes these complexes interesting potential PDT agents.

A series of tpy-terminated Ru(II)—Pt(II) binuclear complexes were designed to study the effect of BL on bioactivity [29]. These complexes were studied to demonstrate DNA binding at the Pt BAS. The complex $[(tpy)RuCl(dpp)PtCl_2]^+$ was also reported to inhibit *in vivo* growth of *E. coli* as a direct result of *cis*-$PtCl_2$ coordination, as the $[(tpy)RuCl(dpp)]^+$ monomer exhibits no antibacterial properties [30]. The binuclear complex $[(tpy)RuCl(dpp)PtCl_2]^+$ more avidly binds to pBluescript DNA than it does to cisplatin. Concentration studies comparing $[(tpy)RuCl(dpp)PtCl_2]^+$ and cisplatin showed complete *E. coli* growth inhibition upon treatment with 0.2, 0.4, and 0.6 mM cisplatin, whereas $[(tpy)RuCl(dpp)PtCl_2]^+$ required 0.4 and 0.6 mM for observable growth inhibition. This *in vivo* study is unusual for supramolecular complexes, as, typically, only DNA gel shift assays are performed.

### 7.4.2 Photosensitizers with a Ru(II) Metal Center Coupled to Rh(III) Bioactive Sites

While *cis*-$Rh^{III}Cl_2$ BASs are structurally similar to *cis*-$Pt^{II}Cl_2$ BAS, $Rh^{III}$—Cl bonds are not labile at room temperature like $Pt^{II}$—Cl. This result provides a unique forum for the development of light-activated halide labilization in Rh-based systems. In addition, the presence of the $Ru^{II}$-(BL) PSs provides for low-energy excitation of these couples' Rh BASs. Also unique to this Ru, Rh forum is the ability to display $O_2$-independent light-induced reactivity with biomolecules, that is, DNA.

#### 7.4.2.1 Trinuclear Complexes with Ru(II), Rh(III), and Ru(II) Metal Centers

A supramolecular architecture that couples two $Ru^{II}$ PSs to a *cis*-$Rh^{III}Cl_2$ BAS was studied through component variation to understand the effects of the TL and BL variation on the orbital energetics and DNA reactivity of this motif [31]. The three complexes, $[\{(bpy)_2Ru(dpp)\}RhCl_2]^{5+}$, $[\{(bpy)_2Ru(bpm)\}_2RhCl_2]^{5+}$, and $[\{(tpy)RuCl(dpp)\}_2RhCl_2]^{3+}$ are shown in Figure 7.6.

Unlike most complexes that utilize Ru-MLCT PSs, the Rh-centered systems display oxygen-independent photoreactions with DNA allowing for their application in treating hypoxic tumors. Photocleavage studies with the three Rh-centered trinuclear complexes were performed to monitor the impact of TL and BL variation on the ability to photomodify pUC18 and pBluescript DNA when excited with irradiation ($\lambda_{irr} = 485$ nm) in the absence of molecular oxygen [31a]. The systems incorporating a dpp BL were shown to photocleave DNA via an oxygen-independent mechanism. When photolyzed in the presence of DNA plasmid, $[\{(bpy)_2Ru(bpm)\}_2RhCl_2]^{5+}$ does not cause cleavage. This result is rationalized by the inaccessible Rh($d\sigma^*$) orbitals in this motif; photoexcitation results in electron localization on the bpm $\pi^*$ orbitals. The gel electrophoresis images from photocleavage studies of $[\{(bpy)_2Ru(dpp)\}_2RhCl_2]^{5+}$ and $[\{(tpy)RuCl(dpp)\}_2RhCl_2]^{3+}$ are shown in Figure 7.6. Both complexes exhibit rapid photocleavage of both pUC18 and pBluescript DNA plasmids in the absence of $O_2$ when irradiated with wavelengths longer than 475 nm. Photocleavage results from reactivity of DNA with the newly generated Rh(II)

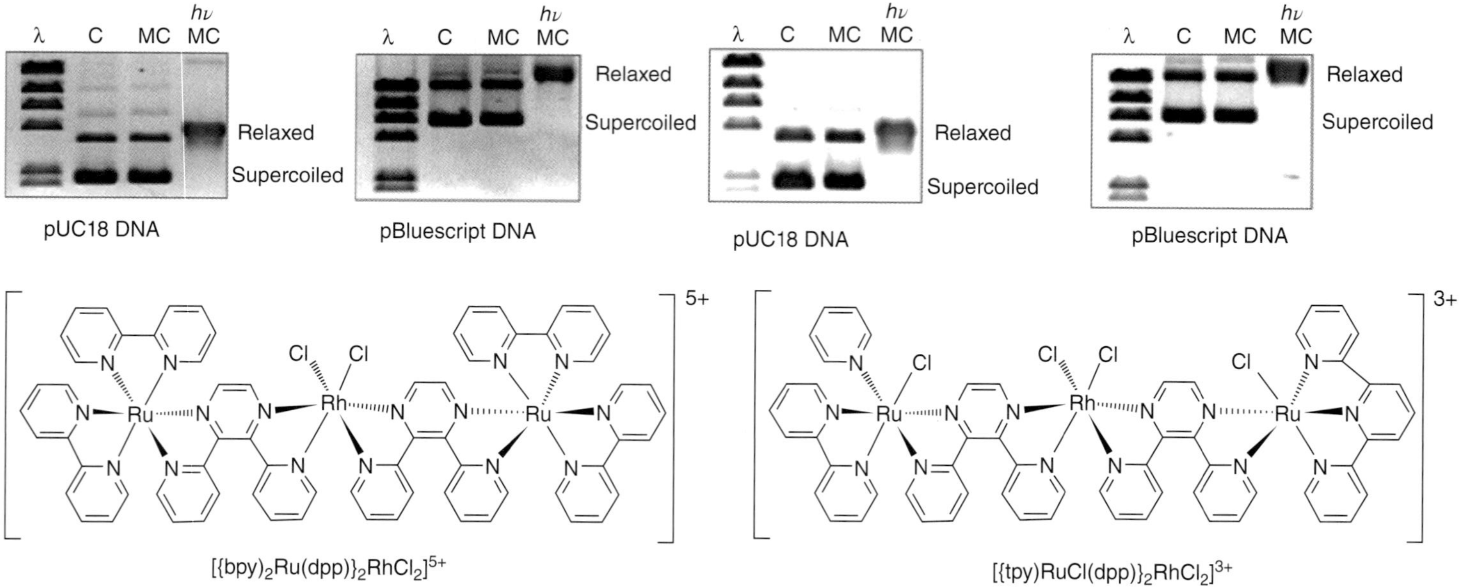


**Figure 7.6** Imaged agarose gel showing the photocleavage of pUC18 and pBluescript plasmid by $[\{(bpy)_2Ru(dpp)\}_2RhCl_2]^{5+}$, and $[\{(tpy)RuCl(dpp)\}_2RhCl_2]^{3+}$ in the absence of molecular oxygen. Lanes $\lambda$ are the $\lambda$ molecular-weight standards, lanes C are the plasmid controls, lanes MC are the dark plasmid controls incubated at 37 °C (2 h) in the presence of the metal complex at a 1 : 5 metal complex:base pair ratio, and lanes $h\nu$, MC are the plasmids irradiated at $\lambda \geq 475$ nm for 20 min in the presence of the metal complex at a 1 : 5 metal complex:base pair ratio. (Reprinted with permission from Holder *et al.* [31a] Copyright 2004 American Chemical Society.)

site upon excitation in the Ru → Rh metal-to-metal charge-transfer (MMCT) photoreactive state. No change in migration is observed when the complexes are incubated with DNA in the absence of light, demonstrating that these complexes likely ionically bind or groove bind DNA prior to excitation and photocleavage, resulting in low dark toxicity.

The trinuclear complex $[\{(bpy)_2Ru(dpp)\}_2RhCl_2]^{5+}$ was reported to inhibit cell growth when Vero cells (African green monkey kidney epithelial cells) were treated with metal complex and photolyzed at $\lambda = 460$ nm with a focused beam for 4 min [31b]. Micrographs of cells treated with $[\{(bpy)_2Ru(dpp)\}_2RhCl_2]^{5+}$ showing cell death where the sample was illuminated are pictured on the left in Figure 7.7. Plots of cell replication inhibition versus metal complex concentration are depicted on the right in Figure 7.7. Cells treated with a metal complex, but not irradiated with visible light, demonstrated normal growth in all concentrations studied [31b]. Upon irradiation, light-induced cell death occurred with the use of $[\{(bpy)_2Ru(dpp)\}_2RhCl_2]^{5+}$. The ability to inhibit cell replication is an important criterion for light-activated anticancer agents. The unique activity of these complexes upon photolysis using an oxygen-independent pathway and lack of dark toxicity makes these systems promising for PDT drug development.

#### 7.4.2.2 Binuclear Complexes with Ru(II) and Rh(III) Metal Centers

Smaller analogues to the Ru–Rh trinuclear complexes, which exhibit better cell-membrane permeability, more efficient binding to DNA, provide a lower cationic charge, reduce the number of stereoisomers, and provide an easily tunable TL on the Rh(III) center, have also been studied and are desirable. A new class of PMDs that couple one Ru(II) PS to a *cis*-$RhCl_2$(TL) BAS was designed [31c, 32]. These new metal systems are designed to display the same $O_2$-independent DNA photocleavage ability while possessing a more sterically accessible Rh site to enhance photoreactivity.

The Ru–Rh dinuclear complexes displayed the $O_2$-independent DNA photocleavage characteristic of the trinuclear complexes, as well as a new DNA photobinding reaction made possible by MMCT labilization of the chloride and a sterically accessible Rh BAS. Photocleavage and photobinding studies were performed with pUC18 DNA and 455-nm excitation of the 5 : 1 BP/MC solutions using $[(bpy)_2Ru(dpp)RhCl_2(phen)]^{3+}$ and $[(bpy)_2Ru(bpm)RhCl_2(phen)]^{3+}$ in the absence of $O_2$ [32]. Binding of the metal complexes to DNA is observed due to the smaller size, leading to a sterically accessible Rh BAS. Conversion of SC DNA to the OC form is observed more rapidly when BL = dpp compared to BL = bpm. Emission studies of $[(bpy)_2Ru(dpp)RhCl_2(phen)]^{3+}$ revealed that the complex emits at 786 nm with a lifetime of 30 ns. No emission was detected for the bpm analogue [32]. The emission is expected to occur at lower energy, which should result in a shorter excited-state lifetime. The slower photocleavage using $[(bpy)_2Ru(bpm)RhCl_2(phen)]^{3+}$ can be attributed to the complex remaining in the excited state for less time relative to the dpp analogue [32]. The activity with DNA does not require molecular oxygen, which is an important property of potential anticancer agents.

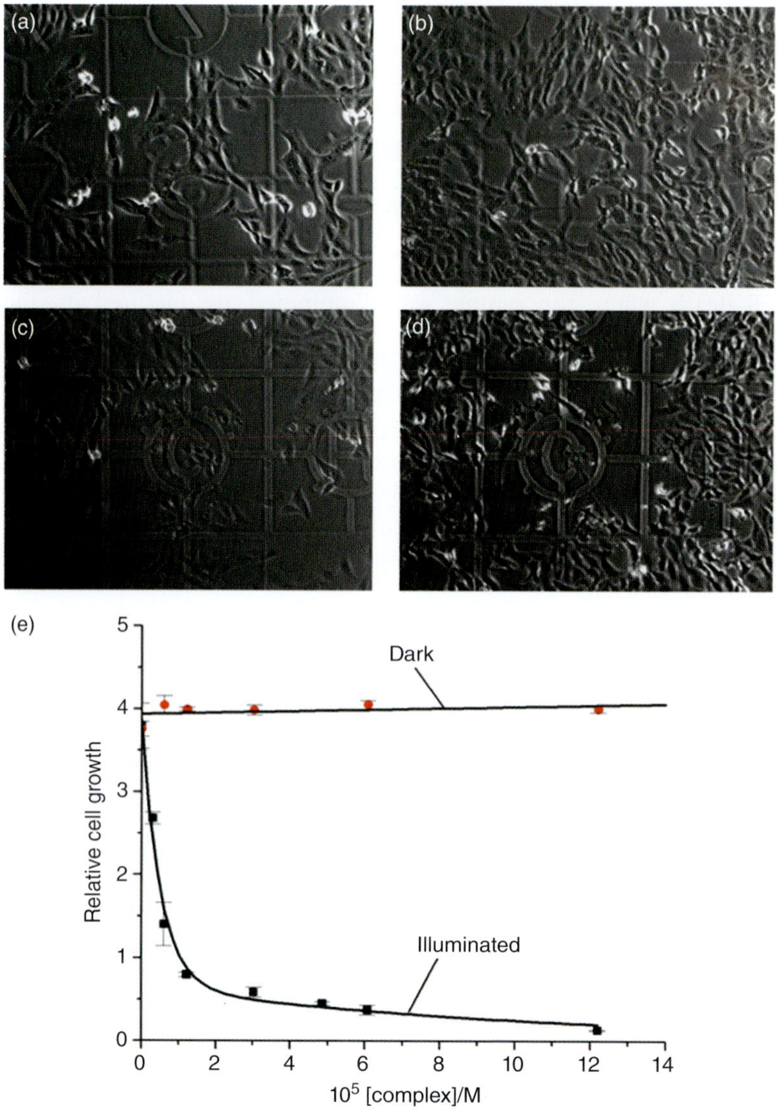


**Figure 7.7** Phase-contrast images of Vero cells after uptake of **8** (a) dark, (b) dark 48 h, (c) preillumination (0 h), (d) postillumination (48 h). (e) Inhibition of Vero cell replication by **8**. (Reprinted with permission from Holder *et al.* 2007 [31b]. Copyright 2007 American Chemical Society.)

### 7.4.3 Photosensitizers with a Ru(II) Metal Cenetr Coupled to Other Bioactive Sites

#### 7.4.3.1 Binuclear Complexes with Ru(II) and Cu

Artifical nuclease-active agents such as Cu(3-Clip-Phen) and also Cu-containing polynuclear coordination compounds have been recently reported [33]. Following these, Reedijk and coworkers reported on a ruthenium-copper terpyridine complex that serves as an efficient DNA cleaver [34]. It is known that Cu(terpy) complexes are capable of cleaving RNA; however [35], there has been no formal report of their ability to cleave DNA. Thus, the ruthenium moiety is employed as a DNA-targeting agent to direct the copper unit to DNA; which subsequently allows the copper unit to cleave DNA. This is the first report of Cu(terpy) complex selectively cleaving DNA. Cleavage activity was found to increase with an increase in the number of copper-active sites. Large fractions of linearized DNA were observed at micromolar concentrations for a complex containing two ruthenium and two copper centers most likely owing to the enhanced targeting ability associated with multiple Ru(II) metal centers. Polyacrylamide gel electrophoresis (PAGE) experiments indicate that DNA cleavage is nonselective [34]. It was speculated that the copper-bound oxidant is responsible for the DNA-cleavage activity, and the antitumor activity of these novel class of compounds is currently under investigation.

#### 7.4.3.2 Binuclear Complexes with Ru(II) and Co(III) Metal Centers

Cobalt(III) complexes have not received as much attention as other systems. However, it does possess some of the same interesting characteristics of metallointercalation and DNA-cleavage properties of $Ru^{II}$ complexes [36]. In addition, being a bioessential transition metal ion, complexes incorporating $Co^{III}$ may find better application at the cellular level in comparison to 4d and 5d metal analogues. Liang *et al.* [37] reported biological application of $[(phen)_2Ru(bpibH_2)Co(phen)_2](ClO_4)_5 \cdot 2H_2O$ (where bpibH2 = 1,4-bis([1,10]phebanthroline-[5,6-*d*]imidazol-2-yl)-benzene}), which includes $Ru^{II}—Co^{III}$ metal centers. The complex shows a higher binding affinity for RNA as opposed to DNA, with both interactions being of an intercalative manner. However, it was noted that this system does not show cytotoxic behavior comparable to cisplatin in either HL-60 or HepG2 cells. However, information from this study can enhance the understanding of the mechanism for binding of metal polypyridyl complexes to nucleic acid [37].

#### 7.4.3.3 Binuclear Complexes with Ru (II) and V(IV) Metal Centers

The bioactivity of vanadium complexes has long been reported since the first report by Lyonnet and Martz Martin in 1899 [38]. Since this time, vanadium-containing compounds have found extensive application in PDT, with many reports appearing on the photocleaving ability of these complexes [39]. Following a report by Sasmal and coworkers [21a], Holder *et al.* [40] reported the synthesis, characterization, and biological properties of two supramolecular $Ru^{II}–V^{IV}$-containing complexes: $[Ru(pbt)_2(tpphz)VO(sal\text{-}L\text{-}tryp)](PF_6)_2 \cdot 6H_2O$ and $[Ru(pbt)_2(phen_2DTT)VO(sal\text{-}L\text{-}tryp)](PF_6)_2 \cdot 5H_2O$ (where pbt = 2-(2′-pyridyl)benzothiazole), tpphz = tetrapyrido[3,2-a:2′,3′-c:3″, 2″-h:2‴,3‴-j]phenazine), sal-L-tryp = *N*-salicylidene-L-tryptophanate, and $phen_2DTT$ = 1,4-bis(1,10-phenanthrolin-5-ylsulfanyl)butane-2,3-diol. This served as the first ever report of the combination of these metal centers for PDT applications. Dark toxicity screenings showed the monomer $[VO(sal\text{-}L\text{-}tryp)(phen)] \cdot H_2O$ to be nonselective as significant activity was observed in both A431 carcinoma cells as well as HFFs (human foreskin fibroblasts) with $IC_{50}$ values comparable to that of cisplatin. In contrast, [Ru(pbt)$_2$(tpphz)VO(sal-L-tryp)]Cl$_2$ was mostly active against A431 carcinoma cells with $IC_{50}$ values comparable to those obtained for cisplatin, while showing significantly less toxicity to noncancerous HFFs. Light toxicity studies were conducted with irradiation of [Ru(pbt)$_2$(tpphz)VO(sal-L-tryp)]Cl$_2$ and [Ru(pbt)$_2$(phen$_2$DTT)VO(sal-L-tryp)]Cl$_2$ at 740 nm and the complexes showed drastic differences in the morphologies of HFF and A431 cells at a concentration of 20 μM, indicating a light-enhanced cytopathic effect in A431 carcinoma cells, as shown in Figure 7.8 [40]. Table 7.1 shows the results of this qualitative study which suggest ruthenium(II)–vanadium(IV) complexes may be potentially good PDT agents.

#### 7.4.3.4 Applications of Ru(II) Metal Centers in Nanomedicine

Hybrid nanoparticles which incorporate multiple functionalities such as luminescence and magnetic properties are paving the way for the growing field of "theranostics." In this field, particles displaying imagining and PDT functions have been recently reviewed [41]. The incorporation of Ru(II) polypyridyl complexes into the inorganic matrix is not uncommon, and often these "hybrid" nanoparticles are employed as luminescent cellular imaging agents [42]. More recently, however, a report by Fransconi and coworkers described a potential theranostic agent obtained from grafting the surface of mesoporous silicon nanoparticles (MSNPs) with a "capping" ruthenium(II) dppz complex [43]. Following irradiation with visible light, the surface-grafted Ru(II) complex is selectively substituted by water, thereby releasing the aqua complex which is then capable of binding to DNA. While unreactive in the dark, the ruthenium complex on the MSNP is activated by light, allowing it to form monoadducts with DNA and act as a DNA-light-switching complex. Spectroscopic and HPLC analysis reveals that when this ruthenium(II) dppz complex is selectively cleaved, the release of an internalized drug molecule, that is, paclitaxel (an anticancer drug), from the porous structure occurs with high release efficiency. Cytotoxic studies against two breast cancer cell lines indicate the loaded functionalized MSNPs showed comparable cytotoxicity when compared with the parent drug molecule following light activation [43]. In a similar study conducted by Lux, Lemercier, and coworkers, gadolinium-based nanoparticles with Ru(II) complexes covalently grafted on the inorganic matrix were created. These hybrid nanoparticles were created to utilize the magnetic resonance imaging (MRI) properties to detect the nanoparticle in disease areas in order to determine the optimal moment to perform irradiation for PDT [44]. Photocytotoxicity studies in HEK293 cells show that the surfaces modified were able to induce cytotoxicity, and this toxicity was enhanced over the free Ru complexes [44]. These examples show the applications of Ru(II) PDT agents in nanomedicine.

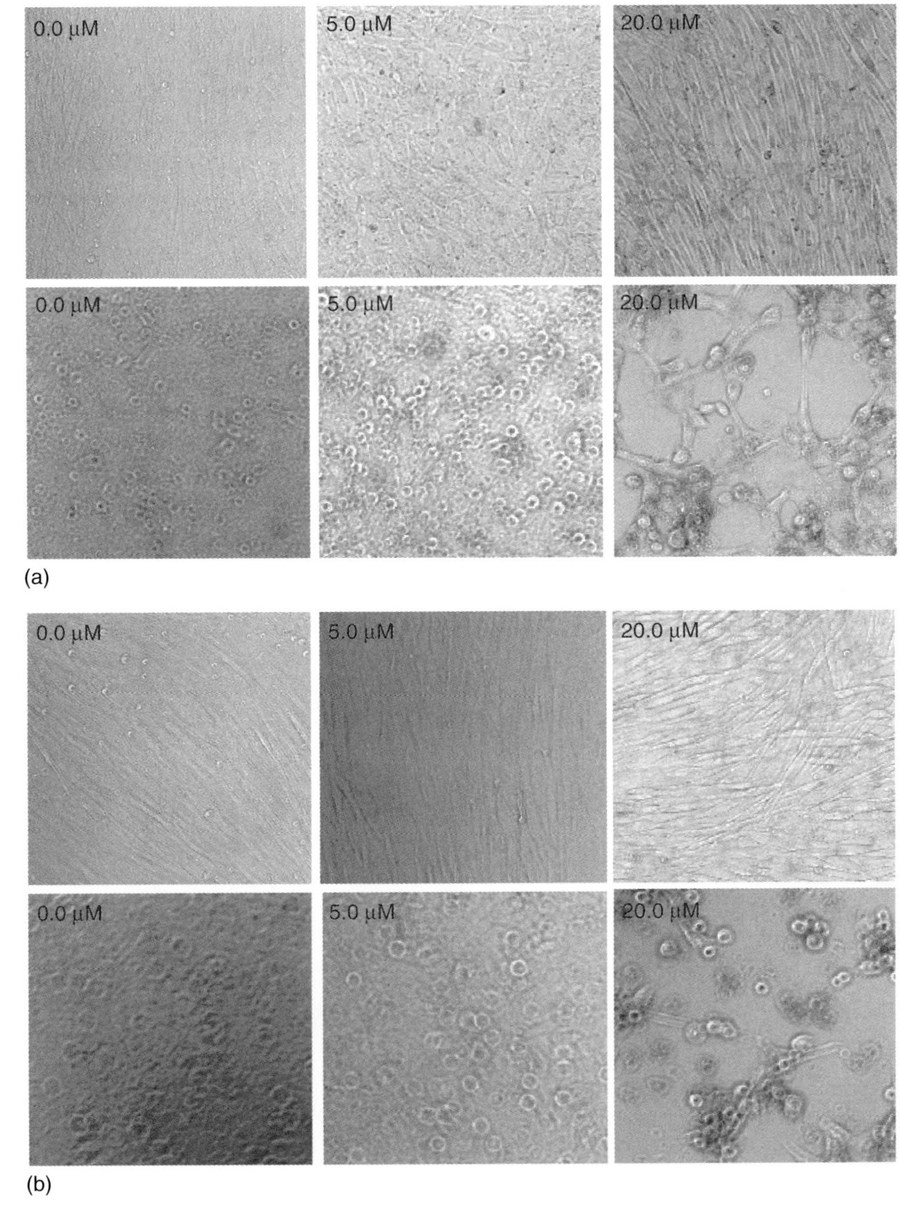


**Figure 7.8** Phase-contrast image of HFF (upper panel) and A431 cells (lower panel). (a) (dark) and (b) (light) toxicity with $[Ru(pbt)_2(tpphz)VO(sal\text{-}L\text{-}tryp)]Cl_2$. (c) (dark) and (d) (light) toxicity with $[Ru(pbt)_2(phen_2DTT)VO(sal\text{-}L\text{-}tryp)]Cl_2$. Magnification = 10×. (Reprinted with permission [40].)

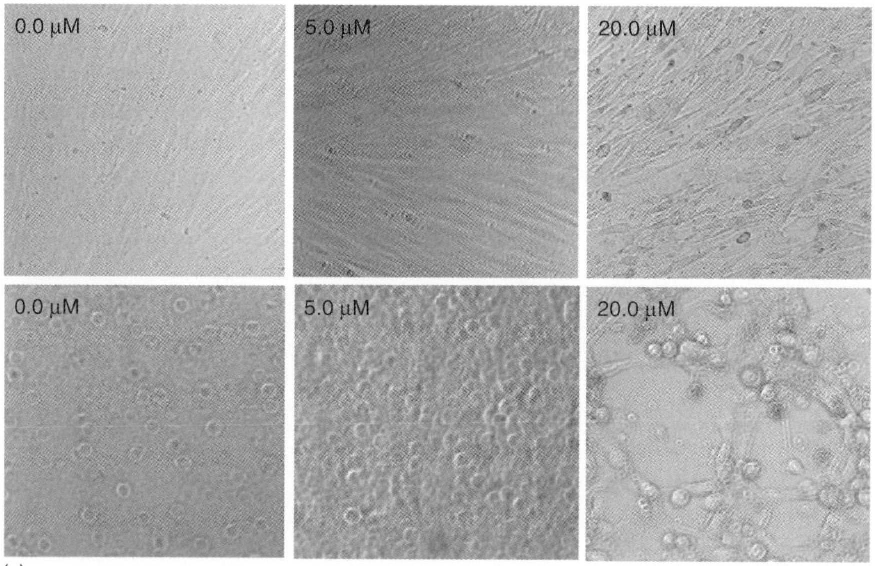


(c)

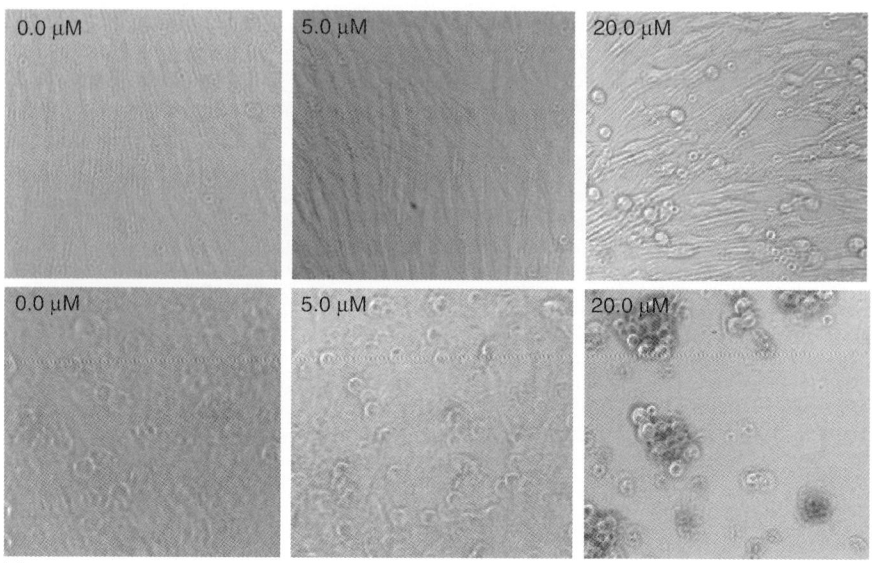


(d)

**Figure 7.8** (*Continued*)

**Table 7.1** Anti-proliferative data obtained for respective complexes in the presence of respective cell lines.

| Species | $IC_{50}$ (μM) | |
|---|---|---|
| | A431 | HFF |
| $[Ru(pbt)_2(tpphz)VO(sal\text{-}L\text{-}tryp)]Cl_2$ | $41.3 \pm 7.6$ | $100.7 \pm 17.7$ |
| $[Ru(pbt)_2(phen_2DTT)VO(sal\text{-}L\text{-}tryp)]Cl_2$ | $48.6 \pm 13.1$ | $204.4 \pm 45.1$ |
| $[VO(sal\text{-}L\text{-}tryp)(phen)] \cdot H_2O$ | $41.6 \pm 5.8$ | $63.1 \pm 28.3$ |
| Cisplatin | $40.1 \pm 11.5$ | $82.0 \pm 8.9$ |

Data expresses as an $IC_{50}$ value in μM.
Source: Reprinted with permission [40].

## 7.5 Summary and Conclusions

Supramolecular complexes, which combine the characteristic traits of individual building blocks into larger assemblies that perform a complex function through contributions of each building block, offer the ability to couple multifunctional units into one molecule providing a type of targeting not possible otherwise and combination-type therapy in one molecular architecture. Early approaches couple one or more $^1O_2$-generating PDT agents or PS to a covalent DNA binder through a BL to target ROS generation to the DNA biomolecule. Platinum and rhodium complexes are reported to thermally or photochemically bind to DNA, making them attractive to incorporate into supramolecular complexes for light-activated DNA modification. Properties of supramolecular complexes can be controlled by subunit variation. Subunits (e.g., TLs, BLs, and BASs) play a specific and vital role in the functioning of supramolecular complexes for DNA modification and tuning excited-state reactions in these new classes of potential PDT drugs.

Selecting the PS metal center is important in the photophysical properties of PDT agents. The Ru(II)–polypyridyl PSs are typically used as they provide rich UV and visible light absorption, relatively long-lived excited states, and photostability. Their $^3$MLCT excited states undergo energy transfer in the presence of molecular oxygen to form ROSs that are potent in cleavage of the DNA phosphate backbone. The selection of TL and BL in these supramolecular architectures impacts properties, such as light absorption, photophysics, and lipophilicity of the metal complex. A new motif for metal complex binding to DNA reported with the binuclear complex $[(Ph_2phen)_2Ru(dpp)PtCl_2]^{2+}$ provides interesting insight into the interactions of supramolecular complexes with biomolecules.

The ability to photocleave DNA in the absence of $O_2$ is a critical goal in developing successful anticancer drugs as the $O_2$ concentration is low in aggressive, hypoxic tumors. The Rh(III)-centered binuclear and trinuclear complexes have been reported to promote DNA photocleavage in deoxygenated conditions. Ruthenium-centered supramolecular complexes offer an exciting and promising stride toward potent anticancer drugs.

While $Pt^{II}$ and $Rh^{III}$ supramolecular complexes continue to dominate the literature regarding mixed-metal complexes for PDT activity, more recent reports of other bioessential elements have appeared including Cu, Co, and V. It is of note that many reports also exist currently where the $Ru^{II}$-PS unit is utilized specifically for cellular localization studies which represent a large area of research. Finally, Ru-PSs are finding applications in nanomedicine as capping units for nanoparticles of various core design.

The ability to tune the properties of supramolecular complexes to afford a molecular device that can target biomolecules to prelocalize a PS agent close to DNA, efficiently absorb low-energy visible light that can pass through the skin, cleave DNA through formation of ROS or through an $O_2$-independent pathway, and exhibit very low dark toxicity to healthy cells, while selectively killing cancer cells when excited, is a forum that holds much potential. Much progress has recently be reported in this new field. Ardent research and dedication in the field must continue in order to build and improve upon the successes of currently utilized drugs for disease treatment including cancer and to develop PDT agents and realize the potential of these supramolecular light-activated anticancer drugs.

## Abbreviations

| | |
|---|---|
| BAS | bioactive site |
| BL | bridiging ligand |
| $bpibH_2$ | 1,4-bis([1,10]phebanthroline-[5,6-*d*]imidazol-2-yl)-benzene |
| bpm | 2,2′-bipyrimidine |
| bpy | 2,2′-bipyridine |
| DNA | deoxyribonucleic acid |
| dpp | 2,3-bis(2-pyridyl)pyrazine |
| dppz | dipyrido[3,2-a:2′,3′-*c*]phenazine |
| dpq | 2,3-bis(2-pyridyl) benzoquinoxaline |
| dpq | dipyrido[3,2-f:2′,3′-h]-quinoxaline |
| HAT | 1,4,5,8,9,12-hexaazatriphenylene |
| $mbpibH_2$ | 1,3-bis([1,10]phenanthroline[5,6-*d*]imidazol-2-yl)benzene |
| $Me_2bpy$ | 4,4′-dimethyl-2,2′-bipyridine |
| $Me_4phen$ | 3,4,7,8-tetramethyl-1,10-phenanthroline |
| MLCT | metal-to-ligand charge transfer |
| MSNP | mesoporous silicon nanoparticle |
| OC | open circular |
| pbt | 2-(2′-pyridyl)benzothiazole) |
| PDT | photodynamic therapy |
| $Ph_2phen$ | 4,7-diphenyl-1,10-phenanthroline |
| phen | 1,10-phenanthroline |
| $phen_2DTT$ | 1,4-bis(1,10-phenanthrolin-5-ylsulfanyl)butane-2,3-diol |
| PMD | photochemical molecular device |
| PS | photosensitizer |

| | |
|---|---|
| ROS | reactive oxygen species |
| sal-L-tryp | *N*-salicylidene-L-tryptophanate |
| SC | supercoiled |
| SOS | 2-mercaptoethyl ether |
| TL | terminal ligand |
| tpphz | tetrapyrido[3,2-a:2′,3′-c:3″,2″-h:2‴,3‴-j]phenazine) |
| tpy | 2,2′:6′,2″-terpyridine |

## References


1 (a) Clarke, M.J. (2002) *Coord. Chem. Rev.*, **232**, 69–93; (b) Clarke, M.J. (2003) *Coord. Chem. Rev.*, **236**, 209–233.

2 (a) Srivastava, S.C., Richards, P., Som, P., Meinken, G., Atkins, H.L., Sewatkar, A., and Ku, T.H. (1980) in *Frontiers in Nuclear Medicine* (eds W. Horst, H. Wagner Jr.,, and J. Buchanan), Springer, Berlin Heidelberg, pp. 123–133; (b) Antonarakis, E. and Emadi, A. (2010) *Cancer Chemother. Pharmacol.*, **66**, 1–9.

3 (a) Gao, F., Chao, H., and Ji, L.-N. (2008) *Chem. Biodivers.*, **5**, 1962–1979; (b) Neidle, S. (2012) *Nat. Chem.*, **4**, 594–595; (c) Pizarro, A.M. and Sadler, P.J. (2009) *Biochimie*, **91**, 1198–1211; (d) Smith, J.A., Collins, J.G., and Keene, F.R. (2009) Groove-Binding Ruthenium (II) Complexes as Probes of DNA Recognition, in *Metal Complex–DNA Interactions*, John Wiley & Sons Ltd., pp. 319–346.

4 (a) Lincoln, P. and Norden, B. (1996) *Chem. Commun.*, 2145–2146; (b) Zou, X.-H., Ye, B.-H., Li, H., Liu, J.-G., Xiong, Y., and Ji, L.-N. (1999) *J. Chem. Soc. Dalton Trans.*, 1423–1428; (c) O'Reill, F.M. and Kelly, J.M. (2000) *J. Phys. Chem. B*, **104**, 7206–7213; (d) Önfelt, B., Lincoln, P., and Nordén, B. (2001) *J. Am. Chem. Soc.*, **123**, 3630–3637; (e) Brodkorb, A., Kirsch-De Mesmaeker, A., Rutherford, T.J., and Keene, F.R. (2001) *Eur. J. Inorg. Chem.*, **8**, 2151–2160; (f) Smith, J.A., Collins, J.G., Patterson, B.T., and Keene, F.R. (2004) *Dalton Trans.*, 1277–1283; (g) Aldrich-Wright, J., Brodie, C., Glazer, E.C., Luedtke, N.W., Elson-Schwab, L., and Tor, Y. (2004) *Chem. Commun.*, 1018–1019.

5 Chao, H., Yuan, Y.-X., Zhou, F., Ji, L.-N., and Zhang, J. (2006) *Transition Met. Chem.*, **31**, 465–469.

6 Rosenberg, B., Van Camp, L., and Krigas, T. (1965) *Nature*, **205**, 698–699.

7 (a) Rosenberg, B. (1973) *Naturwissenschaften*, **60**, 399–406; (b) Rosenberg, B. (1978) *Interdiscip. Sci. Rev.*, **3**, 134; (c) Rosenberg, B., Vancamp, L., Trosko, J.E., and Mansour, V.H. (1969) *Nature*, **222**, 385–386.

8 (a) Nakamura, T. and Miki, T. (2010) *Int. J. Urol.*, **17**, 148–157; (b) Wang, D. and Lippard, S.J. (2005) *Nat. Rev. Drug Discov.*, **4**, 307–320; (c) Cohen, S.M. and Lippard, S.J. (2001) *Prog. Nucleic Acid Res. Mol. Biol.*, **67**, 93–130; (d) Siddik, Z.H. (2003) *Oncogene*, **22**, 7265–7279.

9 (a) Zwelling, L.A. and Kohn, K.W. (1979) *Cancer Treat. Rep.*, **63**, 1439–1444; (b) Reedijk, J. and Lohman, P.H.M. (1985) *Pharm. Weekbl.*, **7**, 173–180; (c) Eastman, A. (2006) The Mechanism of Action of Cisplatin: From Adducts to Apoptosis, in *Cisplatin*, Verlag Helvetica Chimica Acta, pp. 111–134; (d)

Lu, Q.B., Kalantari, S., and Wang, C.R. (2007) *Mol. Pharm.*, **4**, 624–628; (e) Zamble, D.B. and Lippard, S.J. (1995) *Trends Biochem. Sci.*, **20**, 435–439.
10 Alderden, R.A., Hall, M.D., and Hambley, T.W. (2006) *J. Chem. Educ.*, **83**, 728.
11 Richards, A.D. and Rodger, A. (2007) *Chem. Soc. Rev.*, **36**, 471–483.
12 (a) Pabla, N. and Dong, Z. (2008) *Kidney Int.*, **73**, 994–1007; (b) Sullivan, M.J. (2009) *Cancer*, **115**, 5623–5626.
13 Monneret, C. (2011) *Ann. Pharm. Fr.*, **69**, 286–295.
14 Mahnken, R.E., Billadeau, M.A., Nikonowicz, E.P., and Morrison, H. (1992) *J. Am. Chem. Soc.*, **114**, 9253–9265.
15 Muir, M.M. and Huang, W.-L. (1973) *Inorg. Chem.*, **12**, 1831–1835.
16 Crosby, G.A. and Demas, J.N. (1970) *J. Am. Chem. Soc.*, **92**, 7262–7270.
17 (a) Harmon, H.L. and Morrison, H. (1995) *Inorg. Chem.*, **34**, 4937–4938; (b) Billadeau, M.A., Wood, K.V., and Morrison, H. (1994) *Inorg. Chem.*, **33**, 5780–5784.
18 Loganathan, D. and Morrison, H. (2006) *Photochem. Photobiol.*, **82**, 237–247.
19 Verhoeven, J.W. (1988) *Recl. Trav. Chim. Pays-Bas*, **107**, 399.
20 (a) Bertrand, N., Gauthier, M.A., Bouvet, C., Moreau, P., Petitjean, A., Leroux, J.-C., and Leblond, J. (2011) *J. Controlled Release*, **155**, 200–210; (b) Oleksi, A., Blanco, A.G., Boer, R., Usón, I., Aymamí, J., Rodger, A., Hannon, M.J., and Coll, M. (2006) *Angew. Chem.*, **118**, 1249–1253; (c) Rabut, G., Doyle, V., and Ellenberg, J. (2004) *Nat. Cell Biol.*, **6**, 1114–1121; (d) Yu, H., Wang, X., Fu, M., Ren, J., and Qu, X. (2008) *Nucleic Acids Res.*, **36**, 5695–5703; (e) Uerpmann, C., Malina, J., Pascu, M., Clarkson, G.J., Moreno, V., Rodger, A., Grandas, A., and Hannon, M.J. (2005) *Chem. Eur. J.*, **11**, 1750–1756.
21 (a) Sasmal, P.K., Patra, A.K., Nethaji, M., and Chakravarty, A.R. (2007) *Inorg. Chem.*, **46**, 11112–11121; (b) Kolárová, H., Mosinger, J., Lenobel, R., Kejlová, K., Jírová, D., and Strnad, M. (2003) *Toxicol. In Vitro*, **17**, 775–778; (c) Nishiyama, N., Jang, W.-D., and Kataoka, K. (2007) *New J. Chem.*, **31**, 1074–1082.
22 Milkevitch, M., Shirley, B.W., and Brewer, K.J. (1997) *Inorg. Chim. Acta*, **264**, 249–256.
23 Swavey, S. and Brewer, K.J. (2003) in *Comprehensive Coordination Chemistry II* (eds J.A. McCleverty and J. Meyer), Pergamon, Oxford, pp. 135–157.
24 (a) Milkevitch, M., Brauns, E., and Brewer, K.J. (1996) *Inorg. Chem.*, **35**, 1737–1739; (b) Milkevitch, M., Storrie, H., Brauns, E., Brewer, K.J., and Shirley, B.W. (1997) *Inorg. Chem.*, **36**, 4534–4538.
25 Higgins, S.L.H., White, T.A., Winkel, B.S.J., and Brewer, K.J. (2010) *Inorg. Chem.*, **50**, 463–470.
26 (a) Sakai, K., Ozawa, H., Yamada, H., Tsubomura, T., Hara, M., Higuchi, A., and Haga, M.-A. (2006) *Dalton Trans.*, 3300–3305; (b) Sakai, K. and Tsubomura, T. (1997) *J. Inorg. Biochem.*, **67**, 349.
27 Miao, R., Mongelli, M.T., Zigler, D.F., Winkel, B.S.J., and Brewer, K.J. (2006) *Inorg. Chem.*, **45**, 10413–10415.
28 (a) Swavey, S., Fang, Z., and Brewer, K.J. (2002) *Inorg. Chem.*, **41**, 2598–2607; (b) Fang, Z., Swavey, S., Holder, A., Winkel, B., and Brewer, K.J. (2002) *Inorg. Chem. Commun.*, **5**, 1078–1081.

29 Williams, R.L., Toft, H.N., Winkel, B., and Brewer, K.J. (2003) *Inorg. Chem.*, **42**, 4394–4400.
30 Jain, A., Winkel, B.S.J., and Brewer, K.J. (2007) *J. Inorg. Biochem.*, **101**, 1525–1528.
31 (a) Holder, A.A., Swavey, S., and Brewer, K.J. (2004) *Inorg. Chem.*, **43**, 303; (b) Holder, A.A., Zigler, D.F., Tarrago-Trani, M.T., Storrie, B., and Brewer, K.J. (2007) *Inorg. Chem.*, **46**, 4760–4762; (c) Wang, J., Higgins, S.L.H., Winkel, B.S.J., and Brewer, K.J. (2011) *Chem. Commun.*, **47**, 9786–9788.
32 Wang, J., Zigler, D.F., Hurst, N., Othee, H., Winkel, B.S.J., and Brewer, K.J. (2012) *J. Inorg. Biochem.*, **116**, 135–139.
33 (a) Jiang, Q., Xiao, N., Shi, P., Zhu, Y., and Guo, Z. (2007) *Coord. Chem. Rev.*, **251**, 1951–1972; (b) Li, L., Murthy, N.N., Telser, J., Zakharov, L.N., Yap, G.P.A., Rheingold, A.L., Karlin, K.D., and Rokita, S.E. (2006) *Inorg. Chem.*, **45**, 7144–7159; (c) Pitie, M., Sudres, B., and Meunier, B. (1998) *Chem. Commun.*, 2597–2598.
34 van der Steen, S., de Hoog, P., van der Schilden, K., Gamez, P., Pitie, M., Kiss, R., and Reedijk, J. (2010) *Chem. Commun.*, **46**, 3568–3570.
35 Liu, S. and Hamilton, A.D. (1999) *Chem. Commun.*, 587–588.
36 (a) Wang, X.-L., Chao, H., Li, H., Hong, X.-L., Liu, Y.-J., Tan, L.-F., and Ji, L.-N. (2004) *J. Inorg. Biochem.*, **98**, 1143–1150; (b) Sun, Y., Zhou, Q.-X., Chen, J.-R., Hou, Y.-J., Lei, W.-H., Wang, X.-S., and Zhang, B.-W. (2009) *J. Inorg. Biochem.*, **103**, 1658–1665; (c) Sudhamani, C.N., Bhojya Naik, H.S., Girija, D., Sangeetha Gowda, K.R., Giridhar, M., and Arvinda, T. (2014) *Spectrochim. Acta Part A*, **118**, 271–278; (d) Reddy, K.L., Kumar, K.A., and Satyanarayana, S. (2011) *Synth. React. Inorg. Met. Org. Nano-Met. Chem.*, **41**, 182–192; (e) Miao, T.-F., Li, S., Li, J., and Zheng, K.-C. (2012) *J. Inorg. Biochem.*, **109**, 16–25; (f) Kawade, V.A., Kumbhar, A.A., Kumbhar, A.S., Naether, C., Erxleben, A., Sonawane, U.B., and Joshi, R.R. (2011) *Dalton Trans.*, **40**, 639–650; (g) Indumathy, R., Radhika, S., Kanthimathi, M., Weyhermuller, T., and Unni Nair, B. (2007) *J. Inorg. Biochem.*, **101**, 434–443; (h) Arounaguiri, S. and Maiya, B.G. (1996) *Inorg. Chem.*, **35**, 4267–4270; (i) Zhang, Q.-L., Xu, H., Li, H., Liu, J., Liu, J.-Z., Ji, L.-N., and Liu, J.-G. (2002) *Transition Met. Chem.*, **27**, 149–154.
37 Liang, X., Zou, X., Tan, L., and Zhu, W. (2010) *J. Inorg. Biochem.*, **104**, 1259–1266.
38 Lyonnet, B.M. and Martz Martin, E. (1899) *Presse Med.*, **7**, 191.
39 (a) Kwong, D.W.J., Chan, O.Y., Wong, R.N.S., Musser, S.M., Vaca, L., and Chan, S.I. (1997) *Inorg. Chem.*, **36**, 1276–1277; (b) Sasmal, P.K., Saha, S., Majumdar, R., De, S., Dighe, R.R., and Chakravarty, A.R. (2010) *Dalton Trans.*, **39**, 2147–2158; (c) Sasmal, P.K., Saha, S., Majumdar, R., Dighe, R.R., and Chakravarty, A.R. (2010) *Inorg. Chem.*, **49**, 849–859; (d) Guo, H., Lu, J., Ruan, Z., Zhang, Y., Liu, Y., Zang, L., Jiang, J., and Huang, J. (2012) *J. Coord. Chem.*, **65**, 191–204; (e) Balaji, B., Somyajit, K., Banik, B., Nagaraju, G., and Chakravarty, A.R. (2013) *Inorg. Chim. Acta*, **400**, 142–150; (f) Prasad, P., Pant, I., Khan, I., Kondaiah, P., and Chakravarty, A.R. (2014) *Eur. J. Inorg. Chem.*, 2420–2431.

**40** Holder, A.A., Taylor, P., Magnusen, A.R., Moffett, E.T., Meyer, K., Hong, Y., Ramsdale, S.E., Gordon, M., Stubbs, J., Seymour, L.A., Acharya, D., Weber, R.T., Smith, P.F., Dismukes, G.C., Ji, P., Menocal, L., Bai, F., Williams, J.L., Cropek, D.M., and Jarrett, W.L. (2013) *Dalton Trans.*, **42**, 11881–11899.

**41** (a) Couleaud, P., Bechet, D., Vanderesse, R., Barberi-Heyob, M., Faure, A.-C., Roux, S., Tillement, O., Porhel, S., Guillemin, F., and Frochot, C. (2011) *Nanomedicine*, **6**, 995–1009; (b) Brigger, I., Dubernet, C., and Couvreur, P. (2002) *Adv. Drug Deliv. Rev.*, **54**, 631–651; (c) Celli, J.P., Spring, B.Q., Rizvi, I., Evans, C.L., Samkoe, K.S., Verma, S., Pogue, B.W., and Hasan, T. (2010) *Chem. Rev.*, **110**, 2795–2838.

**42** Elmes, R.B.P., Orange, K.N., Cloonan, S.M., Williams, D.C., and Gunnlaugsson, T. (2011) *J. Am. Chem. Soc.*, **133**, 15862–15865.

**43** Frasconi, M., Liu, Z., Lei, J., Wu, Y., Strekalova, E., Malin, D., Ambrogio, M.W., Chen, X., Botros, Y.Y., Cryns, V.L., Sauvage, J.-P., and Stoddart, J.F. (2013) *J. Am. Chem. Soc.*, **135**, 11603–11613.

**44** Truillet, C., Lux, F., Moreau, J., Four, M., Sancey, L., Chevreux, S., Boeuf, G., Perriat, P., Frochot, C., Antoine, R., Dugourd, P., Portefaix, C., Hoeffel, C., Barberi-Heyob, M., Terryn, C., van Gulick, L., Lemercier, G., and Tillement, O. (2013) *Dalton Trans.*, **42**, 12410–12420.

# 8

# Ruthenium Anticancer Agents En Route to the Tumor: From Plasma Protein Binding Agents to Targeted Delivery

*Muhammad Hanif and Christian G. Hartinger*

*School of Chemical Sciences, University of Auckland, Private Bag 92019, Auckland 1142, New Zealand*

## 8.1 Introduction

After the clinical success of platinum anticancer agents, complexes of other metals such as gold, gallium, titanium, iron, rhodium, and ruthenium have been considered for the development of metal-based cancer chemotherapeutics [1]. By replacing platinum with other metal ions, different chemical properties in terms of oxidation state, redox potential, coordination geometry, and ligand exchange kinetics are obtained, which makes it likely that such non-platinum compounds have different mechanisms of action, biodistribution, and toxicities than platinum-based drugs. Therefore, this strategy may provide a means to overcome resistance of tumors to platinum anticancer agents and of their adverse effects observed during chemotherapy.

Among the transition metal complexes studied so far for anticancer activity, ruthenium complexes are the most promising. The $Ru^{III}$ complexes NAMI-A and KP1019/NKP-1339 (Figure 8.1) were developed to clinical trials [2, 3], while RAPTA-C is likely to join them soon [4].

Ruthenium has chemical properties which make it an ideal choice over other metals for the design of therapeutic agents [5, 6]. In addition to the well-developed synthetic coordination chemistry and predictable geometry of ruthenium complexes, they exhibit relatively low ligand exchange rates which depend on the oxidation state of the metal center [7, 8]. Slow ligand exchange is important to ensure that the drug reaches its biological target without being modified. However, very few metallodrugs reach their biological target without undergoing ligand exchange reactions, and some of these interactions are often essential for inducing the desired therapeutic properties while other binding events can cause drug deactivation and detoxification.

Ruthenium is known to have a range of oxidation states accessible under physiological conditions. $Ru^{III}$ complexes are usually more inert than the related $Ru^{II}$ and $Ru^{IV}$ counterparts and the redox potential of a metal complex can be modified by changing the coordinating ligands. Biological reductants such as glutathione, ascorbate, and single-electron transfer proteins are able to reduce $Ru^{III}$

*Ruthenium Complexes: Photochemical and Biomedical Applications*, First Edition.
Edited by Alvin A. Holder, Lothar Lilge, Wesley R. Browne, Mark A.W. Lawrence, and Jimmie L. Bullock Jr.

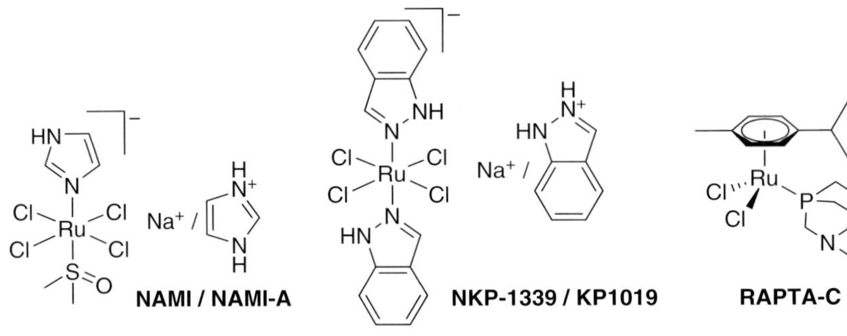


**Figure 8.1** Structures of promising Ru anticancer agents.

to $Ru^{II}$, rendering them reactive to their biological targets ("activation by reduction" hypothesis) [9, 10], while molecular oxygen and cytochrome *c* oxidase can oxidize $Ru^{II}$ in some complexes [11]. The redox potential of ruthenium compounds can be exploited to improve the effectiveness of drugs, as the altered metabolism associated with cancer often results in a lower oxygen concentration and related lower pH values in tumor tissue than in healthy tissue. In addition, elevated levels of glutathione contribute to a reducing environment in these tissues.

A significant number of metal ions are essential in biological processes. Being found in the same group as iron, ruthenium has the ability to mimic iron and may exploit its transport mechanism or replace it in iron-dependent enzymes [12, 13], both processes being of relevance to the modes of action of novel Ru anticancer agents.

One key to success in anticancer drug development is to achieve selective delivery of the pharmacophore to the tumor site. A wide variety of strategies have been developed for anticancer and diagnostic agents which include, among others, the functionalization of the pharmacophore with a vector to target a specific overexpressed receptor and the loading on proteins (see Figure 8.2 for some examples)

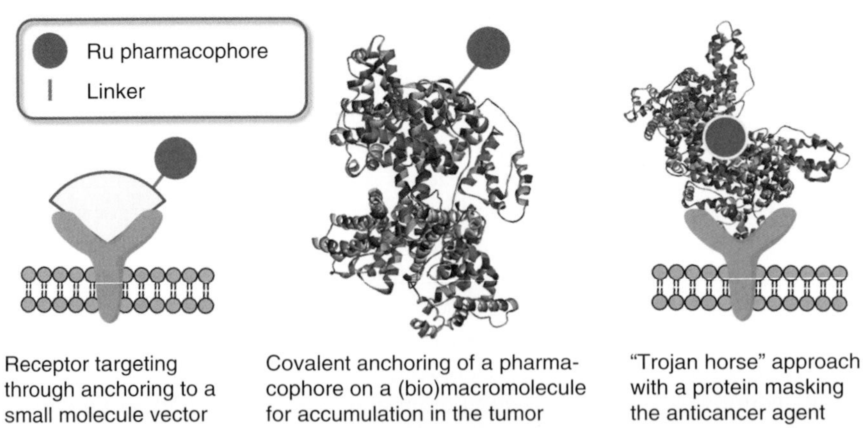


**Figure 8.2** Delivery strategies for Ru anticancer agents to the tumor and into tumor cells.

[14]. These and other delivery strategies have also been employed for Ru-based anticancer agents and a selection is highlighted in this chapter.

## 8.2 Protein Binding Ru$^{III}$ Anticancer Drug Candidates

### 8.2.1 Ru$^{III}$ Anticancer Drug Candidates Targeting Primary Tumors

Since rapidly dividing cancer cells have a greater iron requirement, they overexpress transferrin receptors on their surfaces in order to sequester more of the circulating Fe$^{III}$-loaded transferrin. As mentioned, the chemical properties of Fe and Ru are similar and therefore it is not surprising that at an early stage of development of Ru anticancer agents, interference with the Fe metabolism was suggested as central in the mode of action. This includes the transport and sequestration of Ru into tumor cells mediated by transferrin (Tf) transport and receptor-mediated uptake which has been the design paradigm for a wide variety of Ru$^{III}$ anticancer drugs. The transferrin-receptor–mediated uptake involves the formation of endosomes, which have a lower pH compared to the extracellular space (pH 5.5 vs7.4), and this was suggested to trigger the release of the ruthenium drug inside the cell to exert its pharmacological effects. Such selective transport, in conjunction with the redox activation, may be responsible for the low side effects and the general low toxicity observed for Ru$^{III}$ anticancer agents studied *in vivo* and in clinical trials [15].

It was as early as in 1980 that Clarke reported the anticancer properties of the first ruthenium complexes with the general formula $[Ru(NH_3)_{6-x}Cl_x]^{n+}$ in which several ammine and chlorido ligands were coordinated to either Ru$^{II}$ or Ru$^{III}$ centers. While $[Ru^{II}(NH_3)_5Cl]^+$ and $[Ru^{II}(NH_3)_5(H_2O)]^{2+}$ were not anticancer active, despite forming adducts with DNA in an analogous way to cisplatin, the ruthenium(III) complexes *cis*-$[Ru^{III}(NH_3)_4Cl_2]^+$ and *fac*-$[Ru^{III}(NH_3)_3Cl_3]$ (Figure 8.3)

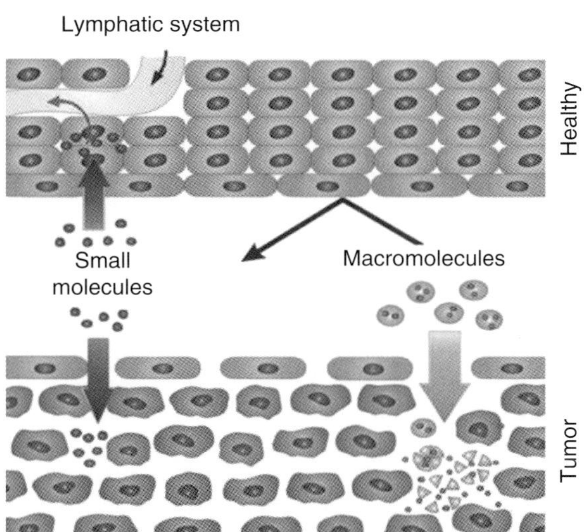


**Figure 8.3** The enhanced permeability and retention effect allows macromolecules to accumulate in tumor tissue. (Adapted from ref. [14] © Wiley 2008.)

exhibited antitumor activity comparable to that of cisplatin in selected cell lines [13]. The *fac*-[$Ru^{III}(NH_3)_3Cl_3$] complex was the most promising among this series of compounds, but further investigations were discontinued due to its poor water solubility [5, 13]. In order to improve the water solubility, anionic complexes with a higher number of halido ligands were designed. Keppler *et al.* developed a series of such complexes and evaluated them for their biological activity. This resulted in the indazole- and imidazole-containing lead compounds indazolium *trans*-[tetrachloridobis(1*H*-indazole)ruthenate(III)] (KP1019) and imidazolium *trans*-[tetrachloridobis(1*H*-imidazole)ruthenate(III)] (KP418). *In vivo* experiments revealed that KP418 showed therapeutic activity against murine P388 leukemia and B16 melanoma [10] as well as caused a significant reduction of tumor burden in rats with autochthonous, chemically induced colorectal cancer [16]. KP1019 showed excellent activity against a rat colon cancer model. In preclinical studies, KP1019 was capable of reducing tumor growth by up to 95% without drug-related mortality and considerable weight loss. In addition, KP1019 was found to have superior activity than 5-fluorouracil, the standard drug used for the treatment of colorectal cancer [17]. Because of poor water solubility of KP1019, the analogous $Ru^{III}$ compound NKP-1339 was prepared with a sodium counterion instead of indazolium. The solubility of NKP-1339 was 35-fold better than that of KP1019 and therefore was recently selected for further clinical development. Interesting effects in combinations with other established drugs were observed and it appears that non-small cell lung cancer may be a target tumor type for this compound [18, 19].

As KP1019 and NKP-1339 are administered intravenously, their reactions in the bloodstream and, in particular, their interactions with serum proteins are of great relevance to their modes of action. Because they consist of the same complex anion, a very similar, if not identical, reaction pattern can be expected. When KP1019 enters the blood stream, it was shown to bind within minutes to the serum proteins Tf and especially human serum albumin (HSA) [20]. Both of these proteins may act as molecular transporters. While Tf may carry the Ru drug into the cancer cell via endocytosis using its receptor in a "Trojan horse"-type approach (see preceding text) [20], HSA accumulates in tumor tissue due to the enhanced permeability and retention (EPR) effect (Figure 8.3). The EPR effect allows macromolecules to penetrate tumors through leaky blood vessels surrounding the tumor and to be retained due to poor lymphatic drainage [21].

The binding ability of the ruthenium(III) drugs to Tf was investigated in several independent studies. Both KP1019 and NKP-1339 reacted with Tf within several minutes in the test tube [22]. Binding studies of KP1019 to apolactoferrin by X-ray crystallography revealed that ruthenium binds to histidine-253, while the two indazole ligands still remained coordinated to the metal center [23]. Furthermore, the exposure of the colon cancer cell line SW480 with KP1019–Fe(III)–transferrin (1 : 0.3 : 1) incubation mixtures resulted in more than 2.5-fold higher ruthenium levels than in studies with KP1019 alone. This indicates that a certain amount of iron loading might be required for optimal Tf-mediated transport, which may be related to the structural modification of the protein to allow for the receptor to detect it [15].

A detailed investigation on serum protein binding of KP1019 by size exclusion chromatography-inductively coupled plasma mass spectrometry (SEC-ICP/MS) under physiologically relevant conditions revealed that Ru was mainly bound to the albumin- and transferrin-containing protein fraction in the 60–80 kDa mass range [24]. It appears that although $Ru^{III}$–Tf adduct formation is kinetically favored, $Ru^{III}$–albumin adducts are thermodynamically more stable [25]. Notably, in competition experiments at physiologically relevant concentrations the total amount of Ru bound to Tf was less than 2%, while the concentration of HSA was about 15 times greater. From these studies, it can be concluded that intravenous administration of such $Ru^{III}$ anticancer agents results in extensive binding to serum proteins with a large majority of the total Ru found at HSA. This may result in passive targeting of the tumor exploiting the EPR effect. At the tumor site, the drug may be released to exert its anticancer effects upon HSA degradation. Alternatively, it has been suggested that HSA may act as a reservoir for Ru and other anticancer agents [21]. The preference of binding to HSA was confirmed by electron paramagnetic resonance spectroscopic studies conducted by Walsby and coworkers. They investigated the binding of KP1019 to HSA, apoTf, and whole human serum. The rapid and noncovalent binding of KP1019 with both HSA and serum showed that initially the Ru drug predominantly binds to HSA through hydrophobic interactions. On prolonged exposure of the drug to serum, the noncovalently bound Ru complexes were converted slowly to coordinatively bound Ru–protein adducts. In addition, Ru complexes maintained in serum their 3+ oxidation state. In contrast, apoTf showed slow binding to Ru exclusively through ligand exchange with the donor atoms of the protein, while noncovalent interactions were not observed [26]. However, the protein binding appears to be related to the low toxic side effects of this compound class and plays a major role in the mode of action. The observations in the test tube were confirmed in the analysis of blood serum samples from cancer patients treated with KP1019 [27].

### 8.2.2 Antimetastatic $Ru^{III}$ Compounds

Contemporaneous to the discovery of KP1019, Sava and coworkers developed the dimethylsulfoxide (DMSO)-containing *cis*- and *trans*-$[Ru^{II}Cl_2(DMSO)_4]$ complexes. In animal tumor models, the *trans* isomer demonstrated a pronounced effect on metastasis in the lung while having a minimal effect on the growth of primary tumors [28]. After surgical removal of the primary tumor in lung-carcinoma–bearing mice, this compound resulted in significant life prolongation [29].

This discovery initiated the development of ruthenium complexes as antimetastatic agents and the $Ru^{III}$ complex Na [*trans*-$Ru^{III}Cl_4$(DMSO)(Him)] (NAMI; Him = 1*H*-imidazole) was selected as a lead compound for further studies [30]. The antimetastatic effects of NAMI were found to be independent of cytotoxicity. Intramuscular administration of the drug compound into mice resulted in a tremendous reduction of both the number and weight of lung metastases, whereas for the bis-DMSO complexes no effect on tumor growth at the site of the primary implant was observed [31]. In order to improve the stability and have access to a more straightforward synthetic route, the sodium counterion of NAMI was replaced with imidazolium to yield $(H_2im)$ [*trans*-$Ru^{III}Cl_4$(DMSO)(Him)] (NAMI-A). NAMI-A was found to have pharmacological properties and efficacy in rodent tumor models equivalent or even superior to NAMI. Preclinical *in vitro* and *in vivo* studies indicated that NAMI-A was able to inhibit the development and growth of pulmonary metastases in several types of metastasizing lung cancer cell lines such as Lewis lung, MCa mammary carcinoma, and TS/A adenocarcinoma. The antimetastatic properties of NAMI-A led to the first clinical trials involving a $Ru^{III}$ complex [3]. As NAMI-A has shown negligible impact on primary tumors, it entered a phase-I/II clinical trial in combination with gemcitabine in patients with non-small cell lung cancer but was found to be less active than gemcitabine alone [32, 33].

Protein binding is generally thought to be detrimental for the anticancer activity of most chemotherapeutics, but its role in the specific anti-metastatic activity of NAMI-A is controversial. Based on animal model studies, Sava and coworkers proposed that the binding of NAMI-A to albumin or transferrin reduced both bioavailability and antimetastatic activity [34]. However, in another study, the activity was increased in the presence of biological reductants such as ascorbate or glutathione, conditions which enhance aquation and hence protein binding of $Ru^{III}$ complexes [35]. Lay and coworkers recently investigated in detail the interaction of $Ru^{III}$ drugs with bovine serum albumin (BSA) by X-ray absorption spectroscopy (XAS) [36]. Based on these investigations, the authors concluded that the anti-metastatic activity of NAMI-A is possibly due to its ability to form adducts with HSA.

## 8.3 Functionalization of Macromolecular Carrier Systems with Ru Anticancer Agents

Macromolecules have been conjugated to cytotoxins through different linker systems to improve the properties of anticancer agents or even allow selective release in tumor tissue. As macromolecules in the size range 10–100 nm [37, 38] can penetrate into tumors (EPR effect) [39], this provides a means of passive targeting that should lead to accumulation of the active Ru unit in the tumor.

### 8.3.1 Proteins as Delivery Vectors for Organometallic Compounds

One highly successful approach investigated is the use of HSA as a delivery vector [21]. HSA is known to accumulate in tumors and has been exploited as the carrier of various anticancer agents such as chlorambucil, doxorubicin, paclitaxel, and platinum compounds [14]. Due to the presence of the highly nucleophilic thiol group of cysteine-34, HSA can be conjugated with cytotoxic agents through a maleimide moiety. Kratz and coworkers investigated this prodrug concept by preparing an acid-sensitive (6-maleimidocaproyl)hydrazone derivative of doxorubicin, that is, INNO-206, which induced rapid and selective binding to the cysteine-34 residue of endogenous albumin. This albumin binding prodrug was developed to clinical trials and showed high plasma stability in its albumin-bound

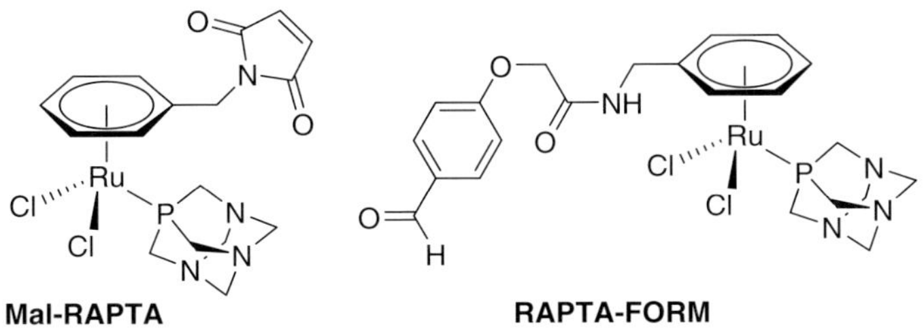


**Figure 8.4** Structures of Ru(arene) compounds designed to utilize HSA as a vector to the tumor.

form, superior efficacy in some murine tumor models, a favorable toxicity profile, and has recently entered phase II [21].

Using HSA as a carrier for organometallic anticancer agents was recently introduced for the RAPTA scaffold. RAPTA compounds, that is, [Ru($\eta^6$-arene)(PTA)$Cl_2$] (PTA = 1,3,5-triaza-7-phosphatricyclodecane; see Figure 8.1 for prototype RAPTA-C with arene = *p*-cymene) have shown antimetastatic activity while exhibiting marginal cytotoxicity in a wide range of cancer cells [40]. Introducing a maleimide functional group led to the development of Mal-RAPTA (Figure 8.4) [41]. The arene moiety was chosen for functionalization, as it was found to have less impact on the overall biological activity than the PTA co-ligand. The reactivity of Mal-RAPTA toward thiol-containing biomolecules, such as L-cysteine, glutathione and HSA, was studied using NMR spectroscopy, electrospray ionization mass spectrometry (ESI-MS), and size exclusion chromatography (SEC) coupled with inductively coupled plasma mass spectrometry (ICP-MS). These experiments demonstrated chemoselective covalent anchoring of Mal-RAPTA to cysteine-34 on HSA without altering significantly the *in vitro* anticancer effect of the RAPTA pharmacophore, as demonstrated in human ovarian (CH1), colon (SW480), and non-small cell lung cancer (A549) cell lines. In aqueous solution, the arene ligand slowly releases the metal center to exhibit the antitumor effect [41].

The concept of introducing maleimide moieties into organometallic anticancer agents has recently been extended to functionalize co-ligands such as the N-donor compounds indazole and pyridine. This yielded in Ru(arene) and Os(arene) organometallics with different mono- and bidentate leaving groups to study the effect of such modification on both the reactivity to thiols and biological activity [42].

In an orthogonal approach, anticancer organometallic compounds were functionalized to conjugate with modified HSA as a carrier protein via acid-labile hydrazone bonds using aldehyde and hydrazine functional groups. This was achieved by derivatizing the RAPTA fragment with an aldehyde to give RAPTA-FORM (Figure 8.4), which can conjugate to recombinant human serum albumin (rHSA) functionalized with hydrazine groups, through formation of a hydrazone bond. *In vitro* cytotoxicity studies in the A2780 ovarian carcinoma cell line to compare rHSA, rHSA modified with the hydrazine linker, RAPTA-FORM, and rHSA conjugated with RAPTA-FORM, revealed that RAPTA-conjugated rHSA was 20-fold more active than RAPTA-FORM, while rHSA and the hydrazine-modified derivative did not significantly affect the cell growth within the tested concentration range [43].

### 8.3.2 Polymers and Liposomes as Delivery Systems for Bioactive Ruthenium Complexes

As mentioned, the Ru complexes NAMI-A, KP1019, and NKP-1339 have demonstrated huge potential in clinical trials. Using formulation strategies that involve nanoparticles, liposomes, and polymers should help improve their stability and also support the targeted delivery of the Ru drug to tumor tissue. In addition to their high stability and favorable biocompatibility and biodegradability, polymeric nanoparticles have high drug loading capacity and may act as a reservoir for the pharmacophore that could be released over prolonged periods. Polymeric nanocarriers have various forms including micelles, liposomes, and dendrimers. Such encapsulation of Ru drugs in liposomes or other co-aggregates will also reduce ligand exchange processes that may occur before reaching the target tissue [44].

Nanoformulation with poly(lactic acid) (PLA) and different surfactants was recently reported as a feasible strategy for improving the stability and biological activity of KP1019 [45]. PLA has been approved by the FDA as a nanoparticle drug delivery system, as it has no systemic toxicity. Preliminary cytotoxicity assays revealed that the activity was increased up to 20-fold compared to KP1019.

Another study reported polymer-based micelles as a nanoparticle formulation of KP1019, which also resulted in significantly enhanced bioactivity due to the high stability of the drug in aqueous solution and facilitated cellular accumulation [46]. Similarly organoruthenium moieties were loaded on poly(ethylene glycol)-*b*-poly(L-glutamic acid) through direct coordination of the Ru center to the carboxylato moieties of the polymer backbone [47]. The micelles of about 60 nm were efficiently taken up by cancer cells and exhibited higher anticancer activity compared to the related organometallic complexes and cisplatin in a mouse xenograft model of cisplatin-resistant human ovarian cancer.

NAMI-A analogues for incorporation into micelles or vesicles were prepared by coupling the $RuCl_4$(DMSO) rest through coordination to a pyrimidine deoxyribo-(thymidine) or ribonucleoside (uridine) scaffold on an amphiphilic nucleolipidic structure comprising two lipophilic C18 chains attached to the C2 and C3 of the ribose and an oligoethylene glycol of variable chain length attached to the C4 (Figure 8.5) [48, 49]. Such structures are capable of forming supramolecular assemblies, thus making them potential carriers for metal-based anticancer drugs loaded onto them. These formulations were stable for months and demonstrated excellent tumor-inhibiting properties against a panel of human and non-human cell lines with significantly higher potency than NAMI-A as the parent compound [49, 50]. The same Ru pharmacophore was also covalently linked to a cholesterol-bearing nucleolipid which allowed incorporating it into liposomes [51]. *In vitro* cytotoxicity experiments revealed high antiproliferative activity for this liposomal formulation against MCF-7 and WiDr adenocarcinoma cells.

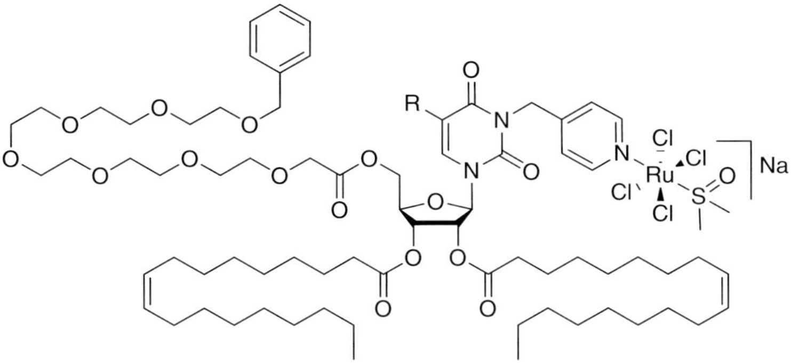


**Figure 8.5** Structure of a NAMI derivative for incorporation in micelles.

Similarly, NAMI-A conjugated to a biocompatible amphiphilic block copolymer showed higher cytotoxicity against human cancer cell lines than NAMI-A. Notably, the trademark antimetastatic activity of NAMI-A was also improved [52]. The same strategy was employed for RAPTA-C as another antimetastatic agent. RAPTA-C was loaded onto the water-soluble polymers poly(2-chloroethyl methacrylate) and poly(2-iodoethyl methacrylate) by alkylation of a PTA-nitrogen atom [53]. Higher cytotoxicity of the copolymer–RAPTA-C conjugate was observed against human cancer cells than for RAPTA-C. A confocal microscopy study confirmed the cell uptake of the macromolecular micelles into the lysosome of the cells, and accumulation was significantly enhanced in comparison to RAPTA-C [54].

### 8.3.3 Dendrimers

Dendrimers are a class of synthetic, highly branched macromolecules and have a well-defined architecture that can easily be functionalized. Moreover, they have been suggested to also be accumulating in tumors because of the EPR effect.

Smith and coworkers prepared an extensive series of tetranuclear and octanuclear ruthenium(II)-arene (arene = *p*-cymene or hexamethylbenzene) metallodendrimers based on poly(propyleneimine) dendritic scaffolds functionalized with different ligand systems featuring *N*,*O* and *N*,*N* chelators [55]. Depending on the nature of the ligands, this afforded neutral or cationic dendritic scaffolds [56]. The antiproliferative potential of these metallodendrimers was evaluated against human ovarian A2780 cancer cells. The metallodendrimers were cytotoxic with $IC_{50}$ values of 20–50 μM which appeared to be dependent on the nuclearity of the dendritic compound. The tetranuclear and octanuclear complexes showed higher cytotoxicities than the monoruthenium counterparts [55, 56].

## 8.4 Hormones, Vitamins, and Sugars: Ruthenium Complexes Targeting Small Molecule Receptors

In addition to macromolecular vectors, low-molecular-weight carriers have been explored to facilitate the delivery of cytotoxins to tumors [57]. As steroid hormones are key contributors in the development of certain breast and prostate cancers, the design of therapeutic agents which have strong affinity to the estrogen receptor (ER) and sex hormone binding globulin (SHBG) appears an appealing approach. The conjugation of cytotoxic drugs with steroids (estrogens and androgens), which act as the shuttle group, may enhance accumulation at the target site by targeting hormone receptors.

About two-thirds of breast cancer cases are hormone-dependent, in which the estrogen receptor is present (ER-positive). Jaouen and coworkers modified the pharmacophore of tamoxifen by replacing a phenyl residue with ruthenocene (Figure 8.6; **1**) or ferrocene [58]. Tamoxifen is the prodrug to the active metabolite hydroxytamoxifen and is given as an anti-estrogen treatment to patients with only ER-positive breast cancer. The introduction of a ferrocene moiety into tamoxifen resulted in a novel anticancer agent termed ferrocifen with interesting redox properties to yield quinonemethide intermediates. Ferrocifen was more active against both hormone-dependent (ER-positive) and hormone-independent breast cancer cell lines [58]. This approach was extended to ruthenium as the heavier congener of iron and a series of ruthenocene derivatives, 1-[4-($O(CH_2)_nN(CH_3)_2$)phenyl]-1-(4-hydroxyphenyl)-2-ruthenocenylbut-1-ene **1** ($n = 2$–5), was prepared on the basis of tamoxifen derivatives. The compounds with $n = 2$ and 3 showed very high relative binding affinity (RBA) values of 85% and 53% to the estrogen receptor α (ERα), respectively, as compared to ca. 40% for the active metabolite of tamoxifen. The cytotoxicity of the ruthenocene derivatives was equal or superior to that of hydroxytamoxifen on ER-positive

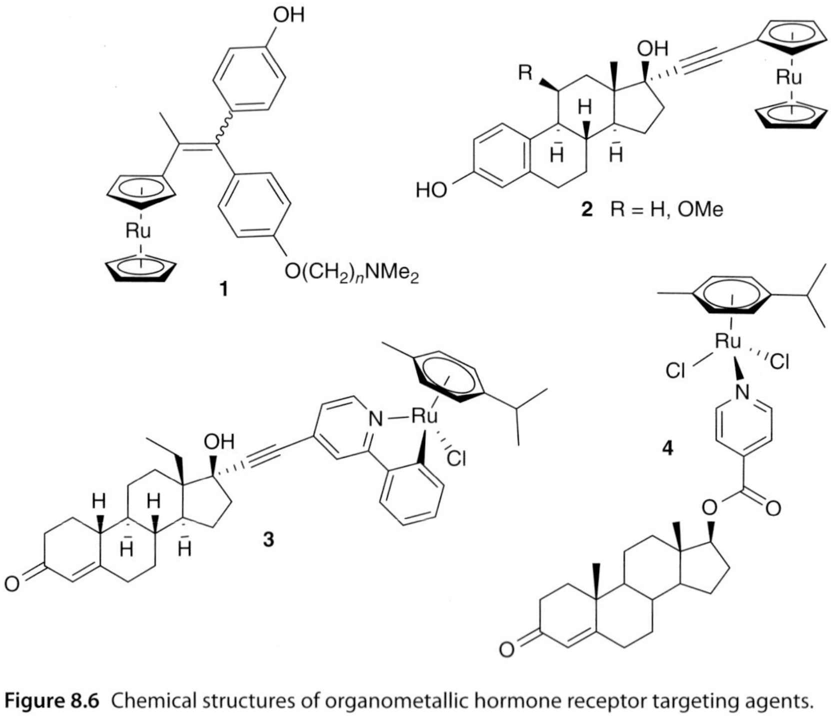


**Figure 8.6** Chemical structures of organometallic hormone receptor targeting agents.

breast cancer cell lines; but, in contrast to the ferrocifen derivatives, they did not show antiproliferative effects on ER-negative breast cancer models, which was explained by different redox chemistry of the Fe and Ru analogues [59].

Using a similar strategy, ruthenocene was also conjugated to 17β-estradiol employing an alkyne-type rigid spacer (**2**; Figure 8.6) or a flexible methylene group at the 17α position of the steroid to give hybrid molecules. While the molecule featuring an alkyne linker gave still moderate RBA values to ERα, the use of a more flexible linker caused the RBA to drop dramatically [60].

Hannon *et al.* prepared organoruthenium derivatives of levonorgestrel, which is structurally related to estradiol. Levonorgestrel is a second-generation progestin which may be used to target steroid hormone receptors. The organoruthenium moiety introduced was the frequently used [Ru($\eta^6$-*p*-cymene)Cl] fragment. Using 2-pyridinephenyl as a chelate for the metal center and an alkyne linker allowed preparation of **3** (Figure 8.6) [61]. The cytotoxicity of this ruthenium bioconjugate was about eightfold superior to cisplatin and its nonsteroidal counterpart in the T47D human breast cancer cell line. Moreover, this compound was more active in cisplatin-resistant A2780 human ovarian carcinoma cells than in the wild-type variant of this cell line. The conjugation of levonorgestrel to a $Ru^{II}$ complex resulted in synergistic effects between the metal center and the steroidal ligand, generating highly potent anticancer agents from inactive constituents.

Modification of androgen and estrogen isonicotinates with [Ru($\eta^6$-*p*-cymene)Cl] moieties to yield **4** (Figure 8.6) resulted in a decline in the binding ability of the conjugates to both ERα and SHBG [62], as compared to the parent compounds. Cytotoxicity assays revealed that the Ru($\eta^6$-*p*-cymene) complexes of 2-substituted estrogens were equally or even more cytotoxic than the steroids against hormone-dependent MCF-7 breast and KB-V1 cervix carcinoma and hormone-independent 518A2 melanoma cells.

As cancer cells require significant amounts of vitamins to sustain their rapid growth, the conjugation of drug compounds to vitamins, such as vitamin $B_{12}$, folic acid, and biotin is a another option to improve tumor targeting [63]. The main biotin uptake system in human intestinal epithelial cells is the sodium-dependent multivitamin transporter (SMVT) system [64, 65]. However, the influence of biotin conjugation is still not unambiguously clarified. For example, the cytotoxicity of a biotinylated RAPTA complex was similar to that of RAPTA-C [66], while biotin-ferrocene conjugates [67] and biotinylated cisplatin-loaded nanoparticles [68] showed enhanced the *in vitro* activity compared with the respective non-biotinylated analogues [69]. Biotinylation of [Ru($\eta^6$-*p*-cymene)] through linkage with a monodentate N-donor ligand such as imidazole or indazole resulted in compounds still able to bind strongly to avidin as a model system (Figure 8.7) [70]. In general, the pyridine derivatives were more cytotoxic than the indazole analogues; however, there was no clear-cut trend in the anticancer activity in cell lines with low or high expression levels of SMVT.

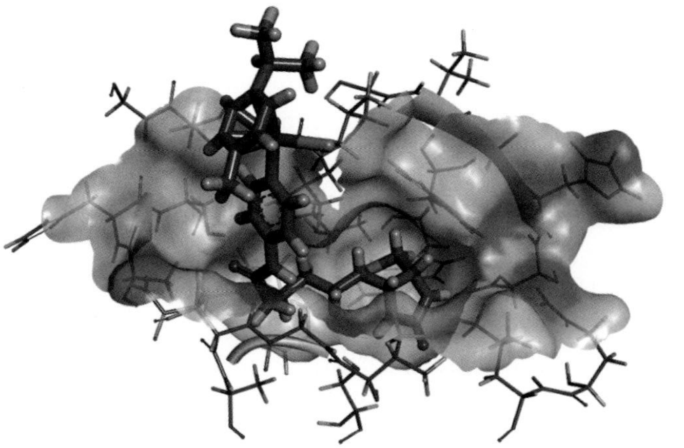

**Figure 8.7** Docking results of a biotinylated Ru(arene) complex to the biotin binding site of streptavidin [70].

Similar to targeting hormone receptors, glucose transporters (GLUT) may provide a means to help accumulate cytotoxins in cancer cells. Rapidly dividing cancer cells often overexpress GLUT in order to meet the high demand for glucose. Therefore, linking pharmacophores to glucose or other carbohydrates may enhance their cellular accumulation via GLUT-mediated uptake [71].

RAPTA analogues were prepared by replacing PTA with glucose-based phosphites as structurally related co-ligands [72–74]. The cytotoxicity of these compounds against cancer cell lines was in the low-to-moderate range and similar to that of the parent RAPTA compounds, an effect which has also been observed for other metal complexes of sugar derivatives [75]. In order to track the cellular uptake and localization of compounds in cancer cells, these Ru sugar–phosphite complexes were further functionalized with a fluorescent anthracene moiety [76]. However, no difference in uptake was observed for the sugar complexes as compared to the PTA analogues, which may be related to the high lipophilicity of the compounds once functionalized with anthracene, overshadowing the contribution of GLUT-mediated cellular accumulation. Glucose-derived phosphite ligands were also coordinated to the Ru centers in triruthenium-carbonyl clusters [77]. This approach resulted in compounds with either anticancer or specific anti-angiogenic activity, depending on the number of sugar-derived co-ligands. While the compounds were tested *in vivo* to demonstrate their anti-angiogenic activity, no studies on the accumulation in tumors were reported. More recently, the half-sandwich organometallic Ru complexes of methyl 2,3-diamino-4,6-*O*-benzylidene-2,3-dideoxy-α-D-hexopyranosides of glucose, mannose, gulose, and talose, and methyl 2-amino-4,6-benzylidene-2,3-dideoxy-3-tosylamido-α-D-glucopyranoside were reported with cytotoxicities in the low micromolar range in different cancer cell types. However, no investigations on the cell uptake were carried out to elucidate the contribution of active transport mechanisms [78].

## 8.5 Peptides as Transporters for Ruthenium Complexes into Tumor Cells and Cell Compartments

Peptides have become increasingly popular over the past years as transport vehicles for anticancer agents. For example, cell-penetrating peptides (CPPs) derived from HIV transactivator of transcription (TAT) protein [79] and structurally related oligoarginine CPPs were investigated as promoters of cellular uptake of proteins, oligonucleotides, plasmids, and others [80].

Ruthenium polypyridyl complexes have demonstrated a significant potential for the development of therapeutic and diagnostic agents. These compounds have shown strong, noncovalent binding to DNA through multiple interactions such as intercalative, electrostatic, and groove binding as well as the ability to induce photocleavage of DNA [81]. However, their poor water solubility and inability to be transported across cell membranes limit their clinical application. Puckett and Barton showed that the cellular uptake using octaarginine conjugates of dipyridophenazine–$Ru^{II}$ complexes was significantly enhanced compared to unconjugated analogues [82]. Endocytosis was identified as the major mechanism of cell entry and trapping in the endosome resulted in a lack of nuclear entry. However, under similar testing conditions, the Ru-peptide conjugate with an appended fluorescein residue was found to accumulate in the nucleus, indicating a cell compartment directing ability. The replacement of the highly positively charged (+8 charge) octaarginine moiety with a shorter oligoarginine sequence (Figure 8.8) further enhanced the nuclear localization [83]. The more positively charged octaarginine may also interact with the negative charge-bearing DNA backbone.

The conjugation of organoruthenium moieties to a series of peptides, such as octreotide targeting the somatostatin receptor which is overexpressed by a variety of tumors, was pioneered by Metzler-Nolte [84–86]. They also loaded

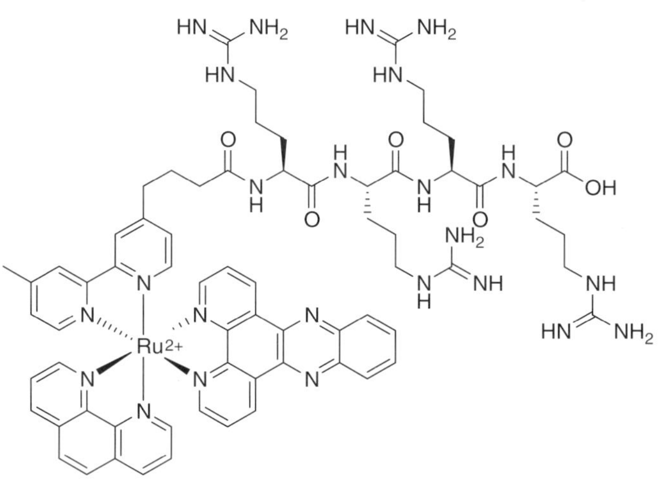


**Figure 8.8** Chemical structure of a Ru complex conjugated to a tetraarginine residue for targeting the cell nucleus of cells.

ruthenocene on a peptide nucleic acid (PNA) dodecamer, which resulted in low cytotoxicity despite a high degree of accumulation in cells [87]. Sadler and coworkers developed a photoactivated Ru(arene) complex conjugated to the dicarba analogue of octreotide and the Arg-Gly-Asp (RGD) tripeptide, both of which can bind to receptors overexpressed on the membranes of tumor cells [88]. The Ru-peptide conjugates were stable in aqueous solution in the dark, but the peptide bound via a pyridyl moiety selectively photodissociated from the Ru center upon irradiation with visible light to form aqua species which reacted with DNA models.

## 8.6 Polynuclear Ruthenium Complexes for the Delivery of a Cytotoxic Payload

Ruthenium coordination cages such as metallaprisms, metallarectangles, and metallacycles have been suggested as carriers for drugs for cancer chemotherapy. Equipped with large conjugated aromatic systems, some of them were shown to be able to encapsulate planar aromatic molecules and also coordination compounds. These Ru cages proved to be promising drug delivery vectors supposedly entering the tumor similar to macromolecules by exploiting the EPR effect. Hexacationic cages such as $[Ru_6(\eta^6\text{-}p\text{-cymene})_6(tpt)_2(dobq)_3]^{6+}$ and $[Ru_6(\eta^6\text{-}p\text{-cymene})_6(tpt)_2(donq)_3]^{6+}$ (tpt = 2,4,6-tris-(pyridin-4-yl)-1,3,5-triazine; dobq = 2,5-dioxydo-1,4-benzoquinonato; donq = 5,8-dioxydo-1,4-naphthoquinonato) [89], were able to encapsulate Pt or Pd acetylacetonato (acac) complexes (Figure 8.9) and planar aromatic compounds of various sizes [90]. The encapsulation of the guest by these complexes

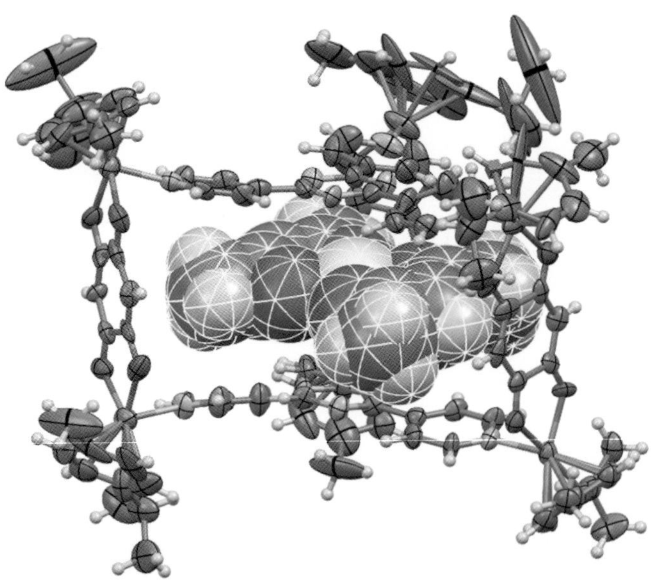

**Figure 8.9** Molecular structure of $Pt(acac)_2$ trapped in a hexanuclear Ru complex [89].

did not impact the physical properties of the host and the guest molecules were stable in these complexes [90]. The biological activity of these metallacages increased with the encapsulation of a guest, suggesting transport and release of the guest once taken up by the cell [90, 91]. A range of guest molecules were evaluated for their delivery to cancer cells. These include delivery of hydrophobic porphin molecules as photosensitizers, pyrenyl-nucleoside derivatives, and pyrenyl-RAPTA analogues. The *in vitro* cytotoxicity of the RAPTA derivatives functionalized with a pyrenyl group and coupled with a water-soluble metallacage was established against a series of cancer cells [92]. The cytotoxicity of these conjugates was at least 10 times higher than that for RAPTA-C as a reference, probably due to higher lipophilic properties induced by the presence of the pyrenyl motif, while the metallacage was 50 times more cytotoxic than RAPTA-C. The uptake of RAPTA derivatives was considerably improved upon encapsulation in the metallacage. Such new derivatives could potentially act as multifunctional drugs, that is, the cytotoxicity may be induced by interactions of the pyrenyl moiety with DNA and the antimetastatic properties are characteristic of RAPTA compounds [6].

## 8.7 Summary and Conclusions

Ru-based pharmacophores have been shown to have potential in the treatment of a variety of cancer types. The ease of modification has led to the development of compounds that are active against primary tumors, inhibit metastasis, or both. With the aim of improving the accumulation as a means of overcoming unwanted side effects, Ru complexes were equipped with tumor targeting properties. A range of strategies have been explored and anticancer-active complexes were modified with functional groups for loading to carrier proteins, cell (or compartment)-specific peptides, nanoparticles, liposomes, micelles, and polymers, while other approaches involve the introduction of low-molecular-weight vectors that allow for interaction with receptors overexpressed by tumor cells. Advantages of such approaches are manifold and include in some cases prolonged half-life in serum and also reduced side reactions, which coordination compounds are often prone of when in aqueous media and in presence of a range of biological ligand systems. However, the number of *in vivo* experiments reported is scarce despite having a range of element-specific methods available that allow detecting trace amounts of Ru in living organisms. This is an issue that needs to be tackled in future to demonstrate the feasibility of the suggested delivery approaches and boost future clinical development.

## Acknowledgments

We thank the organizations and foundations that supported our research efforts in this area, especially the University of Auckland, Genesis Oncology Trust (GOT-1263-RPG), the Austrian Science Fund (Schrödinger fellowship to M.H.), the India-New Zealand Education Council, the New Zealand Education, the India-New Zealand Research Institute, the Royal Society of New Zealand and COST CM1105.

## References

1 Jakupec, M.A., Galanski, M., Arion, V.B., Hartinger, C.G., and Keppler, B.K. (2008) *Dalton Trans.*, 183–194.
2 Hartinger, C.G., Jakupec, M.A., Zorbas-Seifried, S., Groessl, M., Egger, A., Berger, W., Zorbas, H., Dyson, P.J., and Keppler, B.K. (2008) *Chem. Biodiversity*, **5**, 2140–2155.
3 Rademaker-Lakhai, J.M., van den Bongard, D., Pluim, D., Beijnen, J.H., and Schellens, J.H. (2004) *Clin. Cancer Res*, **10**, 3717–3727.
4 Murray, B., Babak, M., Hartinger, C.G., and Dyson Paul, J. (2016) *Coord. Chem. Rev.*, **306**, 86–114.
5 Clarke, M.J., Zhu, F., and Frasca, D.R. (1999) *Chem. Rev.*, **99**, 2511–2533.
6 Dyson, P.J. (2007) *Chimia*, **61**, 698–703.
7 Wang, D. and Lippard, S.J. (2005) *Nat. Rev. Drug Discovery*, **4**, 307–320.
8 Reedijk, J. (2008) *Platinum Met. Rev.*, **52**, 2–11.
9 Reisner, E., Arion, V.B., Keppler, B.K., and Pombeiro, A.J.L. (2008) *Inorg. Chim. Acta*, **361**, 1569–1583.
10 Keppler, B.K. and Rupp, W. (1986) *J. Cancer Res. Clin. Oncol.*, **111**, 166–168.
11 Allardyce, C.S. and Dyson, P.J. (2001) *Platinum Met. Rev.*, **45**, 62–69.
12 Kratz, F., Keppler, B.K., Messori, L., Smith, C., and Baker, E.N. (1994) *Met. Based Drugs*, **1**, 169–173.
13 Clarke, M.J. (2003) *Coord. Chem. Rev.*, **236**, 209–233.
14 Kratz, F., Müller, I.A., Ryppa, C., and Warnecke, A. (2008) *ChemMedChem*, **3**, 20–53.
15 Pongratz, M., Schluga, P., Jakupec, M.A., Arion, V.B., Hartinger, C.G., Allmaier, G., and Keppler, B.K. (2004) *J. Anal. At. Spectrom.*, **19**, 46–51.
16 Garzon, F.T., Berger, M.R., Keppler, B.K., and Schmähl, D. (1987) *Cancer Chemother. Pharmacol.*, **19**, 347–349.
17 Hartinger, C.G., Zorbas-Seifried, S., Jakupec, M.A., Kynast, B., Zorbas, H., and Keppler, B.K. (2006) *J. Inorg. Biochem.*, **100**, 891–904.
18 Peti, W., Pieper, T., Sommer, M., Keppler, B.K., and Giester, G. (1999) *Eur. J. Inorg. Chem.*, 1551–1555.
19 Trondl, R., Heffeter, P., Kowol, C.R., Jakupec, M.A., Berger, W., and Keppler, B.K. (2014) *Chem. Sci.*, **5**, 2925–2932.
20 Sulyok, M., Hann, S., Hartinger, C.G., Keppler, B.K., Stingeder, G., and Koellensperger, G. (2005) *J. Anal. At. Spectrom.*, **20**, 856–863.
21 Kratz, F. and Elsadek, B. (2012) *J. Controlled Release*, **161**, 429–445.
22 Timerbaev, A.R., Rudnev, A.V., Semenova, O., Hartinger, C.G., and Keppler, B.K. (2005) *Anal. Biochem.*, **341**, 326–333.
23 Smith, C.A., Sutherland-Smith, A.J., Kratz, F., Baker, E.N., and Keppler, B.H. (1996) *J. Biol. Inorg. Chem.*, **1**, 424–431.
24 Groessl, M., Zava, O., and Dyson, P.J. (2011) *Metallomics*, **3**, 591–599.

**25** Bytzek, A.K., Boeck, K., Hermann, G., Hann, S., Keppler, B.K., Hartinger, C.G., and Koellensperger, G. (2011) *Metallomics*, **3**, 1049–1055.

**26** Cetinbas, N., Webb, M.I., Dubland, J.A., and Walsby, C.J. (2010) *J. Biol. Inorg. Chem.*, **15**, 131–145.

**27** Groessl, M., Hartinger, C.G., Polec-Pawlak, K., Jarosz, M., and Keppler, B.K. (2008) *Electrophoresis*, **29**, 2224–2232.

**28** Coluccia, M., Sava, G., Loseto, F., Nassi, A., Boccarelli, A., Giordano, D., Alessio, E., and Mestroni, G. (1993) *Eur. J. Cancer*, **29A**, 1873–1879.

**29** Sava, G., Pacor, S., Zorzet, S., Alessio, E., and Mestroni, G. (1989) *Pharmacol. Res.*, **21**, 617–628.

**30** Sava, G., Pacor, S., Bergamo, A., Cocchietto, M., Mestroni, G., and Alessio, E. (1995) *Chem.-Biol. Interact.*, **95**, 109–126.

**31** Sava, G., Capozzi, I., Clerici, K., Gagliardi, G., Alessio, E., and Mestroni, G. (1998) *Clin. Exp. Metastasis*, **16**, 371–379.

**32** Bratsos, I., Gianferrara, T., Alessio, E., Hartinger, C.G., Jakupec, M.A., and Keppler, B.K. (2011) in *Bioinorganic Medicinal Chemistry* (ed. E. Alessio), Wiley-VCH Verlag GmbH & Co. KGaA, Weinheim, Germany, pp. 151–174.

**33** Leijen, S., Burgers, S., Baas, P., Pluim, D., Tibben, M., van Werkhoven, E., Alessio, E., Sava, G., Beijnen, J., and Schellens, J.M. (2015) *Invest. New Drugs*, **33**, 201–214.

**34** Bergamo, A., Messori, L., Piccioli, F., Cocchietto, M., and Sava, G. (2003) *Invest. New Drugs*, **21**, 401–411.

**35** Sava, G., Bergamo, A., Zorzet, S., Gava, B., Casarsa, C., Cocchietto, M., Furlani, A., Scarcia, V., Serli, B., Iengo, E., Alessio, E., and Mestroni, G. (2002) *Eur. J. Cancer*, **38**, 427–435.

**36** Liu, M., Lim, Z.J., Gwee, Y.Y., Levina, A., and Lay, P.A. (2010) *Angew. Chem., Int. Ed. Engl.*, **49**, 1661–1664.

**37** Thorek, D.J., Chen, A., Czupryna, J., and Tsourkas, A. (2006) *Ann. Biomed. Eng.*, **34**, 23–38.

**38** Bhojani, M.S., Van Dort, M., Rehemtulla, A., and Ross, B.D. (2010) *Mol. Pharmaceutics*, **7**, 1921–1929.

**39** Maeda, H. (2010) *Bioconj. Chem.*, **21**, 797–802.

**40** Scolaro, C., Bergamo, A., Brescacin, L., Delfino, R., Cocchietto, M., Laurenczy, G., Geldbach, T.J., Sava, G., and Dyson, P.J. (2005) *J. Med. Chem.*, **48**, 4161–4171.

**41** Hanif, M., Nazarov, A.A., Legin, A., Groessl, M., Arion, V.B., Jakupec, M.A., Tsybin, Y.O., Dyson, P.J., Keppler, B.K., and Hartinger, C.G. (2012) *Chem. Commun.*, **48**, 1475–1477.

**42** Moon, S., Hanif, M., Kubanik, M., Holtkamp, H., Söhnel, T., Jamieson, S.M.F., and Hartinger, C.G. (2015) *ChemPlusChem*, **80**, 231–236.

**43** Ang, W.H., Daldini, E., Juillerat-Jeanneret, L., and Dyson, P.J. (2007) *Inorg. Chem.*, **46**, 9048–9050.

**44** Callari, M., Aldrich-Wright, J.R., de Souza, P.L., and Stenzel, M.H. (2014) *Prog. Polym. Sci.*, **39**, 1614–1643.

**45** Fischer, B., Heffeter, P., Kryeziu, K., Gille, L., Meier, S.M., Berger, W., Kowol, C.R., and Keppler, B.K. (2014) *Dalton Trans.*, **43**, 1096–1104.

46 Heffeter, P., Riabtseva, A., Senkiv, Y., Kowol, C.R., Korner, W., Jungwith, U., Mitina, N., Keppler, B.K., Konstantinova, T., Yanchuk, I., Stoika, R., Zaichenko, A., and Berger, W. (2014) *J. Biomed. Nanotechnol.*, **10**, 877–884.
47 Duan, X., Liu, D., Chan, C., and Lin, W. (2015) *Small*, **11**(32): 3962–3972.
48 Vaccaro, M., Del Litto, R., Mangiapia, G., Carnerup, A.M., D'Errico, G., Ruffo, F., and Paduano, L. (2009) *Chem. Commun.*, 1404–1406.
49 Mangiapia, G., Vitiello, G., Irace, C., Santamaria, R., Colonna, A., Angelico, R., Radulescu, A., D'Errico, G., Montesarchio, D., and Paduano, L. (2013) *Biomacromolecules*, **14**, 2549–2560.
50 Mangiapia, G., D'Errico, G., Simeone, L., Irace, C., Radulescu, A., Di Pascale, A., Colonna, A., Montesarchio, D., and Paduano, L. (2012) *Biomaterials*, **33**, 3770–3782.
51 Vitiello, G., Luchini, A., D'Errico, G., Santamaria, R., Capuozzo, A., Irace, C., Montesarchio, D., and Paduano, L. (2015) *J. Mater. Chem. B*, **3**, 3011–3023.
52 Blunden, B.M., Rawal, A., Lu, H., and Stenzel, M.H. (2014) *Macromolecules*, **47**, 1646–1655.
53 Blunden, B.M., Thomas, D.S., and Stenzel, M.H. (2012) *Polym. Chem.*, **3**, 2964–2975.
54 Blunden, B.M., Lu, H., and Stenzel, M.H. (2013) *Biomacromolecules*, **14**, 4177–4188.
55 Govender, P., Antonels, N.C., Mattsson, J., Renfrew, A.K., Dyson, P.J., Moss, J.R., Therrien, B., and Smith, G.S. (2009) *J. Organomet. Chem.*, **694**, 3470–3476.
56 Govender, P., Renfrew, A.K., Clavel, C.M., Dyson, P.J., Therrien, B., and Smith, G.S. (2011) *Dalton Trans.*, **40**, 1158–1167.
57 Dao, K.-L. and Hanson, R.N. (2012) *Bioconj. Chem.*, **23**, 2139–2158.
58 Nguyen, A., Vessieres, A., Hillard, E.A., Top, S., Pigeon, P., and Jaouen, G. (2007) *Chimia*, **61**, 716–724.
59 Pigeon, P., Top, S., Vessieres, A., Huche, M., Hillard, E.A., Salomon, E., and Jaouen, G. (2005) *J. Med. Chem.*, **48**, 2814–2821.
60 Top, S., El Hafa, H., Vessières, A., Huché, M., Vaissermann, J., and Jaouen, G. (2002) *Chem.–Eur. J.*, **8**, 5241–5249.
61 Ruiz, J., Rodriguez, V., Cutillas, N., Espinosa, A., and Hannon, M.J. (2011) *Inorg. Chem.*, **50**, 9164–9171.
62 Schobert, R., Seibt, S., Effenberger-Neidnicht, K., Underhill, C., Biersack, B., and Hammond, G.L. (2011) *Steroids*, **76**, 393–399.
63 Bildstein, L., Dubernet, C., and Couvreur, P. (2011) *Adv. Drug Delivery Rev.*, **63**, 3–23.
64 Said, H.M. (2009) *J. Nutr.*, **139**, 158–162.
65 Wang, H., Huang, W., Fei, Y.J., Xia, H., Yang-Feng, T.L., Leibach, F.H., Devoe, L.D., Ganapathy, V., and Prasad, P.D. (1999) *J. Biol. Chem.*, **274**, 14875–14883.
66 Babak, M.V., Meier, S.M., Huber, K.V.M., Reynisson, J., Legin, A.A., Jakupec, M.A., Roller, A., Stukalov, A., Gridling, M., Bennett, K.L., Colinge, J., Berger, W., Dyson, P.J., Superti-Furga, G., Keppler, B.K., and Hartinger, C.G. (2015) *Chem. Sci.*, **6**, 2449–2456.
67 Plazuk, D., Zakrzewski, J., Salmain, M., Blauz, A., Rychlik, B., Strzelczyk, P., Bujacz, A., and Bujacz, G. (2013) *Organometallics*, **32**, 5774–5783.

68 Yellepeddi, V.K., Kumar, A., Maher, D.M., Chauhan, S.C., Vangara, K.K., and Palakurthi, S. (2011) *Anticancer Res.*, **31**, 897–906.
69 Vadlapudi, A.D., Vadlapatla, R.K., and Mitra, A.K. (2012) *Curr. Drug Targets*, **13**, 994–1003.
70 Babak, M.V., Plazuk, D., Meier, S.M., Arabshahi, H.J., Reynisson, J., Rychlik, B., Blauz, A., Szulc, K., Hanif, M., Strobl, S., Roller, A., Keppler, B.K., and Hartinger, C.G. (2015) *Chem.–Eur. J.*, **21**, 5110–5117.
71 Medina, R.A. and Owen, G.I. (2002) *Biol. Res.*, **35**, 9–26.
72 Berger, I., Hanif, M., Nazarov, A.A., Hartinger, C.G., John, R.O., Kuznetsov, M.L., Groessl, M., Schmitt, F., Zava, O., Biba, F., Arion, V.B., Galanski, M., Jakupec, M.A., Juillerat-Jeanneret, L., Dyson, P.J., and Keppler, B.K. (2008) *Chem.–Eur. J.*, **14**, 9046–9057.
73 Hanif, M., Meier, S.M., Kandioller, W., Bytzek, A., Hejl, M., Hartinger, C.G., Nazarov, A.A., Arion, V.B., Jakupec, M.A., Dyson, P.J., and Keppler, B.K. (2011) *J. Inorg. Biochem.*, **105**, 224–231.
74 Hanif, M., Nazarov, A.A., Hartinger, C.G., Kandioller, W., Jakupec, M.A., Arion, V.B., Dyson, P.J., and Keppler, B.K. (2010) *Dalton Trans.*, **39**, 7345–7352.
75 Hartinger, C.G., Nazarov, A.A., Ashraf, S.M., Dyson, P.J., and Keppler, B.K. (2008) *Curr. Med. Chem.*, **15**, 2574–2591.
76 Nazarov, A.A., Risse, J., Ang, W.H., Schmitt, F., Zava, O., Ruggi, A., Groessl, M., Scopelitti, R., Juillerat-Jeanneret, L., Hartinger, C.G., and Dyson, P.J. (2012) *Inorg. Chem.*, **51**, 3633–3639.
77 Nazarov, A.A., Baquie, M., Nowak-Sliwinska, P., Zava, O., van Beijnum, J.R., Groessl, M., Chisholm, D.M., Ahmadi, Z., McIndoe, J.S., Griffioen, A.W., van den Bergh, H., and Dyson, P.J. (2013) *Sci. Rep.*, **3**, 1485, 7 pp.
78 Boge, M., Fowelin, C., Bednarski, P., and Heck, J. (2015) *Organometallics*, **34**, 1507–1521.
79 Magzoub, M. and Graslund, A. (2004) *Q. Rev. Biophys.*, **37**, 147–195.
80 Wender, P.A., Mitchell, D.J., Pattabiraman, K., Pelkey, E.T., Steinman, L., and Rothbard, J.B. (2000) *Proc. Natl. Acad. Sci. U. S. A.*, **97**, 13003–13008.
81 Mari, C., Pierroz, V., Ferrari, S., and Gasser, G. (2015) *Chem. Sci.*, **6**, 2660–2686.
82 C. A. Puckett, J. K. Barton, *J. Am. Chem. Soc.* 2009, 131, 8738–8739.
83 Puckett, C.A. and Barton, J.K. (2010) *Bioorg. Med. Chem.*, **18**, 3564–3569.
84 Gross, A., Huesken, N., Schur, J., Raszeja, L., Ott, I., and Metzler-Nolte, N. (2012) *Bioconjugate Chem.*, **23**, 1764–1774.
85 Meier, S.M., Novak, M., Kandioller, W., Jakupec, M.A., Arion, V.B., Metzler-Nolte, N., Keppler, B.K., and Hartinger, C.G. (2013) *Chem.–Eur. J.*, **19**, 9297–9307.
86 Zagermann, J., Kuchta, M.C., Merz, K., and Metzler-Nolte, N. (2009) *J. Organomet. Chem.*, **694**, 862–867.
87 Gross, A. and Metzler-Nolte, N. (2009) *J. Organomet. Chem.*, **694**, 1185–1188.
88 Barragan, F., Lopez-Senin, P., Salassa, L., Betanzos-Lara, S., Habtemariam, A., Moreno, V., Sadler, P.J., and Marchan, V. (2011) *J. Am. Chem. Soc.*, **133**, 14098–14108.

89 Therrien, B., Suess-Fink, G., Govindaswamy, P., Renfrew, A.K., and Dyson, P.J. (2008) *Angew. Chem., Int. Ed.*, **47**, 3773–3776.

90 Therrien, B. (2012) in *Chemistry of Nanocontainers* (eds M. Albrecht and E. Hahn), Springer, pp. 35–55.

91 Schmitt, F., Freudenreich, J., Barry, N.P.E., Juillerat-Jeanneret, L., Suss-Fink, G., and Therrien, B. (2012) *J. Am. Chem. Soc.*, **134**, 754–757.

92 Furrer, M.A., Schmitt, F., Wiederkehr, M., Juillerat-Jeanneret, L., and Therrien, B. (2012) *Dalton Trans.*, **41**, 7201–7211.

# 9

# Design Aspects of Ruthenium Complexes as DNA Probes and Therapeutic Agents

*Madeleine De Beer and Shawn Swavey*

*University of Dayton, Department of Chemistry, 300 College Park, Dayton, OH 45469, USA*

## 9.1 Introduction

The focus of this chapter is ruthenium complexes for use as DNA probes and therapeutic agents. Ruthenium complexes are unique among transition metals due, in part, to their octahedral geometry which allows for a variety of ligand coordination, their tunable and rich redox and photophysical properties, stability under physiological conditions, and their relatively low cytotoxicity. For these reasons, ruthenium complexes have been used extensively over several decades to study diverse scientific questions, ranging from the structure and function of DNA to the development of novel antimicrobial and cancer treatments.

Ruthenium complexes likely mediate their functions through a number of mechanisms. Three broad categories of mechanism of action are physical interactions to disrupt structure of biomolecules, interference of transport of other metals necessary for biological function, and the targeted generation of biologically active chemical compounds such as nitric oxide (NO) and reactive oxygen species (ROS), which have been linked to many disease states. These mechanisms of action are not necessarily mutually exclusive. For example, intercalation could provide the means to physically interact with a biomolecule and thus target the ROS-induced damage to a specific time and place in a biological system. Ruthenium complexes could also cause additive cellular damage by depriving cells of essential components due to targeted disruption of iron, calcium, and other cellular transport systems.

## 9.2 Physical Interaction to Disrupt DNA Structure

Perhaps the most straightforward method of disrupting biological function is to disrupt the overall structure of a biomolecule and consequently prevent additional interactions or conformational changes that are required for function. For example, DNA, the genetic blueprint of all cells, contains specific sequence elements that interact with proteins, and these interactions need to be transient in order for processes like DNA replication and transcription of the DNA into

*Ruthenium Complexes: Photochemical and Biomedical Applications,* First Edition.
Edited by Alvin A. Holder, Lothar Lilge, Wesley R. Browne, Mark A.W. Lawrence, and Jimmie L. Bullock Jr.

RNA to proceed unhindered. Thus, metal complexes that interact with and disrupt the functions of nucleic acids can bind to the DNA or RNA molecule in a number of ways. For example, irreversibly with the nucleic acid structure permanently modified or reversibly through a number of different binding modes, ranging from intercalation between adjacent DNA bases to electrostatic interactions of the cationic metal complex with the negatively charged nucleic acid phosphate backbone, exploiting specific aspects of nucleic acid structure [1].

Several researchers contributed to solving the initial structure of B form DNA in the 1950s [2]. Since then, additional research has focused on high-resolution structural details of DNA and the discovery of alternate forms of DNA such as the A form and Z form. All these DNA forms are double helices comprised of two DNA strands made up of individual nucleotides covalently linked together by phosphodiester bonds between the 3′ hydroxyl group of a ribose sugar of one nucleotide and the 5′ phosphate group of another nucleotide. The two DNA strands are oriented in opposite directions as designated by the ribose moiety of the sugar-phosphate backbone at the end of each single strand and noncovalently paired together by complementary base pairing (A to T and C to G), maximizing the amount of hydrogen bonds that form between the bases. In addition, hydrophobic interactions between the stacked bases found primarily in the interior of the helix contribute to the overall structural stability [3, 4]. Distinctions between the different forms of DNA are based on whether the helices are right-handed (A and B) or left-handed (Z) and differences in the number of base pairs per turn and overall tightness of the helical turns, ranging from 10 base pairs per turn of a right-handed helix for B-DNA to 12 base pairs per turn of a left-handed helix for Z-DNA [4]. Helical turns are important because they define the characteristics of the two types of grooves that form along the length of the double-stranded DNA. The two types of grooves are designated as the major and the minor groove based on their structural characteristics, with the major groove being wider and more shallow and the minor groove being narrower and deeper. These physical characteristics differ among different forms of DNA. For example, the major groove of B-DNA is wider and more accessible compared to the major groove of A-DNA, and, in an even more extreme structural variation, Z-DNA appears to have a distinctly different major groove with a convex surface protruding out of instead of into the helix [5]. B-DNA is thought to be the most stable structure for the majority of double-stranded DNA molecules under physiologically relevant conditions, while the A-DNA is thought to be favored under more hydrophobic conditions [6]. Z-DNA was discovered in the 1980s and tends to form from DNA sequences that have stretches of alternating purine-pyrimidine sequences. Z-DNA forms under high salt conditions *in vitro* and has been observed *in vivo* in specific regions such as telomeres, the ends of linear chromosomes, although the predominant form of DNA remains the B form, originally described in the 1950s [4, 7].

### 9.2.1 Irreversible Covalent Binding

Because disruption of DNA structure can ultimately lead to cell death, DNA is an attractive drug target. For example, structural studies have shown that DNA becomes irreversibly cross-linked in the presence of cisplatin, a small molecule platinum derivative that has become widely used in cancer treatment [8]. The discovery and subsequent success of cisplatin (cis-$Pt(NH_3)_2Cl_2$, as a chemotherapeutic drug revealed the possibilities associated with the use of transition metal complexes as anticancer agents. Upon cellular uptake, cisplatin undergoes ligand exchange to form the diaqua species $Pt(NH_3)_2(OH_2)_2$. The cytotoxic effects of cisplatin result from DNA coordination of the platinum(II) center preferentially at the guanine N-7 position. These covalent interactions cause significant kinking in the DNA, widening of the minor groove, and disruption of overall structure [9]. Because the DNA structure is disrupted, the central processes of replication and transcription that depend on the structural integrity of DNA are halted, ultimately leading to cell death [10]. Although the first line of treatment for the past four decades for a number of cancers, cisplatin is not without some shortfalls. Most notably, cisplatin has nonspecific cytotoxicity leading to deleterous side effects, and certain types of tumors develop a resistance to cisplatin over time. For these reasons, much interest has been focused on finding other coordinate metal complexes that have biological effects [11] (Figure 9.1).

For example, in a 1984 study, a racemic mixture of cis-$Ru(phen)_2Cl_2$ (phen = 1,10-phenanthroline) was incubated with calf thymus DNA [12]. Spectroscopic analysis indicated that only one of the enantiomers of the ruthenium complex, namely, the $\Lambda$-isomer, interacted with the DNA, most likely through coordination to the N-7 position of guanine, the most exposed base in the major groove. The mechanism of this interaction must go through thermal ligand exchange of a chloride in a manner similar to that for cisplatin. The reason for the preference of the $\Lambda$-isomer over the $\Delta$-isomer is a matter of steric favorability associated with the right-handed stereochemistry of the B-DNA. Subsequent studies of an aquaruthenium(II) complex showed even greater stereospecificity, suggesting that ruthenium complexes with labile ligands might serve as alternatives to cisplatin [13]. Although stereospecific DNA binding of ruthenium complexes by thermal ligand exchange has been demonstrated, these complexes do not render tumor-specific cytotoxicity since both tumor and normal cells produce the right-handed stereochemistry of B-DNA.

A more idealized modality would be to develop complexes capable of targeting tumor cells. One method that has received a great deal of attention in the past decade is photodynamic therapy (PDT). This method offers a more targeted approach to treating tumor cells and involves a photosensitizer, light, and molecular oxygen. The photosensitizer is otherwise benign but can be activated by light. Once the photosensitizer is excited, it reacts with molecular oxygen to give either excited singlet oxygen or superoxide. These ROS then react with cellular biomolecules, for example, DNA, ultimately destroying the tumor cell. Unfortunately, many aggressive tumors exist in either a hypoxic environment or low concentrations of molecular oxygen. This, of course, makes PDT less effective for this type of tumor. To circumvent the need for oxygen, researchers examined photoinduced cisplatin like DNA binding of a ruthenium complex, *cis*-$[Ru(phpy)(phen)(CH_3CN)_2]^+$ (phpy = 2-phenylpyridine, phen = 1,10-phenanthroline) [14]. Gel electrophoresis experiments performed on aqueous solutions of linearized ds-DNA and *cis*-$[Ru(phpy)(phen)(CH_3CN)_2]^+$

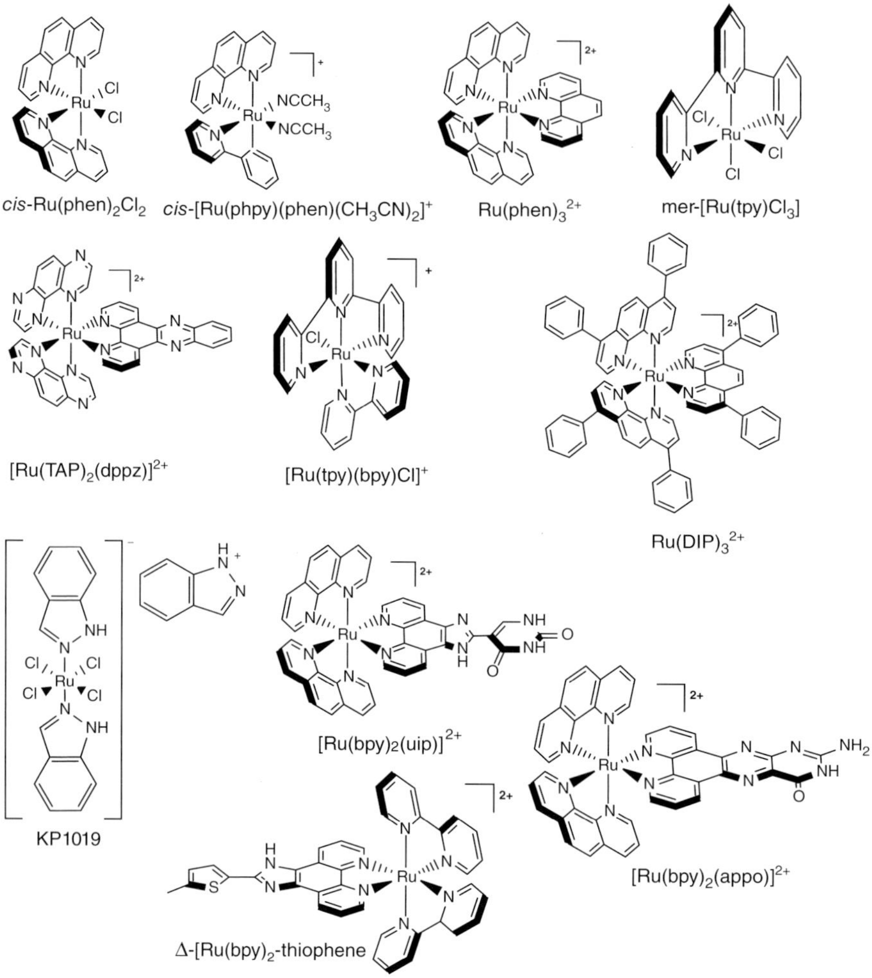


**Figure 9.1** Structures of ruthenium complexes.

after irradiation with low-energy light revealed slowed migration of the DNA through the gel, similar to that observed for cisplatin. It was concluded that irradiation of the complex induced sequential ligand exchange substituting water for the acetonitrile ligands. The diaqua complex coordinates to DNA through loss of the water ligands in a manner similar to the thermal activation of cisplatin. These results offer a new outlook for the use of PDT.

### 9.2.2 Intercalation

Because of ruthenium's chemical and physical properties, much interest has focused on developing ruthenium complexes capable of replicating cisplatin's effects, while circumventing the limitations of toxicity and tumor resistance [15]. However, while some ruthenium complexes form covalent bonds with biomolecules, a significant number appear to function differently from cisplatin due to the fact that they interact primarily through noncovalent interactions with DNA [15]. Although some studies have shown that Ru-based complexes consistently exhibit a higher DNA-binding affinity compared to analogous complexes with different metal centers such as nickel, the type of ruthenium-complex/DNA interaction appears to be primarily determined by the structure of the ligand coordinated with ruthenium [16]. Specifically, ligands that are small and planar such as phenantroline tend to favor interaction by intercalation between the stacked DNA bases of the double helix. Studies by Lerman in the 1960s provide the basis for our understanding of intercalation. Lerman examined the physical and chemical behavior of DNA in the presence of small planar molecules and found that in the presence of these molecules DNA strands became longer and stiffer. These observations were explained by the "intercalation hypothesis," in which small, planar molecules could be found noncovalently bound between the stacked DNA bases [17]. In the 1970s, Lippard and colleagues used several methods to establish that planar-platinum–based compounds are able to intercalate with DNA. For example, they observed that metal complexes with planar aromatic ligands competitively inhibited binding of the known intercalator, ethidium, to DNA. In addition, sedimentation velocity and gel electrophoresis methods indicated that the DNA complexes were partially unwound, which is consistent with intercalators' ability to disrupt the base stacking interactions of double-stranded DNA [18]. Analysis of X-ray diffraction patterns of these metallointercalators established that a metal complex fits between alternating bases, following the "neighbor exclusion principle" [19]. These interactions are primarily hydrophobic and are largely driven by the type of ligand coordinated with the metal center and, unlike cisplatin or Ru complexes that covalently bind to DNA, are not directly coordinated to the DNA molecule [20]. A high-resolution co-crystal structure of a DNA-bound ruthenium complex (Λ-enantiomer of $[Ru(TAP)_2(dppz)]^{2+}$) solved by Hall and coworkers revealed that the dppz ligand intercalates with the DNA while one of the TAP ligands semiintercalates, interacting with only the guanine component of a base pair and within the minor groove of the DNA substrate. Interestingly, this interaction creates a kink in the DNA structure and, together with interactions between the DNA and the second TAP ligand, helps promote noncovalent cross-linking of two distinct double-stranded DNA molecules. These observations indicate that significant structural distortion of DNA can be achieved through noncovalent binding [21].

### 9.2.3 Additional Noncovalent Binding Interactions

In addition to intercalation, the other types of noncovalent interactions observed between metal complexes and DNA are groove-binding and electrostatic interactions. The groove-binding mode may, in fact, overlap with intercalation. Specifically, intercalating ligands are able to access the stacked bases which are found in the core of the DNA helix through interactions with the grooves. In addition, metal–ligand complexes, such as $Ru(phen)_3{}^{2+}$, are capable of simultaneously intercalating into the major groove and using hydrophobic interactions to bind to the minor groove of the DNA molecule [20, 22]. Besides demonstrating dual binding modes for a Ru complex with DNA, Barton's work from the 1980s with $Ru(phen)_3{}^{2+}$ clearly demonstrates the importance of chirality in interactions with DNA since the Δ-enantiomer interacts with DNA primarily through intercalation and the Λ-enantiomer interacts with DNA primarily through hydrophobic interactions with the minor groove [22]. Using more sterically hindered ligands such as diphenylphenanthroline (DIP), enantiomer-specific substrate preferences were further demonstrated, with the Δ-enantiomer of $[Ru(DIP)_3]^{2+}$ able to bind to right-handed B-DNA and the Λ-enantiomer able to bind to the left-handed Z-DNA [23]. Thus, Ru complexes have been actively developed to probe specific structural aspects of DNA *in vitro* and *in vivo* [24].

The overall architecture of the DNA double helix places the hydrophobic bases in the interior of the helix and the negatively charged phosphate groups of the sugar-phosphate backbone on the outside. This structure is stable in an aqueous environment because hydrophobic interactions are maximized in the interior and the negatively charged phosphates can favorably interact with the polar solvent. Furthermore, metal cations such as $Na^+$ and $Mg^{2+}$ are readily soluble in aqueous solution and are found interacting with the negative charges on the exterior of the DNA helix. Under physiological conditions, the positively charged ions interact favorably with the negatively charged phosphate groups, reducing electrostatic repulsion between neighboring phosphate groups and contributing to the overall stability of the double helix [3]. Thus, the propensity for the exterior of double-stranded DNA to bind to positively charged compounds is also often exploited in designing compounds that will be capable of binding DNA, and ruthenium compounds with a 2+ or 3+ charge will have favorable electrostatic interactions with DNA *in vivo* [24, 25].

## 9.3 Biological Consequences of Ru-Complex/DNA Interactions

The well-established fact that metal compounds interact with DNA *in vitro* coupled with the cytotoxic effects observed when cells are exposed to specific metal compounds suggests that DNA is also an important *in vivo* target [26]. For example, cisplatin's primary mechanism of action is thought to be the generation of intrastrand and interstrand DNA cross-links that disrupt DNA structure and prevent the unwinding of the double helix, which is necessary for most DNA-dependent processes. The generation of a large number of cross-links activates the DNA damage response pathway; but if the damage is too extensive, then the cell is targeted for programmed cell death [10]. Because of cisplatin's severe side effects and ability to effectively target a limited range of cancer types, Fruhauf and Zeller sought to discover metal complexes that were effective against a wider range of tumor cells with limited cytotoxic effects for normal cells. They compared cisplatin to five other metal complexes (three platinum, one titanium, and one ruthenium complex, KP1019, currently in clinical trials for cancer treatments) for their DNA and DNA/protein cross-linking ability and cytotoxicity in rat ovarian tumor cells. A positive correlation between DNA cross-linking ability and growth inhibition was observed for the platinum complexes, with cisplatin being the most effective. Interestingly, they observed a positive correlation between the platinum complexes' abilities to cross-link DNA and growth inhibition of the tumor cells. KP1019 was also able to induce DNA and DNA/protein cross-links, but at a significantly reduced ability of approximately 15-fold less and with a different time profile; the maximum number of cross-links was observed at the earliest time point of exposure of 6 h in contrast to the platinum complexes that caused the greatest number of cross-links after 24 h of exposure. This early study suggests that DNA cross-linking ability and extent of DNA damage is positively correlated with cytotoxicity, although other factors also contribute to cell killing ability. For example, the titanium complex did not induce cross-links but was able to inhibit cell growth, and although KP1019 behaved differently from the platinum cross-linking ability and growth inhibition activity complexes, it did exhibit both DNA comparable to the three non-cisplatin platinum complexes and was chosen for the study based on its ability to kill tumor cells, while exhibiting low toxicity toward normal cells in a rat colorectal cancer model. Furthermore, the authors suggest that KP1019's mechanism of action can partly be explained by the relatively stable Ru(III) being converted to the more reactive Ru(II) in the reductive environment of cancer cells [27]. This phenomenon, termed the "activation by reduction" hypothesis, was proposed by Clarke as a means of selective drug activation in cancer cells compared to normal cells, thus providing a plausible explanation for efficient killing of tumor cells while also exhibiting low toxicity toward normal cells [28].

More recent studies on KP1019 have focused on elucidating its mechanism of action in HT29 and SW480 colorectal tumor cells, LT97 cells used to model early stages in cancer development, and the eukaryotic model organism, the unicellular yeast *Saccharomyces cerevisiae*. These studies provide experimental evidence for both direct and indirect effects leading to DNA damage, and characterize the type of DNA damage that is elicited in cells [29]. Initially, Fruhauf and Zeller's work showed that KP1019 was able to induce DNA interstrand cross-links, lesions that severely disrupt DNA structure and function, and other studies have detected Ru in the nuclei and bound to DNA isolated from cancer cells treated with KP1019 [30]. Capillary electrophoresis studies have shown preferential binding of KP1019 to guanosine monophosphate and adenosine monophosphate; thus, KP1019 likely interacts with purines in the context of the DNA molecule [31]. However, KP1019 was significantly less efficient in inducing interstrand DNA cross-links compared to cisplatin, and additional studies from other researchers suggest that the generation of ROS through the "activation by reduction" mechanism is the main mode of inflicting DNA damage *in vivo*.

Specifically, Kapitza and colleagues used a fluorometric assay in which the nonfluorescent intracellular probe, dichlorofluorescin-diacetate (DCFH-DA) can be oxidized to the fluorescent form of dichlorofluorescein (DCF) to detect the formation of the ROS, $H_2O_2$. When cancer cell lines were treated with both DCFH-DA and increasing amounts of KP1019, a corresponding increase in fluorescence was observed which strongly suggests that KP1019 exposure generates ROS *in vitro*. In addition, when the cancer cells were simultaneously treated with KP1019 and *N*-acetylcysteine (NAC), which inhibits $H_2O_2$ formation, fluorescence levels did not increase in a dose-dependent manner. These results were positively correlated with the amount of DNA damage observed, with the simultaneous treatment of KP1019 and NAC resulting in reduction of DNA strand breaks similar to control levels [29]. Thus, the DNA damage observed upon treatment of cancer cells with specific Ru complexes like KP1019 is most likely a result of both damage due to direct binding of the Ru complex to the DNA and indirect damage due to the increased intracellular levels of ROS. Furthermore, experiments performed on SW 480 cells with different DNA repair inhibitors showed a different sensitivity profile between KP1339 (the sodium analogue of KP1019) treatment and cisplatin treatment, with the KP1339-treated cells showing enhanced sensitivity to methoxyamine which inhibits base excision repair and aphidicolin which inhibits nucleotide excision repair. These results suggest that the type of DNA damage resulting from KP1019/KP1339 treatment is different from the DNA damage caused by cisplatin treatment, consistent with KP1019 having a different mode of action for cytotoxicity [28a,b].

It is important to discern the specific DNA damage response pathway that is activated upon drug exposure, because distinct types of DNA lesions are repaired by different mechanisms [32]. For example, damage to single bases can be repaired by the base excision repair (BER) pathway, which uses a family of enzymes called DNA glycosylases to recognize and remove the damaged base [33]. In contrast, significant distortions to the DNA helix due to intra- and interstrand cross-linking is likely targeted by the nucleotide excision repair (NER) pathway which uses a distinct set of proteins to recognize and repair the damage by removing and replacing several bases on either side of the damaged site [34]. Interstrand cross-links as well as single-stranded and double-stranded breaks in the DNA can also be repaired by the recombination pathway [35]. Thus, determining the type of DNA damage caused upon drug treatment will help elucidate a mechanism of action and is important to consider in cancer therapy as different cancers have different genetic profiles and may exhibit markedly distinct DNA repair pathway activities [36].

Studies in yeast have further elucidated the type of DNA damage KP1019 causes in eukaryotic cells. Singh and coworkers initially used native agarose gel electrophoresis to analyze DNA isolated from *in vitro* mononucleosomes (single core chromatin particle of double-stranded DNA wound around the core histone proteins) to examine whether KP1019 would affect any changes to chromatin structure. They observed an increase in free DNA with increasing KP1019 concentration, suggesting that histones were removed from the DNA in the presence of the drug. Further *in vitro* experiments determined that the likely target of KP1019 was histone H3. Specifically, the core histone proteins, H3 and H4, were incubated with KP1019 and protein blot hybridization experiments detected the presence of additional higher molecular weight species in the H3 samples, suggesting that KP1019 interacts with H3. *In vivo* sensitivity assays were then conducted on an array of yeast mutants, and increased drug sensitivity for a histone 3 point mutant (H3K56A) was observed. Serial dilutions of yeast cells were spotted onto plates containing KP1019 and an estimated over 10-fold decrease in viability was observed at 50 μg/ml treatment and almost no detectable growth was observed at the 100 μg/ml treatment compared to a slight decrease in growth of wild-type yeast at the same concentrations [29]. This result is significant because the acetylation of lysine56 on H3 is required to elicit the DNA damage response and maintain genomic integrity; thus, yeast strains that contain the H3K56A mutant are impaired in their ability to respond to DNA damage [37].

Additional experiments examined the cellular response to DNA damage. Specifically, yeast cells treated with increasing concentrations of KP1019 and subsequent protein expression experiments indicated corresponding increased expression of ribonucleotide reductase (RNR) genes which encode enzymes that catalyze the replacement of the 2′hydroxyl group on the ribose ring of a nucleotide with a hydrogen atom, an activity that converts ribonucleotides into deoxyribonucleotides and is upregulated to provide a source of nucleotides during times of increased DNA damage [38]. These results are consistent with an earlier study that showed increased expression of an RNR reporter in the presence of increasing KP1019, suggesting an increase in the amount of DNA damage [29]. In addition, *in vivo* sensitivity assays on an array of yeast mutants defective in different DNA repair pathways indicated that strains defective in single- and double-stranded break repair were the most sensitive, although an increased sensitivity to KP1019 was also observed for NER and BER mutants compared to wild type [29]. These results are consistent with previous studies showing multiple DNA repair pathways (RAD52-dependent recombination, NER and translesion synthesis) significantly contribute to KP1019 resistance [29]. Fluorescence-activated cell sorting (FACS) analysis indicated that the KP1019-treated cells accumulated in the $G_2$/M phase of the cell cycle, which is indicative of DNA damage, since cells will halt their growth until any detected DNA damage is repaired [32]. These results were also observed by Stevens and colleagues, who used microscopy to examine cell morphology of KP1019-treated yeast cells and found a large number of dumbbell-shaped cells with elongated or partially separated nuclei within or close to the bud neck, a yeast cell morphology characteristic of $G_2$/M arrest [29b, 39].

In summary, these studies demonstrate that exposure to KP1019 causes DNA damage which in turn elicits a DNA damage response. It is highly likely that KP1019 is able to directly or indirectly induce various types of DNA damage as the major repair pathways are implicated in conferring drug resistance. These studies highlight the complexity of determining a drug's mechanism of action, while Fruhauf and Zeller's work from the 1990s demonstrates that it is easier to see the contribution of DNA-binding affinity among metal complexes that are more structurally similar.

A number of other studies examining a series of similar Ru compounds have positively correlated the cell killing ability of various Ru complexes with their ability to bind to DNA, observing that the most effective compounds were able to bind DNA with the highest affinity and/or cause the greatest amount of structural disruption [15b, 40]. Several Ru complexes that successfully bind DNA contain polypyridyl ligands, although comparisons of similar Ru complexes have discerned possible connections between DNA binding, extent of DNA damage, and cytotoxicity. For example, Novakova and colleagues synthesized a series of related Ru-polypyridyl complexes differing in the number of chemically labile coordinated chloride ligands. They observed that the most effective cytotoxic agent, mer-[Ru(tpy)$Cl_3$], had the most leaving groups and was the only compound in the series able to form interstrand DNA cross-links. Cellular uptake experiments established that mer-[Ru(tpy)$Cl_3$] was taken into cells at rates similar to that of most of the other compounds studied and about seven times less than [Ru(tpy)(bpy)Cl]Cl, which exhibited no DNA interstrand ability under the conditions tested and no significant levels of cellular cytotoxicity [15b, 40].

In a separate study, Tan and colleagues examined a series of Ru complexes containing bioactive β-carboline alkaloids as ligands. They also observed a positive correlation between DNA-binding affinity *in vitro* and cytotoxic effects on four different human cancer cell lines. Importantly, cellular uptake and intracellular experiments that determined that the most potent cell-killing compound was most efficiently internalized were also performed and targeted to the nucleus where it was able to interact with DNA [41]. Thus, efficient cellular uptake and intracellular localization are also important factors in drug action. In addition, as a basis for experimental comparison, Tan and coworkers included NAMI-A, the first Ru-based compound to enter anticancer clinical trials. NAMI-A exhibited low cytotoxicity even though NAMI-A has been shown to interact with DNA *in vitro* [42]. However, NAMI-A's primary mode of drug action is thought to be the inhibition of metastasis, the spreading of cancer cells away from the primary tumor, resulting in the prevention of additional tumor formation. It has been suggested that NAMI-A interferes with the function of integrins, extracellular proteins that control cell migration and adhesion, processes essential for establishing a cell population in a different environment [43]. This highlights the complexity of a drug's mechanism of action, as DNA binding may be one of several contributing factors.

Because of DNA's essential role as the repository of genetic information in all cells, disruption of DNA structure is detrimental to both prokaryotes and eukaryotes. For example, a study by Mazumder and coworkers has shown that a number of structurally related ruthenium compounds exhibit both anticancer activity, decreasing tumor growth and increasing life span of a mouse cancer model, and antibacterial activity comparable to the control treatment of chloramphenicol for the microorganisms, *Vibrio cholera* 865, *Staphylococcus aureus* 6571, and *Shigella flexneri* [44]. Although DNA-binding studies were not performed as part of the study, the Ru compounds examined were structurally similar to ruthenium tris-chelates proposed to bind DNA via major groove binding and intercalation [45]. Additional studies have shown that a number of Ru complexes with known DNA intercalating ability also exhibit antimicrobial activity [46]. Specifically, a series of ruthenium (II) polypyridyl complexes that have been shown to bind DNA *in vitro* also exhibited antimicrobial activity against the Gram-positive bacteria *Bacillus subtilis* and *S. aureus* in zone inhibition assays. The toxicity and ability to protect against infection were assessed in the eukaryotic model organism, the nematode *Caenorhabditis elegans* [47]. In these experiments, the Ru compounds were able to rescue the nematodes from infection, and the worms exhibited identical survival rates in the presence and absence of the compounds, indicating that the Ru compounds were not toxic. The antimicrobial effects of the Ru compounds positively correlated with increasing size of the aromatic ring of the ligands with $[Ru(2,9\text{-}Me_2phen)_2(dppz)]^{2+}$ being the most active. It has been suggested that increased lipophilicity is an important factor in effective antimicrobials, although the antimicrobial abilities of these Ru compounds also positively correlated with the ability to bind to DNA *in vitro*, consistent with studies in cancer cells discussed earlier, and the development of dinuclear Ru complexes that exhibit both increased DNA-binding affinity and antimicrobial activity [40, 46, 47].

Because of the increased incidence in drug-resistant strains of bacteria, pursuing the development of novel antimicrobials is paramount to preserving human health. Thus, many research groups have focused their efforts on development of a number of Ru-based compounds with antimicrobial activity. Although the interaction with other cellular targets such as RNA and the protein translational machinery also likely contributes to the antimicrobial activity of some Ru compounds, the ability to disrupt DNA structure and function remains an important theme in designing effective drugs [46, 48]. In an effort to discover antibacterials effective against methicillin-resistant *S. aureus* (MRSA), Lam and colleagues synthesized four novel Ru complexes containing *N*-phenyl-substituted diazafluorenes. Two of the complexes exhibited antibacterial effects against MRSA in zone inhibition assays. Additional experiments on the most effective compound, Ru-C7, demonstrated that it was capable of generating significant levels of ROS as measured by a DCFH-DA assay. Although cell localization experiments were not performed, the authors speculate that the increased antimicrobial activity of Ru-C7 could be due to the longer carbon chain increasing hydrophobic character, leading to enhanced cellular uptake. The authors further speculate that the generation of ROS leads to increased levels of DNA damage and, subsequently, cell death. In addition, Ru-C7-treated human skin cells did not exhibit any decrease in cell viability, suggesting that this Ru compound may be a viable topical antibacterial against MRSA skin infections [49].

## 9.4 Effects of Ru Complexes on Topoisomerases and Telomerase

It is reasonable to assume that cytotoxic effects that correlate positively with the extent of DNA damage are due, in part, to the disruption of protein/DNA interactions. Specifically, DNA in prokaryotic and eukaryotic cells needs to be unwound for most of its functions and the normal sequence of DNA bases encodes important information like the start and end sites for transcription of a gene, and the intervening sequence that will ultimately be translated into functional protein. DNA structure and sequence also designate other functional genomic regions such as replication origins in both prokaryotes and eukaryotes and telomeres, the ends of linear chromosomes in eukaryotes. Two families of DNA-binding proteins that have been shown to be specifically targeted by Ru compounds are the topoisomerases which modify DNA topology and telomerase which catalyzes the lengthening of linear chromosomes.

Essential cellular processes such as the unwinding of double-stranded DNA during replication cause changes in overall DNA topology and changes in supercoiling, overwinding or underwinding of the double-stranded DNA molecule, or coiling of the coiled double helix. For example, DNA replication begins at a single site called the origin of replication on the bacterial circular chromosome, and negative supercoiling caused by underwinding is required at this site for the initiator protein, DNAa, to bind. However, the introduction of negative supercoils in one region of the chromosome induces positive supercoils in other regions to relieve the physical stress and prevent DNA breaks. If this process is not regulated, increased supercoiling will eventually lead to breaks in the DNA molecule and cause detrimental DNA damage [50]. Furthermore, both prokaryotic and eukaryotic genomes are organized into topological domains that facilitate the condensation of the genetic material. allowing the DNA molecule to fit in the cell as well as promoting processes like DNA repair and transcription [51].

Topoisomerases are a conserved family of enzymes that promote appropriate supercoiling to prevent unregulated DNA breaks and help maintain the overall chromosomal structural organization and genomic integrity. Topoisomerase activity is conserved in both prokaryotes and eukaryotes and can be divided into two general types depending on the enzymatic activity [50]. Type I topoisomerases are able to catalyze the cleavage of a single strand of DNA of the double-stranded DNA molecule. The other single DNA strand is then passed through the break and the cut site is resealed, resulting in a relaxation or removal of a negative supercoil. Type II topoisomerases are able to catalyze the cleavage of both DNA strands of the double-stranded DNA molecule, the passage of a second double-stranded DNA molecule through the break, and a ligation of the cut sites. This activity is especially important during DNA replication when two replication forks approach each other and supercoiling proceeds prior to fork progression. Topoisomerase activity is also essential to ensure the proper untangling of newly replicated DNA at the completion of S phase and the separation and segregation of chromosomes during M phase. Both types of topoisomerases function by forming a transient covalent bond between a tyrosine residue in the enzyme active site and an electron-deficient phosphorus in the DNA's phosphodiester backbone. Because topoisomerase activity is required to maintain genomic integrity, these enzymes make viable drug targets. In addition, because their activity is dependent on intimately interacting with DNA, disruption in DNA structure can lead to the disruption of topoisomerase function. Three general drug mechanisms are to prevent the initial interaction of topoisomerase with the DNA, inhibit the catalytic activity of the enzyme without resulting in a DNA strand break, or to act as a topoisomerase poison which promotes the formation of a DNA structure that irreversibly binds to the enzyme, consequently preventing its ability to catalyze the religation of DNA strand breaks [52].

Because topoisomerases are promising drug targets, a number of research groups have focused on developing Ru compounds that disrupt their function. Specifically, Gopal and Kondapi examined the ability of KP1019 and the structurally related imidazole-based compound, KP418, to inhibit topoisomerase II activity. [$H^3$]Thymidine incorporation assays, which monitor the rates of newly synthesized DNA, a hallmark of cell proliferation, were performed on a colon cancer cell line (Colo-205) and a breast cancer cell line (ZR-75-1) that had been treated with micromolar concentrations of either drug. Both Ru compounds were able to decrease radiolabeled thymidine incorporation in cancer cells to an extent similar to that of the known topoisomerase II inhibitor, amsacrine (m-ASMA). Because topoisomerases modify DNA topology, their activity can be monitored *in vitro* by gel electrophoretic analysis of plasmid DNA treated with the enzymes. For example, gel electrophoresis results of a plasmid relaxation assay can indicate the extent of topoisomerase's ability to catalyze the removal of supercoils, indicated by an increase in relaxed, slower migrating DNA bands compared to the faster migrating, compacted, and supercoiled plasmid. When plasmid relaxation assays with topoisomerase II were performed in the presence of micromolar amounts of KP1019 and KP418, both compounds were able to inhibit the enzyme's activity in a dose-dependent manner, with KP1019 being the more effective inhibitor [53].

Because the results of *in vitro* plasmid relaxation assays can be complicated by effects of the DNA–drug interactions, which can also alter the DNA topology and migration pattern in gel electrophoresis, additional experiments are needed to confirm the inhibition of topoisomerase activity. Specifically, topoisomerase II activity can be monitored by measuring rates of ATP hydrolysis because the strand passage of one double-stranded DNA molecule through a double-stranded break in a second DNA molecule requires ATP hydrolysis. This ATPase activity of topoisomerase II is greatly stimulated in the presence of negatively supercoiled DNA and can be monitored *in vitro* using a radiolabeled [$\gamma^{32}$P]ATP substrate and quantitating how much $\gamma^{32}$P is generated. A second important assay is the *in vitro* cleavage or strand passage assay which determines if compounds are able to function as topoisomerase poisons and trap the enzyme and the DNA in a ternary drug-dependent complex, and prevent the enzyme from catalyzing the ligation of the DNA strand break. If a compound is able to act as a poison, then when DNA samples are analyzed by gel electrophoresis, linear DNA is observed in DNA samples treated with both the enzyme and the compound [54].

Further experiments on KP1019 and KP418's ability to inhibit topoisomerases indicate that these compounds are able to act as topoisomerase II poisons *in vitro* and *in vivo*. For example, topoisomerase II ATPase assays performed in the presence of the drugs resulted in an 80% inhibition by KP1019 and a 58% inhibition by KP418. These results suggest that both compounds can function as topoisomerase II poisons *in vitro* and are consistent with *in vitro* cleavage assay results which indicate formation of cleavage complexes at 150 micromolar KP1019 and 300 micromolar KP418 [55]. These results are comparable to inhibition seen with the known topoisomerase II inhibitor, m-ASMA and are consistent with work by Stevens and colleagues who explored KP1019's mechanism of drug action in yeast. Specifically, yeast cells that express high levels of topoisomerase II display increased sensitivity to topoisomerase poisons because in the presence of these drugs, more double-stranded DNA breaks will be generated. When yeast cells overexpressing topoisomerase II were treated with KP1019, a slight decrease in growth was observed compared to wild-type yeast, consistent with KP1019 able to function as a topoisomerase II poison *in vivo* [29b].

Studies from the Liang-Nian Ji research group have also focused on developing Ru complexes capable of inhibiting topoisomerase activity. A number of related Ru-bipyridyl (bpy) complexes with the Ru center coordinated to a third bioactive ligand, ranging from nucleotides such as uracil and guanine to neoplastic anthraquinones, have been synthesized and characterized for their cancer drug potential. *In vitro* DNA binding, thermal denaturation studies, and viscosity measurements were used to establish that these Ru complexes bound to DNA primarily through intercalation and generally exhibited DNA-binding affinities in the $10^5\ M^{-1}$ range. These Ru complexes were able to catalyze the photocleavage of DNA in the presence of light at 365 nm. Although this wavelength is outside the window for photodynamic therapy, some of these compounds were able to inhibit the activity of one or both topoisomerases as measured by *in vitro* cleavage assays and effectively kill a number of different cancer cell lines in the absence of any irradiation. In general, a significant inhibition of topoisomerase activity *in vitro* correlated positively with cytotoxic ability, suggesting that topoisomerases may represent important drug targets in cancer cells. For example, two enantiomers of $[Ru(bpy)_2(uip)]^{2+}$ where uip is 2-(5-uracil)-1 *H*-imidazo[4,5-*f*] [1, 10, phenanthroline) were synthesized containing uracil, a nitrogenous base that is a component of RNA and the basis for the commonly used 5-fluorouracil chemotherapeutic agent [56]. Both enantiomers had inhibitory ability comparable to the known topoisomerase poison, Topostatin and the cancer drug, doxorubicin. In addition, both Ru-based drugs were two- to fivefold more cytotoxic in HELA, hepG2, and BEL-7402 cancer cell lines compared to 5-fluorouracil, which interferes with nucleotide synthesis in proliferating cells. Thus, the Ru-based compounds likely exert their cytotoxic effects through multiple mechanisms including the induction of DNA damage and the inhibition of topoisomerase activity [57]. In a separate study, $[Ru(bpy)_2(appo)]^{2+}$ where appo is 11-aminopteridion[6,7-*f*][1, 10, phenanthrolin-13(12*H*)-one was synthesized and shown to have a preference for interacting with GC-rich DNA sequences *in vitro*. This binding preference may primarily be due to the appo ligand containing a group that resembles guanine and is thus able to hydrogen-bond with cytosines in addition to intercalating via the bpy ligands. *In vitro* strand passage (cleavage) assays indicated that this Ru compound was able to inhibit topoisomerase II activity at concentrations below 1 micromolar comparable to doxorubicin, a cancer drug that targets topo II activity [58]. The same research group also characterized the cytotoxic ability of a series of $Ru(bpy)_2$–thiophene complexes, where the thiophene functional group has previously been shown to have cytotoxic effects particularly in liver cells, the site of drug detoxification in mammals [59]. *In vitro* topoisomerase assays demonstrated that these Ru compounds were able to inhibit both topoisomerase I and topoisomerase II activity in the 8–12 micromolar range. *In vitro* DNA-binding and viscosity assays established that these Ru compounds bound to CT-DNA through intercalation. Interestingly, the ability to increase viscosity, a hallmark of the intercalation binding mode, positively correlated with cytotoxicity and cellular uptake. Thus, the ability to interact with DNA and the ability to be internalized into cells are two important factors contributing to cytotoxicity. Microscopy studies exploited the intrinsic fluorescence of the Ru compounds to determine the kinetics of uptake and accumulation, with the Δ enantiomers displaying slightly faster entry into cells compared to the Λ counterparts. Overall, all compounds accumulated in the cell nuclei after 24 h and likely bound to chromosomal DNA. The possibility that chromosomal DNA was the target for the Ru compounds was confirmed by the inability of DAPI stain to enter the nuclei and interact with the DNA at this time point [60]. A suggested drug mechanism is that these Ru complexes are targeted to the nucleus where they interact with both chromosomal DNA and topoisomerases, trapping the enzyme–DNA cleavage complex which results in permanent DNA strand breaks. The presence of DNA strand breaks then elicits a DNA damage response unless the extent of damage is so severe, triggering apoptosis.

Results from additional experiments support these conclusions. Specifically, alkaline single-cell gel electrophoresis or COMET assays were performed on HELA cells treated with 50 micromolar Δ-1 (Cl substituent on the thiophene ring) or Δ-2 (Br substituent on the thiophene ring). COMET assays allow the sensitive detection of DNA breaks in cells by embedding cells in agarose on microscope slides, subjecting them to electrophoresis at high pH, and visualizing the nuclear DNA by staining with a fluorescent dye. Breaks in the DNA result in a decrease of supercoiling and a greater migration of fragmented DNA compared to intact DNA, resulting in structures that resemble comets with a bright circular head corresponding to intact DNA and a diffuse tail containing the DNA with strand breaks [61]. The comparison of these two regions of the stained DNA reveals a qualitative measure of DNA strand breaks. Both Ru compounds tested were able to produce DNA profiles similar to known topoisomerase inhibitors. To confirm that the extent of DNA damage triggered cell death by apoptosis, HELA cells were stained with Annexin which detects the presence of phosphatidyl serine, a phospholipid that is translocated to the outer surface of the plasma membrane during apoptosis [62]. An increase in Annexin staining was observed in a dose-dependent manner, strongly suggesting that the treated cells were undergoing drug-induced apoptosis, which was confirmed by cell cycle analysis experiments indicating an increase in cells with sub-$G_1$ DNA content over the course of a 48-h drug treatment [60]. Cells will progress normally through the cell cycle, unless significant damage is detected and cell cycle arrest occurs, providing the cells with an opportunity to repair any damage or trigger cell death. The sub-$G_1$ DNA content is a marker of cells that have suffered DNA fragmentation and are undergoing apoptosis [63]. Although the studies discussed focused on *in vitro* topoisomerase activity and cytotoxic effects on human cancer cell lines, the conservation of topoisomerase activity in both eukaryotes and prokaryotes and the existence of topoisomerase-inhibiting antimicrobials suggest that topoisomerases are a viable target for the development of novel antimicrobials as well.

In contrast to targeting topoisomerase activity, some researchers have focused efforts on developing Ru complexes that interfere with the activity of another DNA-binding enzyme, telomerase. Telomerase activity is unique to eukaryotes, and the strategy of targeting telomerase is based on observations that this enzymatic activity is increased in approximately 80–90% of all human cancers while not being detectable in most normal cells [64]. The need for telomerase activity becomes apparent when the general structure of the eukaryotic linear chromosome is considered in light of DNA replication which proceeds in the

5′ to 3′ direction. Because DNA replication is unidirectional and template dependent, continuous replication occurs on the 3′ to 5′ strand of the original duplex (the leading strand), while disjointed replication occurs on the 5′ to 3′ strand of the original duplex (the lagging strand) with short Okazaki fragments of DNA being synthesized every 100-200 nucleotides. DNA polymerase activity requires a primer, a short RNA segment that base pairs with part of the DNA strand to be replicated [65]. This primer requirement could present a problem for lagging strand replication at the end of the chromosome because there would be no place for the RNA primer to anneal, causing eukaryotic chromosomes to shorten with each replication event leading to increasing loss of genetic material. However, this usually does not happen for several cell divisions because eukaryotic chromosomes have their ends protected by telomeres, long stretches of species-specific repetitive DNA sequences with single-stranded 3′ overhangs on the G-rich strand to serve as templates for telomerase. For example, the human telomeric repeat is $(TTAGGG)_n$ where n can initially be in the several thousands at the beginning of a normal cell's life span and progressively decrease with each cell division. Reactivation of telomerase in cancer cells enable them to maintain a critical telomere length, one factor which confers unlimited replicative ability [66].

Human genome sequence analysis has determined that G-rich sequences able to form G-quadruplex structures are overrepresented in both the regulatory regions surrounding the DNA sequence that will ultimately get translated into protein and at the telomeres, strongly suggesting that these DNA secondary structures have important roles in controlling gene expression and maintaining functional telomeres [67]. Guanines adjacent to additional guanines have the lowest oxidation potential among the nucleotides [68], and charge transport experiments suggest that the G-quadruplex structure is a more efficient electron trap compared to two adjacent guanines in double-stranded DNA [69], raising the possibility that G-rich regions are more susceptible to oxidative damage compared to other genomic regions. Ru complexes developed to date seem to primarily function by stabilizing G-quadruplex DNA structures, although the development of Ru(III) complexes may be a promising approach to create drugs capable of G-quadruplex stabilization as well as targeted oxidative damage due to generation of ROS at telomeric sites by the "activation by reduction" mechanism. Several studies targeting telomerase activity have focused on developing structurally similar Ru complexes that are capable of stabilizing G-quadruplex DNA structures which in turn inhibit telomerase activity, leading to impaired DNA replication and subsequent cancer cell death [70–74].

## 9.5 Summary and Conclusions

After several decades of research, we are just scratching the surface of the biological and therapeutic uses of ruthenium complexes. Ruthenium's highly tunable redox and photophysical properties coupled to its relative low cytotoxicity toward normal cells, ligand exchange kinetics, high coordination number, and variable ligand coordination make ruthenium unique among the elements. Collaboration between traditional coordination chemists, biochemists, medicinal chemists, and biologists should yield some very exciting new roles of ruthenium complexes as therapeutic agents.

## References


1 Strekowski, L. and Wilson, B. (2007) *Mutat. Res.*, **623**, 3–13.
2 Watson, J.D. and Crick, F.H.C. (1953) *Nature*, **171**, 737.
3 Nelson, D.L. and Cox, M.M. (eds) (2013) *Lehninger Principles of Biochemistry*, 6th edn, W.H. Freeman, New York, pp. 281–291.
4 Zimmerman, S.B. (1982) *Annu. Rev. Biochem.*, **51**, 395–427.
5 Harteis, S. and Schneider, S. (2014) *Int. J. Mol. Sci.*, **15**, 12335–12363.
6 Travis, A. and Muskhelishvili, G. (2015) *FEBS J.*, **282**, 2279–2295.
7 Blackburn, E.H. (1991) *Nature*, **350**, 569–573.
8 Lippard, S.J. (1982) *Science*, **218**, 1075–1081.
9 Gelasco, A. and Lippard, S.J. (1998) *Biochemistry*, **37**, 9230–9239.
10 Jamieson, E.R. and Lippard, S.J. (1999) *Chem. Rev.*, **99**, 2467–2498.
11 Clarke, M.J., Zhu, F., and Frasca, D.R. (1999) *Chem. Rev.*, **99**, 2511–2533.
12 Barton, J.K. and Lolis, E. (1985) *J. Am. Chem. Soc.*, **107**, 708–709.
13 Grover, N., Gupta, N., and Thorp, H.H. (1992) *J. Am. Chem. Soc.*, **114**, 3390–3393.
14 (a) Sears, R.B., Joyce, L.E., Ojami, M., Gallucci, J.C., Thunnel, R.P., and Turro, C. (2013) *J. Inorg. Biochem.*, **121**, 77–87; (b) Singh, T.N. and Turro, C. (2004) *Inorg. Chem.*, **43**, 7260–7262.
15 (a) Zhang, C.K. and Lippard, S.J. (2003) *Curr. Opin. Chem. Biol.*, **7**, 481–489; (b) Brabec, V. and Novakova, O. (2006) *Drug Resist. Updates*, **9**, 111–122.
16 Talib, J., Harman, D.G., Dillon, C.T., Aldrich-Wright, J., Beck, J.L., and Ralph, S.F. (2009) *Dalton Trans.*, **39**, 504–513.
17 (a) Lerman, L.S. (1961) *J. Mol. Biol.*, **3**, 18–30; (b) Berman, H.M. and Young, P.R. (1981) *Annu. Rev. Biophys. Bioeng.*, **10**, 87–114.
18 Howe-Grant, M., Wu, K.C., Bauer, W.R., and Lippard, S.J. (1976) *Biochemistry*, **15**, 4339–4346.
19 Lippard, S.J., Bond, P.J., Wu, K.C., and Bauer, W.R. (1976) *Science*, **194**, 726–727.
20 Zeglis, B.M., Pierre, V.C., and Barton, J.K. (2007) *Chem. Commun.*, **44**, 4565–4579.
21 Hall, J.P., O'Sullivan, K., Naseer, A., Smith, J.A., Kelly, J.M., and Cardin, C.J. (2011) *PNAS*, **108**, 17610–17614.
22 Barton, J.K. (1986) *Science*, **233**, 727–734.
23 Barton, J.K., Basile, L.A., Danishersky, A., and Alexandrescu, A. (1984) *Proc. Nat. Acad. Sci.*, **81**, 1961–1965.
24 Gill, M.R. and Thomas, J.A. (2012) *Chem. Soc. Rev.*, **14**, 3179–3192.
25 Li, F., Collins, J.G., and Keene, F.R. (2015) *Chem. Soc. Rev.*, **10**, 1039–1052.
26 Pizarro, A.M. and Sadler, P.J. (2009) *Biochimie*, **91**, 1198–1211.

27 (a) Fruhauf, S. and Zeller, W.J. (1991) *Cancer Res.*, **51**, 2943–2948; (b) Berger, M.R., Garzon, F.T., Keppler, B.K., and Schmahl, D. (1989) *Anticancer Res.*, **9**, 761–765.

28 (a) Hartinger, C.G., Zorbas-Seifried, S., Jakupec, M.A., Kynast, B., Zorbas, H., and Keppler, B.K. (2006) *J. Inorg. Biochem.*, **100**, 891–904; (b) Hartinger, C.G., Jakupec, M.A., Zorbas-Seifried, S., Groessl, M., Egger, A., Berger, W., Zorbas, H., Dyson, P.J., and Keppler, B.K. (2008) *Chem.Biodivers.*, **5**, 2140–2155; (c) Clarke, M.J. (2003) *Coord. Chem. Rev.*, **236**, 209–233.

29 (a) Kapitza, S., Jakupec, M.A., Uhl, M., Keppler, B.K., and Marian, B. (2005) *Cancer Lett.*, **226**, 115–121; (b) Stevens, S.K., Strehle, A.P., Miller, R.L., Gammons, S.H., Hoffman, K.J., McCarty, J.T., Miller, M.E., Stultz, L.K., and Hanson, P.K. (2013) *Mol. Pharmacol.*, **83**, 225–234; (c) Singh, V., Azad, G.K., Reddy, A., Baranwal, S., and Tomar, R.S. (2014) *Eur. J. Pharmacol.*, **736**, 77–85.

30 Trondl, R., Heffeter, P., Kowol, C.R., Jakupec, M.A., Berger, W., and Keppler, B.K. (2014) *Chem. Sci.*, **5**, 2925–2932.

31 Kung, A., Pieper, T., Wissiack, R., Rosenberg, E., and Keppler, B.K. (2001) *J. Biol. Inorg. Chem.*, **6**, 292–299.

32 (a) Walters, R. (2006) *EMBO Rep.*, **7**, 377–381; (b) Ciccia, A. and Elledge, S.J. (2010) *Mol. Cell Rev.*, **40**, 179–204.

33 Jacobs, A.L. and Schar, P. (2012) *Chromosoma*, **121**, 1–20.

34 (a) de Laat, W.L., Jaspers, N.G.J., and Hoeijmakers, J.H.J. (1999) *Genes Dev.*, **13**, 768–785; (b) Clauson, C., Scharer, O.D., and Niedernhofer, L. (2013) *CSH Perpect.*, **5**, 1–25; (c) Scharer, O.D. (2013) *CHS Perspect.*, **5**, 1–19.

35 Krejci, L., Altmannova, V., Spirek, M., and Zhao, X. (2012) *Nucleic Acids Res.*, **40**, 5795–5818.

36 Ahmed, D., Eide, P.W., Eilertsen, I.A., Danielsen, S.A., Eknaes, M., Hektoen, M., Lind, G.E., and Lothe, R.A. (2013) *Oncogenesis*, **2**, 1–8.

37 (a) Han, J., Zhou, H., Horazdovsky, B., Zhang, K., Xu, R.-M., and Zhang, Z. (2007) *Science*, **315**, 653–655; (b) Driscoll, R., Hudson, A., and Jackson, S.P. (2007) *Science*, **315**, 649–652.

38 Nordlund, P. and Reichard, P. (2006) *Annu. Rev. Biochem.*, **75**, 681–706.

39 Hampsey, M. (1997) *Yeast*, **13**, 1099–1133.

40 Novakova, O., Kasparkova, J., Vrana, O., van Vliet, P.M., Reedijk, J., and Brabec, V. (1995) *Biochemistry*, **34**, 12369–12378.

41 Tan, C., Lai, S., Wu, S., Hu, S., Zhou, L., Chen, Y., Wang, M., Zhu, Y., Lian, W., Peng, W., Ji, L., and Xu, A. (2010) *J. Med. Chem.*, **53**, 7613–7624.

42 Gallori, E., Vettori, C., Alessio, E., Vilchez, F.G., Vilaplana, R., Orioli, P., Casini, A., and Messori, L. (2000) *Arch. Biochem. Biophys.*, **376**, 156–162.

43 Bergamo, A. and Sava, G. (2007) *Dalton Trans.*, **2007**, 1267–1272.

44 Mazumder, U.K., Gupta, M., Karki, S.S., Bhattacharya, S., Rathinasamy, S., and Sivakumar, T. (2005) *Bioorg. Med. Chem.*, **13**, 5766–5773.

45 Tysoe, S.A., Morgan, R.J., Baker, A.D., and Strekas, T.C. (1993) *J. Phys. Chem.*, **97**, 1707–1711.

46 Li, F., Feterl, M., Mulyana, Y., Warner, J.M., Collins, J.G., and Keene, F.R. (2012) *J. Antimicrob. Chemother.*, **67**, 2686–2695.

47 Bohuis, A., Hand, L., Marshall, J.E., Richards, A.D., Rodger, A., and Aldrich-Wright, J. (2011) *Eur. J. Pharm. Sci.*, **42**, 313–317.
48 Li, F., Harry, E., Bottomley, A., Edstein, M.D., Birrell, G.W., Woodward, C.E., Keene, F.R., and Collins, J.G. (2014) *Chem. Sci.*, **5**, 685–693.
49 Lam, P.L., Lu, G.L., Hon, K.M., Lee, K.W., Ho, C.L., Wang, X. *et al.* (2014) *Dalton Trans.*, **43**, 3949–3957.
50 Lodish, H., Berk, A., Zipursky, S.L. *et al.* (2000) *Molecular Cell Biology*, 4th edn, W.H. Freeman, New York, Section 12.3.
51 (a) Postow, L., Hardy, C.D., Arsuaga, J., and Cozzarelli, N.R. (2004) *Genes Dev.*, **18**, 1766–1779; (b) Dixon, J.R., Selvaraj, S., Yue, F., Kim, A., Li, Y., Shen, Y., Hu, M., Liu, J.S., and Ren, B. (2012) *Nature*, **485**, 376–380.
52 Champoux, J.J. (2001) *Annu. Rev. Biochem.*, **70**, 369–413.
53 Gopal, Y.N.V. and Kondapi, A.K. (2001) *J. Biosci.*, **26**, 271–276.
54 Nitiss, J.L., Soans, E., Rogojina, A., Seth, A., and Mishina, M. (2012unit 3.3) *Curr. Protoc. Pharmacol.*, **57**, 1–34.
55 Gopal, Y.N.V., Konuru, N., and Kondapi, A.K. (2002) *Arch. Biochem. Biophys.*, **40**, 53–62.
56 Longley, D.B., Harkin, D.P., and Johnston, P.G. (2003) *Nature*, **3**, 330–337.
57 Gao, F., Chao, H., Yuan, Y.X., Sun, B., Wei, Y.F., Peng, B., and Ji, L.N. (2007) *J. Biol. Inorg. Chem.*, **12**, 1015–1027.
58 Goa, F., Chao, H., Zhou, F., Chen, X., Wei, Y.F., and Ji, L.N. (2008) *J. Inorg. Biochem.*, **102**, 1050–1059.
59 Gramec, D., Masic, L.P., and Doenc, M.S. (2014) *Chem. Res. Toxicol.*, **27**, 1344–1358.
60 Wang, Y.C., Qian, C., Peng, Z.L., Hou, X.J., Wang, L.L., Chao, H., and Ji, L.N. (2014) *J. Inorg. Biochem.*, **130**, 15–27.
61 Collins, A.R. (2004) *Mol. Biotechnol.*, **26**, 249–261.
62 Elmore, S. (2007) *Toxicol. Pathol.*, **35**, 495–516.
63 Kajstura, M., Halicka, H.D., Pryjma, J., and Darzynkiewicz, Z. (2007) *Cytometry A*, **71A**, 125–131.
64 Kim, N.W., Piatyszek, M.A., Prowse, K.R., Harley, C.B., West, M.D., Ho, P., Coviello, G., Wright, W., Weinrich, S., and Shay, J.W. (1994) *Science*, **266**, 2011–2015.
65 Nelson, D.L. and Cox, M.M. (eds) (2013) *Lehninger Principles of Biochemistry*, 6th edn, W.H. Freeman, New York, pp. 1011–1013.
66 (a) Nelson, D.L. and Cox, M.M. (eds) (2013) *Lehninger Principles of Biochemistry*, 6th edn, W.H. Freeman, New York, pp. 984–985, and 1089–1092; (b) Buseman, C.M., Wright, W.E., and Shay, J.W. (2012) *Mutat. Res.*, **730** (1–2), 90–97.
67 (a) Huppert, J.L. and Balasubramanian, S. (2005) *Nucleic Acids Res.*, **33**, 2908–2916; (b) Huppert, J.L., Bugaut, A., Kumari, S., and Balasubramanian, S. (2008) *Nucleic Acids Res.*, **36**, 6260–6268; (c) Huppert, J.L. and Balasubramanian, S. (2007) *Nucleic Acids Res.*, **35**, 406–413; (d) Lipps, H.J. and Rhodes, D. (2009) *Trends Cell Biol.*, **19**, 414–422.
68 Saito, I., Takayama, M., Sugiyama, H., Nakatani, K., Tsuchida, A., and Yamamoto, M. (1995) *J. Am. Chem. Soc.*, **117**, 6406–6407.
69 Delaney, S. and Barton, J.K. (2003) *Biochemistry*, **42**, 14159–14165.

**70** Chen, Z.F., Qin, Q.P., Qin, J.L., Zhou, J., Li, Y.L., Li, N., Liu, Y.C., and Liang, H. (2015) *J. Med. Chem.*, **58**, 4771–4789.
**71** Liao, G., Chen, X., Wu, J., Qian, C., Wang, Y., Ji, L., and Chao, H. (2015) *Dalton Trans.*, **44**, 15145–15156.
**72** Zhang, J., Yu, Q., Li, Q., Yang, L., Chen, L., Zhou, Y., and Liu, J. (2014) *J. Inorg. Biochem.*, **134**, 1–11.
**73** Liao, G., Chen, X., Wu, J., Qian, C., Wang, H., Ji, L., and Chao, H. (2014) *Dalton Trans.*, **43**, 7811–7819.
**74** Xu, L., Liao, G.L., Chen, X., Zhao, C.Y., Chao, H., and Ji, L.N. (2010) *Inorg. Chem. Commun.*, **13**, 1050–1053.

# 10

# Ruthenium-Based Anticancer Compounds: Insights into Their Cellular Targeting and Mechanism of Action

*António Matos[1], Filipa Mendes[2], Andreia Valente[3], Tânia Morais[3,4], Ana Isabel Tomaz[3], Philippe Zinck[5], Maria Helena Garcia[3], Manuel Bicho[6,7], and Fernanda Marques[2]*

[1] *Centro de Microscopia Electrónica e Histopatologia, Centro de Investigação Interdisciplinar Egas Moniz, Monte da Caparica, 2829-511, Caparica, Portugal*
[2] *Centro de Ciências e Tecnologias Nucleares, Instituto Superior Técnico, Universidade de Lisboa, E.N.10, km 139.7, 2695-066, Bobadela LRS, Portugal*
[3] *Centro de Química Estrutural, Faculdade de Ciências, Universidade de Lisboa, Campo Grande, 1749-016, Lisboa, Portugal*
[4] *Instituto de Medicina Molecular, Faculdade de Medicina, Universidade de Lisboa, Av. Prof. Egas Moniz, 1649-028, Lisboa, Portugal*
[5] *UCCS, UMR CNRS 8181, ENSCL, Bât. C7, Cité Scientifique, 59652, Villeneuve d'Ascq Cédex, France*
[6] *Laboratório de Genética, Faculdade de Medicina, Universidade de Lisboa, Av. Prof. Egas Moniz, 1649-028, Lisboa, Portugal*
[7] *Instituto Rocha Cabral, Cç. Bento Rocha Cabral, 1257-047, Lisboa, Portugal*

## 10.1 Introduction

Over the past few decades the field of anticancer medicinal chemistry has progressed toward new and innovative chemotherapeutics. In this way, a large amount of effort has been dedicated to the use of transition-metal–based compounds [1, 2]. Cisplatin, the benchmark drug based on platinum, and its analogues carboplatin and oxaliplatin, are by far the metallodrugs most used worldwide, being employed in 50–70% of chemotherapy regimens for the treatment of ovarian, testicular, lung, and bladder cancers, among others. Although highly efficient, their use is limited by severe side effects and acquired drug resistance, which has motivated the search for alternative metal-based anticancer agents with reduced toxicity, higher selectivity, and enhanced efficacy, especially against tumors resistant to platinum drugs [3, 4].

Transition metals can adopt a wide variety of coordination numbers, geometries, and oxidation states. In addition, they are usually able to switch between several oxidation states under physiological conditions. Thus, based on these different chemical features, it is expected that the mode of action and spectrum of activity will vary in relation to platinum-based drugs [5, 6]. In this frame, a plethora of non-platinum metal complexes based on transition elements namely of groups 8, 9, 10, and 11 such as gold, gallium, titanium, iron, vanadium, ruthenium, osmium, and iridium, have been tested for anticancer activity with

*Ruthenium Complexes: Photochemical and Biomedical Applications*, First Edition.
Edited by Alvin A. Holder, Lothar Lilge, Wesley R. Browne, Mark A.W. Lawrence, and Jimmie L. Bullock Jr.

varying degrees of success depending on the identity of the core metal ion and of the ligands in its coordination sphere [7–13]. Titanocene dichloride and budotitane were the first anticancer metal complexes to enter clinical trials after platinum compounds [7].

Ruthenium is recognized as a highly attractive alternative to platinum since the toxicity of many Ru compounds is lower, and some complexes are quite selective for cancer cells [14]. Ru offers several favorable chemical properties, which make them particularly useful for rational anticancer drug design. These features include a favorable ligand-exchange kinetics and the ability to access oxidation states from +2 to +4 in biological milieu and to mimic iron by binding to plasma proteins [14–18].

Research in the field of ruthenium complexes for cancer therapy has been quite extensive and has been the subject of many reviews [5, 19]. Several families of compounds based on Ru(II) and Ru(III) showing structural diversity have been developed. Among them, the octahedral KP1019 and NAMI-A (Figure 10.1), are by far the most studied compounds that have progressed into phase I clinical trials, and for which the mode of action has been fairly outlined [20–23]. In a different approach, strategies have been employed to stabilize ruthenium in the +2 oxidation state, namely, the introduction of a *N,N* polypyridyl-type

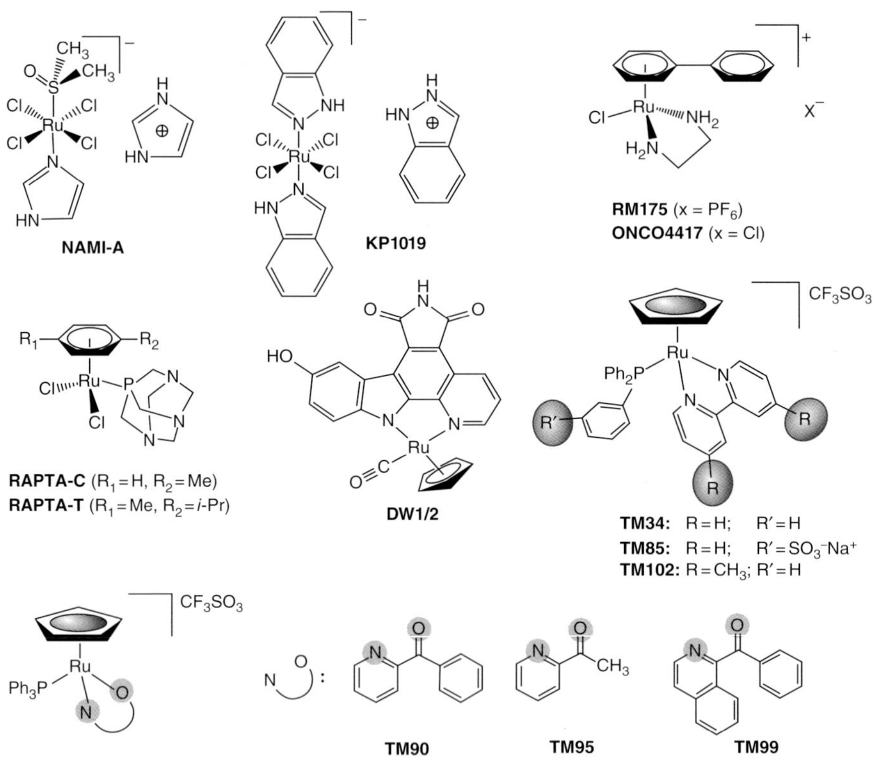


**Figure 10.1** Chemical structures of ruthenium-based compounds reported to have anticancer activity (Ph = phenyl).

ligand (e.g., 2,2′-bipyridyl or 1,10-phenantroline) in octahedral "Ru(NN)$_2$" complexes, or an arene moiety in piano-stool structured 'Ru($\eta^6$-arene)' and 'Ru($\eta^5$-cyclopentadienyl).' The stabilizing scaffolds are relatively inert toward displacement under physiological conditions, providing considerable scope for optimizing the design of anticancer complexes [16, 24–27].

Ru(II) compounds inspired on staurosporine and bearing the "Ru($\eta^5$-$C_5H_5$)" core were successfully explored for the development of complexes as kinase inhibitors (DW1/2, Figure 10.1). *In vivo* studies of DW1/2 or other compounds in this group have not yet been conducted, although its ability to inhibit Pim kinases has been demonstrated *in vitro* [22, 28, 29]. More recently, a new family of "Ru($\eta^5$-$C_5H_5$)" complexes of the type [Ru($\eta^5$-$C_5H_5$)(NN)P]$^+$ and [Ru($\eta^5$-$C_5H_5$)(NO)P]$^+$ (NN/NO being heteroaromatic bidentate ligands coordinated by nitrogen and/or oxygen atoms) (Figure 10.1) has shown increased selectivity and effectiveness. These monocationic compounds present good stability in air and moisture, remarkable cytotoxic properties in a wide range of human cancer cells, and in general lack of cross-resistance with cisplatin [30–32]. *In vivo* evaluation of lead compounds in this family is under progress in order to highlight their potential as prospective metallodrugs.

The most remarkable feature of ruthenium-prospective chemotherapeutics is the efficiency of some of these agents against metastases, placing them in the front stage for the next generation of anticancer drugs. While KP1019 exhibits activity against primary tumors (colon carcinoma and some other primary explanted human tumors) and is practically devoid of side effects, NAMI-A is not cytotoxic and is efficient against metastases.

Complexes with the "Ru($\eta^6$-arene)" scaffold are weakly cytotoxic against tumor cells and are usually free of toxicity toward healthy cells. Importantly, antimetastatic activity was reported in particular for [($\eta^6$-biphenyl)-Ru(ethylenediamine)Cl]X ($X = PF_6^-$ in RM175; $X = Cl^-$ in ONCO4417, Figure 10.1) [22]. A subfamily with the "Ru($\eta^6$-arene)" scaffold, characterized by the presence of 1,3,5-triaza-7-phosphaadamantane (pta), known as RAPTAs, was developed. The water-soluble RAPTA-C ([Ru($\eta^6$-*p*-cymene)(pta)Cl$_2$]) and RAPTA-T ([Ru($\eta^6$-*p*-toluene)(pta)Cl$_2$]) were found to exhibit antimetastatic properties *in vivo* as well (Figure 10.1).

The activity *in vitro* and *in vivo* shown by these compounds supports the hypothesis that numerous targets are involved in their mode of action, and the overall activity is the result of multiple extracellular and intracellular interactions [33]. This is particularly true for NAMI-A- and NAMI-A-type complexes that share the capacity to modify important parameters of metastasis such as tumor invasion, matrix metalloproteinases activity and cell cycle progression [5, 34]. The fact that complexes with completely different structures exhibit similar *in vitro* and *in vivo* effects (and vice versa) challenges the purpose of establishing plain structure–activity relationships and highlights the importance of understanding common features of their action.

In the biological environment, the most likely interactions of Ru complexes that might be relevant to the mechanism of action are with serum proteins and cellular components: (i) human serum albumin (HSA) and serum transferrin (Tf), the former acting as transport system and storage and the latter possibly as a carrier protein; (ii) extracellular matrix and actins on the cell surface, which are likely to be involved in the antimetastatic action of some Ru complexes; (iii) regulatory enzymes within the cell membrane and/or in the cytoplasm; and (iv) DNA in the nucleus [35–39]. In this direction, the cellular targets/pathways that have emerged as highly relevant for the mode of action of ruthenium anticancer compounds are focused on in the following sections.

## 10.2 Cellular Uptake

Cellular uptake refers to the interaction with the cell membrane and represents a considerable barrier to compound transport and accumulation. Identification of the routes across the cell membrane that could affect the rate of uptake and cellular fate is mandatory when evaluating the therapeutic potential of a drug. How a drug distributes and accumulates in the cellular compartment(s) that hosts its targets is a fundamental condition for its activity [40]. Only molecules within a narrow range of molecular weight, charge, and polarity are typically able to cross the plasma membrane by passive diffusion, an energy-independent process. Larger molecules are generally internalized by endocytosis, an energy-dependent process. By endocytosis, molecules are often trapped in endosomes and face degradation by lysosomal enzymes [41, 42]. The extent to which a drug is trapped in lysosomes is important as it can influence its ability to interact with other specific targets [43, 44].

Few mechanistic details are known regarding the uptake of metal complexes. Passive diffusion was considered the main mechanism of cellular uptake for cisplatin. However, recent reports indicated that the intracellular accumulation of cisplatin is modulated by membrane transporters and channels [45]. The cellular uptake properties of Ru complexes challenge these concepts. Different entry mechanisms may be preferred depending on the overall charge, type of ligand set, oxidation state of the metal, and can be cell-type specific [46]. The correlation between charge and uptake is consistent with the plasma membrane potential serving as the driving force for cellular entry.

Uptake mediated by transferrin through the transferrin receptor, which is overexpressed in tumor cells, is an attractive approach for attaining some degree of selectivity. Transferrin enters endosomal compartments within the cell via the transferrin cycle and could thus take ruthenium-based drugs across the cell membrane into an acidic compartment where the low pH would favor disassociation from the protein. In fact, KP1019 is thought to bind to transferrin (after displacement of a labile chloride ligand) and enters the cell by transferrin-receptor–mediated endocytosis [47]. Following cellular entry, the ruthenium complex releases from transferrin and subsequently can escape from endosomes.

The uptake pathways were also studied for $Ru^{II}(\eta^6\text{-}p\text{-cymene})$ complexes using endocytosis and ion-pump modulators [48]. Studies with ouabain, a $Na^+/K^+$-ATPase inhibitor, suggested that the cell membrane potential is a key determinant of uptake of complexes.

Analytical methods (e.g., ICP-MS, inductively coupled plasma mass spectrometry) and pharmacological tools (e.g., endocytosis inhibitors) were used to elucidate the mechanism of uptake of $Ru^{II}(\eta^5\text{-}C_5H_5)$-2,2′bipyridyl complexes (TM34, TM85, and TM102, Figure 10.1) on A2780 ovarian and MDAMB231 breast cancer cells. By ICP-MS, complexes seemed to be retained predominantly at the membrane and cytosol (~90%) with a total Ru content of ~4 nmol Ru/mg protein for TM34 and TM102 [49]. The introduction of a sulphonate group in the triphenylphosphane ligand, with the aim of increasing water solubility (in TM85), led to a substantial decrease (down to ~0.16 nmol Ru/mg protein) in the total cellular uptake. TM34 and TM102 have a different subcellular distribution and cytotoxic profile compared with cisplatin that accumulates preferentially in the cytosol and nucleus, indicating different mechanisms of action [49]. TM34 and TM85 bind to HSA and transferrin and form adducts that apparently do not affect their cytotoxicity in A2780 cells [31, 50]. The uptake was slightly improved for the transferrin-bound TM34 (~50%) and TM85 (~100%) but reduced for the transferrin-bound TM102 (~38%). The cytotoxicity is reduced both in the presence of endocytosis inhibitors (e.g., monodansylcadaverine, choroquine, and amiloride) and at 4 °C, supporting that an energy-dependent process is involved in their cellular uptake mechanism consistent with endocytosis (nevertheless, passive/active transport could not be completely ruled out).

## 10.3 DNA and DNA-Related Cellular Targets

DNA is generally accepted to be the major intracellular target of cisplatin, which triggers its cytotoxicity. Accordingly, it was first assumed by analogy with platinum drugs that the ruthenium complexes would target DNA as well, and the DNA-binding properties of ruthenium compounds have been extensively studied mainly with DNA models [19]. However, the anticancer activity of some ruthenium compounds did not correlate directly with DNA binding or DNA damage. NAMI-A, KP1019, and RAPTA-complexes showed only modest and reversible DNA adducts formation. For KP1019 and NAMI-A and some $Ru^{II}$(arene) complexes, their action is not based only on direct DNA damage [51].

A wide family of Ru(II) complexes based on polypyridyl architectures has been developed to target DNA by an intercalative binding mode taking advantage of the drastic emission enhancement that accounts for their DNA light-switching effect. The activity and luminescence of these compounds is strongly influenced by the set of co-ligands around the ruthenium center. Those complexes with dppz derivatives are suitable for numerous applications ranging from imaging to therapeutics [23, 41, 42]. Studies on $Ru^{II}(\eta^5\text{-}C_5H_5)$-(2,2′bipyridyl) complexes (TM34, TM85, and TM102) suggest that the interactions *in vitro* of plasmid DNA with the complexes are very weak, indicating no formation of adducts which seemed to indicate that the DNA double helix is not a target [49].

The reaction of some $Ru^{III}$ complexes with DNA bases is often influenced by glutathione (GSH) levels and other biological reductants such as ascorbic acid. For $Ru^{III}$ coordination compounds, like KP1019 or NAMI-A, it is assumed that reduction of ruthenium takes place inside the cell after release of the Ru moiety from its biological carrier (e.g., serum proteins), which makes the Ru complex accessible for reduction (and later be reoxidized) in the cellular environment. In the case of $Ru^{II}$(arene) complexes, redox reactions with GSH were also reported, although the $Ru^{II}$ center itself is usually unable to participate in redox reactions. Particularly, the type of interactions with GSH differ between the diverse $Ru^{II}$(arene) complexes according to the type of ligand. A competition between GSH and guanine (N7) bases of DNA was observed for $[Ru^{II}(\eta^6\text{-bip})Ru(en)Cl][PF_6]$ (bip = biphenyl) [52, 53].

KP1019 activity involves disturbance of the cellular redox balance, followed in succession by induction of G2/M cell cycle arrest, blockage of DNA synthesis, and induction of apoptosis via the mitochondrial pathway. In a similar way, RAPTA-C inhibited cell growth effectively by triggering G2/M phase arrest and the mitochondrial apoptotic pathway, resulting in cytochrome c release and caspase-9 activation [18, 23].

Topoisomerase II (DNA girase) is a class of enzymes that controls the topology of DNA at several different steps in replication [54]. Many antitumor Ru complexes act as topoisomerase inhibitors, and these inhibition mechanisms are usually related to DNA binding. Ru(III) polypyridyl complexes bearing 2,2′-bipyridyl and 1,10-phenanthroline, and "$Ru^{II}(\eta^6$-*p*-cymene)" with flavonoid ligands are efficient inhibitors toward topoisomerase II by interference with the DNA religation and direct topoisomerase II binding [55, 56]. Some other related novel complexes revealed multitarget inhibition toward topoisomerase and telomerase at the same time [56].

PARP-1 (poly(ADP-ribose) polymerase-1), a well-known member of the PARP family, is involved in DNA repair mechanisms [57]. RAPTA-T and NAMI-A were tested for their action on this enzyme and showed to be stronger inhibitors ($IC_{50}$ = 28 and 19 μM, respectively) than the control 3-aminobenzamide ($IC_{50}$ = 33 μM) but less effective than cisplatin (($IC_{50}$ = 12 μM)) [58]. TM34 behaves as a strong inhibitor of PARP-1 being, in fact, the strongest ruthenium PARP-1 inhibitor reported so far ($IC_{50}$ = 1.0 μM) [31].

The role of DNA as a primary target for Ru complexes gained controversy since research pointed out that nonnuclear targets are also implicated. The observation that KP1019 was found predominantly in the cytosolic fractions raised the question that other cellular targets could be involved in the mechanism of action, specifically the plasma membrane, the endosomal/lysosomal system, and the Golgi apparatus [49–51, 59, 60]. Insights into how membrane-enclosed intracellular structures and organelles interact and are regulated by antitumor drugs are of vital importance within cell biology and medicinal chemistry areas.

While classical chemotherapy includes drugs interfering with cellular replication and mitotic processes, a more recent strategy involves targeting cellular signaling pathways and enzymes of cancer cells, yielding highly effective cancer treatments with less severe side effects.

## 10.4 Targeting Signaling Pathways

The highly reactive nitric oxide (NO) is known as the second messenger for diverse physiological processes, namely, vascular homeostasis, inflammatory/immune response as well as tumor progression and angiogenesis [61, 62]. Thus, drugs that interfere in the nitric oxide synthase (NOS) pathway could be useful against angiogenesis-dependent tumors. Ru(III) polyaminocarboxylate complexes were found effective as NO scavengers [62]. This NO-scavenging ability was shown to inhibit endothelial cell migration and angiogenesis, especially in the case of NAMI-A. It seems likely that the antiangiogenic activity of NAMI-A might be related to this NO-scavenging activity [62, 63]. Within antiangiogenic therapy, Ru(II) polypyridyl complexes containing benzimidazole and phenanthroline derivatives were developed as anti-vascularized drugs as alternatives to NAMI-A. The antiangiogenic effect results in the activation of antiangiogenic signaling pathways, in particular the activation of VEGF (vascular endothelial growth factor) and VEGFR-2 phosphorylation, which blocked the transmission of the mitogenic signal through protein-serine/threonine kinases Akt and ERK1/2 pathways, thus enhancing cell cycle arrest [64].

Some Ru complexes can also promote the intracellular formation of free radical species, either through reactions with cellular components or through irradiation (as in photodynamic therapy) [65]. The extent to which individual drugs participate in various steps along the metabolic pathways is a crucial factor in determining whether they are mainly antimetastatic or antiproliferative. Ru(II) polypyridyl phenanthroline derivatives enter the cells partially through transferrin-receptor–mediated endocytosis, are translocated from lysosomes to the mitochondria, where it activates mitochondrial dysfunction by regulation of antiapoptotic family proteins Bcl-2, thus leading to intracellular reactive oxygen species (ROS) overproduction. Excess ROS amplified apoptotic signals by activating many downstream pathways such as p53 and mitogen-activated protein kinases (MAPK) to promote cell apoptosis [66].

## 10.5 Targeting Enzymes of Specific Cell Functions

Thioredoxin reductase (TrxR) is an NADPH-dependent selenoenzyme that is upregulated in a number of cancer conditions. It plays a pivotal role in cancer progression and represents an attractive target for anticancer drugs [67]. Ru complexes with labile ligands can undergo ligand-substitution reactions and target proteins with selenocysteine or cysteine at the active site of the enzyme [68]. Ru(II) polypridyl complexes with diimine ligands exhibit antiproliferative activities against A375 human melanoma cells through inhibition of TrxR. The TrxR-inhibitory potency was more effective than auranofin, a positive TrxR inhibitor [66].

The endosomal/lysosomal system and the ER-Golgi network have been recognized as potential targets for drug therapy, providing a window of therapeutic

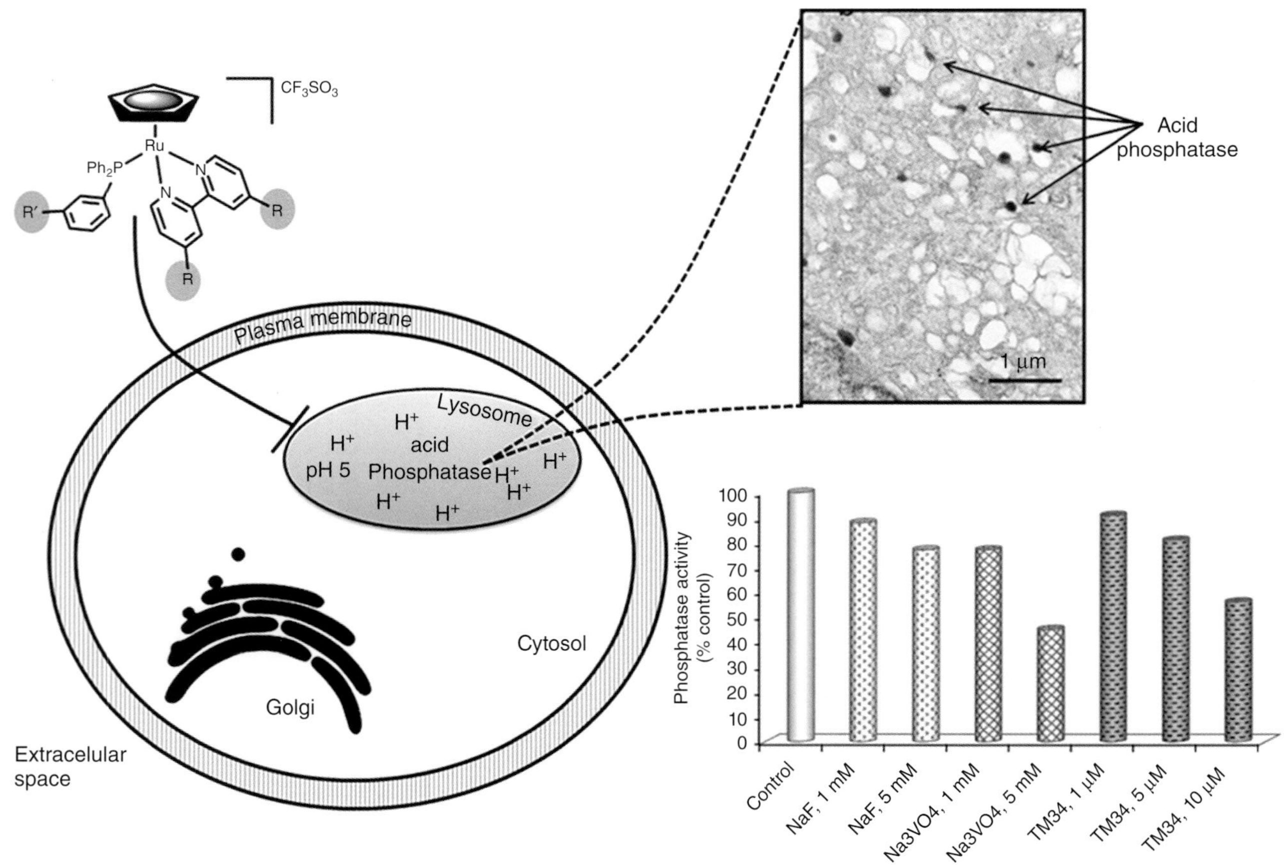


**Figure 10.2** Acid phosphatase (AcP) activity in the cancer cell line MDAMB231. Cytochemical localization of AcP was done by a cerium-based method [44]. The $Ru^{II}(\eta^5\text{-}C_5H_5)$ complex TM34 enters the cells probably by endocytosis, and inhibits AcP activity in a dose-dependent mode. The same trend was observed for NaF and vanadate, respectively noncompetitive and competitive inhibitors of the enzyme [75]. Ultrastructural analysis by TEM showed that the complex also causes disruption and vesiculation of the Golgi apparatus, which suggested the endosomal/lysosomal system as a possible target [44].

opportunities [69]. Lysosomes contain a number of hydrolases that help degrade intracellular and extracellular material delivered. Among hydrolases, cathepsins have a major role [70]. Following their release into the cytosol, they possibly degrade antiapoptotic Bcl-2 proteins, thereby triggering the mitochondrial pathway of apoptosis, with the lysosomal membrane permeabilization the critical step in this pathway. In this scope, a series of RAPTA-type compounds were shown to be effective inhibitors of cathepsin B (cat B), the unique cathepsin involved in pathological degradation of the extracellular matrix [71, 72]. These observations matched *in vivo* data, which showed that cat B could be primarily implicated in the process of metastasis [73].

Acid phosphatases (AcPs) are a class of nonspecific hydrolases that catalyze the hydrolysis of phosphomonoesters at acidic pH. In addition to the mobilization of phosphate, some members of this class are also involved in many essential biological functions [74]. The effect on AcP was evaluated for the prospective antitumor $Ru^{II}(\eta^5\text{-}C_5H_5)$ derivatives. Given the cationic nature of the complexes, AcP was measured to evaluate any lysosomotropism or if any effect would result from trapping into the lysosomes. Using a cerium-based method and by TEM (transmission electron microscopy), new evidence concerning the localization of acid phosphatase activities in the MDAMB231 cells was provided [44]. TM34 inhibition of AcP in breast cancer MDAMB231 cells was found to be a dose-dependent process, and disruption and vesiculation of the Golgi apparatus was observed, which suggested the endosomal/lysosomal system as a possible target (Figure 10.2) [44]. A similar effect, but at a higher dose, was found for its water-soluble analogue TM85, confirming the effect of the complex on the endomembrane system, in particular the Golgi apparatus. Morphological changes in the mitochondria also supported a direct effect on these cellular organelles [50].

## 10.6 Targeting Glycolytic Pathways

The glycolytic phenotype of cancer cells, commonly known as "Warburg effect," is considering a *quasi*-universal trait of all the growing tumors. Therefore, tumor glycolysis could be a logical target for the prospective anticancer compound. Since the metabolism of cancer is different from that of healthy cells, drugs that target this metabolic difference have the potential to selectively kill cancer cells. Glycolytic enzymes and glucose transporters are overexpressed in tumor cells compared with normal cells, thus offering new opportunities for the development of anticancer drugs [76].

Glycolytic cells have been shown to consume oxygen at the cell surface via trans-plasma membrane electron transport (tPMET) systems, a process that utilizes intracellular NADH and supports glycolytic ATP production. Reducing equivalents from cytosolic NADH are transported to extracellular electron acceptors through tPMET. Ferricyanide has been commonly used as an artificial electron acceptor to measure NADH-ferricyanide reductase activity [77, 78]. The observation that $Ru^{II}(\eta^5\text{-}C_5H_5)$ complexes with bidentate N,N-heteroaromatic

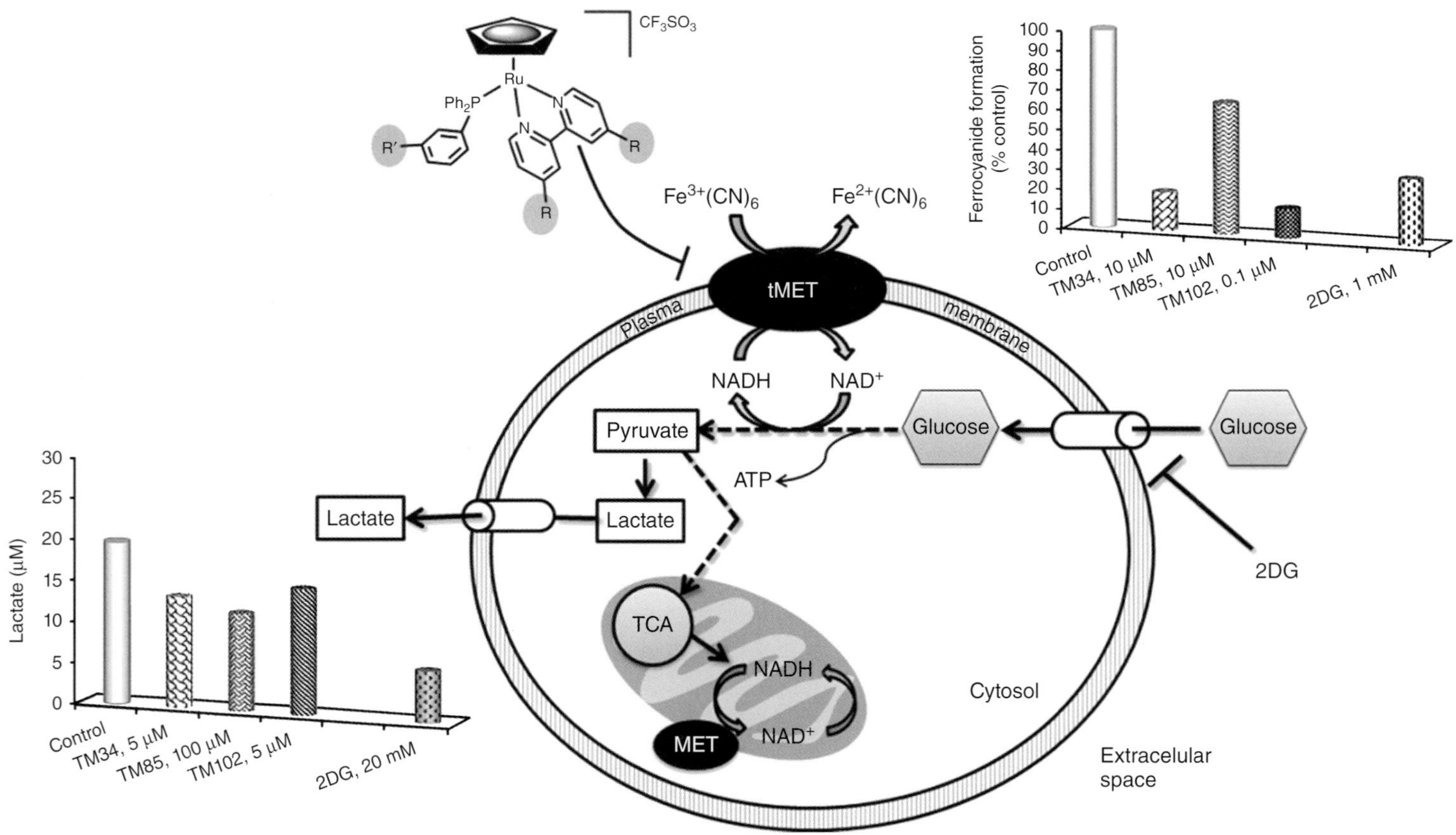


**Figure 10.3** TPMET activity in the high glycolytic breast cancer cells MDAMB231 (simplified proposed model). TPMET activity was measured by reduction of ferricyanide. Mitochondrial NADH production is linked to ferricyanide reduction via the mitochondrial electron transport system (MET). TPMET can ameliorate intracellular reductive stress, which originates from the mitochondrial tricarboxylic acid (TCA) cycle. We propose that the inhibition of this pathway leads to the decrease in the viability of MDAMB231 cells, which rely on tPMET activity for survival. Results indicated that the Ru complexes TM34, TM85 and TM102 can inhibit ferricyanide reductase and lactate production in a manner dependent on their anticancer potential. 2-Deoxyglucose was used as a control inhibitor, as it targets hexokinase which is the first step of glycolysis, and inhibits tPMET by depletion of pyruvate and ATP levels [49].

ligands could be retained preferentially at the membranes prompted exploration of their interaction with the redox system as a possible target. The results from this study indicated that ruthenium complexes can inhibit ferricyanide reductase and pyruvate reduction to lactate depending on the glycolytic phenotype of cancer cells and on the complex antiproliferative potential (Figure 10.3) [49, 79]. The tPMET system can be considered as an alternative way to control the redox status of cells through NADH-mediated signaling pathways and thus be proposed as a target for anticancer metallodrugs [80, 81].

## 10.7 Macromolecular Ruthenium Conjugates: A New Approach to Targeting

The difficulty in achieving selective destruction of tumor cells while sparing normal tissues has motivated the design of new macromolecular compounds based on the well-known EPR ("enhanced permeation and retention") effect [82]. Following on this concept, multiple strategies have been developed on the basis of micelles, [83] coordination cages, [84, 85] lipids, [86, 87] protein conjugates, [88] nanoparticles, [89] and linear ruthenium–polymer conjugates [90].

The synthetic strategy to obtain covalently bound ruthenium drug carriers must fulfill several requirements to achieve high bioactivity which ensures that the drug is still active when bound to the carrier: degradability of the carrier over time without the formation of toxic fragments and *in vivo* protection of the cytotoxic moiety. Most approaches are based on multinuclear drugs with the inclusion of several metal centers per molecule to increase the cytotoxicity of the drug. The mechanism of action for these conjugates is expected to differ from that of their low-molecular-weight models due to their structural differences, where the high molecular weight plays an important role (e.g., increased cellular uptake efficiency probably caused by changes in cell entry mechanism).

RAPTA-C polymeric micelles based on amphiphilic copolymer were prepared to improve the *in vivo* stability and selectivity of RAPTA-C [83]. *In vitro* studies with these macromolecular drugs in ovarian cancer cells (A2780/A2780cis and Ovcar3) revealed an increase in cellular uptake and cytotoxicity when compared to RAPTA-C. The uptake observed into the lysosomes was indicative of endocytic pathway.

Ruthenium-based coordination-cage conjugates have also been explored as molecular carriers able to release their cytotoxic cargo inside cancer cells and to increase cytotoxic activity. The cytotoxicity of the conjugates seemed to be highly dependent on the size, spacer, geometry, solubility, and host–guest properties [84, 85]. To improve the activity of RAPTA-C compounds, pyrenyl moieties of RAPTA-C complexes were encapsulated within the hydrophobic cavity of water-soluble metallacages [84]. These pyrenyl-RAPTA-C derivatives showed higher cytotoxicity compared to RAPTA-C in human cancer cells (A549, A2780/A2780cisR, Me300, and HeLa). The loaded metallacage also exhibited a higher accumulation inside cells which could be partially explained by the increase of lipophilicity induced by the pyrenyl moiety.

Another synthetic strategy to obtain metal-based supramolecular drug carriers and to retard degradation kinetics was to trap ruthenium complexes in liposome bilayers. By exploiting the high affinity of cholesterol with phospholipids to achieve better internalization a "NAMI-A-like" complex (AziRu) was linked to a cholesterol-containing nucleolipid yielding ToThy-CholRu, which was then stabilized by lipid aggregates of POPC (palmitoyl-2-oleoyl-sn-glycero-3-phosphocholine) to delay degradation of NAMI-A [86]. These ruthenium-containing liposomes were more effective than the parent NAMI-A-like complex AziRu in inhibiting the growth of MCF7 cells.

Inspired by NAMI-A and using AziRu as a core molecular scaffold of amphiphilic ruthenium anionic complexes, nanoparticles with DOTAP (the cationic lipid 1,2-dioleyl-3-trimethylammoniumpropane chloride) were analyzed using fluorescent probes to access their uptake and cellular localization in MCF7 cells. The *in vitro* bioactivity profile of these Ru-loaded nanoparticles showed a rapid uptake and a tendency to accumulate in close proximity to the nuclei [87].

With the goal of attaining improved selectivity *in vivo* by the EPR effect (see preceding text) highly cytotoxic "$Ru^{II}(\eta^5-C_5H_5)$" complexes based on the TM34 core were tethered to a polylactide chain incorporating a tumor-targeting moiety (a glucose derivative) (Figure 10.4) [90]. This approach differs from all others previously reported, with expectedly different reactivity patterns and diverse *in vivo* behavior. Compared with other macromolecular ruthenium conjugates, this strategy has the advantages of (i) exact control on the amount of cytotoxic drug in the polymeric chain which is fine-tuned through polymer functionalization (other approaches use a variable percentage of metal per quantity of drug) and (ii) preservation of drug activity even after the polymer hydrolysis [90].

The structural features of this "smart drug" (RuPMC) comprise three components: a ruthenium anticancer fragment, a biodegradable polymer, and a sugarlike molecule at the polymer chain end that could more efficiently target the tumor cells. RuPMC exhibited high cytotoxicity in the A2780, MCF7, and MDAMB231 cells. Its subcellular distribution in MCF7 cells (levels of Ru) showed that this compound was able to enter more efficiently (or be more retained) than its nonpolymeric analogue TM34, and was able to reach the nucleus (levels of Ru in nucleus fractions). DNA could be a possible target

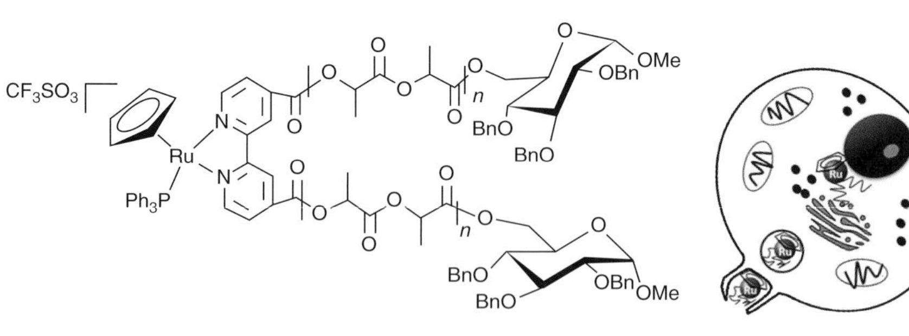


**Figure 10.4** Chemical structure of D-glucose end-capped polylactide ruthenium-cyclopentadienyl complex (RuPMC) and its proposed cellular uptake mechanism (n = 15).

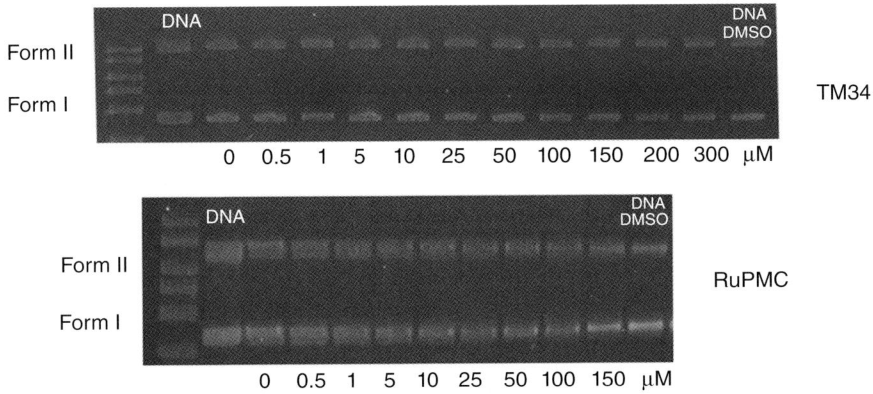


**Figure 10.5** Interaction between supercoiled phiX174 plasmid DNA and the ruthenium complexes TM34 (1 μM) and RuPMC (25 μM) after 24-h incubation at 37 °C in phosphate buffer (pH 7.2). Forms I and II are supercoiled and nicked circular isoforms of DNA, respectively. The concentrations of complex used were equivalent to the $IC_{50}$ values found at 24-h incubation.

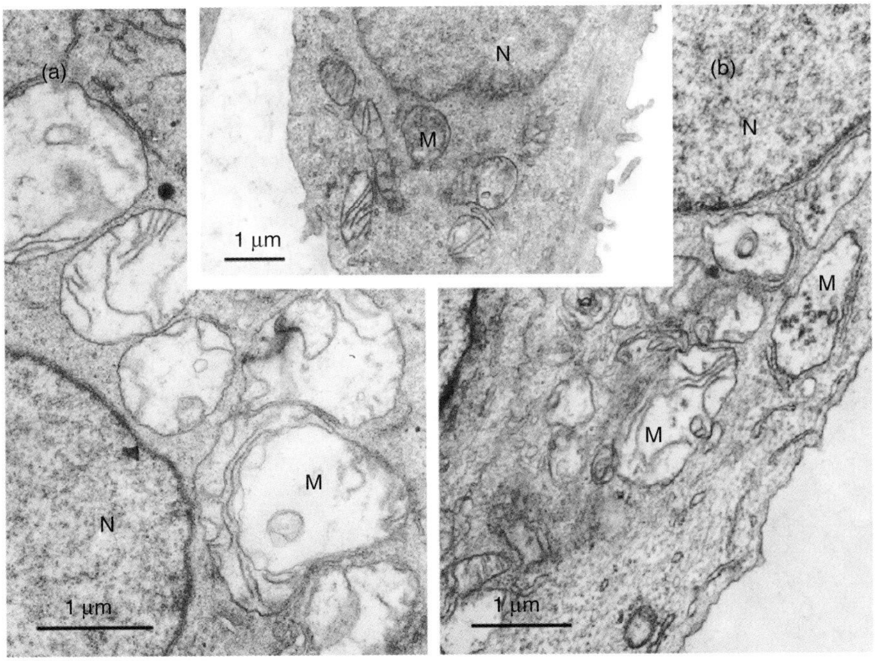


**Figure 10.6** Ultrastructural analysis by TEM of the MCF7 cancer cells after 24-h treatment with RuPMC and TM34: (a) cells treated with RuPMC (25 μM); (b) cells treated with TM34 (1 μM); inset) control cells (no treatment). Mitochondrial alterations were detected in Ru-treated cells, consisting of edema and disorganization of the cristae. In particular, intramitochondrial dense inclusions are present in TM34-treated cells and mitochondrial edema seemed increased in RuPMC-treated cells. M, mitochondria; N, nucleus.

for RuPMC, and its reactivity with DNA *in vitro* was assessed by monitoring drug-induced conformational changes of supercoiled plasmid DNA. As shown in Figure 10.5, the electrophoretic mobilities of both the nicked and closed circular DNA isoforms do not change after incubation with RuPMC (similar behavior was also observed for TM34) [49]. Thus, it can be predicted that even if RuPMC reaches the nucleus, the DNA double helix is not its primordial target.

The mechanism of cell death was evaluated in the MCF7 cells by an apoptosis array assay and ultrastructural analysis by TEM. Results showed that an increase in the levels of heat-shock proteins hsp27 and hsp70, cytochrome C, and, significantly, catalase was promoted by TM34 and RuPMC when compared to controls (untreated cells). In the case of RuPMC, a substantial increase in the hsp60 levels was also observed, indicating proapoptotic stimuli. Although preliminary, results indicated that mitochondria and oxidative stress are involved in the mechanism of cell death of both compounds (unpublished results). Images by TEM (Figure 10.6) clearly indicated mitochondrial alterations caused by exposure of the cells to TM34 and RuPMC (unpublished results).

## 10.8 Conclusions

Platinum-based anticancer drugs such as cisplatin are the only metallodrugs in clinical use for cancer therapy worldwide regardless of their severe side effects. Ruthenium compounds have demonstrated several advantages as metallodrugs in this field such as lower toxicity and the ability to overcome resistance, emerging as new and efficient therapeutic alternatives to platinum drugs. Mechanisms of action different from those attributed to cisplatin are one of the reasons these ruthenium agents are so appealing, although their mode of action is not fully understood.

The primary target for platinum drugs is DNA, although more recent developments have identified other targets. Plain interaction with DNA and DNA models has been observed for several ruthenium compounds; and even though this molecule can be their ultimate target, in most cases the effect is exerted in an indirect manner, for example, through the inhibition of relevant enzymes such as topoisomerase II and PARP-1.

Novel approaches to target one or more phenotypes unique to cancer cells have identified the tumor energy metabolism as promising for developments in this field. How a drug distributes within a tumor and how it is retained are fundamental parameters for its activity. Therefore, even though its cellular uptake is critical, mechanistic details and its intracellular fate are also of vital importance.

Cisplatin enters the cells via passive diffusion and active/facilitated transport. In contrast, most of the Ru complexes are believed to enter the cells by endocytosis. This difference in the cellular uptake implies that compounds are often trapped in the lysosomes and suffer degradation by lysosomal enzymes. Some of these enzymes such as hydrolases (cat B, AcPs) are now emerging as attractive targets for antitumor drugs.

The importance of tPMET in cancer cell survival is indicated by its increased activity in cancer cells, suggesting that tPMET could represent a new target for developing novel anticancer drugs that exploit its unique plasma membrane localization, and may offer new therapeutic options for a wide range of cancer conditions.

A series of Ru(II) piano-stool complexes bearing bidentate heteroaromatic ligands have recently emerged as prospective anticancer drugs and the study on their mode of action is being clarified. In general, this class of complexes showed (i) a broad spectrum of activity against different human tumoral cells; (ii) outstanding cytotoxic efficacy largely surpassing that of cisplatin; (iii) cellular uptake probably facilitated by binding to transferrin; (iv) preferential localization at the cell membrane and cytosol; (v) negligible uptake in the nucleus and no observable direct interaction with DNA; and (vi) interaction with the endosomal/lysosomal system and metabolic redox regulators.

One of the leads of this family TM34 inhibits the activity of AcP and induces disruption and vesiculation of the Golgi apparatus of MDAMB231 cells. In the same cells (known for the high glycolytic phenotype), a plasma membrane redox system (tPMET) was demonstrated as NADH-ferricyanide reductase. This redox system was sensitive to sulfhydryls, metabolic inhibitors, pH, and the Ru complexes TM34, TM85, and TM102 in a dose-dependent mode and apparently in a way related to their cytotoxic potential. The data presented suggest that tPMET could be a possible target and that the inhibition may be fundamental, or at least of prime importance in inducing cytotoxicity.

Altogether, these Ru complexes seemed to exert a pH-dependent anticancer activity through its ability to disturb pH dynamics by two independent mechanisms. In addition, a cross-talk between the two enzymatic systems AcP and tPMET can be foreseen.

These encouraging results prompted the use of TM34 as a structural unit to develop a second generation of "$Ru^{II}(\eta^5\text{-}C_5H_5)$"-polymer conjugates aimed at controlled drug delivery and selective tumor targeting. This new derivative assembles three main moieties: the anticancer core (TM34a-like), a biocompatible and biodegradable polymeric macroligand, and a sugarlike molecule at the polymer chain end (RuPMC). RuPMC showed in relation to TM34: (i) a distinct pattern of cellular distribution with higher uptake in the nucleus; (ii) no observable direct interaction with DNA, and (iii) evidence for the involvement of mitochondria in the mechanism of cell death. Although preliminary, these new findings showed that ruthenium cyclopentadienyl polymeric structures like RuPMC could represent viable candidates as drug delivery vectors.

## References


1 Zang, C.X. and Lippard, S.J. (2003) *Curr. Opin. Chem. Biol.*, **7**, 481.
2 Wang, D. and Lippard, S.J. (2005) *Nat. Rev. Drug Discovery*, **4**, 307.
3 Wheate, N.J., Walker, S., Craig, G.E., and Oun, R. (2010) *Dalton Trans.*, **39**, 8113.
4 Dasari, S. and Tchounwou, P.B. (2014) *Eur. J. Pharmacol.*, **740**, 364.
5 Kostova, I. (2006) *Curr. Med. Chem.*, **13**, 1085.

6 Holder, A.A. (2010) *Annu. Rep. Prog. Chem. Sect. A: Inorg. Chem.*, **106**, 504.
7 Köpf-Maier, P. (1994) *Eur. J. Clin. Pharmacol.*, **47**, 1.
8 Kostova, I. (2009) *Anticancer Agents Med. Chem.*, **9**, 827.
9 Gonçalves, A.C., Morais, T.S., Robalo, M.P., Marques, F., Avecilla, F., Matos, C.P., Santos, I., Tomaz, A.I., and Garcia, M.H. (2013) *J. Inorg. Biochem.*, **129**, 1.
10 Hanif, M., Babak, M.V., and Hartinger, C.G. (2014) *Drug Discovery Today*, **19**, 1640.
11 Geldmacher, Y., Splith, K., Kitanovic, I., Alborzinia, H., Can, S., Rubbiani, R., Nazif, M.A., Wefelmeier, P., Prokop, A., Ott, I., Wölfl, S., Neundorf, I., and Sheldrick, W.S. (2012) *J. Biol. Inorg. Chem.*, **17**, 631.
12 Messori, L., Marcon, G., and Orioli, P. (2003) *Bioinorg. Chem. Appl.*, **1**, 177.
13 Nardon, C., Boscutti, G., and Fregona, D. (2014) *Anticancer Res.*, **34**, 487.
14 Süss-Fink, G. (2010) *Dalton Trans.*, **39**, 1673.
15 Bergamo, A. and Sava, G. (2015) *Chem. Soc. Rev.*, **44**, 8818.
16 Smith, G.S. and Therrien, B. (2011) *Dalton Trans.*, **40**, 10793.
17 Scolaro, C., Bergamo, A., Brescacin, L., Delfino, R., Cocchietto, M., Laurenczy, G., Geldbach, T.J., Sava, G., and Dyson, P.J. (2005) *J. Med. Chem.*, **48**, 4161.
18 Motswainyana, W.M. and Ajibade, P.A. (2015) *Adv. Chem.*, **2015**, 21, Article ID 859730.
19 Clarke, M.J. (2003) *Coord. Chem. Rev.*, **236**, 209.
20 Velders, A.H., Bergamo, A., Alessio, E., Zangrando, E., Haasnoot, J.G., Casarsa, C., Cocchietto, M., Zorzet, S., and Sava, G.J. (2004) *Med. Chem.*, **47**, 1110.
21 Hartinger, C.G., Zorbas-Seifried, S., Jakupec, M.A., Kynast, B., Zorbas, H., and Keppler, B.K. (2006) *J. Inorg. Biochem.*, **100**, 891.
22 Antonarakis, E.S. and Emadi, A. (2010) *Cancer Chemother. Pharmacol.*, **66**, 1.
23 Mazuryk, O., Magiera, K., Rys, B., Suzenet, F., Kieda, C., and Brindell, M. (2014) *J. Biol. Inorg. Chem.*, **19**, 1305.
24 Yan, Y.K., Melchart, M., Habtemariam, A., and Sadler, P.J. (2005) *Chem. Commun. (Camb)*, **14**, 4764.
25 Pizarro, A.M., Melchart, M., Habtemariam, A., Salassa, L., Fabbiani, F.P., Parsons, S., and Sadler, P.J. (2010) *Inorg. Chem.*, **49**, 3310.
26 Egger, A.E., Hartinger, C.G., Renfrew, A.K., and Dyson, P.J. (2010) *J. Biol. Inorg. Chem.*, **15**, 919.
27 Clavel, C.M., Păunescu, E., Nowak-Sliwinska, P., Griffioen, A.W., Scopelliti, R., and Dyson, P.J. (2015) *J. Med. Chem.*, **58**, 3356.
28 Streu, C., Carroll, P.J., Kohli, R.K., and Meggers, E. (2008) *J. Organomet. Chem.*, **693**, 551.
29 Bullock, A.N., Russo, S., Amos, A., Pagano, N., Bregman, H., Debreczeni, J.E., Lee, W.H., von Delft, F., Meggers, E., and Knapp, S. (2009) *PLoS One*, **4**, e7112.
30 Morais, T.S., Silva, T.J., Marques, F., Robalo, M.P., Avecilla, F., Madeira, A.P.J., Mendes, P.J., Santos, I., and Garcia, M.H. (2012) *J. Inorg. Biochem.*, **114**, 65.

31 Tomaz, A.I., Jakusch, T., Morais, T.S., Marques, F., de Almeida, R.F., Mendes, F., Enyedy, E.A., Santos, I., Pessoa, J.C., Kiss, T., and Garcia, M.H. (2012) *J. Inorg. Biochem.*, **117**, 261.

32 Morais, T.S., Santos, F., Côrte-Real, L., Marques, F., Robalo, M.P., Madeira, P.J., and Garcia, M.H. (2013) *J. Inorg. Biochem.*, **122**, 8.

33 Heffeter, P., Böck, K., Atil, B., Reza, M.A., Hoda , Körner, W., Bartel, C., Jungwirth, U., Keppler, B.K., Micksche, M., Berger, W., and Koellensperger, G.J. (2010) *Biol. Inorg. Chem.*, **15**, 737.

34 Sava, G., Zorzet, S., Turrin, C., Vita, F., Soranzo, M., Zabucchi, G., Cocchietto, M., Bergamo, A., DiGiovine, S., Pezzoni, G., Sartor, L., and Garbisa, S. (2003) *Clin. Cancer Res.*, **9**, 1898.

35 Timerbaev, A.R., Hartinger, C.G., Aleksenko, S.S., and Keppler, B.K. (2006) *Chem. Rev.*, **106**, 2224.

36 Kragh-Hansen, U., Chuang, V.T., and Otagiri, M. (2002) *Biol. Pharm. Bull.*, **25**, 695.

37 Pessoa, J.C. and Tomaz, I. (2010) *Curr. Med. Chem.*, **17**, 3701.

38 Levina, A., Mitra, A., and Lay, P.A. (2009) *Metallomics*, **1**, 458.

39 Guo, W., Zheng, W., Luo, Q., Li, X., Zhao, Y., Xiong, S., and Wang, F. (2013) *Inorg. Chem.*, **52**, 5328.

40 Ndolo, R.A., Jacobs, D.T., Forrest, M.L., and Krise, J.P. (2010) *Mol. Cell. Pharmacol.*, **2**, 131.

41 Puckett, C.A. and Barton, J.K. (2008) *Biochemistry*, **47**, 11711.

42 Puckett, C.A., Ernst, R.J., and Barton, J.K. (2010) *Dalton Trans.*, **39**, 1159.

43 Logan, R., Funk, R.S., Axcell, E., and Krise, J.P. (2012) *Expert Opin. Drug Metab. Toxicol.*, **8**, 943.

44 Côrte-Real, L., Matos, A.P., Alho, I., Morais, T.S., Tomaz, A.I., Garcia, M.H., Santos, I., Bicho, M.P., and Marques, F. (2013) *Microsc. Microanal.*, **19**, 1122.

45 Sancho-Martínez, S.M., Prieto-García, L., Prieto, M., López-Novoa, J.M., and López-Hernández, F.J. (2012) *Pharmacol. Ther.*, **136**, 35.

46 Spreckelmeyer, S., Orvig, C., and Casini, A. (2014) *Molecules*, **19**, 15584.

47 Pongratz, M., Schluga, P., Jakupec, M.A., Arion, V.B., Hartinger, C.G., and Allmaier, G. (2004) *J. Anal. Atomic Spectrom.*, **19**, 46.

48 Romero-Canelón, I., Pizarro, A.M., Habtemariam, A., and Sadler, P.J. (2012) *Metallomics*, **4**, 1271.

49 Côrte-Real, L., Mendes, F., Coimbra, J., Morais, T.S., Tomaz, A.I., Valente, A., Garcia, M.H., Santos, I., Bicho, M., and Marques, F. (2014) *J. Biol. Inorg. Chem.*, **19**, 853.

50 Morais, T.S., Santos, F.C., Jorge, T.F., Côrte-Real, L., Madeira, P.J., Marques, F., Robalo, M.P., Matos, A., Santos, I., and Garcia, M.H. (2014) *J. Inorg. Biochem.*, **130**, 1.

51 Stevens, S.K., Strehle, A.P., Miller, R.L., Gammons, S.H., Hoffman, K.J., McCarty, J.T., Miller, M.E., Stultz, L.K., and Hanson, P.K. (2013) *Mol. Pharmacol.*, **83**, 225.

52 Wang, F., Xu, J., Habtemariam, A., Bella, J., and Sadler, P.J. (2005) *J. Am. Chem. Soc.*, **127**, 17734.

53 Clarke, M.J., Zhu, F., and Frasca, D.R. (1999) *Chem. Rev.*, **99**, 2511.

54 Gao, F., Chao, H., and Ji, L.N. (2008) *Chem. Biodivers.*, **5**, 1962.

55 He, X., Jin, L., and Tan, L. (2015) *Spectrochim. Acta A Mol. Biomol. Spectrosc.*, **135**, 101.
56 Liao, G., Chen, X., Wu, J., Qian, C., Wang, Y., Ji, L., and Chao, H. (2015) *Dalton Trans.*, **44**, 15145–15156.
57 Schreiber, V., Dantzer, F., Ame, J.C., and de Murcia, G. (2006) *Nat. Rev. Mol. Cell Biol.*, **7**, 517.
58 Mendes, F., Groessl, M., Nazarov, A.A., Tsybin, Y.O., Sava, G., Santos, I., Dyson, P.J., and Casini, A. (2011) *J. Med. Chem.*, **54**, 2196.
59 Gill, M.R. and Thomas, J.A. (2012) *Chem. Soc. Rev.*, **41**, 3179.
60 Gill, M.R., Cecchin, D., Walker, M.G., Mulla, R.S., Battaglia, G., Smythe, C., and Thomas, J.A. (2013) *Chem. Sci.*, **4**, 4512.
61 Fricker, S.P., Slade, E., Powell, N.A., Vaughan, O.J., Henderson, G.R., Murrer, B.A., Megson, I.L., Bisland, S.K., and Flitney, F.W. (1997) *Br. J. Pharmacol.*, **122**, 1441.
62 Cameron, B.R., Darkes, M.C., Yee, H., Olsen, M., Fricker, S.P., Skerlj, R.T., Bridger, G.J., Davies, N.A., Wilson, M.T., Rose, D.J., and Zubieta, J. (2003) *Inorg. Chem.*, **42**, 1868.
63 Morbidelli, L., Donnini, S., Filippi, S., Messori, L., Piccioli, F., Orioli, P., Sava, G., and Ziche, M. (2003) *Br. J. Cancer*, **88**, 1484.
64 Lai, H., Zhao, Z., Li, L., Zheng, W., and Chen, T. (2015) *Metallomics*, **7**, 439.
65 Holder, A.A., Taylor, P., Magnusen, A.R., Moffett, E.T., Meyer, K., Hong, Y., Ramsdale, S.E., Gordon, M., Stubbs, J., Seymour, L.A., Acharya, D., Weber, R.T., Smith, P.F., Dismukes, G.C., Ji, P., Menocal, L., Bai, F., Williams, J.L., Cropek, D.M., and Jarrett, W.L. (2013) *Dalton Trans.*, **42**, 11881.
66 Luo, Z., Yu, L., Yang, F., Zhao, Z., Yu, B., Lai, H., Wong, K.H., Ngai, S.M., Zheng, W., and Chen, T. (2014) *Metallomics*, **6**, 1480.
67 Jungwirth, U., Kowol, C.R., Keppler, B.K., Hartinger, C.G., Berger, W., and Heffeter, P. (2011) *Antioxid. Redox Signaling*, **15**, 1085.
68 Che, C.M. and Siu, F.M. (2010) *Curr. Opin. Chem. Biol.*, **14**, 255.
69 Wlodkowic, D., Skommer, J., McGuinness, D., Hillier, C., and Darzynkiewicz, Z. (2009) *Leuk. Res.*, **33**, 1440.
70 Linder, S. and Shoshan, M.C. (2005) *Drug Resist. Updat.*, **8**, 199.
71 Chang, W.-S.W., Wu, H.-R., Yeh, C.-T., Wu, C.-W., and Chang, J.-Y. (2007) *J. Cancer Mol.*, **3**, 5.
72 Casini, A., Gabbiani, C., Sorrentino, F., Rigobello, M.P., Bindoli, A., Geldbach, T.J., Marrone, A., Re, N., Hartinger, C.G., Dyson, P.J., and Messori, L. (2008) *J. Med. Chem.*, **51**, 6773.
73 Kos, J., Mitrović, A., and Mirković, B. (2014) *Future Med. Chem.*, **6**, 1355.
74 Bull, H., Murray, P.G., Thomas, D., Fraser, A.M., and Nelson, P.N. (2002) *Mol. Pathol.*, **55**, 65.
75 Gallagher, S.R. and Leonard, R.T. (1982) *Plant Physiol.*, **70**, 1335–1340.
76 Porporato, P.E., Dhup, S., Dadhich, R.K., Copetti, T., and Sonveaux, P. (2011) *Front. Pharmacol.*, **2**, 49.
77 Kennett, E.C. and Kuchel, P.W. (2003) *IUBMB Life*, **55**, 375.
78 Laliberté, J.F., Sun, I.L., Crane, F.L., and Clarke, M.J. (1987) *J. Bioenerg. Biomembr.*, **19**, 69.

79 Philp, A., Macdonald, A.L., and Watt, P.W. (2005) *J. Exp. Biol.*, **208**, 4561.
80 Prata, C., Grasso, C., Loizzo, S., Sega, F.V., Caliceti, C., Zambonin, L., Fiorentini, D., Hakim, G., Berridge, M.V., and Landi, L. (2010) *Leuk. Res.*, **34**, 1630.
81 Liu, Z. and Sadler, P.J. (2014) *Acc. Chem. Res.*, **47**, 1174.
82 Maeda, H. and Matsumura, Y. (1986) *Crit. Rev. Ther. Drug Carrier Syst.*, **46**, 6387.
83 Blunden, B.M., Lu, H., and Stenzel, M.H. (2013) *Biomacromolecules*, **14**, 4177.
84 Furrer, M.A., Schmitt, F., Wiederkehr, M., Juillerat-Jeanneret, L., and Therrien, B. (2012) *Dalton Trans.*, **41**, 7201.
85 Therrien, B. (2015) *CrystEngComm*, **17**, 484.
86 Simeone, L., Mangiapia, G., Vitiello, G., Irace, C., Colonna, A., Ortona, O., Montesarchio, D., and Paduano, L. (2012) *Bioconjugate Chem.*, **23**, 758.
87 Mangiapia, G., Vitiello, G., Irace, C., Santamaria, R., Colonna, A., Angelico, R., Radulescu, A., D'Errico, G., Montesarchio, D., and Paduano, L. (2013) *Biomacromolecules*, **14**, 2549.
88 Stepanenko, I.N., Casini, A., Edafe, F., Novak, M.S., Arion, V.B., Dyson, P.J., Jakupec, M.A., and Keppler, B.K. (2011) *Inorg. Chem.*, **50**, 12669.
89 Süss-Fink, G., Khan, F.-A., Juillerat-Jeanneret, L., Dyson, P.J., and Renfrew, A.K. (2010) *J. Clus. Sci.*, **21**, 313.
90 Valente, A., Garcia, M.H., Marques, F., Miao, Y., Rousseau, C., and Zinck, P. (2013) *J. Inorg. Biochem.*, **127**, 79.
91 Chatterjee, D. and Mitra, A. (2006) *Platinum Met. Rev.*, **50**, 2.

# 11

# Targeting cellular DNA with Luminescent Ruthenium(II) Polypyridyl Complexes

*Martin R. Gill and Jim A. Thomas*

*University of Sheffield, Department of Chemistry, Sheffield, UK*

## 11.1 Introduction

In almost all eukaryotic and prokaryotic organisms, encoded genetic information is stored within the biopolymer deoxyribonucleic acid (DNA). DNA provides the blueprint for the formation of both ribonucleic acid (RNA) and proteins, which are the functional components of cells. The flow of biological information, whereby a section of DNA is transcribed into RNA, which in turn is responsible for the synthesis of specific proteins (translation), is established within "the central dogma of molecular biology," as described by Francis Crick [1]. The central dogma also describes DNA replication, the process by which organisms make copies of their genome before cell division and thus provide the basis of biological inheritance and genetics. Accordingly, small molecules that bind DNA are often able to inhibit cellular growth: either via the inhibition of DNA replication or via the generation of irreparable DNA damage, or a combination of both. As cancer and bacterial cells often display rapid proliferation, molecules that prevent cell growth find use as anticancer and antibacterial agents. Following pioneering studies on the nitrogen mustards, which are indirect DNA alkylating agents [2], small molecules that target – and damage – DNA to inhibit cellular proliferation have formed the basis of chemotherapy used for treating cancers [3]. However, this may be considered a double-edged sword as many agents are themselves mutagenic and carcinogenic. Furthermore, many chemotherapeutics are inherently highly cytotoxic and damage healthy tissue; leading to severe side effects in patients. Hence, there is still great interest in developing novel small molecules that influence DNA structure and function by interacting with it through new binding mechanisms [4]. One particularly relevant example of this is the recent surge of interest in stabilizing noncanonical DNA structures, such as quadruplex DNA, and several new drugs that stabilize this form of DNA are under development, for example, pyridostatin [5]. In addition to targeting DNA for therapeutic reasons, binding substrates that image DNA are also in much demand. Ever since Robert Hooke first visualized a cell in 1665 using a custom-made microscope [6], optical microscopy has facilitated a greater understanding of biological structures and processes. Development of this technique has

*Ruthenium Complexes: Photochemical and Biomedical Applications*, First Edition.
Edited by Alvin A. Holder, Lothar Lilge, Wesley R. Browne, Mark A.W. Lawrence, and Jimmie L. Bullock Jr.

been accompanied by the identification of compounds that target and stain organelles or biomolecules with high specificity. Since these early observations, visualization of cellular DNA – and in particular nuclear DNA – has provided researchers with a range of information, including the discovery of a cell cycle and the identification of highly regulated vital biological processes such as apoptosis.

### 11.1.1 DNA-Binding Modes of Small Molecules

The structure of DNA provides several specific binding modes by which a compound may associate with the biopolymer. This includes direct reactivity, in which the formation of irreversible bonds with the nucleobases or sugar phosphate backbones occurs, and several forms of reversible (noncovalent) associations are also known.

The most basic reversible mode of binding between a small molecule and DNA is electrostatic binding, whereby cationic molecules are simply attracted to the negatively charged double helix; however, two other motifs have more complex driving forces. Groove binding can be within the minor or major groove that runs down the canonical B DNA structure and occurs through a combination of van der Waals' and hydrophobic and hydrogen bonding interactions. Several commercially available and commonly used fluorescent DNA dyes are groove binders, for example, DAPI (4′,6-diamidino-2-phenylindole) – Figure 11.1a.

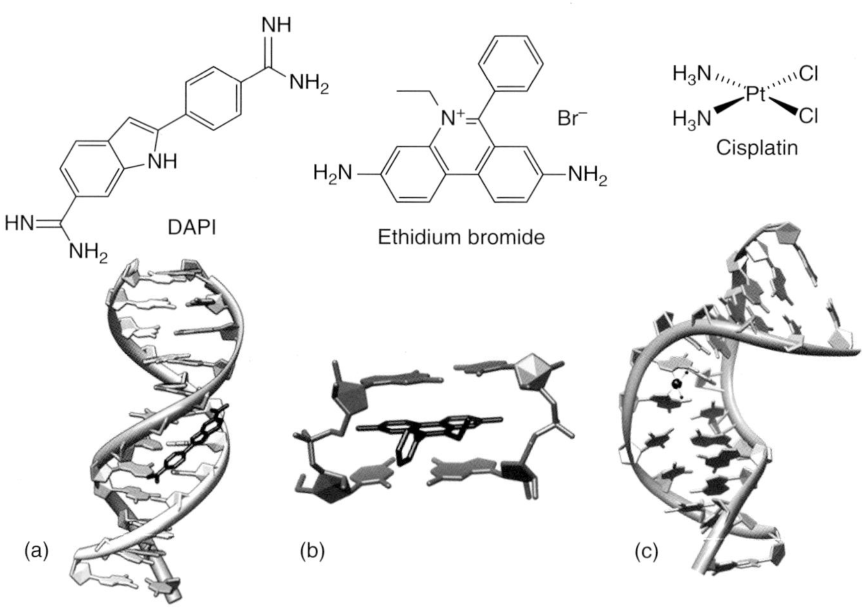


**Figure 11.1** (a) DNA-imaging agent DAPI (top) and X-ray crystal structure of DAPI bound to duplex DNA in minor groove (bottom, PDB: 1D30). (b) Ethidium bromide (EB) and crystal structure of EB intercalating between DNA base pairs (NDB: DRB006). (c) The anticancer drug cisplatin, which forms cisplatin-DNA inter-/intrastrand adducts (PDB: 1A84, [7]).

Another commonly observed reversible DNA-binding interaction is intercalation. First suggested by Lerman in 1961 [8], intercalation involves planar aromatic compounds inserting between adjacent base pairs in the DNA double helix (Figure 11.1b). This interaction involves significant π system overlap between DNA bases and the intercalated molecule which are in van der Waals' contact. Unlike groove binding, which produces minimal disruption of the duplex structure, intercalation unwinds, stiffens, and lengthens the DNA double helix [9]. Examples of intercalating molecules include ethidium bromide and its derivative propidium iodide, which are used as fluorescent nucleic acid stains in applications such as gel electrophoresis and flow cytometry. Several organic intercalators have been developed as anticancer drugs, such as the anthracyclines (e.g., doxorubicin), anthraquinones (e.g., mitoxantrone), or quinolone alkaloids (e.g., Topotecan). Organic intercalators typically act therapeutically via the inhibition/poisoning of topoisomerase I/II enzymes required to unwind DNA during replication, preventing its replication and generating cytotoxic double-strand breaks [10]. Both groove binding and intercalation are commonly accompanied by electrostatic and hydrophobic interactions which act to increase overall binding affinities.

### 11.1.2 Metal Complexes and DNA

While the majority of small molecules that bind and affect the structure and function of DNA are organic, metal coordination and organometallic complexes may also interact with DNA by irreversible or reversible mechanisms. Arguably the most famous example of a metal complex that binds DNA via irreversible coordination is cisplatin (*cis*-diamminedichloroplatinum(II); Figure 11.1c). Developed after Rosenborg and associates noticed that *Escherichia Coli* cell division was inhibited near a platinum electrode [11], cisplatin rapidly became a highly successful anticancer drug. Mechanistically, cisplatin binds DNA through the formation of inter- and intrastrand adducts ("platination;" see Ref. [7] and Figure 11.1c) that inhibit DNA replication, inducing cell cycle arrest and ultimately leading to cell death by activating apoptosis [12]. As cancer cells proliferate rapidly and frequently possess genetic defects within their DNA damage response, cancer may be targeted preferentially by such treatment [13]. Currently, cisplatin is still one of the most heavily used anticancer drugs in the world and is active toward a range of solid tumors, including testicular, ovarian, and bladder cancer [14]. Despite the success of cisplatin, the clinical benefit is limited by its toxicity, particularly nephrotoxicity and acquired or intrinsic tumor resistance. Since the pioneering work by Lippard and coworkers on square planar platinum(II) intercalators in the 1970s [15], there has been growing interest in kinetically inert transition metal complexes that interact with DNA by reversible mechanisms, particularly those able to intercalate DNA, the *metallointercalators* [16]. Work has progressed to exploring the biological activity of metallointercalators as therapeutics, particularly as anticancer therapeutic leads [17]. Parallel efforts have sought to convert their intriguing luminescent properties into DNA-imaging reagents [18]. As we shall see, ruthenium(II) complexes employing polypyridyl ligand architectures offer particular attraction within both categories of agent; specifically their site- and structure-specific binding interactions with DNA combined with fascinating photophysical properties mean that such systems are being investigated as novel imaging and therapeutic candidates.

## 11.2 $[Ru(bpy)_2(dppz)]^{2+}$ and the DNA "Light-Switch" Effect

Early work on the reversible DNA binding of Ru(II) polypyridyl complexes, or RPCs, focused on octahedral Ru(II) tris(phenanthroline) complexes and derivatives [19]. RPCs were chosen because of their chemical stability, high luminescence, and intense metal-to-ligand charge-transfer (MLCT) band in the visible spectrum [20]. The polypyridyl ligands used initially were the bidentate ligands 2,2′-bipyridine (bpy), complex **1.1**, 1,10-phenanthroline (phen), **1.2**, and 4,7-diphenyl-1,10-phenanthroline (DIP), **1.3** (Figure 11.2a) [23]. By examining the variation in photophysical properties in the presence of DNA, **1.1** showed little or no binding, while results for **1.2** and **1.3** suggested two binding modes to duplex DNA. However, the binding affinity of these complexes was low (equilibrium binding constant $K_b \sim 10^3\ M^{-1}$); and Barton and coworkers reasoned that to enhance the binding affinity for DNA, a further increase in ligand surface area was required. In collaboration with J. P. Sauvage, the dipyrido[3,2-*a*:2′,3′-*c*]phenazine (dppz) ligand was developed, which possesses a large aromatic surface area for intercalation. Studies on $[Ru(bpy)_2(dppz)]^{2+}$, **2.1**, revealed that this complex – and its phen analogue $[Ru(phen)_2(dppz)]^{2+}$, **2.2** – binds to DNA with high affinity ($K_b \sim 10^6\ M^{-1}$); Figure 11.2b. Perhaps more significantly, although the luminescence of the unbound complex in water is quenched, it displays intense MLCT-based luminescence on the addition of DNA as a direct result of intercalation (Figure 11.2c,d) [21]. While the exact details of the basis of this effect are much discussed, it is known to be due to solvent-mediated changes in the MLCT excited state of the complex [24]. Through hydrogen bonding interactions with the nitrogens on the phenazine unit of the dppz ligand, the $^3$MLCT excited state is quenched by water; however, when dppz is intercalated between base pairs of DNA, the phenazine nitrogen atoms are shielded from water, luminescence is then activated, and the light is "switched on" [25]. A study by Nordén, Lincoln, *et al.* on the interaction of the resolved Δ and Λ isomers of **2.2** with DNA confirmed that both isomers bind DNA by intercalation; however, this work also revealed that the Δ isomer contributed 85% of the light-switch emission observed for the racemic **2.2** mixture [26]. The binding geometry of Ru(dppz) compounds with DNA may be examined in detail from crystal structures. A clear kinking of the DNA double helix – similar in extent to that induced by cisplatin [7] – is observed due to intercalation of dppz [27]. Intriguingly, this kinking is reversible by dehydration [28]. Furthermore, multiple modes of intercalation are also observed, depending on the specific sequence of DNA employed [29]. For DNA sequences containing a mismatch base pair, side-by-side intercalation and insertion-binding modes are observed for **2.1** (Figure 11.2e) [22]. The discovery of **2.1**'s binding

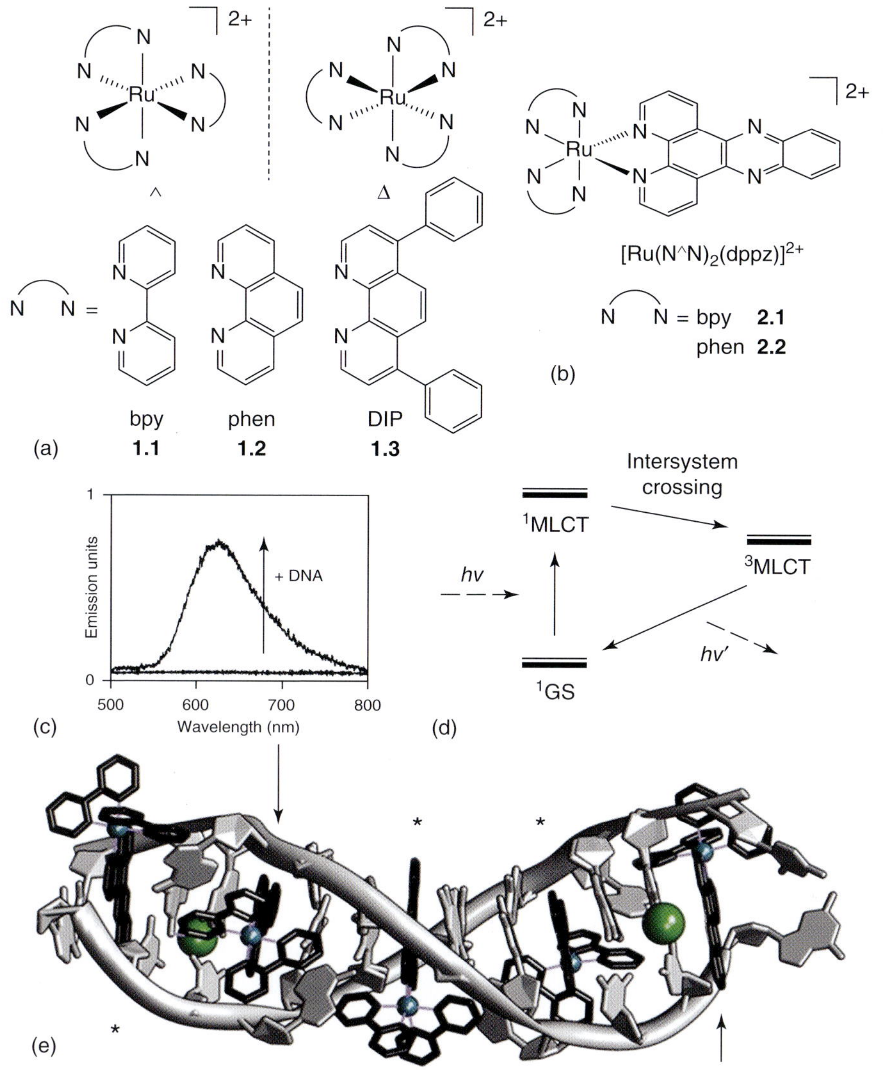


**Figure 11.2** (a) Octahedral Ru(II) tris(phenanthroline) complexes and derivatives, showing stereochemistry of Λ and Δ enantiomers. (b) Structure of Ru(N^N)$_2$(dppz)$^{2+}$ (N^N = bpy or phen) metallointercalators. (c) Increase in luminescence of Ru(bpy)$_2$(dppz)$^{2+}$ with addition of DNA. (Reproduced with permission from [21], © 1990 American Chemical Society (ACS)). (d) Schematic of electronic transitions of RPC MLCT emission (MLCT = metal-to-ligand charge-transfer, GS = ground state). (e) X-ray crystal structure of **2.1** and DNA showing side-by-side intercalation (*) and insertion (→) binding modes (PDB: 4E1U, [22]).

interaction with DNA led to an explosion of interest in the design of RPCs for biological application. While many of these compounds employ the dppz ligand to achieve intercalation, several derivative polypyridyl intercalating ligands have been synthesized and complexes that possess high binding affinities combined with sequence and structure specificities have been developed [30]. In more recent times, attention has been drawn to the biological applications of these systems, where the ability of an RPC to target intracellular DNA successfully is a highly desirable outcome. The next section discusses the cellular uptake and intracellular localization properties of RPCs.

## 11.3 Cellular Uptake of RPCs and Application as DNA-Imaging Agents

The MLCT luminescence that many RPCs possess means fluorescence-based cell biology techniques such as epifluorescence, confocal laser scanning microscopy (CLSM), and flow cytometry are useful tools by which to assess cellular uptake and intracellular localization [31]. In addition, if a "light-switch" complex, MLCT emission may also provide evidence that cellular DNA binding is achieved [30]. In cellular imaging terms, MLCT luminescence possesses several advantages over the current generation of commercial DNA-imaging fluorophores, including photostability, long-lifetime emissions, large Stokes shifts (typically 150–200 nm) and red or far-red emission. However, to enter a live cell and bind DNA, or indeed interact with *any* intracellular organelle or biomolecule, a molecule must first be able to cross the cell membrane. Eukaryotes are surrounded by a phospholipid membrane bilayer, where the hydrophobic interior of the bilayer means that diffusion across the membrane is determined by the polarity and size of the molecule; the more polar or larger the molecule, the more difficult it is to permeate the membrane. If poorly able to enter a cell via passive diffusion, a molecule will require a specific transport mechanism to facilitate cellular uptake. Specific transport mechanisms include carrier proteins embedded within the lipid membrane structure or endocytosis [32]. By their very nature, RPCs are relatively large, hydrophilic, and positively charged molecules; properties which do not encourage passive diffusion. However, this does not preclude cellular uptake. The following section describes the synthetic strategies employed by researchers aimed at promoting cellular uptake and achieving intracellular DNA targeting.

### 11.3.1 Mononuclear Complexes

Despite possessing attractive luminescent properties, the prototype RPCs **1.1** or **1.2** display low levels of cellular uptake, with localization in endosomes or lysosomes typically observed [33]. This is consistent with the complexes displaying low natural membrane permeability, which is also in agreement with their low bioactivity as described in early studies by Dwyer *et al.* [34]. Increased hydrophobicity promoting cellular uptake is well established in organic medicinal chemistry [35] and ruthenium polypyridyl complexes are no exception. Conveniently, any increase in ligand surface area, designed to enhance DNA intercalation, is generally accompanied by a concomitant increase in hydrophobicity. For example, the cellular uptake properties of a series of Ru(dppz) intercalators, including **2.1** and **2.2**, were found to improve uptake with increasing ancillary ligand hydrophobicity [36]. The most hydrophobic complex, $[Ru(DIP)_2(dppz)]^{2+}$, **2.3**, was confirmed to be internalized by passive diffusion;[37] strengthening the hypothesis that an increase in hydrophobicity will increase passive membrane diffusion. However, no nuclear uptake for **2.3** was observed (Figure 11.3a). That poor live-cell uptake is the main barrier to successful nuclear staining by **2.2** was confirmed in a study on fixed cells, which showed that under these conditions the complex is an effective dye for nuclear compartments, including *ex vivo* tissue sections [40]. Interestingly, and in contrast to the behavior demonstrated by **2.3**, $[Ru(DIP)_2(bpy\text{-}COOH)]^{2+}$ functions as a live-cell luminescent DNA dye [41], where DNA binding is presumably achieved via the DIP group. In addition to the ancillary ligands, the intercalating group may be directly modified to increase hydrophobicity. This is demonstrated for a series of complexes where alkyl groups are attached to the phenazine unit of dppz (complexes **3.1-3.3**, Figure 11.3b). As the alkyl chains increase in length, cell uptake is improved [38]. However, as observed for **2.3**, nuclear accumulation *decreases* with increasing hydrophobicity, even in

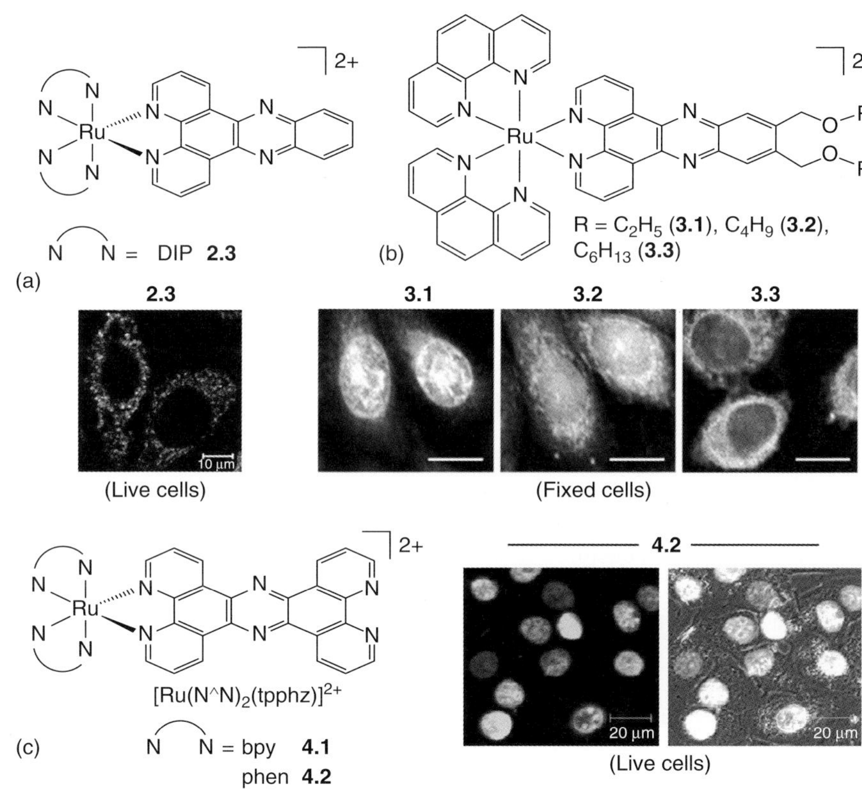


**Figure 11.3** (a) Hydrophobic Ru(dppz) intercalator and intracellular localization in HeLa cervical cancer cells (bottom). (Reproduced with permission from [36], © 2007 ACS). (b) Series of Ru(dppz) complexes substituted with alkyl chains and localization in fixed and permeabilized Chinese hamster ovary (CHO) cells (scale bars = 20 µm). (Reproduced with permission from [38], © 2011 Elsevier). (c) Ru(tpphz) complexes and live-cell uptake by MCF7 human breast cancer cells showing strong nuclear emission (right, see Ref. [39]).

fixed and permeabilized cells. Likewise, intercalating ligands with an increased surface area, such as dppn (benzo[*i*]dipyrido[3,2-*a*:2′,3′-*c*]phenazine)[42] or tpphz (tetrapyrido[3,2-*a*:2′,3′-*c*:3″,2″-*h*:2‴,3‴-*j*]phenazine) results in improved uptake relative to their dppz analogues. In the case of dppn, the light-switch effect is lost as $Ru^{II}$(dppn) systems possess a dppn-based $\pi\pi^*$ excited state [43]. However, both $[Ru(bpy)_2(tpphz)]^{2+}$, **4.1**, and in particular $[Ru(phen)_2(tpphz)]^{2+}$, **4.2**, are effective live-cell nuclear DNA-imaging agents (Figure 11.3c) [39]. In contrast to **2.3**, both these hydrophilic tpphz complexes are internalized by active transport, showing how alternative uptake pathways to passive diffusion may achieve nuclear targeting.

### 11.3.2 Dinuclear Complexes

Several groups have explored the cellular uptake of dinuclear Ru(II) units linked by a variety of bridging ligands (Figure 11.4a). Certain systems have proved to be efficient fixed-cell DNA-imaging agents, for example, $Ru(phen)_2$ units linked by either a rigid PIP-based linking group [46] (PIP = 2-(phenyl) imidazo[4,5-*f*][1,10]phenanthroline) or flexible alkyl chains [47] have both demonstrated success in this regard. Under live-cell conditions, while cytoplasmic accumulation is a common observation [48], several examples of successful DNA targeting have been reported. For example, dinuclear Ru(tpphz)Ru compounds **5.1** and **5.2** (Figure 11.4b) [49], which groove bind DNA (including quadruplex) with high affinity[50], have been developed as cellular imaging agents (Figure 11.4c) [44]. Complex **5.2**, in particular, is an effective DNA-imaging agent for use with CLSM; as demonstrated by overlap with DAPI (Figure 11.4d). Furthermore, two separate emission peaks are observed in **5.2**-stained nuclei, findings which correlate with duplex- and quadruplex-binding events in cell-free studies. Strikingly, nuclear accumulation of **5.2** by live cells is observed 5 min after initial exposure [51], and active transport was implied as the uptake mechanism [44]. When the hydrophobicity of the system is increased by employing DIP ancillary ligands, the resultant lipophilic complex, **5.3**, demonstrates a greater rate of cellular uptake than **5.2**; however, nuclear targeting is lost. Instead, **5.3** localizes in the membrane-dense endoplasmic reticulum, despite possessing a high cell-free DNA-binding affinity ($\sim 10^6\ M^{-1}$) (Figure 11.4c) [45]. Combined with the previously described studies on mononuclear intercalating RPCs, these observations provide evidence that enhanced *cellular* uptake does not necessarily increase *nuclear* targeting as larger hydrophobicity will increase a complex's affinity for membranes and thus promote localization in lipid-dense cellular regions.

### 11.3.3 Cyclometalated Systems

An alternative approach to enhancing uptake is to employ cyclometalated ligands to decrease the overall charge of a complex. For example, Chao *et al.* used the cyclometalated ancillary ligand phpy (phenyl-pyridine) to prepare a monocationic direct structural derivative of **2.1**, $[Ru(phpy)(bpy)(dppz)]^+$, complex **6** (Figure 11.5a) [54]. Strikingly, **6** has a greater level of total cell uptake than **2.1**

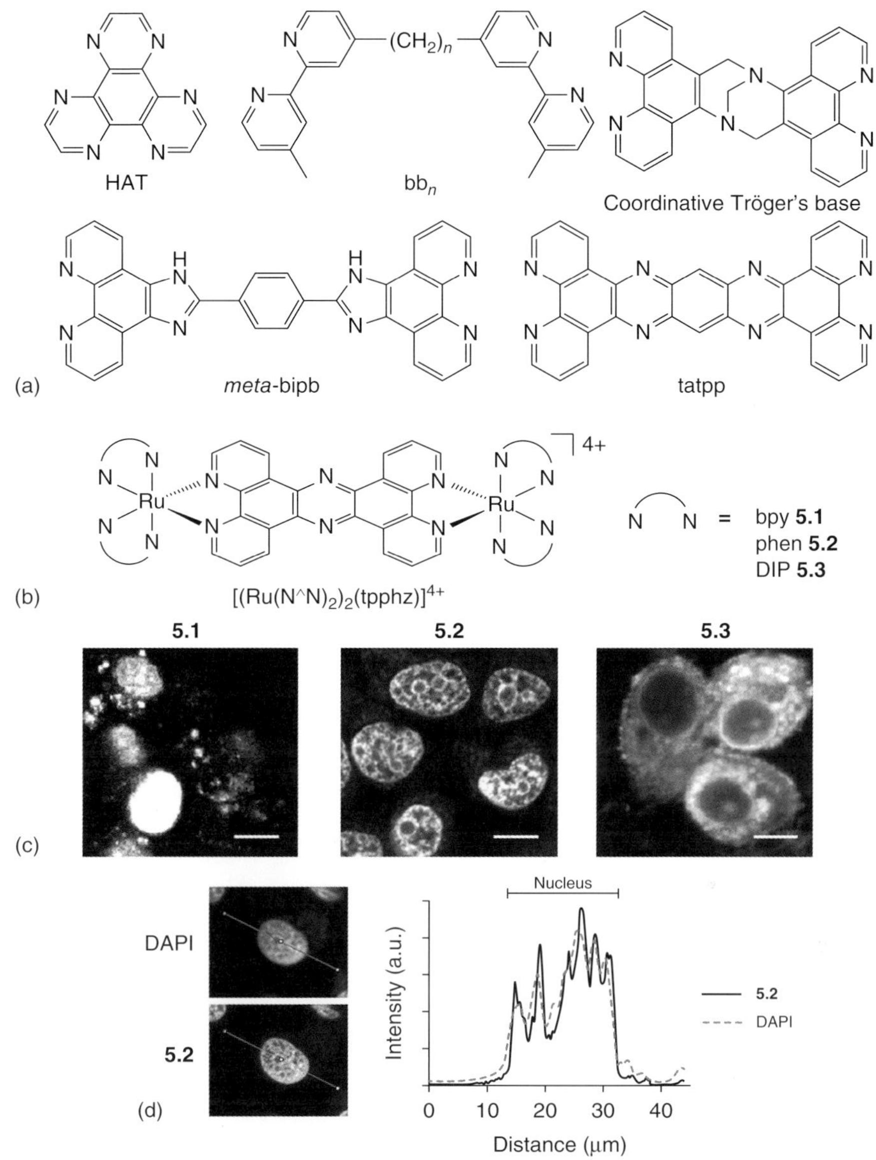


**Figure 11.4** (a) Examples of polypyridyl bridging ligands. (b) Dinuclear Ru-tpphz-Ru complexes. (c) Live-cell uptake of dinuclear Ru(tpphz) complexes by MCF7 human breast cancer cells showing nuclear staining by **5.2** and endoplasmic reticulum (ER) staining by **5.3** (scale bars = 20 μM). (d) Co-staining of MCF7 nucleus with **5.2** and DAPI showing overlap of signals (emission cross section, right). (Adapted from data featured within Refs [44, 45].)

and, more significantly, its nuclear accumulation is high (*vide infra*). No emission details of the complex were reported; however, the compound was demonstrated to inhibit cancer cell proliferation with high potency and inhibition of DNA transcription was implied in these results. A similar approach is involved in the preparation of a series of mixed-metal Ru(II)–Ir(III) complexes **7.1–7.3**

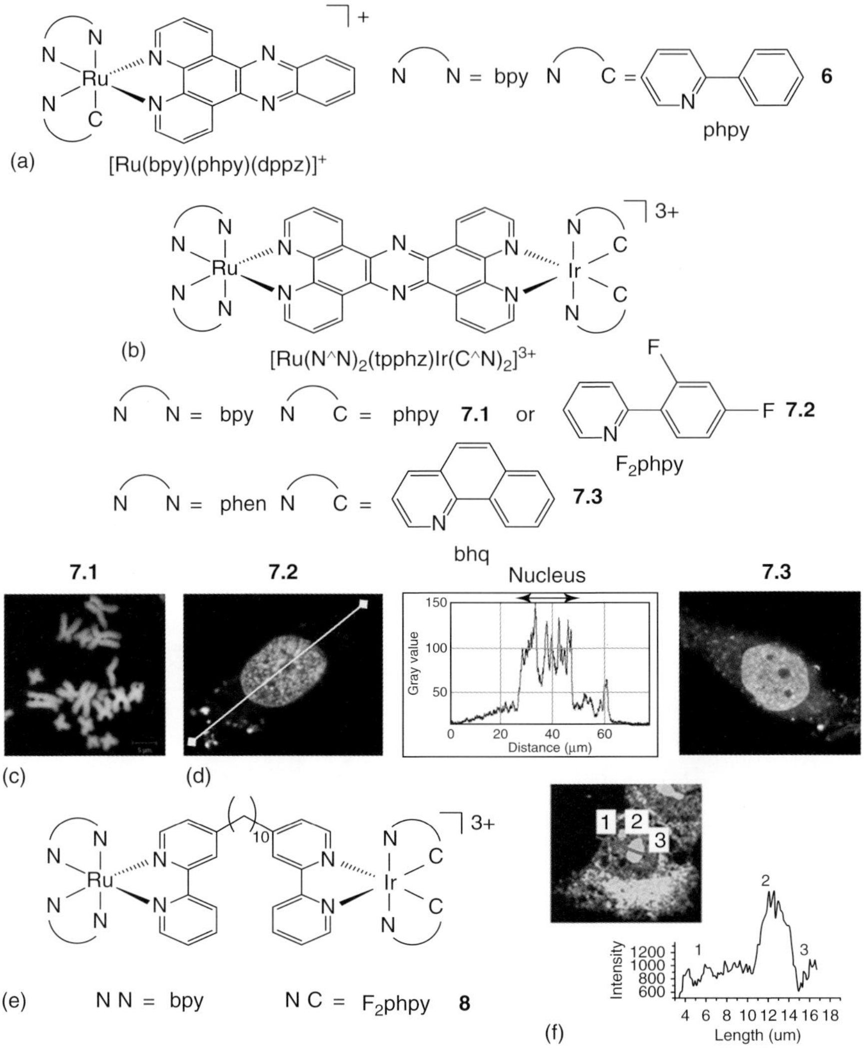


**Figure 11.5** (a) Cyclometallated Ru(dppz) monocation. (b) Hetero bimetallic Ru(II)—Ir(III) tpphz complexes. (c) Isolated HeLa chromosomes stained with **7.1**. (Adapted from Ref. [52a]). (d) HeLa cells incubated with **7.2** or **7.3** showing high nuclear accumulation (emission profile of **7.2**, center, see Refs. [52]). (e) Hetero bimetallic Ru(II)—Ir(III) tpphz complex employing flexible alkyl chain and (f) nucleoli accumulation in MCF7 cells. (Reproduced with permission from [53], © 2014 Royal Society of Chemistry (RSC).)

(Figure 11.5b) [52], which include cyclometalated analogues of the 4+ dinuclear tpphz complexes **5.1** and **5.2** described earlier. Here, a reduced 3+ overall charge is achieved by employing an Ir(III) center incorporating cyclometalated ligands. All complexes function as nuclear DNA-imaging agents in live and fixed cells

(Figure 11.5c,d), and both **7.1** and **7.3** achieve their effects at lower concentrations than their dinuclear ruthenium counterparts, indicating superior rates of cellular uptake. One unexpected consequence is that binding affinity for a common constituent of cell culture, bovine serum albumin (BSA), is increased for the phen analogue **7.3**; thus, inclusion of BSA within cell media strongly inhibits nuclear uptake [52b]. This work also demonstrated that the fluorinated ligands used in complex **7.2** promote cellular uptake while retaining biomolecular targeting, [52a] an approach that has been widely used in medicinal chemistry [55]. Sun *et al.* also explored heterodinuclear Ru(II)–Ir(III) complexes where each metal center was linked by a flexible alkyl chain – for example, complex **8** (Figure 11.5e) [53]. Cellular internalization was observed; however, RNA rather than DNA was the primary target, as evidenced by cell-free binding studies and high accumulation in the RNA-rich nucleolus (Figure 11.5f).

## 11.4 Alternative Techniques to Assess Cellular Uptake and Localization

Of course, steady-state emission microscopy is not the only technique available to researchers. RPCs are compatible with the rapidly emerging technique of Raman microscopy and – due to their long-lifetime emissions – they are also suitable for lifetime-based techniques such as phosphorescence lifetime imaging microscopy (PLIM). Furthermore, they are compatible with two-photon excitation microscopy, a technique which offers deeper tissue penetration than standard confocal microscopy. In addition to these light-based techniques, Ru(II) polypyridyl complexes may also be detected using transmission electron microscopy (TEM) as they possess a second-row transition-metal center. For example, the live cell imaging agent **7.2** is an excellent nuclear stain for two-photon PLIM (Figure 11.6a) [56], while visualizing **7.2**-treated cells post incubation using TEM provides *sub*nuclear localization detail, with heterochromatin accumulation of the complex being evident (Figure 11.6b) [44]. Due to their large Stokes shift values, RPCs are particularly well suited to Raman microscopy. Keyes and coworkers have demonstrated this to great effect by employing a membrane-sensitive RPC–peptide conjugate, **8** (Figure 11.6c) [57]. In combination with CLSM, resonance Raman imaging provides detailed information on the membrane localization of this complex. This is a concept that may, in time, be applied to nuclear-localizing systems.

Cellular metal content may be quantified using analytical techniques such as ICP-MS (inductively coupled plasma mass spectroscopy) or AAS (atomic adsorption spectroscopy). When combined with subcellular fractionization techniques to isolate nuclei and other organelles, detailed profiles of the intracellular distribution of a metal complex may be achieved. Figure 11.6d demonstrates a comparison between ruthenium content in the cytoplasm versus nucleus in cells treated with **2.1**, **2.3**, and **6** for increasing exposure times; showing that the nuclear

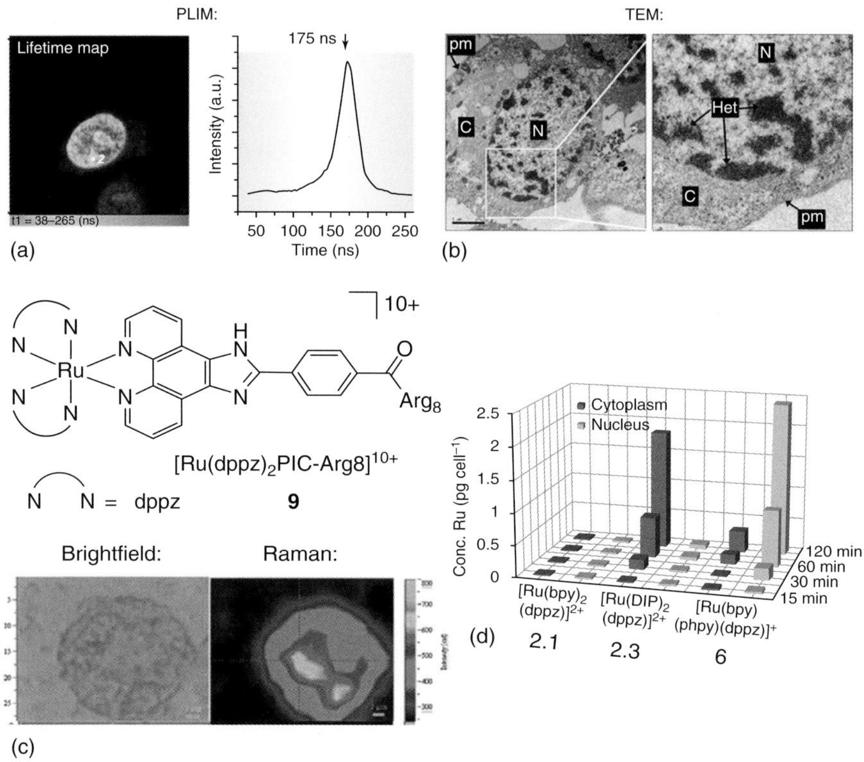


**Figure 11.6** (a) Two-photon PLIM (phosphorescent lifetime emission microscopy) imaging of **7.2**-treated MCF7 cells. (Reproduced with permission from Ref. [56], © 2014 Wiley VCH). (b) Transmission electron micrograph of MCF7 cell pretreated with **7.2**. Key: pm = plasma membrane, C = cytosol, N = nucleus, het = heterochromatin. See Ref. [44]. (c) RPC-labeled cell-penetrating peptide (top) and resonance Raman intensity map of a live myeloma cell treated with complex (bottom right). Reproduced with permission from Ref. [57], © 2010 RSC. (d) Quantification of nuclear and cytoplasmic uptake of a series of RPCs by ICP-MS (inductively coupled plasma mass spectroscopy). (Reproduced with permission from Ref. [54], © 2014 ACS.)

accumulation of **6** is especially significant, thus implying that DNA is the biological target for the complex's antiproliferative properties [54].

## 11.5 Toward Theranostics: luminescent RPCs as Anticancer Therapeutics

It is also possible for DNA-binding molecules to be both emissive and possess antiproliferative properties; indeed, several of the examples discussed in previous sections also display relatively high antiproliferative effects. In such cases, the emissive properties of the complex allow cellular localization to be observed directly, thereby facilitating any exploration into the specific mechanisms of their therapeutic action. This is illustrated by studies on the two tpphz intercalators mentioned previously, **4.1** and **4.2**. In addition to their

DNA-imaging properties, these systems reduce the proliferation of several cancer cell lines, including cisplatin-resistant strains [39]. Their high nuclear uptake and ability to intercalate into DNA suggest that this is their therapeutic target molecule. Further examples are provided in Figure 11.7. This includes work by Gasser and colleagues, who presented a bis(dppz) complex incorporating 2-pyridyl-2-pyrimidinep-4-carboxylic acid as the third bidentate ligand (**10**). Half inhibitory concentrations comparable – or greater – than cisplatin were observed, including in cisplatin-resistant cell lines. Cell death occurs via mitochondrial accumulation, and resultant ROS (reactive oxygen species) generation is observed [58]. This finding is similar to previous work on a series of antiproliferative luminescent ruthenium(II) β-carboline complexes (**11.1–11.3**) [59]. Several ruthenium(II) complexes with substituted PIP ligands demonstrate high mitochondrial accumulation and induction of cell death via the intrinsic (i.e., mitochondrial-mediated) pathway, including in mouse xenograft models (general structure: **13**) [60]. Mechanisms range from the direct induction

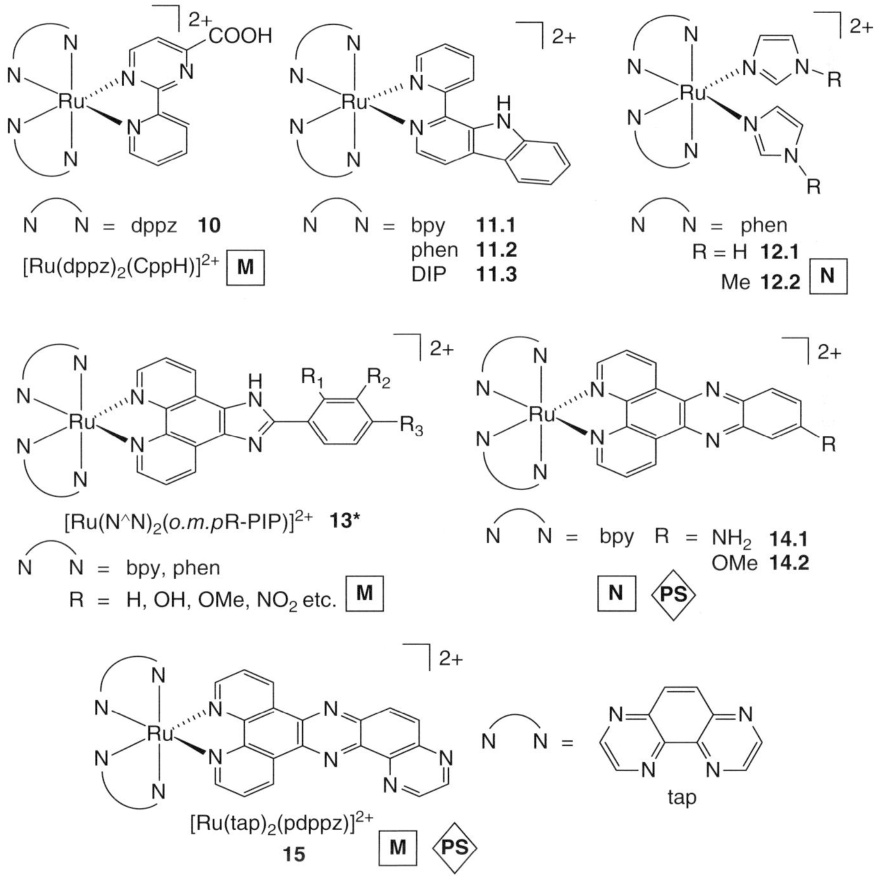


**Figure 11.7** Theranostic ruthenium(II) polypyridyl complexes. Key: N = nuclear uptake, M = mitochondrial uptake. PS = photosensitizer.

of apoptosis via mitochondrial accumulation and ROS generation [61], to affecting cell motility and growth via suppressing FAK (focal adhesion kinase) signaling [62]. These complexes have been designed to target DNA, and systems thought to stabilize G-quadruplex DNA, and thus inhibit telomerase [63], or function as topoisomerase I/II poisons[64] have been discussed; however, these conclusions are largely based on cell-free DNA-binding studies. At this point, it should also be noted that direct induction of cell death (cytotoxicity) is not the only way to inhibit cell proliferation. For example, the luminescent complex $[Ru(phen)_2(ImH)_2]^{2+}$, **12.1**, and its methylated derivative $[Ru(phen)_2(IMe)_2]^{2+}$, **12.2** (ImH = imidazole and IMe = 1-methylimidazole), demonstrate high potency via a cytostatic (i.e., growth inhibition) mechanism of action [65]. Although DNA as a molecular target was not explored, nuclear uptake of **12.2** was suggested.

An alternative method to target cancer cells specifically is to apply the theranostic as a photosensitizer. This relies on the activated state being the cytotoxic group and inducing intracellular damage via the generation of singlet oxygen, other ROS, or by direct damage of DNA and/or protein structures. This approach aims to exploit long-lived MLCT – or even intraligand[66] – states. To date, several luminescent RPCs that show potential in this regard have been reported, and the reader is directed to a recent review of progress in this area [67]. Promising examples of RPC photodynamic therapy (PDT) sensitizers include a bimetallic Ru(II)-V(IV) tpphz compound [68], nuclear-localizing Ru(II) complexes containing dppz ligands substituted with NH2 (**14.1**) or OMe (**14.2**)[69] or a lipophilic dppz derivative, pdppz ([2,3-*h*] dipyrido[3,2-*a*:2′,3′-*c*]phenazine), **15** [70]. In the latter case, phototoxicity is achieved via the incorporation of 1,4,5,8-tetraazaphenanthrene (tap) ancillary ligands combined with mitochondrial accumulation of the complex.

One interesting application of mixed-metal agents is combining ruthenium DNA-binding systems with a platinum therapeutic. In this case, the RPC unit may provide an imaging role by way of MLCT emission while simultaneously influencing the site or sequence of DNA targeted. We direct readers to the pioneering work of Brewer and colleagues [71], and, more recently, a Ru(II)-Pt(II) conjugate that inhibits cell growth via upregulation of the cyclin-dependent kinase inhibitor p27 has been reported [72]. Interestingly, this complex does not demonstrate a mechanism of action involving the generation of cytotoxic DNA damage and is therefore in a different class of agent to conventional platinum drugs, a factor that likely explains why no cross-resistance to a cisplatin-resistant ovarian cancer cell line is observed.

## 11.6 Summary and Conclusions

As this chapter illustrates, recent years have seen significant advances in the understanding of the cellular uptake and biomolecular targeting properties of ruthenium(II) polypyridyl complexes, both in terms of providing cellular DNA-imaging agents for a range of microscopy techniques and developing therapeutic candidates able to halt cancer cell growth. It is anticipated that greater understanding of cellular internalization pathways and pharmacological mechanisms of action will provide the basis for isolate complexes with novel bioactivities compared to existing agents, where the flexible synthetic coordination chemistry will allow rapid optimization of “first in series” complexes for the desired biological application(s). If work continues apace, it is clear that there is a bright future for ruthenium(II) polypyridyl complexes within biomedical science.

## References

1 Crick, F.H.C. (1970) *Nature*, **227**, 561–563.

2 Gilman, A. and Philips, F.S. (1946) *Science*, **103**, 409–436.

3 Chabner, B.A. and Roberts, T.G. (2005) *Nat. Rev. Cancer*, **5**, 65–72.

4 Hurley, L.H. (2002) *Nat. Rev. Cancer*, **2**, 188–200.

5 Rodriguez, R., Miller, K.M., Forment, J.V., Bradshaw, C.R., Nikan, M., Britton, S., Oelschlaegel, T., Xhemalce, B., Balasubramanian, S., and Jackson, S.P. (2012) *Nat. Chem. Biol.*, **8**, 301–310.

6 Hooke, R. (1665) *Micrographia*, J. Martyn and J. Allestry, London.

7 Takahara, P.M., Rosenzweig, A.C., Frederick, C.A., and Lippard, S.J. (1995) *Nature*, **377**, 649–652.

8 Lerman, L.S. (1961) *J. Mol. Biol.*, **3**, 18–30.

9 Long, E.C. and Barton, J.K. (1990) *Acc. Chem. Res.*, **23**, 271–273.

10 Pommier, Y., Leo, E., Zhang, H., and Marchand, C. (2010) *Chem. Biol.*, **17**, 421–433.

11 (a) Rosenberg, B., Van Camp, L., and Krigas, T. (1965) *Nature*, **205**, 698–699; (b) Mansour, V.H., Rosenberg, B., Vancamp, L., and Trosko, J.E. (1969) *Nature*, **222**, 385–386.

12 Siddik, Z.H. (2003) *Oncogene*, **22**, 7265–7279.

13 Lord, C.J. and Ashworth, A. (2012) *Nature*, **481**, 287–294.

14 Cancer Research UK General Cancer Information http://www.cancerresearchuk.org/about-cancer/cancers-in-general/treatment/cancer-drugs/cisplatin (accessed 08 July 2017).

15 (a) Jennette, K.W., Lippard, S.J., Vassiliades, G.A., and Bauer, W.R. (1974) *Proc. Natl. Acad. Sci. U.S.A.*, **71**, 3839–3843; (b) Lippard, S., Bond, P., Wu, K., and Bauer, W. (1976) *Science*, **194**, 726–728.

16 (a) Metcalfe, C. and Thomas, J.A. (2003) *Chem. Soc. Rev.*, **32**, 215–224; (b) Erkkila, K.E., Odom, D.T., and Barton, J.K. (1999) *Chem. Rev.*, **99**, 2777–2796; (c) Zeglis, B.M., Pierre, V.C., and Barton, J.K. (2007) *Chem. Commun.*, 4565–4579.

17 Gasser, G. and Metzler-Nolte, N. (2012) *Curr. Opin. Chem. Biol.*, **16**, 84–91.

18 (a) Fernandez-Moreira, V., Thorp-Greenwood, F.L., and Coogan, M.P. (2010) *Chem. Commun.*, **46**, 186–202; b) Lo, K.K.-W., Choi, A.W.-T., and Law, W.H.-T. (2012) *Dalton Trans.*, **41**, 6021–6047; c) Baggaley, E., Weinstein, J.A., and Williams, J.A.G. (2012) *Coord. Chem. Rev.*, **256**, 1762–1785; d) Zhao, Q., Huang, C., and Li, F. (2011) *Chem. Soc. Rev.*, **40**, 2508–2524.

19 Barton, J. (1986) *Science*, **233**, 727–734.

20 Barton, J.K., Danishefsky, A., and Goldberg, J. (1984) *J. Am. Chem. Soc.*, **106**, 2172–2176.

21 (a) Friedman, A.E., Chambron, J.C., Sauvage, J.P., Turro, N.J., and Barton, J.K. (1990) *J. Am. Chem. Soc.*, **112**, 4960–4962; (b) Hartshorn, R.M. and Barton, J.K. (1992) *J. Am. Chem. Soc.*, **114**, 5919–5925; (c) Jenkins, Y., Friedman, A.E., Turro, N.J., and Barton, J.K. (1992) *Biochemistry*, **31**, 10809–10816.

22 Song, H., Kaiser, J.T., and Barton, J.K. (2012) *Nat. Chem.*, **4**, 615–620.

23 Kumar, C.V., Barton, J.K., and Turro, N.J. (1985) *J. Am. Chem. Soc.*, **107**, 5518–5523.

24 (a) Olson, E.J.C., Hu, D., Hormann, A., Jonkman, A.M., Arkin, M.R., Stemp, E.D.A., Barton, J.K., and Barbara, P.F. (1997) *J. Am. Chem. Soc.*, **119**, 11458–11467; (b) Brennaman, M.K., Alstrum-Acevedo, J.H., Fleming, C.N., Jang, P., Meyer, T.J., and Papanikolas, J.M. (2002) *J. Am. Chem. Soc.*, **124**, 15094–15098; (c) Olofsson, J., Önfelt, B., and Lincoln, P. (2004) *J. Phys. Chem. A*, **108**, 4391–4398.

25 McKinley, A.W., Lincoln, P., and Tuite, E.M. (2011) *Coord. Chem. Rev.*, **255**, 2676–2692.

26 Hiort, C., Lincoln, P., and Norden, B. (1993) *J. Am. Chem. Soc.*, **115**, 3448–3454.

27 Hall, J.P., O'Sullivan, K., Naseer, A., Smith, J.A., Kelly, J.M., and Cardin, C.J. (2011) *Proc. Natl. Acad. Sci. U.S.A.*, **108**, 17610–17614.

28 Hall, J.P., Sanchez-Weatherby, J., Alberti, C., Quimper, C.H., O'Sullivan, K., Brazier, J.A., Winter, G., Sorensen, T., Kelly, J.M., Cardin, D.J., and Cardin, C.J. (2014) *J. Am. Chem. Soc.*, **136**, 17505–17512.

29 Niyazi, H., Hall, J.P., O'Sullivan, K., Winter, G., Sorenson, T., Kelly, J.M., and Cardin, C.J. (2012) *Nat. Chem.*, **4**, 621–628.

30 Gill, M.R. and Thomas, J.A. (2012) *Chem. Soc. Rev.*, **41**, 3179–3192.

31 Puckett, C.A., Ernst, R.J., and Barton, J.K. (2010) *Dalton Trans.*, **39**, 1159–1170.

32 Alberts, B., Johnson, A., Lewis, J., Raff, M., Roberts, K., and Walter, P. (2007) *Molecular Biology of the Cell*, 5th edn, Garland Science, New York.

33 Huntosova, V., Stroffekova, K., Wagnieres, G., Novotova, M., Nichtova, Z., and Miskovsky, P. (2014) *Metallomics*, **6**, 2279–2289.

34 Dwyer, F.P., Gyarfas, E.C., Rogers, W.P., and Koch, J.H. (1952) *Nature*, **170**, 190–191.

35 Lipinski, C.A., Lombardo, F., Dominy, B.W., and Feeney, P.J. (2001) *Adv. Drug Delivery Rev.*, **46**, 3–26.

36 Puckett, C.A. and Barton, J.K. (2007) *J. Am. Chem. Soc.*, **129**, 46–47.

37 Puckett, C.A. and Barton, J.K. (2008) *Biochemistry*, **47**, 11711–11716.

38 a) Svensson, F.R., Matson, M., Li, M., and Lincoln, P. (2010) *Biophys. Chem.*, **149**, 102–106; b) Matson, M., Svensson, F.R., Nordén, B., and Lincoln, P. (2011) *J. Phys. Chem. B*, **115**, 1706–1711.

39 Gill, M.R., Derrat, H., Smythe, C., Battaglia, G., and Thomas, J.A. (2011) *ChemBioChem*, **12**, 877–880.

40 Rajendiran, V., Palaniandavar, M., Periasamy, V.S., and Akbarsha, M.A. (2010) *J. Inorg. Biochem.*, **104**, 217–220.

41 Musatkina, E., Amouri, H., Lamoureux, M., Chepurnykh, T., and Cordier, C. (2007) *J. Inorg. Biochem.*, **101**, 1086–1089.

42 Schatzschneider, U., Niesel, J., Ott, I., Gust, R., Alborzinia, H., and Wölfl, S. (2008) *ChemMedChem*, **3**, 1104–1109.

43 Foxon, S.P., Alamiry, M.A.H., Walker, M.G., Meijer, A.J.H.M., Sazanovich, I.V., Weinstein, J.A., and Thomas, J.A. (2009) *J. Phys. Chem. A*, **113**, 12754–12762.

44 Gill, M.R., Garcia-Lara, J., Foster, S.J., Smythe, C., Battaglia, G., and Thomas, J.A. (2009) *Nat. Chem.*, **1**, 662–667.

45 Gill, M.R., Cecchin, D., Walker, M.G., Mulla, R.S., Battaglia, G., Smythe, C., and Thomas, J.A. (2013) *Chem. Sci.*, **4**, 4512–4519.

46 Svensson, F., Andersson, J., Åmand, H., and Lincoln, P. (2012) *J. Biol. Inorg. Chem.*, **17**, 565–571.

47 (a) Li, X., Gorle, A.K., Ainsworth, T.D., Heimann, K., Woodward, C.E., Grant Collins, J., and Richard Keene, F. (2015) *Dalton Trans.*, **44**, 3594–3603; (b) Onfelt, B., Gostring, L., Lincoln, P., Norden, B., and Onfelt, A. (2002) *Mutagenesis*, **17**, 317–320.

48 (a) Xu, W., Zuo, J., Wang, L., Ji, L., and Chao, H. (2014) *Chem. Commun.*, **50**, 2123–2125; (b) Elmes, R.B.P., Erby, M., Bright, S.A., Williams, D.C., and Gunnlaugsson, T. (2012) *Chem. Commun.*, **48**, 2588–2590.

49 Bolger, J., Gourdon, A., Ishow, E., and Launay, J.-P. (1996) *Inorg. Chem.*, **35**, 2937–2944.

50 (a) Rajput, C., Rutkaite, R., Swanson, L., Haq, I., and Thomas, J.A. (2006) *Chem. Eur. J.*, **12**, 4611–4619; (b) Lutterman, D.A., Chouai, A., Liu, Y., Sun, Y., Stewart, C.D., Dunbar, K.R., and Turro, C. (2008) *J. Am. Chem. Soc.*, **130**, 1163–1170; (c) Wilson, T., Costa, P.J., Félix, V., Williamson, M.P., and Thomas, J.A. (2013) *J. Med. Chem.*, **56**, 8674–8683.

51 Tian, X., Gill, M.R., Cantón, I., Thomas, J.A., and Battaglia, G. (2011) *ChemBioChem*, **12**, 548–551.

52 (a) Wragg, A., Gill, M.R., Turton, D., Adams, H., Roseveare, T.M., Smythe, C., Su, X., and Thomas, J.A. (2014) *Chem. Eur. J.*, **20**, 14004–14011; (b) Wragg, A., Gill, M.R., McKenzie, L., Glover, C., Mowll, R., Weinstein, J.A., Su, X., Smythe, C., and Thomas, J.A. (2015) *Chem. Eur. J.*, **21**, 11865–11871.

53 Sun, S., Wang, J., Mu, D., Wang, J., Bao, Y., Qiao, B., and Peng, X. (2014) *Chem. Commun.*, **50**, 9149–9152.

54 Huang, H., Zhang, P., Yu, B., Chen, Y., Wang, J., Ji, L., and Chao, H. (2014) *J. Med. Chem.*, **57**, 8971–8983.

55 Purser, S., Moore, P.R., Swallow, S., and Gouverneur, V. (2008) *Chem. Soc. Rev.*, **37**, 320–330.

56 Baggaley, E., Gill, M.R., Green, N.H., Turton, D., Sazanovich, I.V., Botchway, S.W., Smythe, C., Haycock, J.W., Weinstein, J.A., and Thomas, J.A. (2014) *Angew. Chem. Int. Ed.*, **53**, 3367–3371.

57 Cosgrave, L., Devocelle, M., Forster, R.J., and Keyes, T.E. (2010) *Chem. Commun.*, **46**, 103–105.

58 Pierroz, V., Joshi, T., Leonidova, A., Mari, C., Schur, J., Ott, I., Spiccia, L., Ferrari, S., and Gasser, G. (2012) *J. Am. Chem. Soc.*, **134**, 20376–20387.

59 Tan, C., Lai, S., Wu, S., Hu, S., Zhou, L., Chen, Y., Wang, M., Zhu, Y., Lian, W., Peng, W., Ji, L., and Xu, A. (2010) *J. Med. Chem.*, **53**, 7613–7624.

60 Wang, J.-Q., Zhang, P.-Y., Ji, L.-N., and Chao, H. (2015) *J. Inorg. Biochem.*, **146**, 89–96.
61 (a) Chen, T., Liu, Y., Zheng, W.-J., Liu, J., and Wong, Y.-S. (2010) *Inorg. Chem.*, **49**, 6366–6368; (b) Wang, J.-Q., Zhang, P.-Y., Qian, C., Hou, X.-J., Ji, L.-N., and Chao, H. (2014) *J. Biol. Inorg. Chem.*, **19**, 335–348; (c) Du, Y., Fu, X., Li, H., Chen, B., Guo, Y., Su, G., Zhang, H., Ning, F., Lin, Y., Mei, W., and Chen, T. (2014) *ChemMedChem*, **9**, 714–718.
62 Cao, W., Zheng, W., and Chen, T. (2015) *Sci. Rep.*, **5**, 9157.
63 (a) Zhang, Z., Wu, Q., Wu, X.-H., Sun, F.-Y., Chen, L.-M., Chen, J.-C., Yang, S.-L., and Mei, W.-J. (2014) *Eur. J. Med. Chem.*, **80**, 316–324; (b) Yu, Q., Liu, Y., Wang, C., Sun, D., Yang, X., Liu, Y., and Liu, J. (2012) *PLoS One*, **7**, e50902.
64 Zhang, P., Wang, J., Huang, H., Qiao, L., Ji, L., and Chao, H. (2013) *Dalton Trans.*, **42**, 8907–8917.
65 Cardoso, C.R., Lima, M.V.S., Cheleski, J., Peterson, E.J., Venâncio, T., Farrell, N.P., and Carlos, R.M. (2014) *J. Med. Chem.*, **57**, 4906–4915.
66 Lincoln, R., Kohler, L., Monro, S., Yin, H., Stephenson, M., Zong, R., Chouai, A., Dorsey, C., Hennigar, R., Thummel, R.P., and McFarland, S.A. (2013) *J. Am. Chem. Soc.*, **135**, 17161–17175.
67 Mari, C., Pierroz, V., Ferrari, S., and Gasser, G. (2015) *Chem. Sci.*, **6**, 2660–2686.
68 Holder, A.A., Taylor, P., Magnusen, A.R., Moffett, E.T., Meyer, K., Hong, Y., Ramsdale, S.E., Gordon, M., Stubbs, J., Seymour, L.A., Acharya, D., Weber, R.T., Smith, P.F., Dismukes, G.C., Ji, P., Menocal, L., Bai, F., Williams, J.L., Cropek, D.M., and Jarrett, W.L. (2013) *Dalton Trans.*, **42**, 11881–11899.
69 Mari, C., Pierroz, V., Rubbiani, R., Patra, M., Hess, J., Spingler, B., Oehninger, L., Schur, J., Ott, I., Salassa, L., Ferrari, S., and Gasser, G. (2014) *Chem. Eur. J.*, **20**, 14421–14436.
70 Cloonan, S.M., Elmes, R.B.P., Erby, M., Bright, S.A., Poynton, F.E., Nolan, D.E., Quinn, S.J., Gunnlaugsson, T., and Williams, D.C. (2015) *J. Med. Chem.*, **58**, 4494–4505.
71 (a) Milkevitch, M., Storrie, H., Brauns, E., Brewer, K.J., and Shirley, B.W. (1997) *Inorg. Chem.*, **36**, 4534–4538; (b) Williams, R.L., Toft, H.N., Winkel, B., and Brewer, K.J. (2003) *Inorg. Chem.*, **42**, 4394–4400; (c) Jain, A., Wang, J., Mashack, E.R., Winkel, B.S.J., and Brewer, K.J. (2009) *Inorg. Chem.*, **48**, 9077–9084; (d) Higgins, S.L.H., Tucker, A.J., Winkel, B.S.J., and Brewer, K.J. (2012) *Chem. Commun.*, **48**, 67–69.
72 Ramu, V., Gill, M.R., Jarman, P.J., Turton, D., Thomas, J.A., Das, A., and Smythe, C. (2015) *Chem. Eur. J.*, **21**, 9185–9197.

# 12

# Biological Activity of Ruthenium Complexes With Quinoline Antibacterial and Antimalarial Drugs

*Jakob Kljun and Iztok Turel*

*University of Ljubljana, Faculty of Chemistry and Chemical Technology, Večna pot 113, SI-1000, Ljubljana, Slovenia*

## 12.1 Introduction

Finding new uses for approved (off-patent) drugs as well as drugs withdrawn from clinical practice is a growing area of research due, in part, to the cost of developing new drugs from scratch. Several examples of approved drugs that have subsequently found novel applications are known such as sildenafil (Viagra), which was developed for heart disease and is now used to treat erectile dysfunction; gemcitabine (Gemzar), which was developed as an antiviral and is now used as an anticancer agent; raloxifene (Evista), which was developed as a birth control drug and is now used as an osteoporosis treatment that can also prevent breast cancer; and acetylsalicylic acid (Aspirin), which was developed for the treatment of inflammation and pain and is now also used as an antiplatelet agent. Such molecules can also serve as viable lead compounds for simple structural modifications involving limited synthetic steps. The fact that they were or still are in production at industrial scale significantly reduces the amount of funding and research needed to move a therapeutic agent toward clinical practice.

From the point of view of an inorganic or organometallic chemist, such structural modifications (Figure 12.1) can involve coordination of a molecule to a metal ion/moiety (formation of a *coordination compound*), derivatization of a molecule by inclusion of a metal-binding moiety through a spacer via a cleavable (amide, ester) or noncleavable covalent bond with subsequent formation of a metal complex (formation of a metal *conjugate*), or an inclusion of a chemically inert metal-containing substituent bound through a covalent bond, most commonly a metallocene (formation of a metal-containing *analogue*).

Generally, the evaluation of biological properties of such complexes, derivatives, and analogues proceeds in two directions – on the one hand, the novel metallodrug can be investigated as an alternative or complementary agent to the original molecule with a similar mode of action but different pharmacokinetics and pharmacodynamics; on the other hand the binding of the organic drug to the metal moiety can be viewed as a means of transferring these

*Ruthenium Complexes: Photochemical and Biomedical Applications*, First Edition.
Edited by Alvin A. Holder, Lothar Lilge, Wesley R. Browne, Mark A.W. Lawrence, and Jimmie L. Bullock Jr.

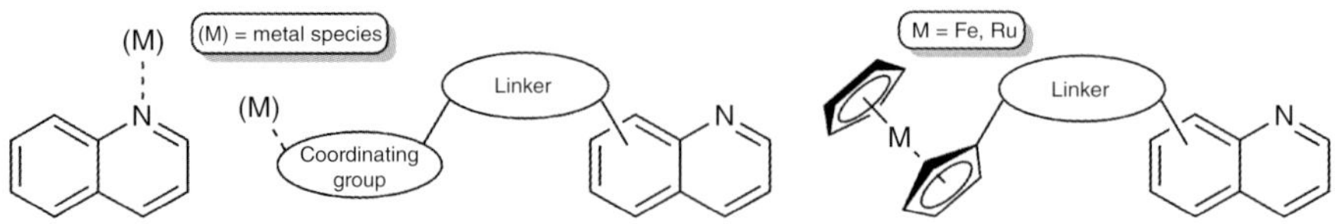


**Figure 12.1** Structural modifications for the inclusion of metal-containing moieties on a quinoline scaffold.

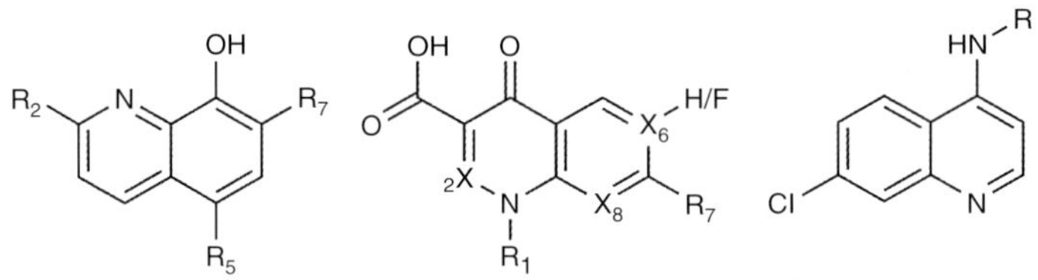


**Figure 12.2** General structure of the three classes of quinoline antibacterial or antimalarial agents. (a–c) 8-hydroxyquinoline, 4-quinolone, and 4-aminoquinoline.

same properties to a metal-containing molecule while retaining the biological activity of the metal species.

Antibacterial and antimalarial (antiprotozoal) drugs include numerous groups of both naturally occurring and synthetic compounds which comprise from simple derivatives of basic heterocycles to compounds with oligosaccharide, oligopeptide, and steroid frameworks, and much more. Among these groups of compounds, the quinoline scaffold occurs in three distinct and numerously represented classes of compounds with either antibacterial or antimalarial properties.

This chapter thus reviews the current state in the chemistry of ruthenium complexes, conjugates, and analogues of the members of antibacterial and antimalarial agents based on the quinoline scaffold (Figure 12.2) as well as their biological properties.

## 12.2 Antibacterial (Fluoro)quinolones

Quinolones are a group of synthetic antibacterial agents containing a 4-oxo-1,4-dihydroquinoline-3-carboxylic acid scaffold (Figure 12.2b). The first compound from this group introduced into clinical practice in the early 1960s was nalidixic acid (nalH; Figure 12.3a). It was serendipitously discovered in an attempt to devise a novel synthesis of antimalarial agent chloroquine. A by-product of the reaction was found to exhibit antibacterial activity. Synthesis of similar compounds and assessment of their activity led to the discovery of nalidixic acid. An expanded "Discovery Story" of this important class of drugs was recently published [1]. A huge number of structurally related molecules have been isolated; and altogether approximately 30 members are (or were) in clinical use for the treatment of a variety of diseases including infections of the urinary tract, skin, soft tissue, respiratory tract, gastrointestinal infections, bone/joint

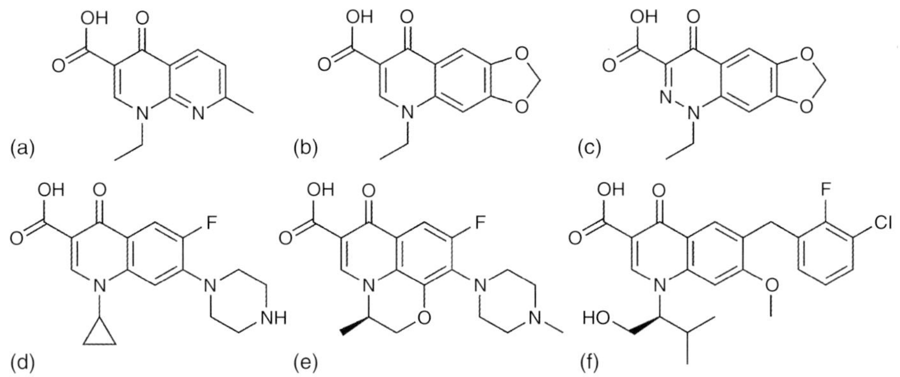


**Figure 12.3** (a–c) Antibacterial agents nalidixic acid, oxolinic acid, cinoxacin; (d–f) antibacterial agents ciprofloxacin, levofloxacin, and HIV integrase inhibitor elvitegravir.

infections, sexually transmitted diseases, prostatitis, community-acquired pneumonia, and more [2]. Nowadays, the quinolones are among the most used antibacterial agents.

Interestingly, some quinolones (e.g., ciprofloxacin, levofloxacin; Figure 12.3d,e) are also used as drugs of choice for treating victims infected by *Bacillus anthracis*—the etiologic agent of anthrax [3]. After terrorist attacks in the United States in 2001, envelopes containing *Bacillus anthracis* spores were mailed to news companies and government officials, which led to the first bioterrorism-related cases of anthrax in the country and to a substantially increased demand for quinolones.

### 12.2.1 Quinolones and Their Interactions with Metal Ions

Soon after the introduction of quinolones in clinical practice, it was discovered that concurrent administration of metal-ion–containing medications or food supplements provokes a near-complete loss of activity of the drug in serum [4]. It was quickly proposed and also confirmed that quinolones interact with metal cations through the 4-oxo and adjacent carboxyl group. These functional groups are also essential for the antibacterial activity of the quinolones. Many quinolone metal complexes were prepared and characterized, [5] and it was found, on the one hand, that metal–quinolone interactions are unfavorable because the absorption of these drugs is reduced (also due to the formation of sparingly soluble metal complexes), but on the other hand it is believed that metal ions are needed for the biological activity of quinolones [6]. More precisely, magnesium ions mediate quinolone interactions with the topoisomerase enzyme, which is crucial for the inhibition process [7]. It was later established that some metal–quinolone complexes might be considered as potential new drugs as they exert different kinds of biological activities.

### 12.2.2 Ruthenium and Quinolones

The first reported ruthenium–quinolone complex was the organometallic complex [($\eta^6$-*p*-cymene)RuCl(oflo)] (Figure 12.4) with ofloxacin (ofloH) [8].

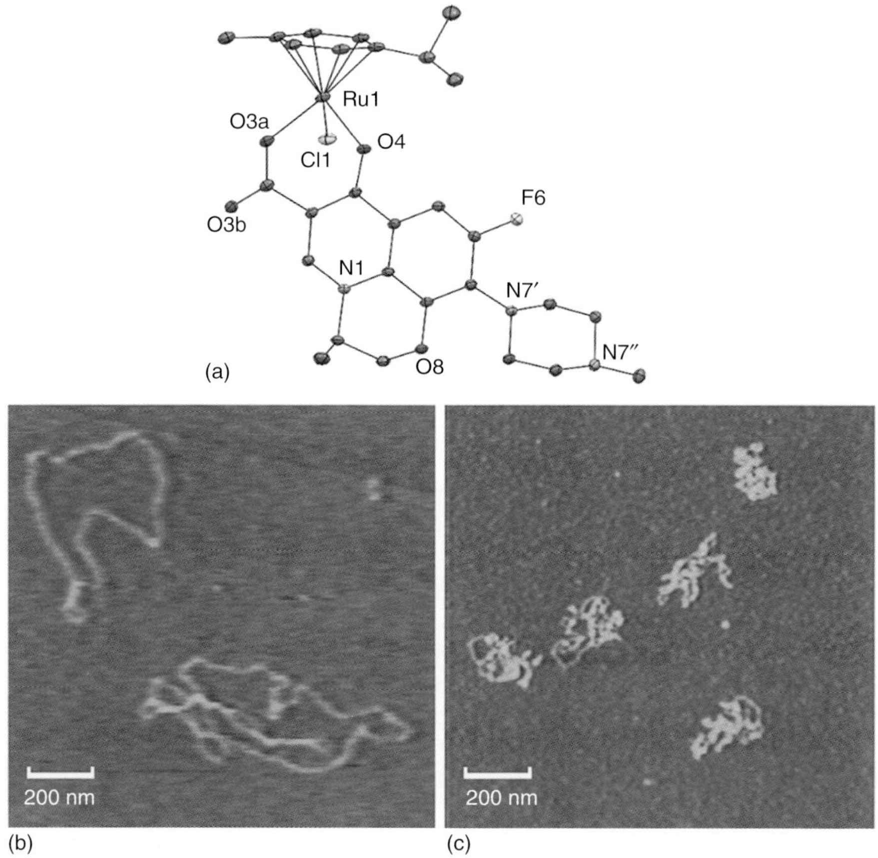


**Figure 12.4** Crystal structure of the organoruthenium complex of quinolone ofloxacin (a). Circular DNA pRS in 5 mM NaCl on mica before (b) and after (c) exposure to 60 μM concentration of complex [($\eta^6$-*p*-cymene)RuCl(oflo)]. (Image adapted from [8] Supporting Information file with permission of ACS. Copyright 2010.)

Its crystal structure revealed that the quinolone is bidentately coordinated to the metal through the pyridone oxygen atom and one of the carboxylic oxygen atoms, forming a so-called piano-stool complex.

Its important advantage over other metal–quinolone complexes was the much higher aqueous solubility. Many metal–quinolone complexes are only sparingly soluble, which impairs the study of biological activity. As one of the possible targets, the interactions with DNA were studied by various spectroscopic methods (electronic, fluorescence, circular dichroism (CD)) and atomic force microscopy (AFM). The ruthenium complex interacts with DNA and binding occurs even if guanine *N7* is protonated. Interestingly, it was also revealed that cisplatin prevents binding of ruthenium complex to DNA in competitive binding experiments and vice versa. AFM experiments have shown that ruthenium complex provokes DNA shrinkage. Rather, uniform compressed structures were observed after adding metal complex to solution of this macromolecule (Figure 12.4).

DNA topoisomerases are enzymes involved in the regulation of DNA supercoiling, which is essential for the DNA transcription and replication. On the one hand, quinolones are selective bacterial-type topoisomerase inhibitors; on the other hand, eukaryotic-type topoisomerase inhibitors are potential anticancer agents and some quinolone derivatives with the extended aromatic ring system, such as the quinobenzoxazines, were also found to exhibit antitumor properties [9]. The enzyme inhibition tests (human topoisomerase IIα) have been performed with ruthenium–ofloxacin complex but no improved activity in comparison to free ligand ofloH was found. This compound was also tested in *in vitro* tests against various microorganisms that cause tropical diseases and in cytotoxicity experiments with rat skeletal myoblasts. Moderate activity in these tests was found against *Trypanosoma b. rhodesiense, Trypanosoma cruzi,* and *Plasmodium falciparum.*

Further work included the development of new organoruthenium complexes with other ligands of the quinolone family (nalidixic acid, cinoxacin; Figure 12.3) in order to elucidate the influence of the structure and substitution pattern of the ligands on the physicochemical and anticancer activity of this compound class [10]. Crystal structures revealed the same coordination around the central metal ion as described for ruthenium-ofloxacin compound. The stability in aqueous solution and the interaction with biological target molecules were studied. It was found that all tested compounds undergo a quick activation step by releasing the chlorido ligand and replacing it by a labile water molecule. Ruthenium complexes with nalidixic acid and cinoxacin are significantly more stable than the corresponding ofloxacin analogue and undergo only minor conversion to the unreactive and biologically inactive hydrolytic product $[(cym)Ru(\mu\text{-}OH)_3Ru(cym)]^+$ species (cym = $\eta^6$-*p*-cymene), the formation of which was previously observed in structurally similar pyr(id)one ligands [11]. However, the hydrolysis rate can be easily monitored and controlled through appropriate choice of ligands, modification of the metal binding sites, and control of the pH values of the utilized media.

Human serum albumin (HSA) is the most abundant protein in blood plasma. It is involved in the transport of hormones, fatty acids, and other compounds. Therefore, it is also important to consider its interactions with drugs. Binding to this protein might change the biological properties of the original drug, or provide paths for drug transport. It is also known that ruthenium complexes have a high affinity for HSA and other serum proteins, which may contribute to their selective accumulation within tumor cells [12]. Capillary zone electrophoresis-inductively coupled plasma mass spectrometry (CZE)-ICP-MS experiments confirmed that in the presence of HSA, ruthenium complexes with ofloxacin, nalidixic acid, and cinoxacin form adducts with this protein within 20 min of incubation, which confirms that HSA is an important protein for their transport. Cytotoxicity assays with three cancer cell lines (human non–small cell lung carcinoma A549, colon adenocarcinoma SW480, and human ovarian cancer CH1) *in vitro* were performed and the complexes were proved to mostly be nontoxic, with the exception of the ofloxacin complex in CH1 ovarian cancer cells. The nontoxic nature of the complexes, in general, should not discourage further studies of metal-based drugs. Most of the commonly used cell lines in screening experiments are derived from primary tumors, while one of the two ruthenium drugs in advanced clinical studies is highly active against metastases *in vivo* but it is not toxic *in vitro* [13]. Therefore, the high $IC_{50}$ values alone are not a sufficient reason to discard a compound as a potential drug candidate.

Further efforts were directed toward the design of a thionated quinolone, where the pyridone oxygen atom is substituted by a sulfur atom as such transformation has previously proved beneficial in ruthenium complexes with pyrone and pyridone ligands in terms of stability in aqueous solution and cytotoxicity [14]. Indeed, the aqueous stability of the complex was increased and only a minor degree of thionalidixicato ligand dissociated within a week. Derivatization also caused changes in biological activity and we have found that the antibacterial activity of both thionalidixic acid and its ruthenium complex against *Escherichia coli* decreased (vs nalidixic acid). It was also reported that ruthenium complex with ofloxacin $[Ru(oflo)_2Cl(H_2O)]_2CI$ was prepared and its activity against three Gram-positive and four Gram-negative bacteria as well as two fungi was tested. Results have shown that antimicrobial activity was comparable to that of free fluoroquinolone [15]. It is therefore clear that changes in the β-ketocarboxylate functionality of the quinolone affects the biological activity, although attributing the loss of antibacterial activity due to the lower affinity towards magnesium ions in the case of the thionated analogue and the blocked chelation site in the case of the ruthenium complexes would be fair assumptions.

On the other hand, the described chemical transformations on the quinolone ligands proved favorable concerning the intrinsic biological properties of the metallic moiety as the organoruthenium complex of thionalidixic acid displayed a toxicity of an order of magnitude higher toward three cancer cell lines in comparison to the parent complex as well as a significant increase of the inhibitory potency against two enzymes from the cathepsin family (Cat B and Cat S). Obviously, the replacement of the pyridone oxygen atom in the quinolone ligand by a sulfur atom affects the inhibition of the enzyme substantially. Cysteine cathepsins are a group of lysosomal proteases, which were found to have an important role in a number of diseases including cancer, rheumatoid arthritis, atherosclerosis, osteoporosis, and others [16]. Moreover, a possible approach to treat tropical parasitic diseases such as malaria comprises the targeting of parasite cysteine proteases with small molecules. The rationale behind this approach is the similarity of parasite cysteine proteases to the human cathepsins. It is assumed that inhibitors originally developed for human cathepsins might be a source of promising lead compounds for the development of antiparasitic agents targeting parasite cysteine proteases [17]. In general, the search for molecules that inhibit cathepsins *in vivo* is interesting and is also an active strategy, especially in the search of novel anticancer therapeutics [18]. It was reported that organoruthenium complexes strongly inhibit Cat B, suggesting that metal-based drugs may represent a novel class of cathepsin inhibitors [19].

A study by Turel *et al.* has therefore focused on complexes bearing the face-capping sulfur macrocycle 1,4,7-trithiacyclononane ($[9]aneS_3$) [20]. Four complexes with the general formula $[Ru([9]aneS_3)(dmso\text{-}\kappa S)(quinolonato)](PF_6)$ with nalidixic acid, levofloxacin, oxolinic acid, and cinoxacin as quinolone ligands (Figure 12.3) are more stable in aqueous solution than their organoruthenium analogues since substitution of the dmso ligand is slow and not quantitative. DNA binding of all four ruthenium complexes was studied using UV/vis absorption spectroscopy, cyclic voltammetry, and viscosity measurements. These experiments suggest that the ruthenium complexes might interact with DNA via intercalation. Further experiments also confirmed that all complexes show good binding affinity to bovine serum albumin and human serum albumin (BSA and HSA) with relatively high binding constants. These results provide additional confirmation that ruthenium complexes have multiple targets in biological systems. All compounds were also tested for biological activity (cytotoxicity against two cell lines; inhibition of cathepsin enzymes), but only weak-to-moderate effect was observed.

An important step in pharmacological characterization of potential drugs is the study of their pharmacokinetics in serum. Speciation of the organoruthenium complex of nalidixic acid [($\eta^6$-*p*-cymene)RuCl(nal)] in spiked human serum was thus investigated by the use of conjoint liquid chromatography on monolithic disks with UV and post-column isotope-dilution inductively coupled plasma mass spectrometry (ID-ICP-MS) detection [21]. A mixture of standard serum proteins (HSA, immunoglobulin (IgG), and transferrin (Tf)) was also used in this analysis. Serum proteins exhibit several important functions and it is therefore important to know how drugs interact with them. Immunoglobulins are large glycoproteins used by the immune system to identify and neutralize foreign objects (e.g., bacteria, viruses). On the other hand, transferrins are iron-binding plasma glycoproteins that control the level of iron in biological fluids [22]. Moreover, it is also well known that ruthenium complexes bind to transferrin and that transferrin promotes their transport into the cell [23]. Ruthenium complex with nalidixic acid is a neutral complex that hydrolyzes into the positively charged aqua species, which allowed successful application of the techniques mentioned earlier. It was clearly demonstrated that the tested compound binds to all three standard serum proteins (HSA, IgG, and Tf), and it was possible to investigate the kinetics of its interactions with serum proteins and the study of the distribution of ruthenium species in serum samples [21].

The presented ruthenium-quinolone case study clearly shows that converting clinically used agents by complexation of a metallic fragment is a viable route for the development of new metal-based drugs. However, the physicochemical properties and biological activities can be drastically altered (often detrimentally) by even minor structural changes, especially if these involve the primary coordination sphere of the metal ion.

### 12.2.3 Ruthenium and HIV Integrase Inhibitor Elvitegravir

The HIV-1 integrase (IN) is the most important enzyme in the replication mechanism of retroviruses, and integrase inhibitors are a new class of drugs used in the treatment of HIV. Elvitegravir (Figure 12.3) is an integrase inhibitor from the quinolone family. There are two magnesium(II) ions in the active site of IN, which are important for the activity of the enzyme. These metal ions are the target of inhibitors such as raltegravir [24]. Carcelli *et al.* have prepared novel elvitegravir model molecules and their ruthenium complexes [25]. They have found that the ligands and the complexes show good inhibition potency (in the sub-/low-micromolar concentration range) in anti-HIV-1 integrase enzymatic assays. However, they have reported that Ru(II) arene moiety did not contribute significantly to the enhancement of the intrinsic activity of the parent model ligands. They have postulated that ruthenium complexes can inhibit the activity of HIV-1 IN enzyme by carrying the ligand inside the active site of the enzyme, where the ligand is released and can interact with the magnesium metal cofactors.

## 12.3 Antibacterial 8-Hydroxyquinolines

### 12.3.1 Mode of Action of 8-Hydroxyquinoline Agents

8-Hydroxyquinoline (8-hqH; Figure 12.5) and its derivatives have been studied for a long time for various medical applications. These compounds have demonstrated activity as antimicrobial, anticancer agents and very encouraging results were also reported in the treatment of Alzheimer's [26] and Parkinson's [27] diseases. 8-Hydroxyquinoline and its derivatives are well-known bidentate chelating agents that readily form metal complexes in the presence of metal-containing species in biological systems. Such complexes may exert various biological effects. They can affect metal homeostasis, [28] act as ionophores, [29] inhibit metal-dependent enzymes, [30] influence metal-dependent protein aggregation, [31] and more.

Metal chelating properties and medicinal applications of 8-hydroxyquinolines were thoroughly reviewed recently [32]. It was reported that the inhibition of the malaria-causing *P. falciparum* (see next paragraph on 4-aminoquinolines) and the chelating ability of the ligands were directly correlated [33]. Parent compound 8-hqH is itself a potent chelator, which is known to possess an antimalarial effect against the intracellular stage of malaria in red blood cells [34].

It is also known that substitutions in different positions of the quinoline ring can substantially change the activity in this family of drugs [32]. Interestingly, it was also established that some anticancer drugs exert antimalarial activity [35]. Moreover, it was reported that antiparasitic and anticancer activities of 8-hqH derivatives are improved upon complexation with certain metal ions such as copper, zinc, and also antimony [36].

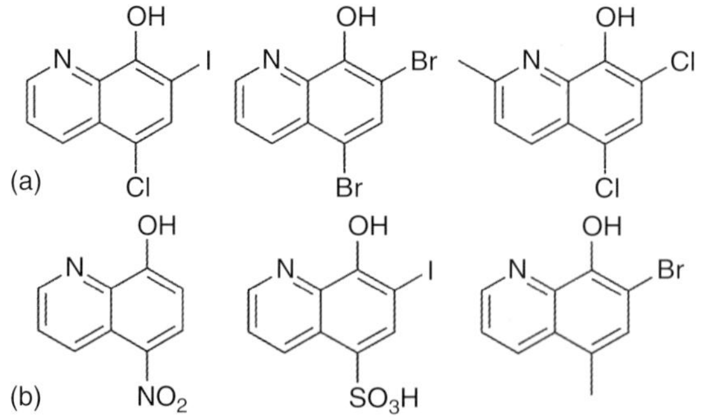


**Figure 12.5** Hydroxyquinolines used in clinical practice: (a) clioquinol (cqH), broxyquinoline, and chloroquinaldol; (b) nitroxoline (nxH), chiniofon, and tilbroquinol.

There are several papers dealing with ruthenium-hydroxyquinoline complexes; however, these compounds were studied for their physicochemical and catalytic properties and only few studies exist on their biological properties [37].

### 12.3.2 Ruthenium and 8-Hydroxyquinolines

Clioquinol (cqH) exerts a wide range of biological activities and was used as an antimicrobial agent for a long time. The interest in this drug has grown after promising use in the treatment of Alzheimer's and Parkinson's diseases, as mentioned [26, 27]. Activity in neurodegenerative diseases was ascribed to its ability to cross the blood–brain barrier and also to its chelation of divalent metal ions (e.g., Cu(II) and Zn(II)). These ions might be associated with protein aggregation in the brain [38]. An organoruthenium–clioquinol complex (Figure 12.6) was recently reported to possess selective toxicity toward leukemia cell lines; and, contrary to the free ligand cqH, it is independent of the concentration of the copper ions present in the cell culture medium. This complex showed proteasome-independent inhibition of the NFκB signaling pathway and results suggest that the mechanism of action is different from that of clioquinol alone [39].

The group of E. C. Glazer reported that coordination of various hydroxyquinolines to a ruthenium bis-dimethylphenanthroline scaffold increased their cytotoxic potential substantially, with values reaching nanomolar concentrations in several examples. They have prepared a small library of nine ruthenium complexes containing different hydroxyquinoline ligands to explore structure – activity relationships [40]. More precisely, in these complexes two methyl-substituted phenanthroline ligands (2,9-dimethyl-1,10-phenanthroline) or two 2,2′-bipyridine ligands (bipy) and one hydroxyquinoline ligand were attached to the central ruthenium ion. They have carefully studied how substituents on hydroxyquinoline ring system affect the activity and they have found that the presence of halogens at the 5- and 7-positions resulted in the most active compounds. They have also reported that the co-ligands (*N*,*N*-ligands of phenanthroline or bipyridine type) are also very important for activity of such ruthenium complexes and the best results were obtained with 2,9-dimethyl-1,10-phenanthroline. In contrast to previous observations

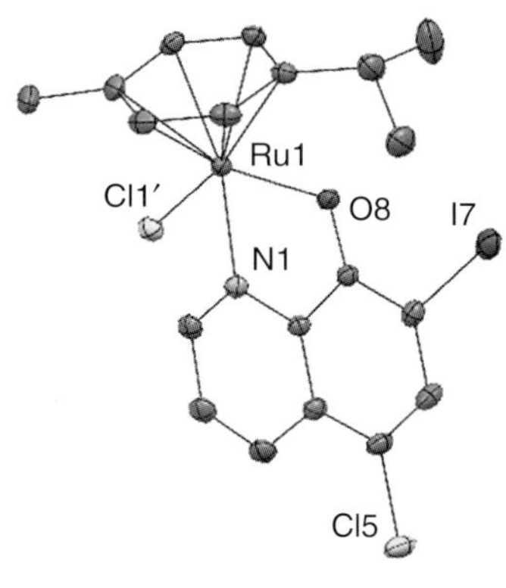


**Figure 12.6** Crystal structure of the organoruthenium complex of clioquinol [39].

for clioquinol that acted as proteasome inhibitor [41], such an effect was not found for ruthenium complexes containing *N,N*-ligands of phenanthroline or bipyridine type as in the case of the abovementioned organoruthenium complex [39]. One of the hydroxyquinoline ligands used in the study of Glazer *et al.* was nitroxoline. This compound is a clinically used antibiotic which is effective in combating biofilm infections [42]. Moreover, it is also studied as a potential anticancer agent [43]. It was shown that nitroxoline interacts with metal ions and functions by chelating iron(II) and zinc(II) ions. However, it was established that incorporation of electron-rich substituents (such as nitro group in nitroxoline) at the 5-position has substantially reduced cytotoxic activity [40].

Thangavel *et al.* prepared several mixed-ligand ruthenium complexes among others also with phen and 7-halo (bromo or iodo) nitroxoline [44]. These complexes were tested for activity in *in vivo* anticancer tests against a transplantable murine tumor cell line, Ehrlich's ascitic carcinoma (EAC). In addition, *in vitro* antibacterial activity against several Gram-positive and Gram-negative bacterial strains was also tested. Prolonged life span of EAC-bearing mice as well as decreased tumor volume was observed for some ruthenium complexes, and it was also reported that all tested complexes exhibited mild-to-moderate antibacterial activity.

## 12.4 Antimalarial 4-Aminoquinolines

Malaria is the most occurring human parasitic disease. The World Health Organization reports approximately 250 million cases annually which are responsible for almost one million deaths, half of them in African children below the age of 5 [45, 46]. The parasites from the *Plasmodium* species, most commonly *P. falciparum,* are able to infect human host and until now, no effective vaccine or reliable means for the prevention of the spreading of the parasitic infection have been developed.

### 12.4.1 Mechanism of Action of Antimalarial Quinoline Agents

4-Aminoquinolines are a class of potent antimalarial agents with two members currently being used in clinical practice, namely, chloroquine (chq) and amodiaquine (Figure 12.7a). After the appearance of the chloroquine-resistant parasitic strains, several other antimalarial agents were developed such as mefloquine, halofantrine, and lumefantrine (Figure 12.7). The 4-aminoquinoline scaffold is still being intensively investigated for novel antimalarial drugs despite the fact that these compounds have been in clinical use for more than 60 years [47]. Progress is, however, slow and it has been nearly 20 years since the introduction of the last antimalarial drug (artemether-lumefantrine in 1998) into clinical practice [48]. In the past decades, most of the advances were reported in the field of artemisinin-based combination therapies, which increase the effectiveness and reduce the risk of resistance development [49]. However, in 2010, Dondorp *et al.* published the first observed case of resistance to artemisinin-family agent artesunate [50].

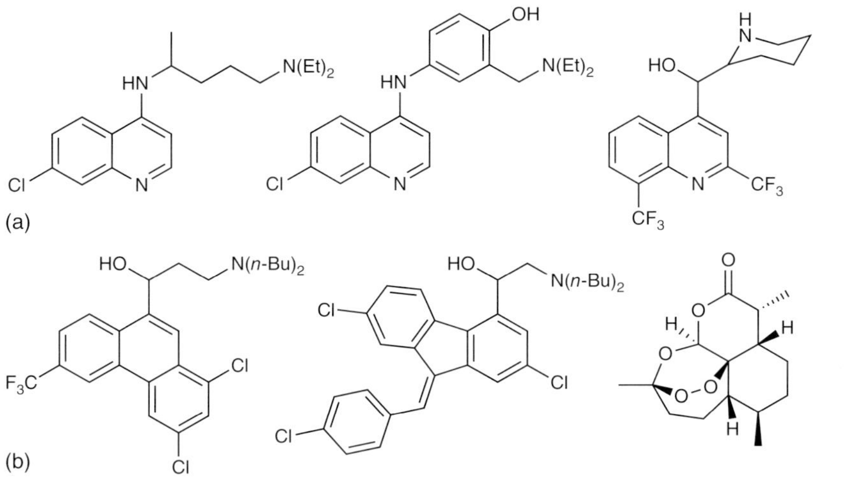


**Figure 12.7** Antimalarial drugs in clinical use: (a) chloroquine, amodiaquine, and mefloquine; (b) halofantrine, lumefantrine, and artemisinin.

The antimalarial mechanism of action of 4-aminoquinolines is well known [51]. Malarial parasites inhabit the red blood cells where they degrade hemoglobin and absorb the essential amino acids required for their growth and metabolism. The heme is however not degraded and, as a highly toxic species, removed by biocrystallization to hemozoin which is then stored in the form of insoluble crystals in the digestive vacuoles of the parasites, giving them distinct pigmentation. Aminoquinolines enter the cells by diffusion and are subsequently doubly protonated to their dicationic form due to the low pH in digestive vacuoles. In this state, the aminoquinoline efflux is prevented and it accumulates inside the parasite where it coordinates the heme-bound iron and prevents the biocrystallization of hemozoin.

## 12.5 Metallocene Analogues of Chloroquine

One of the first reports of the possible use of ruthenium compounds in antimalarial chemotherapy reaches back to 1996 where Sanchez-Delgado reported the first dinuclear ruthenium complex with chloroquine (chq) in which the antimalarial agent acts as a bridging ligand through the quinoline and diethylamine N atoms [52]. The dimeric complex $[RuCl_2(chq)]_2$ was tested for its antimalarial activity against *P. berghei* together with a rhodium-chloroquine complex and was found to be three- to fivefold more active than chloroquine alone, while the rhodium complex showed no increase in activity. Moreover, the complex was proved to be highly active against chloroquine-resistant *P. falciparum* and its activity is attributed to the higher lipophilicity of the complex in comparison to the free chloroquine, while its action is exerted through the inhibition of the heme aggregation as the complex binds to hematin and prevents its aggregation to β-hematin [53]. The binding of hematin is facilitated by the hydrolysis of the complex to its active species $[RuCl(OH_2)_3(chq)]_2Cl_2$ which occurs in aqueous solution, especially at lower pH values.

In 2009, Sanchez-Delgado extended his studies to organoruthenium arene complexes in which chloroquine is bound through the quinoline *N1* atom as well as an unusual complex where the chloroquine molecule is π-bonded to the ruthenium moiety (Figure 12.8) [54]. For such compounds, a clear correlation emerges between the lipophilicity and the heme aggregation inhibition ability at the water/*n*-octanol surface and the overall antimalarial potency. In addition to their antimalarial activity, the compounds were found to induce apoptosis on human lymphoid cell lines while being relatively nontoxic to several normal mammalian cell lines [55]. Moreover, the complexes bind covalently to human serum albumin and apotransferrin and can be reversibly released upon decrease in pH values [56]. As such, they display good transport properties, which makes them better candidates for further drug development.

The same group has subsequently developed chelating analogues of chloroquine, which allow the binding of metal species at the "tail" part of the chloroquine derivative (leaving free the *N1* quinoline site) [57]. Upon complexation of the ligands possessing an *N,N*-binding moiety, which results in monocationic ruthenium complexes, the antimalarial activity decreased. However, the use of an *N,O*-negatively charged binding moiety, which upon complexation gives a neutral ruthenium complex, resulted in the increase of the antimalarial activity in comparison with the free ligands in both chloroquine-sensitive and chloroquine-resistant *P. falciparum* strains. Unfortunately, the correlation between heme aggregation inhibition activity or lipophilicity and the antimalarial activity was not studied due to too big structural diversity and small number of compounds.

The most recent study by Smith *et al.* reports a series of silicon-containing derivatives of chloroquine in which the terminal dialkylamino group is

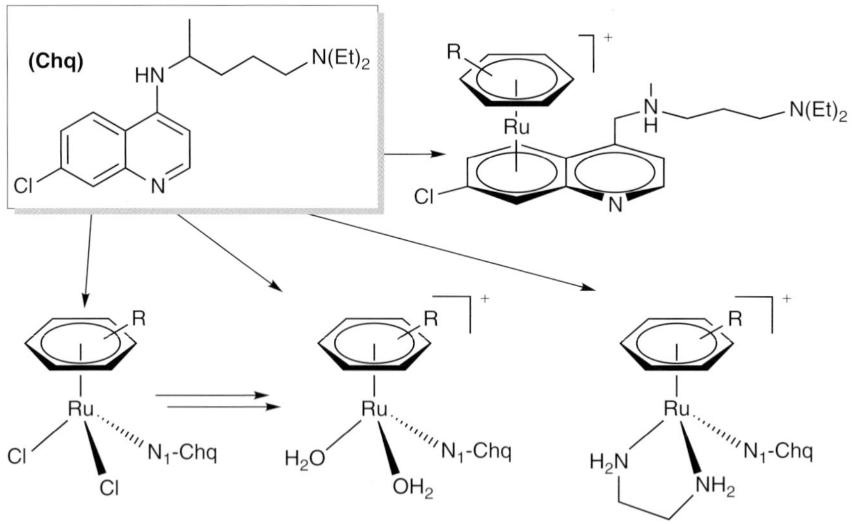


**Figure 12.8** Organoruthenium complexes of antimalarial drug chloroquine.

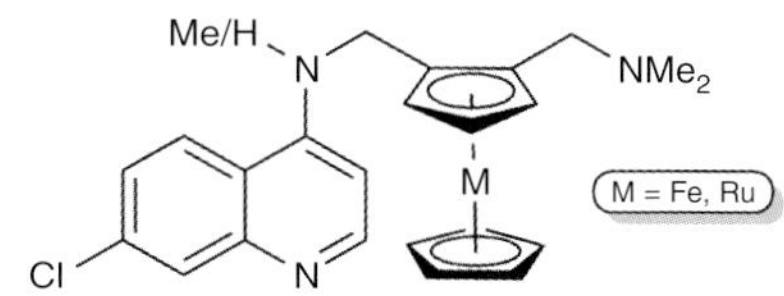


**Figure 12.9** Chemical structure of (methyl)ferroquine and (methyl)ruthenoquine.

substituted by the trimethylsillyl functionality. Several ruthenium(II), rhodium(I), and rhodium(III) complexes in which the metal species is bound at the *N1* quinoline position were prepared and evaluated for their antimalarial, anticancer, and antibacterial properties [58]. The compounds were generally less active toward *P. Falciparum* than the parent compound chloroquine while displaying similar values of β-hematin inhibition. Moreover, they displayed low micromolar $IC_{50}$ values for esophageal carcinoma cell lines (WHCO1) and were widely ineffective against mycobacterium *M. tuberculosis.*

In 2002, Beagley *et al.* reported the synthesis of metallocene analogues of chloroquine in which ferrocene and ruthenocene moieties are used as spacers in the side chain of the 4-aminoquinoline structure [59], building on the breakthrough results of Biot *et al.* [60] who first reported a significant improvement in the biological activity of a metallocene-containing analogue (ferroquine; Figure 12.9) of a clinically used drug which is to date the organometallic compound in the most advanced stage of clinical trials as it completed phase II trials and phase IIb trials were planned to start in late 2014 [46]. Website clinicaltrials.gov however presently (site accessed July 2017) reports only a second phase II study by Sanofi with a single dose regimen of ferroquine and artefenomel for treatment of uncomplicated Plasmodium falciparum malaria which is in its recruitment stage. While Beagley's compounds did not show any improved effectiveness toward chloroquine-sensitive strains of *P. falciparum,* the chloroquine-resistant strain was 1–2 orders of magnitude more susceptible to the metallocene analogues. Surprisingly, the parasites reacted similarly to iron- and ruthenium-containing analogues.

A recent study shows further advancement in the understanding of the mechanism of action of ferrocene and ruthenocene derivatives of chloroquine as it shows that the constrained conformation of the spacer (1,2-disubstituted metallocene) favors the formation of the intramolecular hydrogen bonding between the two amine groups in the absence of water, which facilitates the passage through the cell membrane [61]. This hypothesis was confirmed by a decrease of the $IC_{50}$ values by 1 order of magnitude in the ferroquine and ruthenoquine derivatives, which were methylated at the *N4* position and thus unable to form the intramolecular bonds. All four compounds were able to block the hemozoin crystal growth process. Interestingly the nonmethylated molecules showed no cross-resistance with chloroquine; on the other hand, the methylated analogues did. Not unexpectedly, in contrast to the ruthenium-containing compounds, the iron-containing compounds were found to generate hydroxyl radicals which results in the breakdown of the parasite digestive vacuole.

A further evolution of both lines of work was reported by Li *et al.*, where organoruthenium and organorhodium complexes of syllylated ferroquine derivatives were studied; however, this series of compounds did not show remarkable improvement in hematin formation inhibition and antiplasmodial activity in comparison with parent drugs chloroquine, amodiaquine, ferroquine, and its previously reported derivatives and complexes [58, 62].

## 12.6 Conclusions

As demonstrated in the cases discussed, drug design strategies which expand their horizons to include metallic functionalities offer new possibilities. The advantages of such an approach are, for example, novel geometries of building block arrangements, interactions with target macromolecules beside covalent/hydrogen bonding, and tunable redox and electronic properties. With each opportunity, there are often (multiple) downfalls. The development of metal-based drugs is in its early stages in comparison with classical organic/pharmaceutical chemistry. Commonly used biological screening experiments as well as computational methods are very often found unsuitable due to the specific chemical and physical properties of such newly developed drug candidates. Promising examples such as ferroquine might pave the way for future generations of scientists to use all the tools at their disposal - elements of the periodic table - to "boldly go where no man has gone before" and achieve remarkable advancements in the field of medicinal chemistry.

## References


1 Bisacchi, G.S. (2015) *J. Med. Chem.*, **58**, 4874–4882.

2 (a) Andriole, V.T. (2000) *The Quinolones*, 3rd edn, Academic Press, San Diego, CA; (b) Hooper, D.C. and Rubinstein, E. (2003) *Quinolone antibacterial agents*, 3rd edn, ASM Press, Washington, DC; (c) Ronald, A.R. and Low, D.E. (2003) in *Milestones in Drug Therapy* (eds M.J. Parnham and J. Bruinvels), Birkhäuser Verlag, Basel.

3 Cieslak, T.J. and Eitzen, E.M. Jr., (1999) *Emerging Infect. Dis.*, **5**, 552–555.

4 Hoffken, G., Borner, K., Glatzel, P.D., Koeppe, P., and Lode, H. (1985) *Eur. J. Clin. Microbiol. Infect. Dis.*, **4**, 345.

5 (a) Turel, I. (2002) *Coord. Chem. Rev.*, **232**, 27–47; (b) Uivarosi, V. (2013) *Molecules*, **18**, 11153–11197.

6 Mitscher, L.A. (2005) *Chem. Rev.*, **105**, 559–592.

7 Wohlkonig, A., Chan, P.F., Fosberry, A.P., Homes, P., Huang, J.Z., Kranz, M., Leydon, V.R., Miles, T.J., Pearson, N.D., Perera, R.L., Shillings, A.J., Gwynn, M.N., and Bax, B.D. (2010) *Nat. Struct. Mol. Biol.*, **17**, 1152–1153.

8 Turel, I., Kljun, J., Perdih, F., Morozova, E., Bakulev, V., Kasyanenko, N., Byl, J.A.W., and Osheroff, N. (2010) *Inorg. Chem.*, **49**, 10750–10752.

9 Kwok, Y., Zeng, Q.P., and Hurley, L.H. (1999) *J. Biol. Chem.*, **274**, 17226–17235.

10 Kljun, J., Bytzek, A.K., Kandioller, W., Bartel, C., Jakupec, M.A., Hartinger, C.G., Keppler, B.K., and Turel, I. (2011) *Organometallics*, **30**, 2506–2512.

11 (a) Kandioller, W., Hartinger, C.G., Nazarov, A.A., Bartel, C., Skocic, M., Jakupec, M.A., Arion, V.B., and Keppler, B.K. (2009) *Chem. Eur. J.*, **15**,

12283–12291; (b) Kandioller, W., Hartinger, C.G., Nazarov, A.A., Kuznetsov, M.L., John, R.O., Bartel, C., Jakupec, M.A., Arion, V.B., and Keppler, B.K. (2009) *Organometallics*, **28**, 4249–4251.
12 Timerbaev, A.R., Hartinger, C.G., Aleksenko, S.S., and Keppler, B.K. (2006) *Chem. Rev.*, **106**, 2224–2248.
13 Sava, G., Capozzi, I., Clerici, K., Gagliardi, G., Alessio, E., and Mestroni, G. (1998) *Clin. Exp. Metastasis*, **16**, 371–379.
14 (a) Kandioller, W., Kurzwernhart, A., Hanif, M., Meier, S.M., Henke, H., Keppler, B.K., and Hartinger, C.G. (2011) *J. Organomet. Chem.*, **696**, 999–1010; (b) Hudej, R., Kljun, J., Kandioller, W., Repnik, U., Turk, B., Hartinger, C.G., Keppler, B.K., Miklavcic, D., and Turel, I. (2012) *Organometallics*, **31**, 5867–5874.
15 Golcu, A. (2014) *Synth. React. Inorg., Met.-Org., Nano-Met. Chem*, **44**, 1509–1520.
16 (a) Mohamed, M.M. and Sloane, B.F. (2006) *Nat. Rev. Cancer*, **6**, 764–775; (b) Vasiljeva, O., Reinhackel, T., Peters, C., Turk, D., Turk, V., and Turk, B. (2007) *Curr. Pharm. Des.*, **13**, 387–403.
17 Ang, K.K.H., Ratnam, J., Gut, J., Legac, J., Hansell, E., Mackey, Z.B., Skrzypczynska, K.M., Debnath, A., Engel, J.C., Rosenthal, P.J., McKerrow, J.H., Arkin, M.R., and Renslo, A.R. (2011) *PLoS Negl. Trop. Dis.*, **5**, e1023.
18 Repnik, U., Stoka, V., Turk, V., and Turk, B. (2012) *Biochim. Biophys. Acta*, **1824**, 22–33.
19 Casini, A., Gabbiani, C., Sorrentino, F., Rigobello, M.P., Bindoli, A., Geldbach, T.J., Marrone, A., Re, N., Hartinger, C.G., Dyson, P.J., and Messori, L. (2008) *J. Med. Chem.*, **51**, 6773–6781.
20 Kljun, J., Bratsos, I., Alessio, E., Psomas, G., Repnik, U., Butinar, M., Turk, B., and Turel, I. (2013) *Inorg. Chem.*, **52**, 9039–9052.
21 Martinčič, A., Vidmar, M.R.J., Turel, I., Keppler, B.K., and Ščančar, J. (2014) *J. Chromatogr. A*, **1371**, 168–176.
22 Crichton, R.R. and Charloteauxwauters, M. (1987) *Eur. J. Biochem.*, **164**, 485–506.
23 Pongratz, M., Schluga, P., Jakupec, M.A., Arion, V.B., Hartinger, C.G., Allmaier, G., and Keppler, B.K. (2004) *J. Anal. At. Spectrom.*, **19**, 46–51.
24 Pendri, A., Meanwell, N.A., Peese, K.M., and Walker, M.A. (2011) *Exp. Opin. Ther. Pat.*, **21**, 1173–1189.
25 Carcelli, M., Bacchi, A., Pelagatti, P., Rispoli, G., Rogolino, D., Sanchez, T.W., Sechi, M., and Neamati, N. (2013) *J. Inorg. Biochem.*, **118**, 74–82.
26 Adlard, P.A., Cherny, R.A., Finkelstein, D.I., Gautier, E., Robb, E., Cortes, M., Volitakis, I., Liu, X., Smith, J.P., Perez, K., Laughton, K., Li, Q.X., Charman, S.A., Nicolazzo, J.A., Wilkins, S., Deleva, K., Lynch, T., Kok, G., Ritchie, C.W., Tanzi, R.E., Cappai, R., Masters, C.L., Barnham, K.J., and Bush, A.I. (2008) *Neuron*, **59**, 43–55.
27 Kaur, D., Yantiri, F., Rajagopalan, S., Kumar, J., Mo, J.O., Boonplueang, R., Viswanath, V., Jacobs, R., Yang, L., Beal, M.F., DiMonte, D., Volitaskis, I., Ellerby, L., Cherny, R.A., Bush, A.I., and Andersen, J.K. (2003) *Neuron*, **37**, 899–909.

28 Tardito, S., Barilli, A., Bassanetti, I., Tegoni, M., Bussolati, O., Franchi-Gazzola, R., Mucchino, C., and Marchio, L. (2012) *J. Med. Chem.*, **55**, 10448–10459.
29 Ding, W.Q. and Ling, S.E. (2009) *IUBMB Life*, **61**, 1013–1018.
30 Jacobsen, J.A., Fullagar, J.L., Miller, M.T., and Cohen, S.M. (2011) *J. Med. Chem.*, **54**, 591–602.
31 Matlack, K.E.S., Tardiff, D.F., Narayan, P., Hamamichi, S., Caldwell, K.A., Caldwell, G.A., and Lindquist, S. (2014) *Proc. Nat. Acad. Sci. U.S.A.*, **111**, 4013–4018.
32 Prachayasittikul, V., Prachayasittikul, S., Ruchirawat, S., and Prachayasittikul, V. (2013) *Drug Des. Devel. Ther.*, **7**, 1157–1178.
33 Scheibel, L.W. and Adler, A. (1982) *Mol. Pharmacol.*, **22**, 140–144.
34 Scheibel, L.W. and Adler, A. (1980) *Mol. Pharmacol.*, **18**, 320–325.
35 (a) Denton, H., Roberts, C.W., Alexander, J., Thong, K.W., and Coombs, G.H. (1996) *FEMS Microbiol. Lett.*, **137**, 103–108; (b) McCarthy, S.M. and Davis, C.D. (2003) *J. Parasitol.*, **89**, 886–894; (c) Pino, P., Foth, B.J., Kwok, L.Y., Sheiner, L., Schepers, R., Soldati, T., and Soldati-Favre, D. (2007) *PLoS Pathog.*, **3**, 1092–1108.
36 (a) Ding, W.Q., Liu, B.L., Vaught, J.L., Yamauchi, H., and Lind, S.E. (2005) *Cancer Res.*, **65**, 3389–3395; (b) Reis, D.C., Pinto, M.C.X., Souza-Fagundes, E.M., Rocha, L.F., Pereira, V.R.A., Melo, C.M.L., and Beraldo, H. (2011) *Biometals*, **24**, 595–601.
37 (a) Thai, T.T., Therrien, B., and Suss-Fink, G. (2009) *J. Organomet. Chem.*, **694**, 3973–3981; (b) Bhattacharya, S. (1993) *Polyhedron*, **12**, 235–239; (c) Sears, R.B., Joyce, L.E., and Turro, C. (2010) *Photochem. Photobiol.*, **86**, 1230–1236; (d) Warren, J.T., Chen, W., Johnston, D.H., and Turro, C. (1999) *Inorg. Chem.*, **38**, 6187–6192; (e) Zhao, H.C., Fu, B.L., Schweinfurth, D., Harney, J.P., Sarkar, B., Tsai, M.K., and Rochford, J. (2013) *Eur. J. Inorg. Chem.*, **25**, 4410–4420; (f) Powell, P. (1974) *J. Organomet. Chem.*, **65**, 89–92; (g) Slugovc, C., Koppitz, A., Pogantsch, A., and Stelzer, F. (2005) *Inorg. Chim. Acta*, **358**, 2718–2724.
38 Mot, A.I., Wedd, A.G., Sinclair, L., Brown, D.R., Collins, S.J., and Brazier, M.W. (2011) *Expert Rev. Neurother.*, **11**, 1717–1745.
39 Gobec, M., Kljun, J., Sosič, I., Mlinarič-Raščan, I., Uršič, M., Gobec, S., and Turel, I. (2014) *Dalton Trans.*, **43**, 9045–9051.
40 Heidary, D.K., Howerton, B.S., and Glazer, E.C. (2014) *J. Med. Chem.*, **57**, 8936–8946.
41 (a) Daniel, K.G., Chen, D., Orlu, S., Cui, Q.C., Miller, F.R., and Dou, Q.P. (2005) *Breast Cancer Res.*, **7**, R897–R908; (b) Chen, D., Cui, Q.C., Yang, H.J., Barrea, R.A., Sarkar, F.H., Sheng, S.J., Yan, B., Reddy, G.P.V., and Dou, Q.P. (2007) *Cancer Res.*, **67**, 1636–1644.
42 Pelletier, C., Prognon, P., and Bourlioux, P. (1995) *Antimicrob. Agents Chemother.*, **39**, 707–713.
43 Jiang, H.C., Taggart, J.E., Zhang, X.X., Benbrook, D.M., Lind, S.E., and Ding, W.Q. (2011) *Cancer Lett.*, **312**, 11–17.
44 Mazumder, U.K., Gupta, M., Bhattacharya, S., Karki, S.S., Rathinasamy, S., and Thangavel, S. (2004) *J. Enzyme Inhib. Med. Chem.*, **19**, 185–192.

45 WHO (2014) WHO Global Malaria Programme: World Malaria Report 2014, Geneva.
46 Biot, C., Nosten, F., Fraisse, L., Ter-Minassian, D., Khalife, J., and Dive, D. (2011) *Parasite*, **18**, 207–214.
47 (a) Mushtaque, M. and Shahjahan (2015) *Eur. J. Med. Chem.*, **90**, 280–295; (b) Bray, P.G., Park, B.K., Asadollaly, E., Biagini, G.A., Jeyadevan, J.P., Berry, N.G., Ward, S.A., and O'Neill, P.M. (2006) *Curr. Top. Med. Chem.*, **6**, 479–507.
48 Olliaro, P. and Wells, T.N. (2009) *Clin. Pharmacol. Ther.*, **85**, 584–595.
49 (a) White, N.J. (1999) *Philos. Trans. R. Soc. London, Ser. B*, **354**, 739–749; (b) Nosten, F. and White, N.J. (2007) *Am. J. Trop. Med. Hyg.*, **77**, 181–192.
50 Dondorp, A.M., Yeung, S., White, L., Nguon, C., Day, N.P., Socheat, D., and Von Seidlein, L. (2010) *Nat. Rev. Microbiol.*, **8**, 272–280.
51 Rosenthal, P.J. (2001) *Antimalarial chemotherapy: mechanism of action, resistance and new directions in drug discovery*. Humana Press, Totowa.
52 Sanchez-Delgado, R.A., Perez, N.M.H., and Urbina, J.A. (1996) *J. Med. Chem.*, **39**, 1095–1099.
53 Martinez, A., Rajapakse, C.S.K., Naoulou, B., Kopkalli, Y., Davenport, L., and Sanchez-Delgado, R.A. (2008) *J. Biol. Inorg. Chem.*, **13**, 703–712.
54 (a) Martinez, A., Rajapakse, C.S.K., Jalloh, D., Dautriche, C., Sanchez-Delgado, R.A., and Biol, J. (2009) *Inorg. Chem.*, **14**, 863–871; (b) Rajapakse, C.S.K., Martinez, A., Naoulou, B., Jarzecki, A.A., Suarez, L., Deregnaucourt, C., Sinou, V., Schrevel, J., Musi, E., Ambrosini, G., Schwartz, G.K., and Sanchez-Delgado, R.A. (2009) *Inorg. Chem.*, **48**, 1122–1131.
55 Martinez, A., Rajapakse, C.S.K., Sanchez-Delgado, R.A., Varela-Ramirez, A., Lema, C., and Aguilera, R.J. (2010) *J. Inorg. Biochem.*, **104**, 967–977.
56 Martinez, A., Suarez, J., Magliozzo, R.S., and Sanchez-Delgado, R.A. (2011) *J. Inorg. Biochem.*, **105**, 39–45.
57 Glans, L., Ehnbom, A., de Kock, C., Martinez, A., Estrada, J., Smith, P.J., Haukka, M., Sanchez-Delgado, R.A., and Nordlander, E. (2012) *Dalton Trans.*, **41**, 2764–2773.
58 Li, Y., de Kock, C., Smith, P.J., Chibale, K., and Smith, G.S. (2014) *Organometallics*, **33**, 4345–4348.
59 Beagley, P., Blackie, M.A.L., Chibale, K., Clarkson, C., Moss, J.R., and Smith, P.J. (2002) *J. Chem. Soc., Dalton Trans.*, **23**, 4426–4433.
60 Biot, C., Glorian, G., Maciejewski, L.A., and Brocard, J.S. (1997) *J. Med. Chem.*, **40**, 3715–3718.
61 Dubar, F., Egan, T.J., Pradines, B., Kuter, D., Ncokazi, K.K., Forge, D., Paul, J.-F., Pierrot, C., Kalamou, H., Khalife, J., Buisine, E., Rogier, C., Vezin, H., Forfar, I., Slomianny, C., Trivelli, X., Kapishnikov, S., Leiserowitz, L., Dive, D., and Biot, C. (2011) *ACS Chem. Biol.*, **6**, 275–287.
62 Li, Y., de Kock, C., Smith, P.J., Guzgay, H., Hendricks, D.T., Naran, K., Mizrahi, V., Warner, D.F., Chibale, K., and Smith, G.S. (2013) *Organometallics*, **32**, 141–150.

# 13

# Ruthenium Complexes as NO Donors: Perspectives and Photobiological Applications

*Loyanne C.B. Ramos*[1], *Juliana C. Biazzotto*[1], *Juliana A. Uzuelli*[1], *Renata G. de Lima*[2], *and Roberto S. da Silva*[1]

[1] *Universidade de São Paulo, School of Pharmaceutical Sciences of Ribeirão Preto, Departamento de Física e Química, Faculdade de Ciências Farmacêuticas de Ribeirão Preto, Avenida do Café, Ribeirão Preto, SP 14040-903, Brazil*
[2] *Federal University of Uberlandia, Faculty of Integrated Sciences Pontal, Ituiutaba, MG 38304-402, Brazil*

## 13.1 Introduction

Phototherapy is a clinical therapeutic technique based on the use of light irradiation and consists of exposing the patient to specific wavelengths of light for a prescribed amount of time. The use of phototherapy in medicine has a long history, and, currently, several diseases can be treated with this clinical approach, including psoriasis [1], acne [2], Alzheimer's disease [3], and arthritis [4]. Among the types of phototherapy, photodynamic therapy (PDT) has become clinically widespread because of its use in cancer treatment [5, 6]. PDT essentially depends on the photosensitizer, light dose, and oxygen concentration used, and the correct combination of these parameters will promote the production of reactive oxygen species (ROS) and the exertion of their diverse biological effects, especially on DNA, resulting in cell death [7, 8]. Although this technique is already used in the clinic, several factors have been researched to improve PDT, and substantial efforts have been devoted to improving the subcellular localization of photosensitizers [9, 10], which should significantly enhance the cytotoxicity of PDT. However, hypoxic areas in tumors are insensitive to PDT, which has driven scientists to search for new radical molecules. Among them, nitric oxide (NO) has been implicated in both promoting and preventing cancer in a concentration-dependent manner [11, 12]. The key factor in the production of NO in biological media appears to be related to the stimulus of the NO precursor. This function involves four processes:

1. Spontaneous NO release
2. Effect of pH
3. Reduction process
4. Light irradiation

The well-controlled thermodynamic and kinetic processes of coordination compounds have made this process essential in biological experiments related

*Ruthenium Complexes: Photochemical and Biomedical Applications*, First Edition.
Edited by Alvin A. Holder, Lothar Lilge, Wesley R. Browne, Mark A.W. Lawrence, and Jimmie L. Bullock Jr.

to NO production. Photochemical controlled release of NO may be used in clinical therapy and may enhance PDT. We discuss this subject from the photochemical and photobiological perspectives relating to nitrogen oxide ruthenium complexes in this chapter.

## 13.2 Photochemical Processes of Some Nitrogen Oxide Derivative–Ruthenium Complexes

Metal coordination-nitrogen oxide–derivative complexes have been postulated as NO donor agents [13–15] for several different processes. One possibility is light irradiation as external stimulation [16, 17]. Among all the coordination compounds studied as NO donor agents, ruthenium complexes are known to produce NO by UV-visible photolysis [16, 18, 19]. Three photochemical pathways, including metal-ligand charge transfer $d_\pi(Ru^{II})—\pi^*(NO^+)$ (i), nitrite cleavage (ii), and photoinduced electron transfer (ii), have been proposed as light-irradiation-induced mechanisms for these complexes (Scheme 13.1):

### 13.2.1 Metal-Ligand Charge-Transfer Photolysis of $\{Ru\text{-}NO\}^6$

Photoredox reactions of the metal-to-ligand charge-transfer (MLCT) excited states are one of the processes observed in the photochemistry of nitrosyl ruthenium complexes. The reactive species originate from the photolysis of ruthenium(II)-nitrosyl complexes, which activates the nitrosyl ligand and thus produces NO. The electronic spectra and NO-sensor measurements are described to elucidate the mechanistic details.

The MLCT band arises from the back-bonding of the $d_\pi$ orbital of the Ru(II) and $\pi^*$ $(NO^+)$, as depicted in Figure 13.1.

The energy of the MLCT transition is dependent on the electron density of the fragment $\{Ru^{II}L_5\}$. If $[Ru(NH_3)_5NO]^{3+}$ is used as a model for $[Ru^{II}L_5(NO)^+]^{n+}$, the MLCT band appears near 350 nm, and photolysis in the MLCT region of the ruthenium(II)-nitrosyl complex in aqueous solution generates the Ru(III) analog $[Ru(NH_3)_5H_2O]^{3+}$ and NO. The electron transfer is postulated to occur by an MLCT process. The assumption that the lowest unoccupied molecular

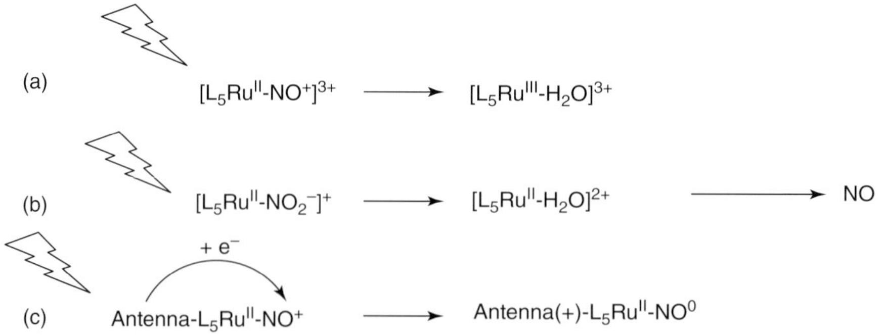


**Scheme 13.1**

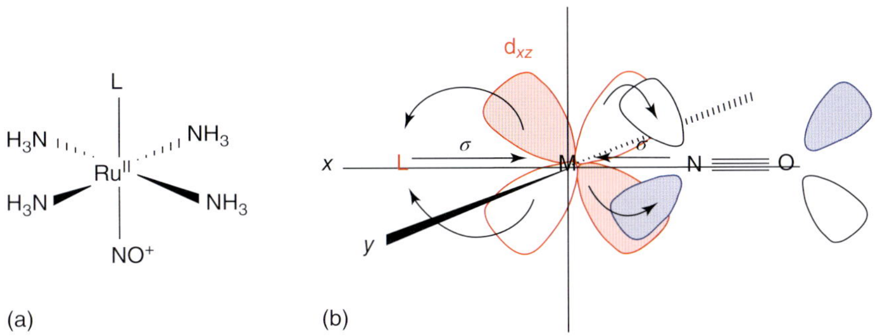


**Figure 13.1** *Trans* configuration of nitrosyl ruthenium complexes (a) and $d_\pi(Ru^{II})$—$\pi^*(NO^+)$ back-bonding (b).

$$trans\text{-}[Ru^{II}(NH_3)_4L(NO^+)]^{3+} \xrightarrow[H_2O]{h\nu} \{trans\text{-}[Ru^{III}(NH_3)_4L(NO^0)]^{3+}\}^* \longrightarrow trans\text{-}[Ru^{III}(NH_3)_4L(H_2O)]^{3+} + NO^0$$

**Scheme 13.2**

orbital (LUMO) of the nitrosyl ligand has a large contribution suggests that the photoaquation takes place from the electron excitation of an orbital with essentially dπ character (highest occupied molecular orbital, HOMO) to the excited-state LUMO. This prediction is confirmed for *trans*-$[RuNO(NH_3)_4L]^{3+}$, in which the HOMO and LUMO energies were influenced by the *trans* effect of "L." Studies have shown that photolysis of the MLCT band affords high NO quantum yield [20–24]. The photochemical pathway is described in Scheme 13.2 for *trans*-$[Ru^{II}NO(NH_3)_4L]^{3+}$ complexes.

Similar photochemical studies have been reported for *trans*-[Ru(NO)Cl (tetraazamacrocycle)]$^{2+}$, where tetraazamacrocycle is cyclam = 1,4,8,11-tetrazacyclotetradecano or [15]ane$N_4$ = 1,4,8,12-tetrazacyclopentadecane (Figure 13.2).

The tetraazamacrocyclic ligand chelator increased the stability of the ruthenium complex, which is beneficial for photobiological studies and avoids extraneous photochemical reactions. The photolysis of both macrocyclic ruthenium

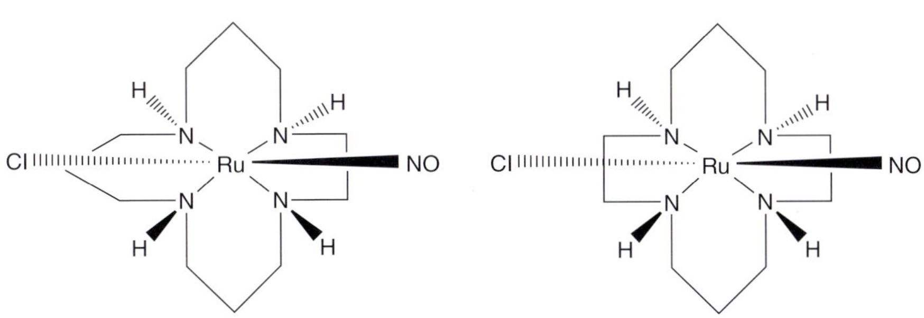


*trans*-$[RuCl([15]aneN_4)NO]^{2+}$ *trans*-$[RuCl(cyclam)NO]^{2+}$

**Figure 13.2** Schematic structures of nitrosyl macrocyclic ruthenium complex.

complexes, *trans*-$[RuCl(cyclam)(NO)]^{2+}$ and *trans*-$[RuCl([15]aneN_4)(NO)]^{2+}$, was carried out with 355-nm light irradiation in buffer solution at pH 7.4 [13]. The resulting UV-visible spectral changes were consistent with the substitution of NO by a solvent molecule, shown by the appearance of a new band at 348 nm attributed to the *trans*-$[RuCl(OH)$ (tetraazamacrocycle)$]^+$ photoproduct at pH 7.4 and the *trans*-$[Ru(H_2O)$(tetraazamacrocycle)$Cl]^{2+}$ photoproduct at pH = 1.

The observed quantum yields for NO release ($\phi_{NO}$) from *trans*-$[RuCl(cyclam)(NO)]^{2+}$ and *trans*-$[RuCl([15]aneN_4)(NO)]^{2+}$ during photolysis were calculated using *in situ* NO determination. The $\phi_{NO}$ of *trans*-$[RuCl(cyclam)(NO)]^{2+}$ (0.16 mol Einstein$^{-1}$) at pH 7.4 is four times smaller than the quantum yield observed for *trans*-$[RuCl([15]aneN_4)(NO)]^{2+}$ (0.61 mol Einstein$^{-1}$) under similar experimental conditions [13, 25, 26]. These results were attributed to the different contributions of the nitrosyl ligand to the LUMO orbital [26].

The effect of the π-acceptor character of the co-ligands "L" in $[RuL_5NO]^{n+}$ has also been explored using N-ligands for L. A representative series of polypyridine ruthenium complexes consists of the following: *cis*-$[Ru(bpy)_2L_a(NO)]^{n+}$ (bpy = 2,2′-bipyridine, $L_a$ = Cl, pyridine (py), 4-acetylpyridine (4-acpy), pyrazine (pz), or 4-picoline (4-pic)) and $[Ru(tpy)L_bNO]^{3+}$ (tpy = 2,2′:6′,2″-terpyridine, $L_b$ = bipyridine (bpy) or 3,4-diiminobenzoic acid (NH·NHq)) complexes as photoactivatable ruthenium-based prodrugs [27–33].

In general, UV-visible spectra of the mononuclear nitrosyl ruthenium complexes described as *cis*-$[Ru(bpy)_2L(NO)]^{n+}$ and $[Ru(tpy)(L)(NO)]^{n+}$ exhibit a shoulder in the region of 330 nm in aqueous solution at pH ≤ 5.0. Based on DFT calculations, this shoulder can be ascribed to the $t_2$-π* MLCT transition because the π* level presents substantial nitrosyl character. The HOMO of the *cis*-$[Ru(bpy)_2L(NO)]^{n+}$ and $[Ru(tpy)(L)(NO)]^{n+}$ complexes does not depend on the pyridine ligands and has an essentially metal character. In addition, the LUMO displays nitrosyl character for all the studied complexes, which is consistent with MLCT in the region of 330 nm [32]. The photolysis at 355-nm light irradiation in aqueous solution of *cis*-$[Ru(bpy)_2L(NO)]^{n+}$ and $[Ru(tpy)(L)(NO)]^{n+}$ complexes resulted in the formation of $\{Ru^{III}\text{-}NO^0\}^{3+}$ species, in accordance with the classical description of light irradiation on the MLCT transition of ruthenium compounds. UV-visible spectroscopy and high-performance liquid chromatography (HPLC) revealed that the main product is $\{Ru^{II}\text{-}H_2O\}$. We have supposed that a secondary photochemical process may explain the reduction of the primary photoproduct $\{Ru^{III}\text{-}H_2O\}$ or maybe ligand field transition was also accessed by 355-nm light irradiation (Scheme 13.3) [16, 33].

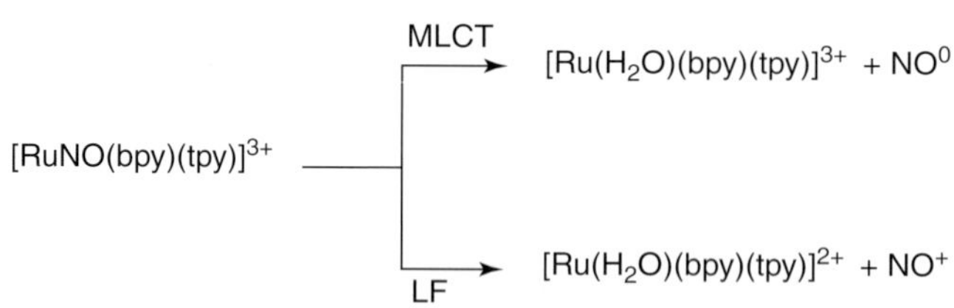


**Scheme 13.3**

**Table 13.1** Polypyridine nitrosyl ruthenium complex quantum yields ($\phi_{NO}$) after flash photolysis at 355 nm in trifluoroacetate buffer solution (pH = 2.01).

| Compounds | $\phi_{NO}$ (mol Einstein$^{-1}$) | LUMO | Refs |
|---|---|---|---|
| *cis*-$[Ru^{II}Cl(bpy)_2(NO)]^{2+}$ | 0.98[a)] | −10.04983863 | [27] |
| *cis*-$[Ru^{II}(bpy)_2(py)(NO)]^{3+}$ | 0.16 | −13.18563427 | [16] |
| *cis*-$[Ru^{II}(bpy)_2(4\text{-pic})(NO]^{3+}$ | 0.17 | −13.05393303 | [16] |
| *cis*-$[Ru^{II}(bpy)_2(4\text{-acpy})(NO)]^{3+}$ | 0.07 | −13.17692675 | [16] |
| $[Ru^{II}(tpy)(bpy)(NO)]^{3+}$ | 0.14 | – | [17] |
| $[Ru^{II}(tpy)(NH{\cdot}NHq)(NO)]^{3+}$ | 0.47 | – | [17] |

a) pH = 5.7 phosphate-buffered solution.

**Table 13.2** Polypyridine nitrosyl ruthenium complex quantum yields ($\phi_{NO}$) after flash photolysis at 355 nm in phosphate-buffered solution (pH = 7.4).

| Compounds | $\phi_{NO}$ (mol Einstein$^{-1}$) [33] |
|---|---|
| *cis*-$[Ru^{II}NO_2(bpy)_2(py)]^+$ | 0.007 |
| *cis*-$[Ru^{II}NO_2(bpy)_2(4\text{-pic})]^+$ | 0.009 |
| *cis*-$[Ru^{II}NO_2(bpy)_2(pz)]^+$ | 0.037 |
| $[Ru^{II}NO_2(tpy)(bpy)]^+$ | 0.036 |

The NO quantum yield appears to rise as the nitrosyl character of the LUMO wavefunction increases [32]. Table 13.1 shows good correlations between the quantum yield and the LUMO energy.

Despite the higher $\phi_{NO}$ under low-intensity UV light, the utility of the *cis*-$[Ru(bpy)_2L(NO)]^{n+}$ and $[Ru(tpy)(L)(NO)]^{n+}$ as NO donor agents is limited in buffer solution at physiological pH. Dissolution of the *cis*-$[Ru(bpy)_2L(NO)]^{n+}$ and $[Ru(tpy)(L)(NO)]^{n+}$ in aqueous solution at pH 7.4 results in conversion of the NO group to $NO_2^-$ (nitrosyl to nitrite ligand). The nitro complexes *cis*-$[RuNO_2(bpy)_2L]^{n+}$ and $[RuNO_2(tpy)(L)]^{n+}$ released NO in buffer solution at pH 7.4 upon exposure to 355-nm light irradiation, as determined using a NO sensor (Table 13.2).

### 13.2.2 Nitrosyl Ruthenium Complexes: Visible-Light Stimulation

Although studies have shown that ruthenium-bipyridine complexes can function as NO donors by photolysis, this method cannot be used in clinical therapy because of restrictions on the use of light in the photoactive regions of those complexes. In this context, the development of methods to sensitize the photochemical release of an active biomolecule is clinically interesting.

Hence, one interesting subclass of nitrosyl ruthenium complexes is binuclear systems, such as $[Ru_A{}^{II}L(NH_3)_4(pz)Ru_B{}^{II}(bpy)_2(NO)](PF_6)_5$ (where L is $NH_3$, pyridine (py), or 4-acetylpyridine (4-acpy), and pz = pyrazine) (Figure 13.3),

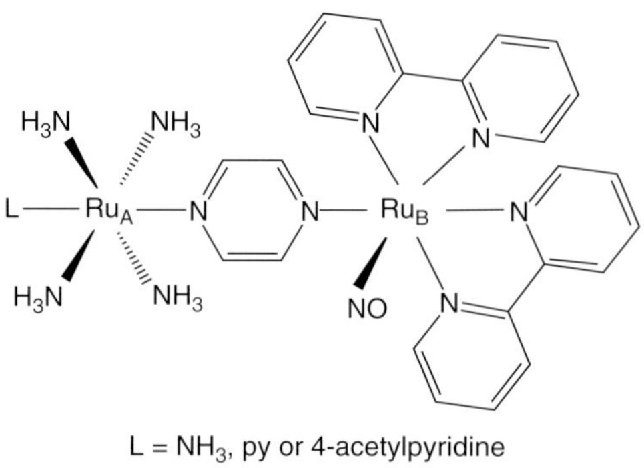


**Figure 13.3** Schematic representation of binuclear ruthenium complex.

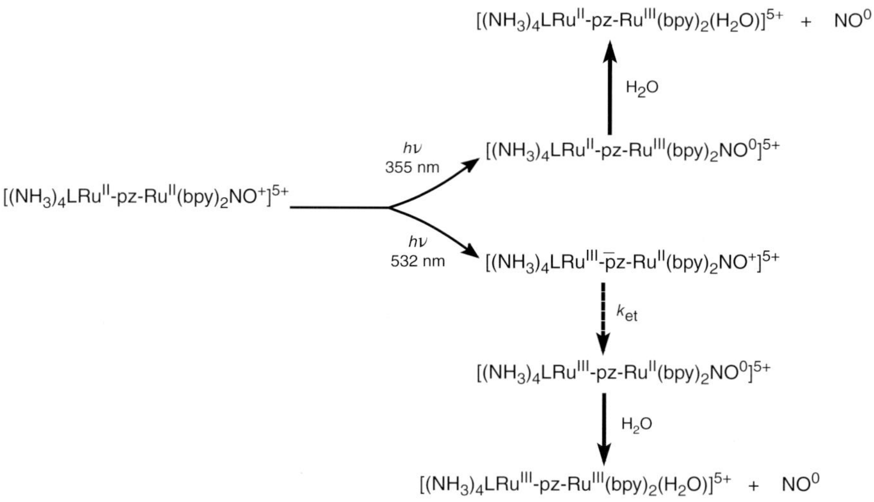


**Scheme 13.4**

which exhibits additional UV-visible features because of the electronic characteristics of the second ruthenium ($Ru_A$) fragment [18, 34]. The design of binuclear complexes with the subunit $\{RuL(NH_3)_4pz\}$ ($L = NH_3$, py, or 4-acpy) with different conjugation and electron donation strengths has enabled us to systematically study the interrelationship between the molecular structure and the photoinduced electron transfer process.

Photolysis of $[Ru_A{}^{II}L(NH_3)_4(pz)Ru_B{}^{II}(bpy)_2(NO)]^{5+}$ complexes under visible or UV light irradiation generates NO, as described in Scheme 13.4 [18, 34]. Photoinduced electron transfer has been postulated as the photochemical pathway resulting from 532-nm light irradiation on these binuclear species. The excitation of $[Ru_A{}^{II}L(NH_3)_4(pz)Ru_B{}^{II}(bpy)_2(NO)]^{5+}$ complexes with visible-light radiation gives the ruthenium complex in the excited state (Scheme 13.4).

There is a substantial contribution from the $\{Ru_A{}^{II}L(NH_3)_4(pz)\}^*$ moiety, which rapidly transfers an electron to the $\{Ru_B{}^{II}\text{-}NO^+\}$ fragment [18]. The thermodynamics of photoinduced electron transfer depends on the driving force operating between the excited reducing species and the $NO^{+/0}$

processes. This may explain the lowest NO quantum yield achieved for the $[Ru_A{}^{II}(NH_3)_5(pz)Ru_B{}^{II}(bpy)_2(NO)]^{5+}$ complex, which presents the smallest $E_{1/2}$ value for the electrochemical process involving the nitrosyl moiety of the ruthenium complex [18, 34].

Two other complexes, *cis*-$[Ru(bpy)_2L_xNO]^{3+}$ and $[Ru(tpy)L_yNO]^{3+}$ (where $L_x$ = quinolone and $L_y$ = 3,4-diiminobenzoic acid), show visible absorption in the electronic spectra in the 500-nm region [17, 30]. Under visible-light irradiation, both compounds release NO with a quite interesting amount for clinical application ($\phi_{NO}$ ca. 0.1 mol Einstein$^{-1}$). Unfortunately, those complexes fail to properly perform the photolysis in the therapeutic window. To improve this process, our group investigated some phthalocyanine ruthenium complexes as NO donors [35–37]. This class of compounds is characterized by intense absorption in the therapeutic window region, shows high stability, and can be bonded to the nitrosyl ligand. These complexes also undergo a synergistic effect resulting from the simultaneous production of NO and singlet oxygen.

A supramolecule of nitrosyl ruthenium and phthalocyanine was taken as the primary compound to be used to produce NO and singlet oxygen [35–38]. As depicted in Figure 13.4, *cis*-$[Ru(Hdcbpy_{-})_2ClNO]$ ($H_2$-dcbpy = 4,4′-dicarboxy-2,2′-bipyridine) and $Na_4[Tb(TsPc)(acac)]$ (TsPc = tetrasulfonated phthalocyanines and aca = acetylacetone) were dissolved in aqueous solution to generate the supramolecule [38].

The NO concentration and $^1O_2$ quantum yield for light irradiation in the $\lambda > 600$ nm region for this system were 1.21 μmol L$^{-1}$ and $\phi_\Delta = 0.41$ mol Einstein$^{-1}$, respectively.

A similar system has been developed using [Zn(phthalocyanine)] and $[Ru(tpy)(NH{\cdot}NHq)NO]^{3+}$ entrapped in a liposomal system. The zinc-phthalocyanine effectively worked as an antenna for the electron transfer process to the nitrosyl ruthenium complexes, which led to the production of NO and $^1O_2$ during photolysis at 675 nm [39]. Although significant

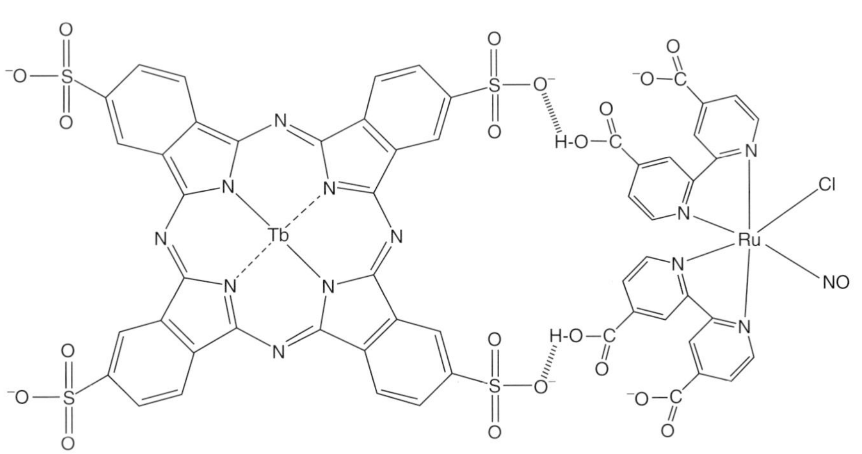


**Figure 13.4** Schematic representation of the conjugate Tb(phthalocyanine) – nitrosyl ruthenium complex.

new information has been obtained in previous experiments, the associated compounds still require a procedure for use in biological experiments. A better system should involve direct interactions between the nitrogen oxide ligand and the ruthenium-phthalocyanine core. We observed this issue in two systems: *trans*-[$RuNO_2$(phthalocyanine)NO] **(I)** and [Ru(phthalocyanine)$(pz)_2${$Ru(bpy)_2NO$}$_2$]$(PF_6)_6$ **(II)** complexes, which were synthesized for use in such photochemical studies (Figure 13.5) [36, 37].

Species **(I)** could produce NO through a reduction process, as shown by the spectroelectrochemical experiments [35]. Under 670-nm light irradiation, NO

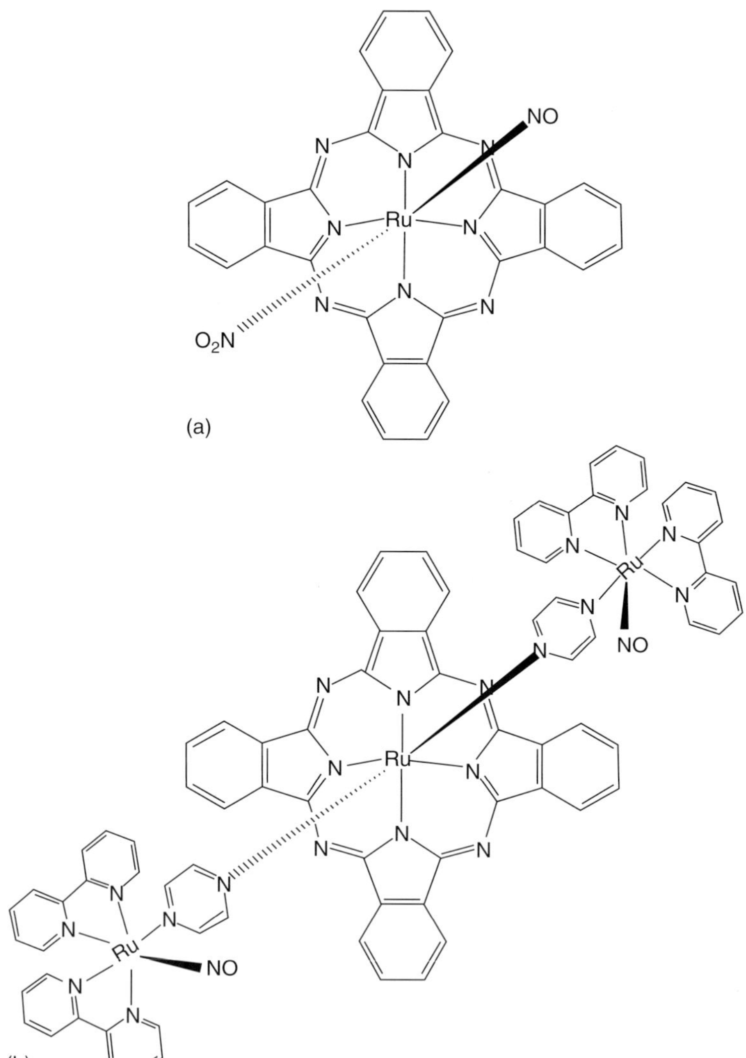


**Figure 13.5** Chemical structures of the nitrosyl phthalocyanine ruthenium complexes [$RuNO_2$(phthalocyanine)NO] (a) and [Ru(phthalocyanine)$(pz)_2${$Ru(bpy)_2NO$}$_2$]$^{6+}$ (b).

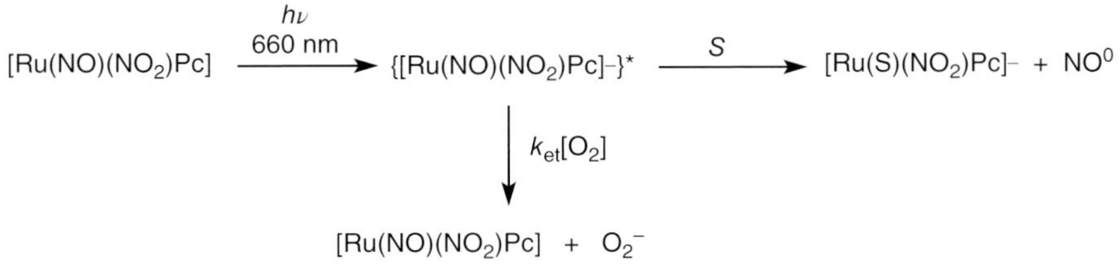

Scheme 13.5

and $^1O_2$ are produced. Molecular orbital calculations were used to explain the contributions of nitrosyl and phthalocyanine to the LUMO orbital resulting in NO photorelease and singlet oxygen production (Scheme 13.5). In the trinuclear species, the photolysis at 670 nm produced only singlet oxygen. In certain circumstances, this method avoids side reactions involving NO and $^1O_2$. For both species, photolysis with UV light results in NO quantum yield between 0.2 and 0.8 mol Einstein$^{-1}$, which is consistent with light irradiation at the MLCT from the dπ(Ru)—π*(NO$^+$) transition.

## 13.3 Photobiological Applications of Nitrogen Oxide Compounds

### 13.3.1 Photovasorelaxation

Nitrovasodilators, such as nitroglycerine and sodium nitroprusside (SNP), are therapeutically used to decrease systemic blood pressure in clinical hypertensive emergencies. These drugs induce vascular relaxation promoted by NO production. NO is well known to be involved in numerous biological process, including relaxing vascular smooth muscle [40–44]. In this case, endogenous NO is synthesized in vascular endothelial cells through the oxidation of L-arginine to L-citrulline mediated by the enzyme NO synthase. NO activates the enzyme guanylate cyclase (GC), which promotes the conversion of the guanosine triphosphate (GTP) to cyclic guanosine monophosphate (cGMP), resulting in a decrease in the intracellular calcium [$Ca^{2+}$]c, which is related to the activation of the vascular smooth muscle [45, 46].

Although different nitrovasodilators are available for clinical use, their diverse side effects, such as toxic cyanide release, nitrate tolerance, and tachyarrhythmia [47, 48], have driven the development of new NO donors as potential therapeutic drugs and for the study of vasodilation pathways. Metalonitrosyl complexes can effectively release NO by a reduction process [17, 26, 49, 50] or by light stimulation [13, 16]. More specifically, complexes in which NO is coordinated to the ruthenium metal have attracted attention because they can potentially deliver NO to specific sites in a controlled manner. In addition, such complexes are thermally stable, have low toxicity, and can be stable at physiological pH [13, 29]. Such nitrosyl ruthenium compounds are relatively scarce in the field of photovasodilators. Furthermore, we have found that some nitrosyl

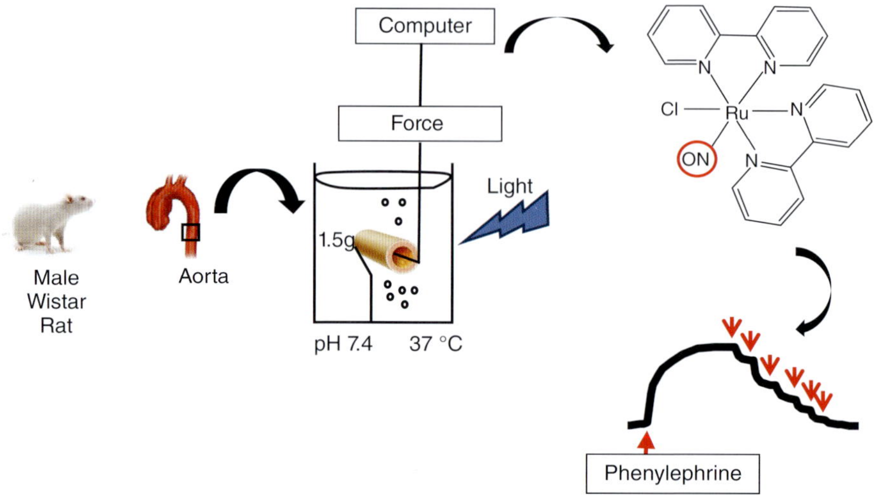


**Scheme 13.6** Vasodilation experiment: the thoracic aorta was isolated from male Wistar rat, cut into rings and placed between two stainless-steel stirrups and connected to an isometric force transducer. The responses were recorded using a computerized system to measure tension in the preparations. The tracheal rings were placed in a 10-ml organ chamber containing Krebs solution at pH 7.4, and gassed with 95% $O_2$ and 5% $CO_2$ at 37 °C. The rings were initially stretched to a basal tension of 1.5 g (optimal basal tone, previously determined by length–tension relationship experiments) before allowing them to equilibrate in the bathing medium. Aortic rings were precontracted with phenylephrine, a concentration that produced half-maximal contraction (EC50). When the contraction to phenylephrine had reached a plateau the nitrosyl ruthenium complex was added cumulatively.

ruthenium complexes exhibit remarkable structural diversity and have relevant vasorelaxation activities induced by light irradiation [15, 51, 52].

As thoroughly described previously, *trans*-$[Ru^{II}Cl(cyclam)NO]^{2+}$ **(I)**, *trans*-$[Ru^{II}Cl([15]aneN_4)NO]^{2+}$ **(II)**, and *cis*-$[Ru^{II}Cl(bpy)_2NO]^{2+}$ **(III)** produce NO under 355-nm light stimulation at pH = 7.4 [18, 33].

This knowledge drove us to study of photovasodilation using those nitrosyl ruthenium complexes as NO delivery agents, as described in Scheme 13.6.

The experiments performed with the nitrosyl macrocyclic ruthenium compounds, for example, clearly showed the effect caused by the NO quantum yield on vasodilation. Although the molecular structures of these complexes are similar, the lowest NO quantum yield for the cyclam species did not show any vasorelaxation activity when light irradiation experiments were performed in rat aorta [26]. The low amount of NO generated was likely insufficient to activate the biological pathway required to promote muscle relaxation. In contrast, the photorelaxation of **(II)** observed in denuded-endothelium aortic ring contracted with prostaglandin (PGF2α) was 99.1 ± 2.1% ($n = 6$) with a time course of 50 s after the endogenous photoactivatable NO stores were depleted [52]. Furthermore, the trans-$[RuCl([15]aneN_4)NO]^{2+}$ complex did not show toxicity against the vascular smooth muscle cell at the concentration needed to induce maximum relaxation of denuded rat aorta.

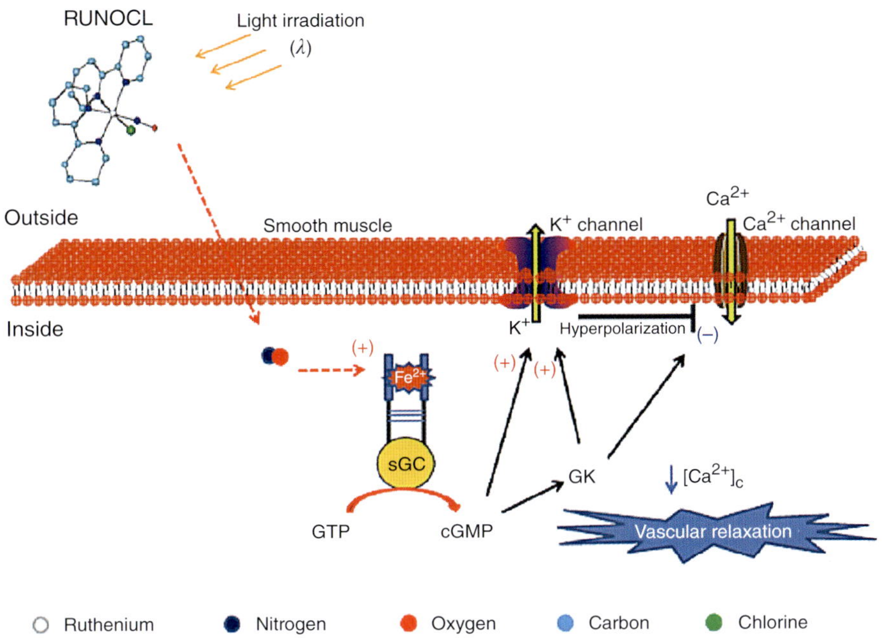


**Figure 13.6** Proposed nitric oxide release and vasodilation mechanism elicited by *cis*-$[Ru^{II}Cl(bpy)_2NO]^{2+}$ (RUNOCL). The colored circles represent atoms in the chemical structure. The hydrogen atom has been omitted in the structure. sGC, soluble guanylyl cyclase; GK, G kinase protein; $Ca^{2+}$, calcium; $K^+$, potassium; $[Ca^{2+}]c$, cytosolic calcium concentration.

At physiological pH (pH = 7.4), *cis*-$[Ru^{II}Cl(bpy)_2NO]^{2+}$ **(III)** is also photochemically active under UV and visible photolysis [53]. The vasodilating effect in rat aortic ring occurs only because of light irradiation. Using light at $\lambda > 380$ nm, the compound induces a maximum effect of 101.2% ± 3.7%, with pD2: 6.62 ± 0.16 ($n = 7$) and a time course of 1630 s to reach the maximum vasodilation at high concentration. When the aortic rings were pretreated with the soluble guanylyl cyclase inhibitor ODQ, the vascular relaxation induced by **(IIIcp** was completely suppressed. These findings indicate that vascular relaxation is strictly associated with the activation of a NO-cGMP pathway. The vascular relaxation caused by the nitrosyl ruthenium complexes could be described as shown in Figure 13.6 [15].

A study of vasodilation was also performed using $[Ru(NH{\cdot}NHq)(tpy)NO]^{3+}$ entrapped in a sol–gel matrix or silicone membrane [53]. The complex was found to be stable inside these matrices and released NO at 532-nm light irradiation. In addition, the amount of NO produced was observed to depend on the amount of entrapped ruthenium complex. The $[Ru(NH{\cdot}NHq)(tpy)NO]^{3+}$ caused maximum relaxation within 100 s in denuded rat aortic rings contracted with phenylephrine under visible-light irradiation. The NO must be delivered in the extracellular medium because no relaxation was observed after preincubation with the NO scavenger oxyhemoglobin [53]. The vasodilation effect was $E_{max} = 29.9 \pm 3.6\%$ for the complex in sol–gel matrix. Compared with the results obtained in solution at physiological pH, the entrapped $[Ru(NH{\cdot}NHq)(tpy)NO]^{3+}$ was found to be less effective because the membrane releases NO extracellularly, and, as a result, less NO may be accessible within the cell for promoting guanylyl-cyclase activation.

The nitrosyl ruthenium compounds *cis*-$[RuL(bpy)_2(NO)]^{3+}$ (L = pyridine, 4-picoline and pyrazine) and $[Ru(bpy)(tpy)NO]^{3+}$ could be converted into the corresponding nitro species at pH 7.4 (Scheme 13.2) and can produce NO upon 355-nm light irradiation [33]. Pharmacological studies were conducted in denuded rat aortic rings contracted with KCl (60 mmol $l^{-1}$) using the nitro complex *cis*-$[RuNO_2L(bpy)_2]^+$ and $[Ru(NO_2)(bpy)(tpy)]^+$ in water/oil (W/O) microemulsion [33]. The NO quantum yield for $[Ru(NO_2)(bpy)(tpy)]^+$ and *cis*-$[Ru(NO_2)(bpy)_2pz]^+$ in solution are similar and greater than that of *cis*-$[Ru(NO_2)L(bpy)_2]^+$ (L = py and 4-pic] (Table 13.2). Nevertheless, in the microemulsions, the time to reach maximum relaxation was observed to be longer for the $[Ru(NO_2)(bpy)(tpy)]^+$ complex (ca. 50 min, $n = 6$) than for the cis-$[Ru(NO_2)L(bpy)_2]^+$ with L = py and 4-pic complex (ca. 28 min, $n = 6$) and pz complex (ca. 24 min, $n = 5$). The authors believe that this difference resulted from the combination of NO and hydroxo-ruthenium complex inside the microemulsion, producing the nitro ruthenium species [33].

## References


1 Kogan, N.N., Chaparro, E., Veira, R., Crespo, A., Rosati, O., and Gusis, S. (2013) *J. Eur. Acad. Dermatol.*, **27**, 65.
2 Elman, M. and Lebzelter, J. (2004) *Dermatol. Surg.*, **30**, 139.
3 Colenda, C.C., Cohen, W., McCall, W.V., and Rosenquist, P.B. (1997) *Alzheimer Dis. Assoc. Disord.*, **11**, 175.
4 Oliveira, P., Santos, A.A., Rodrigues, T., Tim, C.R., Pinto, K.Z., Magri, A.M.P., Fernandes, K.R., Mattiello, S.M., Parizotto, N.A., and Anibal, F.F. (2013) *J. Biomed. Opt.*, **18**, 128004.
5 Manisova, B., Binder, S., Malina, L., Jiravova, J., Langova, K., and Kolarova, H. (2015) *Anticancer Res.*, **35**, 3943.
6 Robertson, C.A., Hawkins, E.D., and Abrahamse, H. (2009) *J. Photochem. Photobiol. B Biol.*, **96**, 1.
7 Dolmans, D.E.J.G.J., Fukumura, D., and Jain, R.K. (2003) *Nat. Rev. Cancer*, **3**, 380.
8 Macdonald, I. and Dougherty, T.J. (2001) *J. Porphyrins Phthalocyanines*, **05**, 105.
9 Ali, S.M. and Olivo, M. (2002) *Int. J. Oncol.*, **21**, 531.
10 Ji, Z., Yang, G., Vasovic, V., Cunderlikova, B., Suo, Z., Nesland, J.M., and Peng, Q.J. (2006) *Photochem. Photobiol. B Biol.*, **84**, 213.
11 Ridnour, L.A., Thomas, D.D., Switzer, C., Flores-Santana, W., Isenberg, J.S., Ambs, S., Roberts, D.D., and Wink, D.A. (2008) *Nitric Oxide*, **19**, 73.
12 Wink, D.A., Vodovotz, Y., Laval, J., Laval, F., Dewhirst, M.W., and Mitchel, J.B. (1998) *Carcinogenesis*, **19**, 711.
13 Oliveira, F.S., Togniolo, V., Pupo, T.T., Tedesco, A.C., and da Silva, R.S. (2004) *Inorg. Chem. Commun.*, **7**, 160.

14 Lunardi, C.N., da Silva, R.S., and Bendhack, L.M. (2009) *Braz. J. Med. Biol. Res.*, **42**, 87.
15 de Lima, R.G., Silva, B.R., da Silva, R.S., and Bendhack, L.M. (2014) *Molecules*, **19**, 9628.
16 Sauaia, M.G., Oliveira, F.S., Tedesco, A.C., and da Silva, R.S. (2003) *Inorg. Chim. Acta*, **355**, 191.
17 de Lima, R.G., Sauaia, M.G., Bonaventura, D., Tedesco, A.C., Bendhack, L.M., and da Silva, R.S. (2006) *Inorg. Chim. Acta*, **359**, 2543.
18 Sauaia, M.G., de Lima, R.G., Tedesco, A.C., and da Silva, R.S. (2005) *Inorg. Chem.*, **44**, 9946.
19 Ghosh, K., Kumar, S., and Kumar, R. (2011) *Inorg. Chem. Commun.*, **14**, 146.
20 Tfouni, E., Doro, F.G., Gomes, A.J., da Silva, R.S., Metzker, G., Benini, P.G.Z., and Franco, D.W. (2010) *Coord. Chem. Rev.*, **254**, 355.
21 Carlos, R.M., Ferro, A.A., Silva, H.A.S., Gomes, M.G., Borges, S.S.S., Ford, P.C., Tfouni, E., and Franco, D.W. (2004) *Inorg. Chim. Acta*, **357**, 1381.
22 Gomes, M.G., Davanzo, C.U., Silva, S.C., Lopes, L.G.F., Santos, P.S., and Franco, D.W. (1998) *Dalton Trans.*, 601.
23 Toledo, J.C., Lopes, L.G.D., Alves, A.A., da Silva, L.P., and Franco, D.W. (2002) *J. Inorg. Biochem.*, **89**, 267.
24 Marcondes, F.G., Ferro, A.A., Souza-Torsoni, A., Sumitani, M., Clarke, M.J., Franco, D.W., Tfouni, E., and Krieger, M.H. (2002) *Life Sci.*, **70**, 2735.
25 Bonaventura, D., Oliveira, F.S., Togniolo, V., Tedesco, A.C., da Silva, R.S., and Bendhack, L.M. (2004) *Nitric Oxide*, **10**, 83.
26 Oliveira, F.S., Ferreira, K.Q., Bonaventura, D., Bendhack, L.M., Tedesco, A.C., Machado, S.P., Tfouni, E., and da Silva, R.S. (2007) *J. Inorg. Biochem.*, **101**, 313.
27 Togniolo, V., da Silva, R.S., and Tedesco, A.C. (2001) *Inorg. Chim. Acta*, **316**, 7.
28 Sauaia, M.G. and da Silva, R.S. (2003) *Trans. Metal Chem.*, **28**, 254.
29 Tfouni, E., Doro, F.G., Figueiredo, L.E., Pereira, J.C.M., Metzker, G., and Franco, D.W. (2010) *Curr. Med. Chem.*, **17**, 3643.
30 de Carvalho, A.N., Fornari, E.C., Gomes, W.R., Araújo, D.M.S., Machado, A.E.H., and Nikolaou, S. (2011) *Inorg. Chim. Acta*, **370**, 444.
31 Rose, M.J. and Mascharak, P.K. (2008) *Coord. Chem. Rev.*, **252**, 2093.
32 da Silva, R.S., de Lima, R.G., and Machado, S.P. (2015) *Adv. Inorg. Chem.*, **67**, 265.
33 de Lima, R.G., Sauaia, M.G., Bonaventura, D., Tedesco, A.C., Lopez, R.F.V., Bendhack, L.M., and da Silva, R.S. (2005) *Inorg. Chim. Acta*, **358**, 2643.
34 Sauia, M.G., de Lima, R.G., Tedesco, A.C., and da Silva, R.S. (2003) *J. Am. Chem. Soc.*, **125**, 14718.
35 da Rocha, Z.N., de Lima, R.G., Doro, F.G., Tfouni, E., and da Silva, R.S. (2008) *Inorg. Chem. Commun.*, **11**, 737.
36 Carneiro, Z.A., de Moraes, J.C., Rodrigues, F.P., de Lima, R.G., Curti, C., da Rocha, Z.N., Paulo, M., Bendhack, L.M., Tedesco, A.C., Formiga, A.L., and da Silva, R.S. (2011) *J. Inorg. Biochem.*, **105**, 1035.
37 Heinrich, T.A., Tedesco, A.C., Fukuto, J.M., and da Silva, R.S. (2014) *Dalton Trans.*, **43**, 4021.

38 Cicillini, S.A., Prazias, A.C.L., Tedesco, A.C., Serra, O.A., and da Silva, R.S. (2009) *Polyhedron*, **28**, 2766.
39 Maranho, D.S., de Lima, R.G., Primo, F.L., da Silva, R.S., and Tedesco, A.C. (2009) *Photochem. Photobiol.*, **85**, 705.
40 Plane, F., Wiley, K.E., Jeremy, J.Y., Cohen, R.A., and Garland, C.J. (1998) *Braz. J. Pharmacol.*, **123**, 1351.
41 Bredt, D.S., Hwang, P.M., and Snyder, S.H. (1990) *Nature*, **347**, 768.
42 Garthwaite, J. (1995) *Trends Neurosci.*, **18**, 51.
43 Rees, D.D., Palmer, R.M., and Moncada, S. (1986) *Proc. Natl. Acad. Sci.*, **86**, 3375.
44 Biel, M., Zong, X., Ludwig, A., Sautter, A., and Hofmann, F. (1999) *Rev. Physiol. Biochem. Pharmacol.*, **135**, 151.
45 Li, P.L., Lee, H.C., Nelson, M.T., Meininger, G.A., and Van Breemen, C. (2003) *Acta Physiol. Scand.*, **179**, 339.
46 Robin, E.D. and McCauley, R. (1992) *Chest*, **102**, 1842.
47 Laursen, J.B., Mulsch, A., Boesgaard, S., Mordvintcev, P., Trautner, S., Gruhn, N., Nielsen-Kudsk, J.E., Busse, R., and Aldershvile, J. (1996) *Circulation*, **94**, 2241.
48 Karidi, K., Garoufis, A., Tsipis, A., Hadjiliiadis, N., den Dulk, H., and Reedijk, J. (2005) *Dalton Trans.*, 1176.
49 Lopes, L.G.F., Catellano, E.E., Ferreira, A.G., Davanzo, C.U., Clarke, M.J., and Franco, D.W. (2005) *Inorg. Chim. Acta*, **358**, 2883.
50 Lunardi, C.N., Cacciari, A.L., da Silva, R.S., and Bendhack, L.M. (2006) *Nitric Oxide*, **15**, 252.
51 Ferezin, C.Z., Oliveira, F.S., da Silva, R.S., Simioni, A.R., Tedesco, A.C., and Bendhack, L.M. (2005) *Nitric Oxide*, **13**, 170.
52 Lunardi, C.N., Vercesi, J.A., da Silva, R.S., and Bendhack, L.M. (2007) *Vasc. Pharmacol.*, **47**, 139.
53 de Lima, R.G., Sauaia, M.G., Ferezin, C.Z., Pepe, I.M., José, N.M., Bendhack, L.M., da Rocha, Z.N., and da Silva, R.S. (2007) *Polyhedron*, **26**, 4620.

# 14

# Trends and Perspectives of Ruthenium Anticancer Compounds (Non-PDT)

*Michael A. Jakupec*[1], *Wolfgang Kandioller*[1], *Beatrix Schoenhacker-Alte*[2], *Robert Trondl*[1], *Walter Berger*[2], *and Bernhard K. Keppler*[1]

[1] *University of Vienna, Faculty of Chemistry, Institute of Inorganic Chemistry & Research Cluster "Translational Cancer Therapy Research", Währinger Straße 42, 1090 Vienna, Austria*
[2] *Medical University of Vienna, Institute of Cancer Research & Comprehensive Cancer Center, Department of Medicine I, Borschkegasse 8a, 1090 Vienna, Austria*

## 14.1 Introduction

Since the beginnings of modern medicinal cancer therapy, research was concerned with the limited tumor selectivity of the cytotoxic effects exerted by anticancer drugs (limiting their tolerable doses) on the one hand, and resistance phenomena either intrinsically present in tumor cells or acquired under drug exposure (limiting the response rates and duration of response to therapy, respectively) on the other. Fueled by the widespread availability of modern molecular methods in the past few decades, an ever increasing number of factors that are differentially expressed by tumor cells compared to normal cells are being recognized, opening up entirely new avenues for pharmacological interventions. As a consequence, a new era of drug development was heralded, which is governed by target-specific compound design and screening, with the promise of high cancer selectivity and reduced side effects. These efforts have resulted in considerable progress in the treatment of many forms of cancer (most notably in renal cell cancer, melanoma, and hematological malignancies, but to a certain extent also in the more prevalent cancers of breast, lung, and colon/rectum), while having little impact on others so far. Even in this era of "targeted therapeutics," certain ruthenium compounds remain attractive for a variety of aspects, such as preferential transport to and activation in tumor tissue, as well as novel targets and modes of action, as will be exemplified subsequently.

In general, two main lines of research have evolved in the field of ruthenium anticancer compounds concurrently, based on different concepts of how to make use of the two most relevant oxidation states of ruthenium complexes (+II and +III) and their implications. Depending on the redox potential, which is influenced by the coordination sphere, ruthenium(III) complexes may be reduced to ruthenium(II) species under physiological conditions. As ruthenium(II) tends to be more reactive, ruthenium(III) complexes can be developed as prodrugs

*Ruthenium Complexes: Photochemical and Biomedical Applications*, First Edition.
Edited by Alvin A. Holder, Lothar Lilge, Wesley R. Browne, Mark A.W. Lawrence, and Jimmie L. Bullock Jr.

that have to be reduced in the reductive environment of solid tumors to fully unfold their pharmacological activities. This idea was developed by Clarke on the basis of pentaammineruthenium(III) complexes [1, 2], but may be applicable to other ruthenium(III) complexes as long as their redox potential lies in the proper (physiologically accessible) range, which can be achieved by choosing a set of ligands with adequate net electron-donor properties [3]. The other main concept is based on the stabilization of the more reactive ruthenium(II) state by arene ligands that are relatively inert to exchange reactions. Still, exchange rates of the other (non-arene) ligands in ruthenium(II)–arene complexes vary broadly, offering ample scope for tuning the reactivity by structural variation [4].

The potential of each approach with respect to the goal of targeted cancer therapy has been discussed before by other authors [5]. Full recognition and appreciation of the potential will clearly require a breakthrough in the clinical setting, and the following sections should also be read as an up-to-date reassessment of how close we have actually come to this point.

## 14.2 Ruthenium(III) Compounds

All three ruthenium complexes that have been studied in clinical trials so far (NAMI-A, KP1019, NKP-1339), and are thus at the forefront of drug development, are based on ruthenium(III) with an octahedral coordination geometry. For all of them, molecular targets or combinations of targets that are not already covered by long-established anticancer drugs have been recognized. While KP1019 and NKP-1339 are based on the same complex anion (Figure 14.1) and, though maybe not exactly bioequivalent, they most likely share the same purported mechanism of action [6] involving interference with cellular responses to endoplasmic reticulum (ER) stress and with histone-related processes, NAMI-A is a distinctly different molecular entity that lacks cytotoxicity in relevant concentrations but interferes with cell–matrix interactions, cell motility, invasion and angiogenesis, and therefore was developed as a metastasis inhibitor [7].

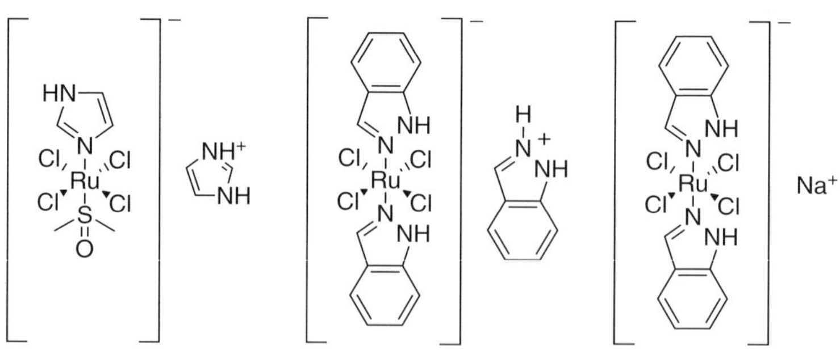


(a) **NAMI-A** (b) **KP1019** (c) **NKP-1339**

**Figure 14.1** Structural formulas of ruthenium(III) complexes NAMI-A (a), KP1019 (b), and NKP-1339 (c).

### 14.2.1 NAMI-A

Among the rational considerations for producing complexes with dimethyl sulfoxide (DMSO) as a ligand were a high capacity for permeating membranes, and a strong *trans* effect induced by coordination through sulfur (labilizing the opposite bond), and mild π-acceptor properties facilitating the reduction of ruthenium. Central to the recognition of DMSO complexes as inhibitors of the metastatic process was the observation that Na[*trans*-$Ru^{III}Cl_4$(dmso)(Him)] (NAMI; Him = imidazole) exerts stronger antimetastatic effects in mice when tumors are implanted intramuscularly and metastasize to the lung than when lung colonies are induced by intravenous injection of tumor cells – a process not involving the early steps of metastasis (detachment and intravasation of tumor cells) [8]. Preference was given to the imidazolium analogue NAMI-A (Figure 14.1) because of favorable chemical properties including higher stability in the solid state, with similar toxicological properties and at least equivalent efficacy *in vivo* [9].

#### 14.2.1.1 Biotransformation

NAMI-A is transformed under biological conditions into different species by hydrolysis and reduction, but their individual contributions to the pharmacological effects remain elusive. Hydrolysis proceeds rapidly by the exchange of two chlorides for water molecules and their partial deprotonation. Finally, hydrolysis results in formation of oxo-bridged oligo/polymeric products [10]. The biological relevance of poly-oxo species is unclear, though, since their formation will be considerably reduced in the presence of N-donor ligands reacting rapidly with early hydrolysis products. Reduction of NAMI-A to the corresponding dianionic ruthenium(II) species can be rapidly accomplished by biological reductants such as ascorbic acid, glutathione, or cysteine. Chloride hydrolysis is tremendously accelerated upon reduction [11], but since NAMI-A is unstable under physiological conditions and hydrolysis products have higher redox potentials, hydrolysis is likely to occur prior to reduction [12]. Still, reduction is fairly facile (due to the influence of the DMSO ligand) and, in fact, slightly advantageous for metastasis inhibition *in vivo* [13], although it seems not an absolute necessity for biological activity.

In serum, NAMI-A binds to proteins such as albumin or transferrin to a high extent. Albumin interactions are of a hydrophobic kind first and are converted into coordinative binding within minutes, which prevents reduction by ascorbic acid and oligomerization, according to electron spin resonance (ESR) spectroscopic measurements [14]. Altered pharmacokinetics of reduced NAMI-A *in vivo* suggest that the original drug is not rapidly reduced in the bloodstream [15]. However, experiments using albumin- and transferrin-bound NAMI-A initially led to rejection of the idea of tumor targeting mediated by these proteins and argued for bolus application associated with higher levels of non-protein-bound drug [16]. More recent studies suggesting that *in vitro* hallmarks of antimetastatic activity remain unaffected by albumin binding despite a complete replacement of ligands (according to X-ray absorption spectroscopy), might argue for a reassessment of this question [17].

#### 14.2.1.2 Antimetastatic Activity

*In vivo*, NAMI-A reduces the number and size of metastases but has little impact on growth of primary tumors in murine tumor models regularly metastasizing to the lungs [9, 18, 19]. The antimetastatic properties are independent of the route of administration (intraperitoneal, oral, or aerosolic) despite quite different ruthenium concentrations achieved in the lung, whereas growth of the primary tumor is only significantly affected in a high-dose, dose-dense setting resulting in high ruthenium levels in the tumor [20]. Experiments using intratumoral injection of NAMI-A confirm that the compound is not completely devoid of activity on primary tumors, but this activity requires higher local concentrations than those usually achieved [21].

Whether NAMI-A inhibits primarily the metastatic process or targets established metastases cannot be answered unequivocally because both effects seem to contribute to overall activity. On the one hand, histological examination of MCa mammary carcinoma in NAMI-A-treated mice revealed higher proportions of connective tissue in tumor parenchyma, around blood vessels, and around the tumor mass, which may constitute barriers against dissemination [9]. In addition, there seems to be a reduction of the metastatic potential of the tumor cell population itself, as evidenced by the reduced formation of lung metastases after transplantation of cells from the primary tumor of treated mice to mice not receiving the drug [22]. On the other hand, accumulation in collagen-rich regions and long retention in lung tissue have fueled speculation that NAMI-A may also affect metastatic cells already located in the lung. The latter is supported by the fact that antimetastatic effects in H460M2 xenografts and murine B16 melanoma are observed independently of whether NAMI-A is given before or after surgical removal of the primary tumor [19, 22]. If accumulation in lung tissue contributes to the antimetastatic effects, selective activation by reduction in lung metastases is deemed unlikely [23].

#### 14.2.1.3 Mode of Action

*In vitro*, NAMI-A is only modestly cytotoxic (except for surprising cytotoxicity in leukemia cells [24]) and is hardly taken up into cells [25], but it has a strong impact on cell motility and on interactions of tumor cells with the extracellular matrix. In particular, it rapidly induces long-lasting actin-dependent cell adhesion (even in usually nonadherent cells) based on a mechanism involving integrin activation [26, 27], associated with cell spreading and extrusion of filopodia [26]. Furthermore, NAMI-A inhibits degradation of the extracellular matrix by reducing the release of matrix metalloproteinases (MMPs), accompanied by direct inhibition of these enzymes, which are associated with metastatic aggressiveness [21]. Consistent with this finding, MMP-2 and MMP-9 levels in plasma of NAMI-A-treated mice are decreased [19]. The observed reduction of the CD44-positive fraction of tumor cells probably has implications for metastatic potential as well, as this adhesion protein is involved in cell motility and a low CD44 level is associated with reduced formation of invadopodia, cellular protrusions capable of degrading the extracellular matrix [21]. As a result of these effects, NAMI-A reduces cell invasiveness and migration [21, 22, 28]. A possible role of KCa 3.1 ($Ca^{2+}$-dependent $K^{+}$ channel) inhibition, which seems to cause the cytotoxicity in leukemia cells, remains to be investigated [24]. Overall, tumor cells adopt a less malignant phenotype with a lower propensity for dissemination. Remarkably, NAMI-A even seems to exert selection effects on a genomically heterogeneous tumor cell population (MCa cells), reducing the cell fraction with high ploidy, which suggests elimination of cell clones with high metastatic potential [21].

In endothelial cells, inhibition of proliferation, chemotactic behavior and MMP secretion has been observed, and in the chorioallantoic membrane assay spreading of new blood vessels is inhibited, suggesting an antiangiogenic component in the mode of action [29]. This probably involves inhibition of the MEK/ERK pathway (a signaling cascade involved in angiogenesis and metastatic tumor growth) [30, 31]. Consistent with the affinity of ruthenium for nitric oxide, experimental evidence suggests that the antiangiogenic effects are caused by interference with NO signaling [32]. Under physiological-like conditions, NAMI-A is indeed capable of scavenging NO upon hydrolysis, but not of releasing it again [33].

Enhanced infiltration of lymphocytes into primary tumors after NAMI-A treatment was observed in various settings [9, 21, 34]. The majority of infiltrating lymphocytes were characterized as T cells, with a decreased helper/suppressor cell ratio [34]. NAMI-A seems to promote extravasation of peripheral blood lymphocytes and production of nitric oxide by tumor-infiltrating lymphocytes [21]. In co-cultures of metastatic tumor cells and splenocytes, NAMI-A stimulates lymphocytes to adhesion onto metastatic cells, resulting in a reduced number of metastatic cells, suggesting that NAMI-A, in contrast to many classic anticancer drugs that are cytotoxic to immune cells, may be capable of activating tumor-suppressing properties in immune cells [35].

#### 14.2.1.4 Clinical Studies and Perspectives

In a clinical phase I study in 24 patients with advanced solid tumors who received NAMI-A infusions on 5 consecutive days every 3 weeks, 300 mg $m^{-2}$ $day^{-1}$ [1] were determined as a safe dose. Higher doses resulted in poorly healing, painful blisters on hands and feet. Premedication with dexamethasone was required to prevent hypersensitivity reactions, and pre- and post-hydration was applied to minimize renal toxicity, which was reversible and not dose-limiting. Vomiting occurred frequently, but could be controlled with antiemetic medication. No objective response was observed, but one patient with non-small cell lung cancer experienced disease stabilization for 21 weeks. In plasma, most of the ruthenium was bound to proteins, associated with a small apparent volume of distribution, low clearance, and long terminal half-life [36].

A subsequent phase I/II study in patients with non-small cell lung cancer who were given a combination of NAMI-A (300–450 mg $m^{-2}$ (see footnote 1) once a week for 2–3 weeks in 3- to 4-week cycles) and gemcitabine as second-line therapy indicated only moderate tolerability due to neutropenia, transient nephrotoxicity, nausea, vomiting, and diarrhea. In the dose escalation part of the study, one patient showed a partial response (with brain metastases remaining progressive,

1 For dosing per square meter, which is common in cancer chemotherapy, the body surface area is assessed on the basis of body height and weight, for example, by the use of nomograms.

though). However, no objective response (yet about 60% disease stabilization) was observed in 15 evaluable patients in the phase II part, suggesting a lower efficacy than could be expected for gemcitabine alone [37].

The authors of this study advocated a future role of NAMI-A only if ways were found to reduce toxicity and increase efficacy. A solution might be sought in innovative drug delivery strategies like the recently reported conjugation to micelle-forming block copolymers [38]. The nanosized NAMI-A-containing micelles were shown to inhibit tumor cell invasion through Matrigel more strongly than the parent drug, and evaluation *in vivo* was held out in prospect.

### 14.2.2 KP1019/NKP-1339

From the class of tetrachloridobisazoleruthenate(III) complexes, KP1019 was originally selected for clinical development because of its superior efficacy in preclinical colorectal carcinoma models, but as development progressed it was replaced by the analogue NKP-1339 (KP1339, IT-139), differing only by the counterion (sodium instead of indazolium), because of the manifold higher solubility of the latter. This modification solved the administration issues and enabled easy application of higher doses, which became possible also due to the high tolerability of the compounds. The pharmacological profile of NKP-1339 turned out to be distinctly different from known anticancer drugs, as detailed subsequently. Recently, alternative approaches to circumvent the administration and stability issues encountered with KP1019 were pursued by production of KP1019-loaded micelle-forming PEGylated polymers [39] and poly(lactic acid) nanoparticles with different surfactants (see Section 2.2.2) [40], both of which proved technically feasible and showed enhanced biological activity *in vitro* but are still at an experimental stage.

#### 14.2.2.1 Tumor Targeting Mediated by Plasma Proteins

One of the greatest challenges to the development of systemically applied anticancer therapies is to target the tumor effectively. KP1019 and NKP-1339 target solid tumors by taking advantage of transport with serum proteins to which they have a strong affinity [41, 42]. The main proteins involved, albumin and transferrin, are both attractive drug carriers [43], even though the ways by which they deliver drugs to tumors differ. Transferrin is specifically endocytosed via transferrin receptors, which are frequently overexpressed by tumor cells to sustain their highly iron-dependent altered metabolism. The suitability of albumin is based on its accumulation in malignant tissues due to (i) leaky blood capillaries and deficient lymphatic drainage in these tissues – known as the "enhanced permeability and retention (EPR) effect" [44]; (ii) albumin-binding proteins mediating transcytosis through tumor endothelia and subsequent trapping in the tumor interstitium; and (iii) incorporation of albumin by tumor cells as an amino acid source [45]. With the approval of nanoparticle-formulated, albumin-bound paclitaxel, the concept of albumin-based drug targeting has become fully established in clinical cancer therapy.

*In vitro* [46, 47] and *in vivo* studies [42] alike have demonstrated that KP1019 binds mainly to the serum protein fraction of 60–80 kDa containing both albumin and transferrin. Interactions with each of these proteins were also addressed specifically *in vitro*: KP1019 reacts with transferrin very fast, and reaction is complete within several minutes [48]. X-ray diffraction analysis of KP1019 bound to apolactotransferrin suggests that the indazole ligands remain bound to the ruthenium center after binding to histidine-253 of the protein [49]. Support for transferrin-mediated cellular uptake came from cell culture experiments showing that accumulation of KP1019 in cancer cells may be favored by prior binding to partially iron-loaded transferrin [50] and that ruthenium levels and sensitivity are increased in transferrin-receptor-overexpressing leukemia cells upon exposure to KP1019 [51].

According to experiments in model systems, binding to transferrin is kinetically favored, but albumin adducts are more stable and likely to prevail in human serum due to the much higher abundance of albumin [48, 52]. This conclusion is supported by analytical studies that accompanied the KP1019 clinical phase I trial, showing that ruthenium in patient plasma was preferably bound to albumin, whereas transferrin seemed to play only a minor role [42]. Binding of KP1019 and NKP-1339 to human serum albumin involves hydrophobic cavities in subdomains IIA and IIIA of the protein with similar and moderately strong affinities, according to displacement experiments with specific site markers, implying that bilirubin and the ruthenium complexes compete for the same binding site [53]. Fast interactions with albumin are mainly hydrophobic, but are converted to coordinative binding over time [54]. Based on the assumption that noncoordinative interactions with albumin might be favorable for anticancer efficacy, modification of the complex by exchange of indazole for pyridine-based ligands with more hydrophobic moieties has been proposed [55], but the consequences *in vivo* have not yet been reported.

#### 14.2.2.2 Activation by Reduction

As outlined in the introduction, ruthenium(III) complexes may serve as bioreductive prodrugs, provided that their redox potential lies in the physiologically accessible range. For NKP-1339 in phosphate buffer, a redox potential of 0.00 to +0.03 V versus NHE (normal hydrogen electrode), independent of pH value, was reported from cyclic voltammetry measurements [56]. This is considered sufficiently high for reduction in a biological environment, as potentials reported for the most important cellular reductants are considerably lower: between −0.25 and −0.20 V for glutathione (GSSG/2GSH), and between −0.4 and −0.3 V for $NAD^+$/NADH and $NADP^+$/NADPH [57]. On the other hand, the redox potential of KP1019/NKP-1339 is not so high that instant reduction in biological liquids (as suggested for the more easily reducible $[Ru^{III}Cl_2(ind)_4]^+$ complex) has to be assumed [58].

The native compound KP1019 can be reduced in the presence of ascorbic acid or glutathione under buffered conditions within minutes or hours, respectively, resulting in increased reactivity at least in the case of low glutathione concentrations (whereas an excess of glutathione decreases reactivity probably due to ruthenium–glutathione coordination) [56]. Activating effects of high ascorbic acid concentrations could be verified in cell culture experiments [59]. However, reduction is prevented as long as the complex is bound to serum proteins, according to capillary electrophoresis studies under simulated extracellular conditions in the presence of ascorbic acid (the most important extracellular reductant) [60]. Under simulated intracellular conditions as well as in diluted cytosolic extracts, incubation of KP1019-transferrin adducts with glutathione was shown to yield low-molecular-weight ruthenium(II) species [61]. Thus, bioreductive activation seems much more likely inside the cell concurrent with drug release from the protein, which is thought to contribute to tumor selectivity in concert with the mechanisms mentioned in the section above.

To address the question whether the complexes are indeed reduced *in vivo*, attempts were made by application of X-ray absorption spectroscopy to tumor and liver samples from murine tumor models (homologous sarcoma 180 treated with KP1019 or NKP-1339, as well as xenografted SW480 carcinoma treated with NKP-1339). Since changes of oxidation state on the one hand and exchange of donor atoms around the metal center on the other have opposite effects on the resulting shift of edge energy, the measured spectra are consistent with two different average coordination patterns, $Ru(III)Cl_3N_2(O/N)$ and $Ru(II)ClN_2(O/N)_3$, for all *in vivo* settings alike [62]. Another, independently reported XAS study only included *in vitro* samples of KP1019-treated hepatoblastoma cells, where shifts were interpreted as the result of ruthenium(III) binding to S-donor ligands [63].

Surfactant-mediated binding of KP1019 to poly(lactic acid) nanoparticles surprisingly yielded biologically active, stabilized ruthenium(II) species as a result of reduction by the surfactant (Tween 80). While this approach is still based on accumulation in tumor tissue by the EPR effect, it hence turned out as a shortcut not only with regard to protein-mediated delivery but also to activation by reduction, obviating both of them [40].

#### 14.2.2.3 Mode of Action

Several unique characteristics of KP1019/NKP-1339 make the precise elucidation of the mode of action of these ruthenium compounds challenging. The double prodrug nature (serum protein binding and activation by reduction) together with the strong reactivity of the resulting Ru(II) species with diverse biomolecules including proteins and nucleic acids necessitate a precise spatial and temporal dissection of activation processes. As soon as KP1019/NKP-1339 is accumulated within the malignant tissue due to the EPR effect, additional factors have to be taken into account regulating actual anticancer activity: uptake of the drug from interstitium into tumor cells, efflux of the drug by export pumps, and chemical reduction and interaction with specific intracellular targets. With regard to uptake of these Ru(III) prodrug compounds, endocytosis/macropinocytosis by the malignant cells in protein-bound form is considered to be dominant (compare Section 14.2.2.1). Concerning active efflux mechanisms, it has been shown that KP1019 is a weak substrate of ABCB1 (P-glycoprotein), resulting in lower intracellular drug accumulation in ABCB1-overexpressing cells [51]. However, the impact of this classic drug resistance mechanism is distinctly limited by uptake in albumin-bound form obviously reducing recognition by the ABC-transporter efflux pump. Accordingly, acquired resistance to KP1019 does not depend on an ABC-transporter-mediated phenotype [51]. Furthermore, sensitivity toward KP1019 treatment does not correlate with intracellular drug accumulation

[46, 51]. Together, these findings suggest that the key factors guiding activity are rather interactions of highly bioactive Ru(II) species (compare Section 14.2.2.2) with specific intracellular targets than altered drug accumulation [46, 64–66]. Consequently, the different responsiveness to these compounds might be explained either by the presence/absence of targets or differences in the ability to repair the induced damage.

Several studies concerning the impact of KP1019/NKP-1339 on cancer cells indicate a combined antiproliferative and cell-death-inducing activity [51, 67–69]. In colorectal cancer cell models, both ruthenium compounds induce apoptosis by the so-called intrinsic pathway dependent on mitochondrial membrane depolarization, cytochrome *c* release into the cytoplasm, and activation of initiator caspase 9 [46, 70]. Especially in hypersensitive cell models, however, additional activation of the extrinsic apoptosis pathway involving cell death receptor signals and initiator caspase 8 has been observed [71]. Besides cell death at higher KP1019/NKP-1339 concentrations, accumulation of cells in G2/M phases of the cell cycle has been reported in human and yeast cells [69, 71–73]. As comparable changes are induced by cisplatin treatment [74, 75], the initial assumption was that these ruthenium compounds might also target primarily DNA. However, more recent studies suggest that protein targets might be more important than DNA and that the mode of action of these ruthenium complexes distinctly differs from that of classic platinum drugs [2, 46, 72, 76].

With respect to DNA as cellular target, cell-free assays demonstrated binding of KP1019/NKP-1339 to nucleotides and DNA probes [64, 66, 77–79]. In addition, studies in human cancer cells and yeast have proposed that the ruthenium compounds interact with DNA, induce double-strand DNA breaks, and lead to formation of DNA–DNA and DNA–protein cross-links [70, 72, 74, 76]. *Saccharomyces cerevisiae* mutants indicate that double-strand break and nucleotide excision repair but not base excision and mismatch repair protect against KP1019 [74]. However, it is so far unclear whether indeed DNA adducts, which are sparse as compared to cisplatin and the minority as compared to DNA–protein adducts [76], are involved in the DNA strand breaks induced by KP1019/KP1339 in rat and human cells [70]. In addition, several studies suggest that KP1019/NKP-1339 DNA damage might be secondary to massive redox disturbance and consequent reactive oxygen species (ROS) formation in cancer cells [39, 70]. In line with this hypothesis, depletion of intracellular glutathione by buthionine-sulfoximine (BSO) sensitizes against KP1019/NKP-1339, while the glutathione precursor *N*-acetylcysteine (NAC) exerts protective activity [70]. In addition, KP1019/NKP-1339-induced DNA damage might be secondary to protein interaction. Hence, the DNA damage response by KP1019 in yeast cells described by Singh *et al.* [72] was, in contrast to suggestions by Stevens *et al.* [74], attributed to specific interaction with histones (mainly H3) leading to histone eviction from nucleosomes. This indicates that even at the level of chromatin primarily histones rather than DNA itself might be targeted. Accordingly, we found a major signature of altered histone gene expression by NKP-1339 in hypersensitive cell models (unpublished observation). These data correspond to papers describing a key role of DNA-binding proteins in DNA damage. Generally, such proteins react with DNA-adduct-forming species, thereby shielding DNA from drug interaction. Nevertheless, the altered peptides may in turn cause damage to adjacent nucleotides or even DNA strand breaks [80, 81]. However, the question remains whether KP1019/NKP-1339 generally reach chromatin in the nucleus at sufficient concentrations or primarily interact with protein components in the cell membrane or the cytoplasm. Stronger nuclear accumulation of NKP-1339 as compared to KP1019 was described [46]. This argues against dominant nuclear targets, as KP1019 exerts a relatively higher cytotoxic activity *in vitro*. In addition, size-exclusion chromatography-inductively coupled plasma mass spectrometry (SEC-ICP-MS) demonstrated that following KP1019/NKP-1339 exposure ruthenium is initially bound mainly to high-molecular-weight cellular protein complexes >700 kDa, probably representing protein clusters in the cell membrane. At prolonged recovery, ruthenium is redistributed to a protein fraction <40 kDa. Whether this indicates drug metabolism or breakdown of damaged proteins needs to be determined. Anyway, this binding pattern does not at all resemble that of cisplatin, strongly supporting a very different mode of action [46].

Based on X-ray fluorescence imaging of single cells, recently co-localization of KP1019 and iron as well as intracellular relocalization of iron in response to the ruthenium compound has been reported [25]. In agreement, mild interference with erythropoiesis was detected in toxicological studies [82]. These alterations might be caused by altered iron uptake via the transferrin receptor in the presence of KP1019. However, based on the similarity between ruthenium and iron, a direct impact on intracellular iron homeostasis also seems likely. In addition, Ru(III) might participate in Fenton-like reactions, probably interfering with redox processes involving iron. Accordingly, preliminary data from our group demonstrated an enhanced activity of KP1019/NKP-1339 when combined with $FeCl_3$ but a reduced one when combined with an iron chelator. These data indicate that not iron deprivation *per se* but a direct interaction with iron-involving processes is participating in the anticancer activity of KP1019/NKP-1339.

The ability of KP1019/NKP-1339 to bind proteins also offers new applications, as has been recently demonstrated by the modulation of Aβ-peptide aggregation in Alzheimer's disease [83]. In concordance, glucose-regulated protein of 78 kDa (GRP78) as well as other parts of the ER-stress sensing and signaling machinery are downregulated by NKP-1339 especially in hypersensitive cell models [71, 84]. This is remarkable as tumor cells are generally characterized by increased levels of protein damage and are therefore in need of unfolded protein response. This may explain, at least in part, the higher sensitivity of malignant cells to KP1019/NKP-1339-induced cell death. The involvement of ER-stress-related effects in the anticancer activity of NKP-1339 was substantiated by application of protein translation and ER-stress inhibitors [71, 85].

In addition, the ruthenium compounds have been demonstrated to target cellular stress and survival signaling pathways. For example, we have demonstrated recently that NKP-1339 strongly synergizes with the clinically approved multikinase inhibitor sorafenib (Nexavar) both *in vitro* and *in vivo*. Remarkably, sorafenib blocks NKP-1339-induced stress signals such as protective p38 mitogen-activated protein kinase (MAPK) activation and subsequent G2/M cell cycle arrest [69]. The influence of KP1019/NKP-1339 on stress-induced

MAPK pathways is not limited to human cancer cells but was also described in *S. cerevisiae*, where increased phosphorylation of the osmolarity stress sensor MAPK Hog1 and consequently upregulation of protein ubiquitination was detected [73]. Part of these KP1019/NKP-1339-induced stress signals might be a consequence of the interaction with the sarco-ER $Ca^{2+}$-ATPase (SERCA), recently described as a specific feature of KP1019 [86].

In addition to its cytotoxic activity in malignant cells, KP1019/NKP-1339 might interfere with the tumor–stroma interaction or components of the microenvironment. As an example, KP1019 was demonstrated to reduce migration and invasion of mammary carcinoma cells [87]. Furthermore, NKP-1339 – like NAMI-A – is able to block several key steps of neoangiogenesis such as endothelial cell proliferation and migration stimulated by vascular endothelial growth factor (VEGF) or $NO^{\bullet}$ donor drugs. Accordingly, these compounds were demonstrated to efficiently react with $NO^{\bullet}$, a key signal mediator in inflammatory and immune responses as well as vascular homeostasis [32].

#### 14.2.2.4 Clinical Studies and Perspectives

KP1019 was studied in one clinical phase I trial in patients with therapy-refractory advanced solid tumors. The drug was infused twice per week over 3 weeks in an absolute dose range of 25–600 mg, with each of eight patients in total receiving only one dose level. Adverse effects were generally mild. No dose-limiting toxicities were observed, but further dose escalation was not feasible because of the large infusion volume required. Six patients were evaluable for response, of whom five (two colorectal, one cholangiocellular, one endometrial carcinoma, one head/neck cancer) showed disease stabilization for 8–10 weeks, apparently irrespective of whether a second therapy cycle was given or not [68]. Consistent with extensive binding to plasma proteins, ultrafiltrable ruthenium and renal excretion were very low, and total ruthenium was associated with a small volume of distribution and long half-life [88].

The safety and maximum tolerated dose of NKP-1339 (IT-139) were studied in a phase I trial in patients with advanced solid tumors, all of them progressive after failure of established prior therapies. The drug was administered in 4-week cycles consisting of an intravenous infusion once a week for 3 weeks followed by 1 week without. The dose was escalated from $20\,mg\,m^{-2}$ (see footnote 1) to a maximum of $780\,mg\,m^{-2}$ where dose-limiting toxicity was observed. Thus, $625\,mg\,m^{-2}$ was considered the maximum tolerated dose. The most frequent treatment-related adverse events were nausea/vomiting and fatigue; mild fever/chills could be controlled with steroid premedication [89a]. Remarkably, one long-lasting partial remission (with 98 weeks of disease control) and 2 cases of disease stabilization for $\geq$19 weeks in patients with carcinoid neuroendocrine tumors (NETs) as well as a further 7 cases of disease stabilization in other malignancies (such as non-small cell lung cancer and colorectal carcinoma) out of 38 evaluable patients were reported [89b]. Thus, the tolerability of the drug is quite high, but even more encouraging are the evident signs of therapeutic efficacy, most notably in NETs, which are poorly responsive to established medicinal treatments. Phase II single-agent studies and phase I combination studies will serve to delineate more precisely the clinical activity profile.

## 14.3 Organoruthenium(II) Compounds

Ruthenium(II) is prone to oxidation, and stabilization of the metal center is essential to investigate its antiproliferative potential. One method of stabilizing the Ru(II) in the highly reactive oxidation state + II is the facial coordination of aromatic systems, for example, arene or cyclopentadienyl derivatives, to the metal center leading to so-called piano-stool complexes, where the aromatic ring forms the "seat" and three remaining coordination sites the "legs." The attachment of aromatic moieties increases the lipophilicity of the compounds and therefore facilitates the passive cellular uptake through the cell membrane. Furthermore, modification of the arene part allows for tuning of the lipophilicity, and in addition tumor targeting vectors can be easily attached, enabling selective metallodrug accumulation in the tumor tissue. The remaining three coordination sites can be occupied either by mono- or polydentate ligand scaffolds, which have a massive impact on the pharmacological and physicochemical properties. These parameters can be easily fine-tuned due to the synthetic versatility of this compound class. In most cases, at least one coordination site is occupied by a halido ligand, which can be easily replaced by a water molecule in aqueous solution. This important activation step leads to the more reactive aqua species which can interact with the biological targets of the respective metallodrug.

### 14.3.1 Ruthenium(II)–Arene Compounds in Preclinical Development

[Ru(II)($\eta^6$-benzene)(dmso)$Cl_2$] was one of the first reported piano-stool complexes with anticancer activity and showed topoisomerase IIα inhibitory properties. However, the lack of stability hindered its further development. The two best studied classes of Ru(II) organometallics with anticancer activity are the so-called RAPTA complexes [Ru(II)($\eta^6$-arene)(pta)$X_2$] (pta = 1,3,5-triaza-7-phosphatricyclo[1,1,3,3]decane, X = halido) developed by Dyson and [Ru(II)($\eta^6$-arene)(en)X]$^+$ (en = ethylenediamine, X = halido) organometallics established by Sadler (Figure 14.2). Both are currently in an advanced preclinical stage.

For the complex class of [Ru(II)(arene)(en)]$^+$, the *N,N*-chelate ethylenediamine has been found to be superior with regard to stability and cytotoxicity because organometallic Ru(II) complexes bearing monodentate *N*-donor molecules are prone to fast ligand exchange reactions, whereas complexes with tridentate *N*-donor scaffolds are highly stable but insufficiently cytotoxic [90, 91]. A broad range of [Ru(II)($\eta^6$-arene)(en)X]$^+$ derivatives with varying

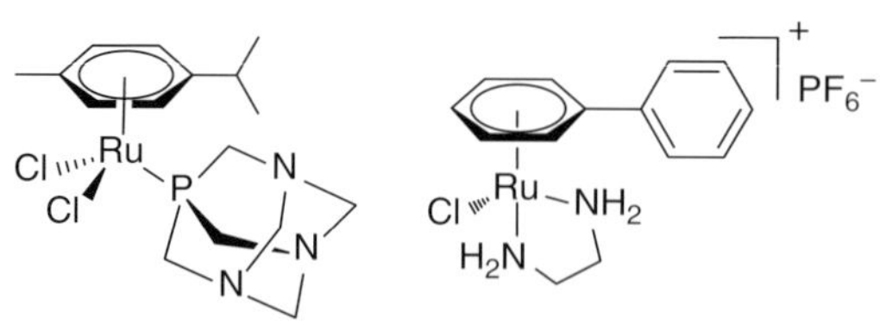


(a) **RAPTA-C** (b) **RM175**

**Figure 14.2** Two representatives of organoruthenium complexes in an advanced preclinical stage of development: RAPTA-C (a) and RM175 (b).

arene and leaving groups were studied for their antiproliferative potential in different human cancer cell lines. It was found that $[Ru(II)(\eta^6\text{-biphenyl})(en)Cl]^+$ (RM175) and $[Ru(II)(\eta^6\text{-}p\text{-cymene})(en)Cl]^+$ exhibit activity in the range of carboplatin ($IC_{50} \sim 6\ \mu M$), and the more lipophilic tetrahydroanthracene analogue $[Ru(II)(\eta^6\text{-tha})(en)Cl]^+$ showed even cytotoxicity in the range of cisplatin ($IC_{50} \sim 0.6\ \mu M$) in A2780 human ovarian cancer cells. The size of the attached arene has a remarkable impact on the cytotoxicity of these compounds [91] due to the ability to intercalate into DNA, the supposed main target of these substances. The complexes hydrolyze rapidly in aqueous solutions, yielding the respective aqua and hydroxido complexes; however, aquation can be suppressed by high chloride concentrations. Therefore, it can be assumed that, similar to cisplatin, the Sadler-type compounds are activated by hydrolysis within the cytosol at low chloride concentrations. Similar to the mode of action of cisplatin, these ruthenium organometallics possess a high affinity to the *N*7 of guanine, but in addition the formed DNA adducts are stabilized by intercalation of the arene ligand and hydrogen bond formation [92]. It seems that the formation of GSH adducts also contributes to their anticancer activity. GSH is responsible for detoxification and resistance processes within the cell; however, the rapidly formed Ru–GSH intermediates are prone to oxidation yielding to the respective sulfenates, which can be replaced by guanine [93]. *In vivo* experiments revealed the potency of RM175 to reduce tumor growth in A2780 ovarian carcinoma xenografts [90]. In further animal experiments, the antitumor and antimetastatic properties against the MCa mammary carcinoma were reported [94].

The general structure of RAPTA complexes allows for a broad range of synthetic possibilities and a remarkable compound library with modifications at the arene, the leaving group, and pta has been reported. RAPTA-C $[Ru(II)(\eta^6\text{-}p\text{-cymene})(pta)Cl_2]$ is the best studied representative of this compound class so far and was shown to hydrolyze quickly in aqueous solution to the corresponding monoaqua complex. In contrast to the Sadler-type organometallics, RAPTA-C showed only moderate activity *in vitro*; however, pronounced antimetastatic and antiangiogenic properties (similar to NAMI-A) in MCa mammary, Ehrlich ascites mammary-bearing mice and A2780 xenografts were observed *in vivo* [95–97]. This different behavior can be explained by the supposed mode of action of RAPTA complexes, which is very different from clinically applied platinum drugs or RM175. The supposed biological target is not DNA, but it seems that interactions with proteins (such as histones) are responsible for the observed biological activity [98]. The observed pronounced antimetastatic properties of this ruthenium complex can be explained by the interference with the underlying mechanisms of metastasis formation. Hydrolysis of the leaving group is an important parameter for metallodrugs; however, too fast aquation can also hamper their bioactivity. Undesired side reactions can be inhibited by coordination of a more stable leaving group; and in the case of RAPTA complexes, the replacement of the chloride ligands by oxalate [99], curcumine [100], or β-diketonates [101] resulted in enhanced stability and cytotoxicity. Selectivity for cancer cells is a major issue in modern chemotherapy and linking of targeting moieties to the "piano-stool" motif might circumvent major drawbacks of classic platinum-based drugs. Recently, maleimide-functionalized

Ru(II)–arene complexes bearing RAPTA or sugar phosphite ligands were shown to rapidly bind to the free thiol group of human serum albumin. The formed HSA adduct is able to selectively accumulate in tumor tissue due to the EPR effect [102]. Also, tethering biotin to the coordinated arene introduces a targeting vector which possibly enhances tumor accumulation [103].

#### 14.3.1.1 Organoruthenium Complexes Bearing Bioactive Ligand Scaffolds

The promising results of Sadler and Dyson had a strong impact on the field of organoruthenium anticancer agents, and a broad range of novel cytotoxic structures within this research field have been reported over the past decade (see Figure 14.3), exploiting the synthetic versatility of this compound class.

A promising approach for the development of novel organometallic anticancer agents is the coordination of bioactive ligand scaffolds to the metal center with the aim of designing metal complexes with novel modes of action, able to interact with multiple biological targets and to exploit arising synergistic effects. Protein kinases were identified as promising targets for cancer therapy, and Meggers and coworkers developed organometallic complexes which are able to mimic the kinase inhibitor staurosporine and show a 100-fold lower $IC_{50}$ value and higher selectivity for the protein kinase Pim-1 than the parent compound [104]. Flavones are well known for their broad range of biological properties and can act as *O,O*-chelates. The respective organoruthenium complexes were found to be highly cytotoxic, and the most potent representative (KP1796) exhibits cytotoxicity in the high nanomolar range. These potential multitargeted compounds are accumulated rapidly in the ER and are both potent inhibitors of topoisomerase IIα and able to bind covalently to DNA. Furthermore, a photoinduced CO release

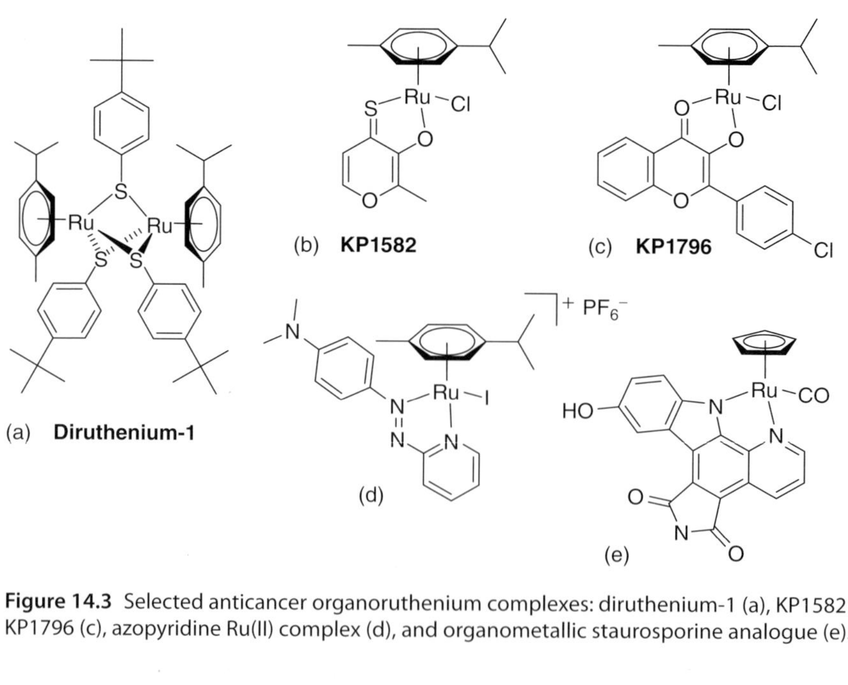


**Figure 14.3** Selected anticancer organoruthenium complexes: diruthenium-1 (a), KP1582 (b), KP1796 (c), azopyridine Ru(II) complex (d), and organometallic staurosporine analogue (e).

was observed; however, the obtained products upon irradiation were found to be inactive *in vitro* [105]. Investigation of a compound library based on this ligand scaffold revealed that the flavone moiety is the cytotoxicity-determining factor and variation of the leaving group or the arene scaffold does not impact relevantly its anticancer activity. Another bioactive *O,O*-chelate is lapachol, which was already investigated in clinical trials for its anticancer potential. It is assumed that the mode of action of lapachol is related to ROS production due to the redox activity of the naphthoquinone backbone. The respective cytotoxic organoruthenium complex is able to induce ROS and induces apoptosis at a higher level than the free ligand. Other reported examples for cytotoxic metal complexes bearing bioactive ligand scaffolds are organometallics with attached paullones [106], indoloquinolines [107], sulfonamides [108], quinolone-based antibiotics [109], or oxicams [110].

The donor atoms of the ligand have a pronounced impact on the stability, ligand exchange kinetics, biological activity, and, generally, on the mode of action of these compounds. For example, thiomaltol, the thionation product of the food additive maltol (E 636), was first reported in the early 2000s and is able to coordinate by its *S,O* donor atoms in a bidentate manner [111]. The respective Ru(II)–arene complex shows remarkable *in vitro* cytotoxicity in the low micromolar range, which is attributed to its enhanced stability under physiological conditions and in the presence of biomolecules compared to its maltol analogue, which exhibits only poor antiproliferative activity [112, 113].

#### 14.3.1.2 Cytotoxic Organoruthenium Complexes without Activation by Aquation

Another class of highly interesting organometallics with *N,N* coordination motif has been developed in the Sadler group. Azopyridine-based ruthenium complexes exhibit pronounced cytotoxicity in several human cancer cell lines and reduce tumor growth in xenograft experiments. Their mode of action differs substantially from the parent ethylenediamine complex because this compound class does not undergo activation by hydrolysis. The mode of action has been proposed to be related to the generation of ROS due to facilitated reduction upon metal coordination of the ligand scaffold by GSH [114].

Recently, Süss-Fink and coworkers reported exciting *in vitro* and *in vivo* results of the dimeric thiolato-bridged ruthenium complex *diruthenium-1* with the lowest reported $IC_{50}$ values (30 nM) so far for ruthenium organometallics. Diruthenium-1 and its analogues show cytotoxicity in both cisplatin-resistant and cisplatin-sensitive cell lines, and interaction studies revealed that these complexes are potent catalysts for oxidation of GSH [115]. However, this first assumption on the mode of action could not be confirmed *in vitro*, and it seems that diruthenium-1 more likely inhibits cellular aerobic metabolism [116].

Overall, the field of anticancer organoruthenium complexes is flourishing with a broad range of novel structures exhibiting promising *in vivo* and *in vitro* activities, and the scientific community is anticipating the first compound of this class to be studied in the clinical setting.

## References


1 Kelman, A.D., Clarke, M.J., Edmonds, S.D., and Peresie, H.J. (1977) *J. Clin. Hematol. Oncol.*, **7**, 274–288.
2 Clarke, M.J. (2003) *Coord. Chem. Rev.*, **236**, 209–233.
3 Reisner, E., Arion, V.B., Keppler, B.K., and Pombeiro, A.J.L. (2008) *Inorg. Chim. Acta*, **361**, 1569–1583.
4 Yan, Y.K., Melchart, M., Habtemariam, A., and Sadler, P.J. (2005) *Chem. Commun.*, **2005**, 4764–4776.
5 Ang, W.H. and Dyson, P.J. (2006) *Eur. J. Inorg. Chem.*, **2006**, 4003–4018.
6 Trondl, R., Heffeter, P., Kowol, C.R., Jakupec, M.A., Berger, W., and Keppler, B.K. (2014) *Chem. Sci.*, **5**, 2925–2932.
7 Bergamo, A. and Sava, G. (2007) *Dalton Trans.*, **2007**, 1267–1272.
8 Sava, G., Pacor, S., Mestroni, G., and Alessio, E. (1992) *Clin. Exp. Metastasis*, **10**, 273–280.
9 Sava, G., Capozzi, I., Clerici, K., Gagliardi, G., Alessio, E., and Mestroni, G. (1998) *Clin. Exp. Metastasis*, **16**, 371–379.
10 Sava, G., Alessio, E., Bergamo, A., and Mestroni, G. (1999) in *Metallopharmaceuticals I* (eds M.J. Clarke and P.J. Sadler), Springer, Berlin, pp. 143–169.
11 Ravera, M., Baracco, S., Cassino, C., Zanello, P., and Osella, D. (2004) *Dalton Trans.*, **2004**, 2347–2351.
12 Brindell, M., Stawoska, I., Supel, J., Skoczowski, A., Stochel, G., and van Eldik, R. (2008) *J. Biol. Inorg. Chem.*, **13**, 909–918.
13 Sava, G., Bergamo, A., Zorzet, S., Gava, B., Casarsa, C., Cocchietto, M., Furlani, A., Scarcia, V., Serli, B., Iengo, E., Alessio, E., and Mestroni, G. (2002) *Eur. J. Cancer*, **38**, 427–435.
14 Webb, M.I. and Walsby, C.J. (2011) *Dalton Trans.*, **40**, 1322–1331.
15 Novohradský, V., Bergamo, A., Cocchietto, M., Zajac, J., Brabec, V., Mestroni, G., and Sava, G. (2015) *Dalton Trans.*, **44**, 1905–1913.
16 Bergamo, A., Messori, L., Piccioli, F., Cocchietto, M., and Sava, G. (2003) *Invest. New Drugs*, **21**, 401–411.
17 Liu, M., Lim, Z.J., Gwee, Y.Y., Levina, A., and Lay, P.A. (2010) *Angew. Chem. Int. Ed.*, **49**, 1661–1664.
18 Bergamo, A., Gagliardi, R., Scarcia, V., Furlani, A., Alessio, E., Mestroni, G., and Sava, G. (1999) *J. Pharmacol. Exp. Ther.*, **289**, 559–564.
19 Gava, B., Zorzet, S., Spessotto, P., Cocchietto, M., and Sava, G. (2006) *J. Pharmacol. Exp. Ther.*, **317**, 284–291.
20 Cochietto, M., Zorzet, S., Sorc, A., and Sava, G. (2003) *Invest. New Drugs*, **21**, 55–62.
21 Pacor, S., Zorzet, S., Cocchietto, M., Bacac, M., Vadori, M., Turrin, C., Gava, B., Castellarin, A., and Sava, G. (2004) *J. Pharmacol. Exp. Ther.*, **310**, 737–744.
22 Sava, G., Zorzet, S., Turrin, C., Vita, F., Soranzo, M.R., Zabucchi, G., Cocchietto, M., Bergamo, A., DiGiovone, S., Pezzoni, G., Sartor, L., and Garbisa, S. (2003) *Clin. Cancer Res.*, **9**, 1898–1905.
23 Bergamo, A. and Sava, G. (2011) *Dalton Trans.*, **40**, 7817–7823.

24 Pillozzi, S., Gasparoli, L., Stefanini, M., Ristori, M., D'Amico, M., Alessio, E., Scaletti, F., Becchetti, A., Arcangeli, A., and Messori, L. (2014) *Dalton Trans.*, **43**, 12150–12155.

25 Aitken, J.B., Antony, S., Weekley, C.M., Lai, B., Spiccia, L., and Harris, H.H. (2012) *Metallomics*, **4**, 1051–1056.

26 Sava, G., Frausin, F., Cocchietto, M., Vita, F., Podda, E., Spessotto, P., Furlani, A., Scarcia, V., and Zabucchi, G. (2004) *Eur. J. Cancer*, **40**, 1383–1396.

27 Frausin, F., Scarcia, V., Cochietto, M., Furlani, A., Serli, B., Alessio, E., and Sava, G. (2005) *J. Pharmacol. Exp. Ther.*, **313**, 227–233.

28 Zorzet, S., Bergamo, A., Cocchietto, M., Sorc, A., Gava, B., Alessio, E., Iengo, E., and Sava, G. (2000) *J. Pharmacol. Exp. Ther.*, **295**, 927–933.

29 Vacca, A., Bruno, M., Boccarelli, A., Coluccia, M., Ribatti, D., Bergamo, A., Garbisa, S., Sartor, L., and Sava, G. (2002) *Br. J. Cancer*, **86**, 993–998.

30 Sanna, B., Debidda, M., Pintus, G., Tadolini, B., Posadino, A.M., Bennardini, F., Sava, G., and Ventura, C. (2002) *Arch. Biochem. Biophys.*, **403**, 209–218.

31 Pintus, G., Tadolini, B., Posadino, A.M., Sanna, B., Debidda, M., Bennardini, F., Sava, G., and Ventura, C. (2002) *Eur. J. Biochem.*, **269**, 5861–5870.

32 Morbidelli, L., Donnini, S., Filippi, S., Messori, L., Piccioli, F., Orioli, P., Sava, G., and Ziche, M. (2003) *Br. J. Cancer*, **88**, 1484–1491.

33 Oszajca, M., Kulis, E., Stochel, G., and Brindell, M. (2014) *New J. Chem.*, **38**, 3386–3394.

34 Magnarin, M., Bergamo, A., Carotenuto, M.E., Zorzet, S., and Sava, G. (2000) *Anticancer Res.*, **20**, 2939–2944.

35 Bacac, M., Vadori, M., Sava, G., and Pacor, S. (2004) *Cancer Immunol. Immunother.*, **53**, 1101–1110.

36 Rademaker-Lakhai, J.M., van den Bongard, D., Pluim, D., Beijnen, J.H., and Schellens, J.H.M. (2004) *Clin. Cancer Res.*, **10**, 3717–3727.

37 Leijen, S., Burgers, S.A., Baas, P., Pluim, D., Tibben, M., van Werkhoven, E., Alessio, E., Sava, A., Beijnen, J.H., and Schellens, J.H.M. (2015) *Invest. New Drugs*, **33**, 201–214.

38 Blunden, B.M., Rawal, A., Lu, H., and Stenzel, M.H. (2014) *Macromolecules*, **47**, 1646–1655.

39 Heffeter, P., Riabtseva, A., Senkiv, Y., Kowol, C.R., Körner, W., Jungwirth, U., Mitina, N., Keppler, B.K., Konstantinova, T., Yanchuk, I., Stoika, R., Zaichenko, A., and Berger, W. (2014) *J. Biomed. Nanotechnol.*, **10**, 877–884.

40 Fischer, B., Heffeter, P., Kryeziu, K., Gille, L., Meier, S.M., Berger, W., Kowol, C.R., and Keppler, B.K. (2014) *Dalton Trans.*, **43**, 1096–1104.

41 Kratz, F., Hartmann, M., Keppler, B.K., and Messori, L. (1994) *J. Biol. Chem.*, **269**, 2581–2588.

42 Sulyok, M., Hann, S., Hartinger, C.G., Keppler, B.K., Stingeder, G., and Koellensperger, G. (2005) *J. Anal. At. Spectrom.*, **20**, 856–863.

43 Kratz, F. and Elsadek, B. (2011) *J. Controlled Release*, **161**, 1–16.

44 Maeda, H., Wu, J., Sawa, T., Matsumura, Y., and Hori, K. (2000) *J. Controlled Release*, **65**, 271–284.

45 Kratz, F. (2014) *J. Controlled Release*, **190**, 331–336.

**46** Heffeter, P., Böck, K., Atil, B., Reza Hoda, M.A., Körner, W., Bartel, C., Jungwirth, U., Keppler, B.K., Micksche, M., Berger, W., and Koellensperger, G. (2010) *J. Biol. Inorg. Chem.*, **15**, 737–748.
**47** Groessl, M., Zava, O., and Dyson, P.J. (2011) *Metallomics*, **3**, 591–599.
**48** Timerbaev, A.R., Rudnev, A.V., Semenova, O., Hartinger, C.G., and Keppler, B.K. (2005) *Anal. Biochem.*, **341**, 326–333.
**49** Smith, C.A., Sutherland-Smith, A.J., Kratz, F., Baker, E.N., and Keppler, B.K. (1996) *J. Biol. Inorg. Chem.*, **1**, 424–431.
**50** Pongratz, M., Schluga, P., Jakupec, M.A., Arion, V.B., Hartinger, C.G., Allmaier, G., and Keppler, B.K. (2004) *J. Anal. At. Spectrom.*, **19**, 46–51.
**51** Heffeter, P., Pongratz, M., Steiner, E., Chiba, P., Jakupec, M.A., Elbling, L., Marian, B., Körner, W., Sevelda, F., Micksche, M., Keppler, B.K., and Berger, W. (2005) *J. Pharmacol. Exp. Ther.*, **312**, 281–289.
**52** Polec-Pawlak, K., Abramski, J.K., Semenova, O., Hartinger, C.G., Timerbaev, A.R., Keppler, B.K., and Jarosz, M. (2006) *Electrophoresis*, **27**, 1128–1135.
**53** Dömötör, O., Hartinger, C.G., Bytzek, A.K., Kiss, T., Keppler, B.K., and Enyedi, E.A. (2013) *J. Biol. Inorg. Chem.*, **18**, 9–17.
**54** Cetinbas, N., Webb, M.I., Dubland, J.A., and Walsby, C.J. (2010) *J. Biol. Inorg. Chem.*, **15**, 131–145.
**55** Webb, M.I., Wu, B., Jang, T., Chard, R.A., Wong, E.W.Y., Wong, M.Q., Yapp, D.T.T., and Walsby, C.J. (2013) *Chem. Eur. J.*, **19**, 17031–17042.
**56** Schluga, P., Hartinger, C.G., Egger, A., Reisner, E., Galanski, M., Jakupec, M.A., and Keppler, B.K. (2006) *Dalton Trans.*, 1796–1802.
**57** Schafer, F.Q. and Buettner, G.R. (2001) *Free Radical Biol. Med.*, **30**, 1191–1212.
**58** Jakupec, M.A., Reisner, E., Eichinger, A., Pongratz, M., Arion, V.B., Galanski, M., Hartinger, C.G., and Keppler, B.K. (2005) *J. Med. Chem.*, **48**, 2831–2837.
**59** Bartel, C., Egger, A.E., Jakupec, M.A., Heffeter, P., Galanski, M., Berger, W., and Keppler, B.K. (2011) *J. Biol. Inorg. Chem.*, **16**, 1205–1215.
**60** Timerbaev, A.R., Foteeva, L.S., Rudnev, A.V., Abramski, J.K., Polec-Pawlak, K., Hartinger, C.G., Jarosz, M., and Keppler, B.K. (2007) *Electrophoresis*, **28**, 2235–2240.
**61** Matczuk, M., Przadka, M., Aleksenko, S.S., Czarnocki, Z., Pawlak, K., Timerbaev, A.R., and Jarosz, M. (2014) *Metallomics*, **6**, 147–153.
**62** Hummer, A.A., Heffeter, P., Berger, W., Filipits, M., Batchelor, D., Büchel, G.E., Jakupec, M.A., Keppler, B.K., and Rompel, A. (2013) *J. Med. Chem.*, **56**, 1182–1196.
**63** Levina, A., Aitken, J.B., Gwee, Y.Y., Lim, Z.J., Liu, M., Singharay, A.M., Wong, P.F., and Lay, P.A. (2013) *Chem. Eur. J.*, **19**, 3609–3619.
**64** Egger, A., Arion, V.B., Reisner, E., Cebrian-Losantos, B., Shova, S., Trettenhahn, G., and Keppler, B.K. (2005) *Inorg. Chem.*, **44**, 122–132.
**65** Küng, A., Pieper, T., Wissiack, R., Rosenberg, E., and Keppler, B.K. (2001) *J. Biol. Inorg. Chem.*, **6**, 292–299.
**66** Groessl, M., Tsybin, Y.O., Hartinger, C.G., Keppler, B.K., and Dyson, P.J. (2010) *J. Biol. Inorg. Chem.*, **15**, 677–688.

67 Kapitza, S., Pongratz, M., Jakupec, M.A., Heffeter, P., Berger, W., Lackinger, L., Keppler, B.K., and Marian, B. (2005) *J. Cancer Res. Clin. Oncol.*, **131**, 101–110.
68 Hartinger, C.G., Jakupec, M.A., Zorbas-Seifried, S., Groessl, M., Egger, A., Berger, W., Zorbas, H., Dyson, P.J., and Keppler, B.K. (2008) *Chem. Biodivers.*, **5**, 2140–2155.
69 Heffeter, P., Atil, B., Kryeziu, K., Groza, D., Koellensperger, G., Körner, W., Jungwirth, U., Mohr, T., Keppler, B.K., and Berger, W. (2013) *Eur. J. Cancer*, **49**, 3366–3375.
70 Kapitza, S., Jakupec, M.A., Uhl, M., Keppler, B.K., and Marian, B. (2005) *Cancer Lett.*, **226**, 115–121.
71 Schoenhacker-Alte, B., Mohr, T., Pirker, C., Kryeziu, K., Kuhn, P. S., Buck, A., Hofmann, T., Gerner, C., Hermann, G., Koellensperger, G., Keppler, B. K., Berger, W., and Heffeter, P. (2017) *Cancer Lett.*, **404**, 79–88.
72 Singh, V., Azad, G.K., Mandal, P., Reddy, M.A., and Tomar, R.S. (2014) *FEBS Lett.*, **588**, 1044–1052.
73 Singh, V., Azad, G.K., Reddy, M.A., Baranwal, S., and Tomar, R.S. (2014) *Eur. J. Pharmacol.*, **736**, 77–85.
74 Stevens, S.K., Strehle, A.P., Miller, R.L., Gammons, S.H., Hoffman, K.J., McCarty, J.T., Miller, M.E., Stultz, L.K., and Hanson, P.K. (2013) *Mol. Pharmacol.*, **83**, 225–234.
75 Kelland, L. (2007) *Cancer*, **7**, 573–584.
76 Frühauf, S. and Zeller, W.J. (1991) *Cancer Res.*, **51**, 2943–2948.
77 Kratz, F., Keppler, B.K., Messori, L., Smith, C., and Baker, E.N. (1994) *Met.-Based Drugs*, **1**, 169–173.
78 Trynda-Lemiesz, L., Karaczyn, A., Keppler, B.K., and Kozlowski, H. (2000) *J. Inorg. Biochem.*, **78**, 341–346.
79 Piccioli, F., Sabatini, S., Messori, L., Orioli, P., Hartinger, C.G., and Keppler, B.K. (2004) *J. Inorg. Biochem.*, **98**, 1135–1142.
80 Zhou, C. and Greenberg, M.M. (2014) *J. Am. Chem. Soc.*, **136**, 6562–6565.
81 Prestwich, E.G., Roy, M.D., Rego, J., and Kelley, S.O. (2005) *Chem. Biol.*, **12**, 695–701.
82 Hartinger, C.G., Zorbas-Seifried, S., Jakupec, M.A., Kynast, B., Zorbas, H., and Keppler, B.K. (2006) *J. Inorg. Biochem.*, **100**, 891–904.
83 Jones, M.R., Mu, C., Wang, M.C., Webb, M.I., Walsby, C.J., and Storr, T. (2015) *Metallomics*, **7**, 129–135.
84 Dickson, N.R., Jones, S.F., Burris, H.A., Ramanathan, R.K., Weiss, G.J., Infante, J.R., Bendell, J.C., McCulloch, W., and von Hoff, D.D. (2011) *J. Clin. Oncol.*, **29** Suppl., Abstr., 2607.
85 Flocke, L.S., Trondl, R., Jakupec, M.A., and Keppler, B.K. (2016) *Invest. New Drugs*, **34**, 261–268.
86 Sadafi, F.Z., Massai, L., Bartolommei, G., Moncelli, M.R., Messori, L., and Tadini-Buoninsegni, F. (2014) *ChemMedChem*, **9**, 1660–1664.
87 Bergamo, A., Masi, A., Jakupec, M.A., Keppler, B.K., and Sava, G. (2009) *Met.-Based Drugs*, **2009**, 681270.

88 Lentz, F., Drescher, A., Lindauer, A., Henke, M., Hilger, R.A., Hartinger, C.G., Scheulen, M.E., Dittrich, C., Keppler, B.K., and Jaehde, U. (2009) *AntiCancer Drugs*, **20**, 97–103.

89 (a) Thompson, D.S., Weiss, G.J., Jones, S.F., Burris, H.A., Ramanathan, R.K., Infante, J.R., Bendell, J.C., Ogden, A., and Von Hoff, D.D. (2012) *J. Clin. Oncol.*, **30**, Suppl., Abstr., 3033; (b) Burris, H.A., Bakewell, S., Bendell, J.C., Infante, J., Jones, S.F., Spigel, D.R., Weiss, G.J., Ramanathan, R.K., Ogden, A., and Von Hoff, D. (2016) *ESMO Open*, **1**, e000154.

90 Aird, R.E., Cummings, J., Ritchie, A.A., Muir, M., Morris, R.E., Chen, H., Sadler, P.J., and Jodrell, D.I. (2002) *Br. J. Cancer*, **86**, 1652–1657.

91 Babak, M.V., Meier, S.M., Legin, A.A., Adib Razavi, M.S., Roller, A., Jakupec, M.A., Keppler, B.K., and Hartinger, C.G. (2013) *Chem. Eur. J.*, **19**, 4308–4318.

92 Chen, H., Parkinson, J.A., Robert, E., Morris, R.E., and Sadler, P.J. (2003) *J. Am. Chem. Soc.*, **125**, 173–186.

93 Wang, F., Xu, J., Wu, K., Weidt, S.K., Logan Mackay, C., Langridge-Smith, P.R.R., and Sadler, P.J. (2013) *Dalton Trans.*, **42**, 3188–3195.

94 Bergamo, A., Masi, A., Peacock, A.F.A., Habtemariam, A., Sadler, P.J., and Sava, G. (2010) *J. Inorg. Biochem.*, **104**, 79–86.

95 Bergamo, A., Masi, A., Dyson, P.J., and Sava, G. (2008) *Int. J. Oncol.*, **33**, 1281–1289.

96 Chatterjee, S., Kundu, S., Bhattacharyya, A., Hartinger, C.G., and Dyson, P.J. (2008) *J. Biol. Inorg. Chem.*, **13**, 1149–1155.

97 Weiss, A., Berndsen, R.H., Dubois, M., Müller, C., Schibli, R., Griffioen, A.W., Dyson, P.J., and Nowak-Sliwinska, P. (2014) *Chem. Sci.*, **5**, 4742–4748.

98 Wu, B., Ong, M.S., Groessl, M., Adhireksan, Z., Hartinger, C.G., Dyson, P.J., and Davey, C.A. (2011) *Chem. Eur. J.*, **17**, 3562–3566.

99 Ang, W.H., Daldini, E., Scolaro, C., Scopelliti, R., Juillerat-Jeannerat, L., and Dyson, P.J. (2005) *Inorg. Chem.*, **45**, 9006–9013.

100 Pettinari, R., Marchetti, F., Condello, F., Pettinari, C., Lupidi, G., Scopelliti, R., Mukhopadhyay, S., Riedel, T., and Dyson, P.J. (2014) *Organometallics*, **33**, 3709–3715.

101 Sersen, S., Kljun, J., Kryeziu, K., Panchuk, R., Alte, B., Körner, W., Heffeter, P., Berger, W., and Turel, I. (2015) *J. Med. Chem.*, **58**, 3984–3996.

102 Hanif, M., Nazarov, A.A., Legin, A., Groessl, M., Arion, V.B., Jakupec, M.A., Tsybin, Y.O., Dyson, P.J., Keppler, B.K., and Hartinger, C.G. (2012) *Chem. Commun.*, **48**, 1475–1477.

103 Babak, M.V., Plażuk, D., Meier, S.M., Arabshahi, H.J., Reynisson, J., Rychlik, B., Błauż, A., Szulc, K., Hanif, M., Strobl, S., Roller, A., Keppler, B.K., and Hartinger, C.G. (2015) *Chem. Eur. J.*, **21**, 5110–5117.

104 Debreczeni, J., Bullock, A.N., Atilla, G.E., Williams, D.S., Bregman, H., Knapp, S., and Meggers, E. (2006) *Angew. Chem. Int. Ed.*, **45**, 1580–1585.

105 Saraf, S.L., Fish, T.J., Benninghoff, A.D., Buelt, A.A., Smith, R.C., and Berreau, L.M. (2014) *Organometallics*, **33**, 6341–6351.

106 Schmid, W.F., John, R.O., Arion, V.B., Jakupec, M.A., and Keppler, B.K. (2007) *Organometallics*, **26**, 6643–6652.

**107** Filak, L.K., Mühlgassner, G., Jakupec, M.A., Heffeter, P., Berger, W., Arion, V.B., and Keppler, B.K. (2010) *J. Biol. Inorg. Chem.*, **15**, 903–918.
**108** Monnard, F.W., Heinisch, T., Nogueira, E.S., Schirmer, T., and Ward, T.R. (2011) *Chem. Commun.*, **47**, 8238–8240.
**109** Kljun, J., Bytzek, A.K., Kandioller, W., Bartel, C., Jakupec, M.A., Hartinger, C.G., Keppler, B.K., and Turel, I. (2011) *Organometallics*, **30**, 2506–2512.
**110** Aman, F., Hanif, M., Siddiqui, W.A., Ashraf, A., Filak, L.K., Reynisson, J., Söhnel, T., Jamieson, S.M.F., and Hartinger, C.G. (2014) *Organometallics*, **19**, 5546–5553.
**111** Lewis, J.A., Puerta, D.T., and Cohen, S.M. (2003) *Inorg. Chem.*, **42**, 7455–7459.
**112** Kandioller, W., Hartinger, C.G., Nazarov, A.A., Kuznetsov, M.L., John, R., Bartel, C., Jakupec, M.A., Arion, V.B., and Keppler, B.K. (2009) *Organometallics*, **28**, 4249–4251.
**113** Kandioller, W., Hartinger, C.G., Nazarov, A.A., Bartel, C., Skocic, M., Jakupec, M.A., Arion, V.B., and Keppler, B.K. (2009) *Chem. Eur. J.*, **15**, 12283–12291.
**114** Dougan, S.J., Habtemariam, A., McHale, S.E., Parsons, S., and Sadler, P.J. (2008) *Proc. Natl. Acad. Sci. U.S.A.*, **105**, 11628–11633.
**115** Giannini, F., Süss-Fink, G., and Furrer, J. (2011) *Inorg. Chem.*, **50**, 10552–10554.
**116** Tomšik, P., Muthná, D., Řezáčová, M., Mičuda, S., Ćmielová, J., Hroch, M., Endlicher, R., Červinková, Z., Rudolf, E., Hann, S., Stíbal, D., Therrien, B., and Süss-Fink, G. (2015) *J. Organomet. Chem.*, **782**, 42–51.

# 15

# Ruthenium Complexes as Antifungal Agents

*Claudio L. Donnici[1], Maria H. Araujo[1], and Maria A. R. Stoianoff[2]*

[1] *Universidade Federal de Minas Gerais, Departamento de Química, Instituto de Ciências Exatas, Av. Antônio Carlos, 6627 – Belo Horizonte, Minas Gerais 31270-901, Brazil*
[2] *Universidade Federal de Minas Gerais, Departamento de Microbiologia, Instituto de Ciências Biológicas, Av. Antônio Carlos, 6627 – Belo Horizonte, Minas Gerais 31270-901, Brazil*

## 15.1 Introduction

Despite the common current control measurements of many diseases, several incidents in the world show that the infectious diseases are not disappearing, but they are reemerging and increasing, and, moreover, new diseases or modifications of the existing diseases have appeared: the emerging infectious diseases (EIDs) [1]. In addition to the loss of hundreds of thousands of lives, these diseases cause massive pain, suffering, and disability, and include a constant feeling of fear and panic in society, as in the case of influenza type A, caused by the H1N1 virus, or in the more recent Ebola case [2]. A new classification of pathogenic species known as "species jump," [3] which initiates when a first human infection caused by a new agent is brought about by a novel or unusual physical contact between potential pathogen and human. Such contacts usually occur because of cultural, social, behavioral, or technological changes on the part of humans. The spread of these new infectious diseases depends on several factors, including environmental or social factors, such as susceptibility to infection, climate, weather, environment, travelling, international trade, poverty, social inequality, and lack of political will. Based on data since 1940, the main sites from the origin of the emergence of infectious diseases have been mapped for zoonotic infections of wild and domestic animals, infections from vectors, and infections from drug-resistant organisms [3, 4].

The reemergence of infectious diseases is also associated with the development of resistance to the usually well-known therapeutic agents and the interruption of previously available measures for infection control [5]. This resistance occurrence, discovered since the 1980s, showed that the EID caused by drug-resistant microorganisms represent about 21% (Figure 15.1) [2b]. In 1997, infectious diseases were the leading cause of death worldwide (33%), among other anthropomorphic causes being the extensive and inappropriate use of antibiotics [2b, 6].

*Ruthenium Complexes: Photochemical and Biomedical Applications,* First Edition.
Edited by Alvin A. Holder, Lothar Lilge, Wesley R. Browne, Mark A.W. Lawrence, and Jimmie L. Bullock Jr.

Although the viral and bacterial diseases are the most known ones, the invasive fungal infections (IFIs) [7], investigated by mycologists, have been a major cause of morbidity and mortality in immunocompromised patients. It should be remembered that the fungal infectious diseases have shown a significant risk to the health of the world's population throughout history [8], and the frequency of IFIs due to opportunistic fungi has increased significantly in the past two decades [9]. It is noteworthy that an opportunistic pathogen is usually harmless (saprobes or saprophytes) in their normal habitat, but they may become pathogenic due to host opportunistic infections, such as several kinds of immunosuppressed patients [10, 11].

In the past 30 years, a significant increase in the incidence of opportunistic fungal infections in hospitals has been noticed worldwide. Serious and fatal infections have been reported, with an increasing range of pathogens causing systemic fungal diseases [12]: isolated yeast of the genus *Candida, Cryptococcus, Trichosporon, Rhodotorula,* and *Pichia;* filamentous fungi of the genus *Aspergillus, Acremonium, Fusarium, Penicillium, Scedosporium, Paecilomyces, Scopulariopsis, Alternaria, Bipolaris, Curvularia, Dactylaria, Exophiala,* and *Phialophora;* other fungi of great interest in the medical field: *Absidia sp., Mucor* sp., *Rhizopus* sp., *Rhizomucor* sp., *Cunninghamella* sp., and *Saksenaea* sp.; causative agents of endemic mycoses that can also cause disease in immunocompromised patients, particularly represented in Brazil by *Histoplasma capsulatum, Coccidioides immitis, Sporothrix schenckii,* and *Paracoccidioides brasiliensis.*

In fact, the incidence of major fungal infections has increased, and among the most serious IFIs stands out candidiasis with an incidence of almost 400%, during the 1980s, in hospitals in the United States. Yeast infections or candidiasis are frequently caused by *Candida albicans* and it may occur as vulvovaginal candidiasis [12g, 13]. The public health problem about IFIs has become alarming, also due to the increase of resistant species of fungi, mainly caused by the widespread use of broad-spectrum antibacterial agents and increase in genetic mutations that lead to resistant species with new mechanisms of action. The most serious and well-known IFIs are candidiasis [14], aspergillosis [15], sporotricosis [16], and paracoccidioidomycosis [17]. Considering the gender of the main species mentioned, the majority of deaths (95%) related to IFIs is associated with *Candida* sp, *Aspergillus* sp. and *Cryptococcus* sp. [18]. Nowadays, indeed, there is no antifungal agent without some restrictions about its utilization, due to the reported toxicity, fungal resistance, or cross-resistance [19]. Amphotericin B (AMB), the bedrock of systemic antifungal therapy [20], and nystatin, another very often used polyene antifungal agent [21], have showed resistance especially in cases of candidiasis. The imidazoles and triazoles ("azoles"), the largest class of antifungal agents in clinical use [22], and even fluconazole, considered during the 1990s "the gold standard" for the treatment of fungal infections, have presented cases of resistance [23]. Itraconazole is also very often used, especially for fluconazole nonsusceptible fungal infections, but this agent is very hydrophobic; toxic and drug-resistant cases have also been observed, since 1997

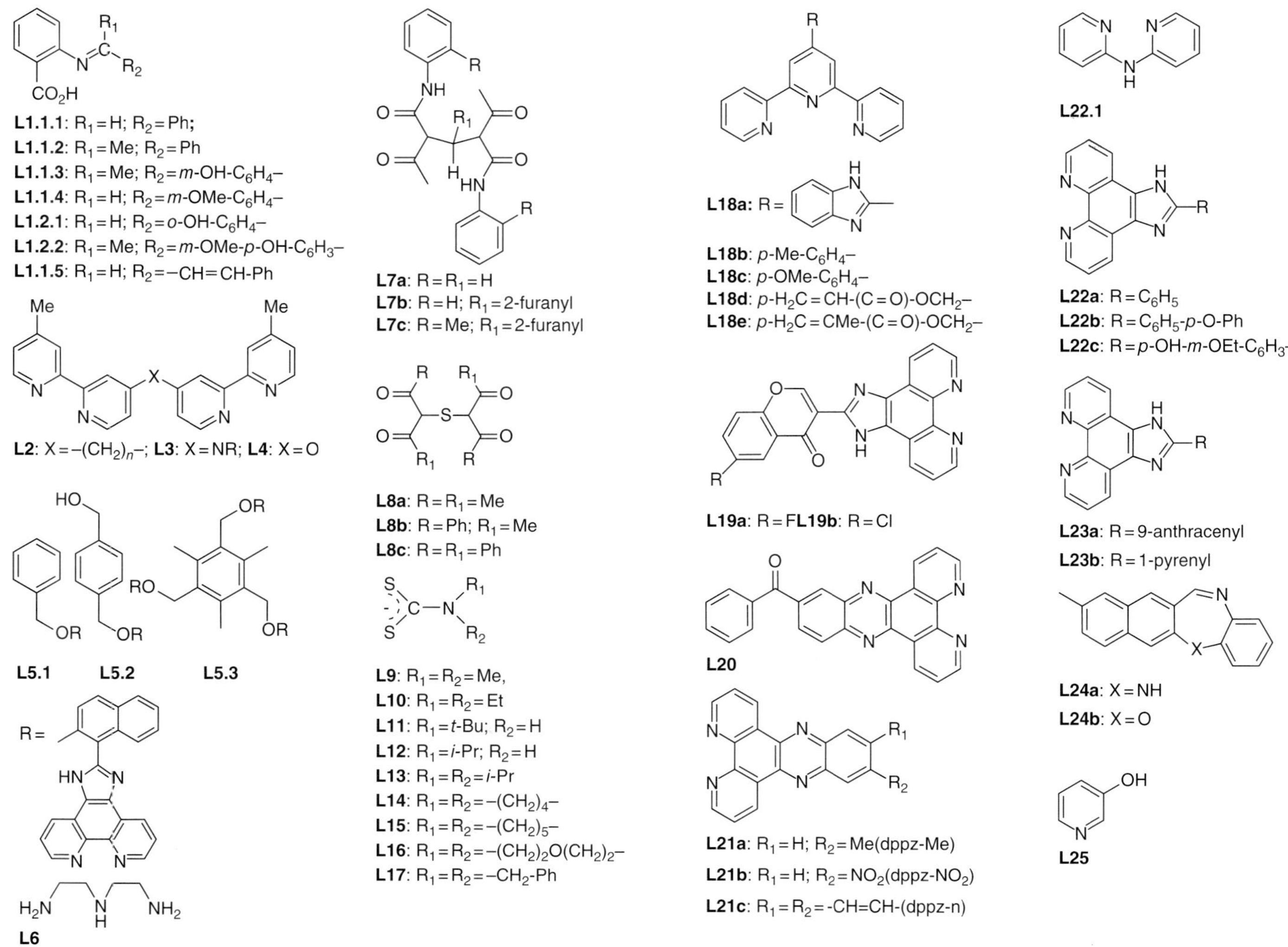


**Figure 15.1** General structures of the pharmacophore ligands.

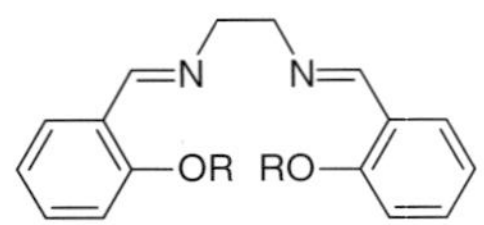

**L26a**: R = H
**L26b**: R = Me

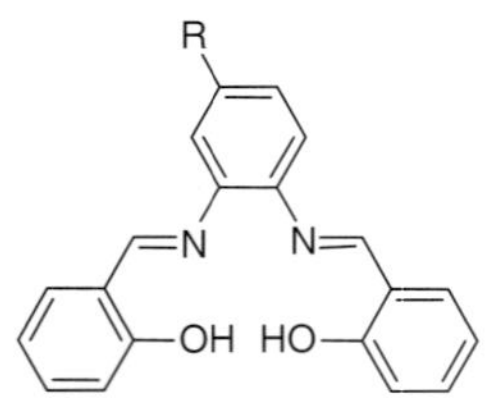

**L26c**: R = H;
**L26d**: R = Me;
**L26e**: R = $NO_2$

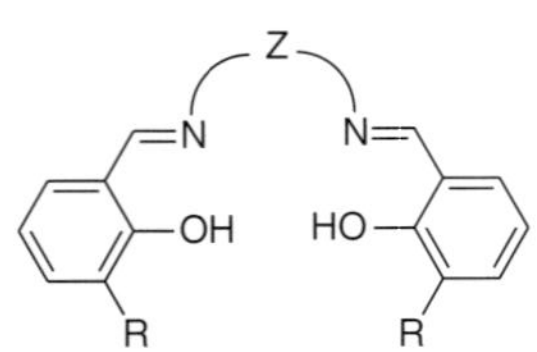

**L26f.1**: R = H, Z = $-(CH_2)_4-$
**L26f.2**: R = H, Z = $-(CH_2CHMe)-$
**L26f.3**: R = OMe, Z = $-(CH_2)_2-$
**L26f.4**: R = OMe, Z = $-(CH_2)_4-$
**L26f.5**: R = OMe, Z = $-(CH_2CHMe)-$
**L26f.6**: R = OMe, Z = *o*-$C_6H_4-$

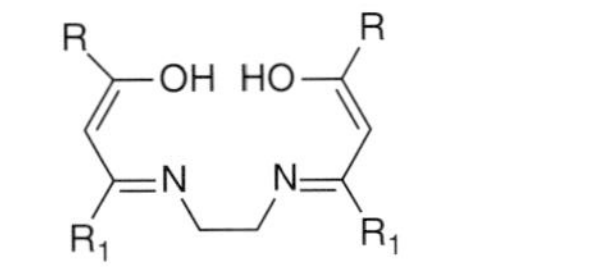

**L27a**: R = $NHC_6H_5$, $R_1$ = Me
**L27b**: R = *o*-(NMe)-$C_6H_4-$, $R_1$ = Me
**L27c**: R = OEt; $R_1$ = Me

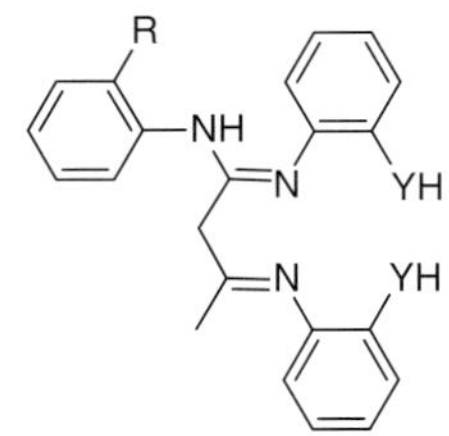

**L28a**: R = H, Y = O;
**L28b**:R = H, Y = S
**L28c**: R = H, Y = $CO_2$;
**L28d**: R = Me, Y = O
**L28e**: R = Me, Y = S;
**L28f**: R = Me, Y = $CO_2$

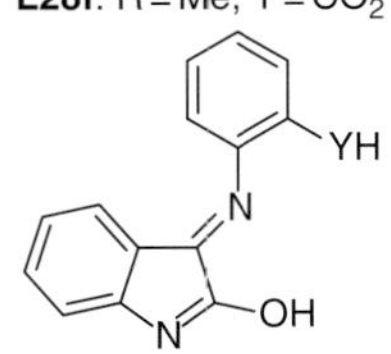

**L29a**: Y = O; **L29b**: Y = S;
**L29c**: Y = $CO_2$

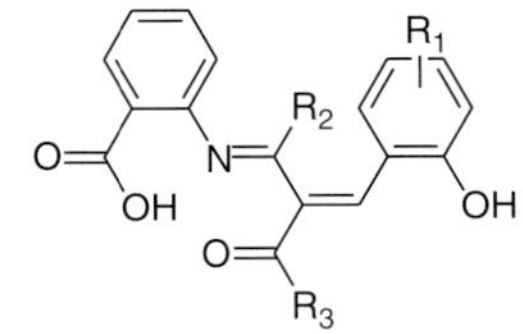

**L30a**: $R_1$ = H, $R_2$ = Me, $R_3$ = OEt
**L30b**: $R_1$ = $-C_4H_4-$, $R_2$ = Me, $R_3$ = OEt

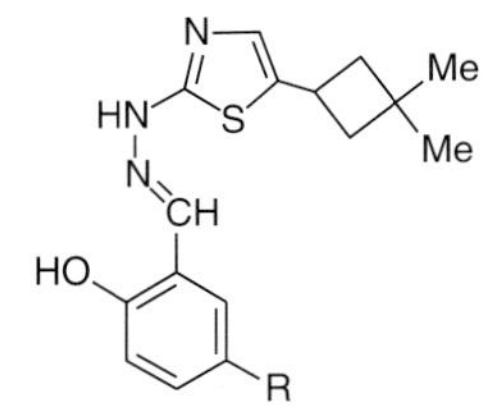

**L31a**: R = H, **L31b**: R = Br

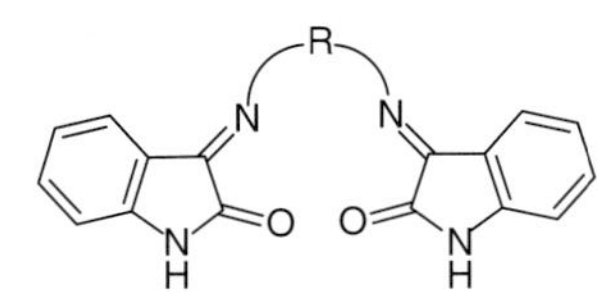

**L32a**: R = $-(CH_2)_2-$; **L32b**: R = $-(CH_2)_3-$;
**L32c**: R = $-(CH_2)_4-$; **L32d**: R = *o*-$C_6H_4-$;
**L32e**: R = *p*-$C_6H_4-$

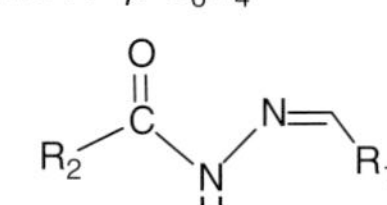

**L33.1a**: $R_1$ = 4-py; $R_2$ = 2-furanyl
**L33.1b**: $R_1$ = 4-py; $R_2$ = 4-py
**L33.1c**: $R_1$ = 4-py; $R_2$ = 2-thienyl
**L33.1d**: $R_1$ = 2-py; $R_2$ = 2-furanyl
**L33.1e**: $R_1$ = 2-py; $R_2$ = 4-py
**L33.1f**: $R_1$ = 2-py; $R_2$ = 2-furanyl
**L33.2a**: $R_1$ = *o*-OH-$C_6H_4-$;
$R_2$ = 2-furanyl
**L33.2b**: $R_1$ = *o*-OH-$C_6H_4-$; $R_2$ = 4-py
**L33.2c**: $R_1$ = *o*-OH-$C_6H_4-$; $R_2$ = 2-thienyl

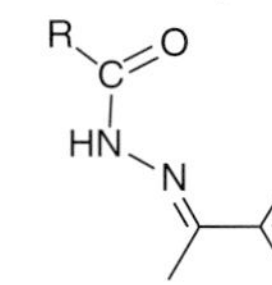

**L34a**: R = Ph
**L34b**: R = 2-py
**L34c**: R = 4-py
**L34d**: R = *o*-Cl-$C_6H_4-$
**L34e**: R = *p*-Cl-$C_6H_4-$
**L34f**: R = *p*-OMe-$C_6H_4-$
**L34g**: R = *p*-$NO_2$-$C_6H_4-$
**L34h**: R = *p*-Me-$C_6H_4-$

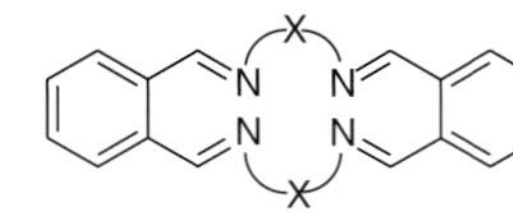

**L35.1a:** X = $-(CH_2)_2-$
**L35.1b:** X = *o*-$C_6H_4-$
**L35.1c:** X = 2,3-py
**L35.1d:** X = 1,8-naphthyl

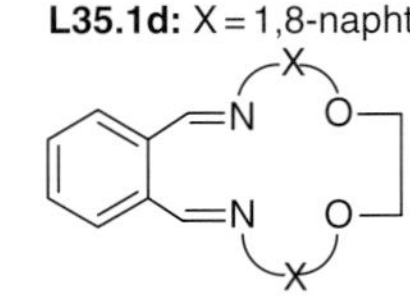

**L35.2b:** X = *o*-$C_6H_4-$
**L35.2c:** X = 2,3-py
**L35.2e:** X = 2,3-naphthyl

**Figure 15.1** (*Continued*)

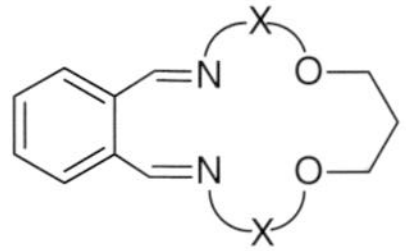

**L35.3b:** X = *o*-$C_6H_4$–
**L35.3c:** X = 2,3-py
**L35.3e:** X = 2,3-naphthyl

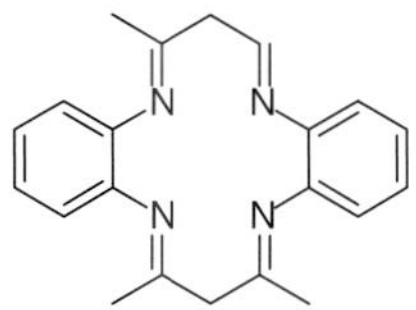

**L36**

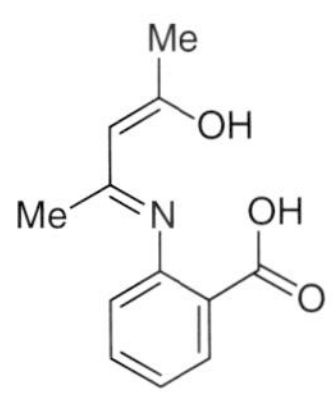

**L37.1**

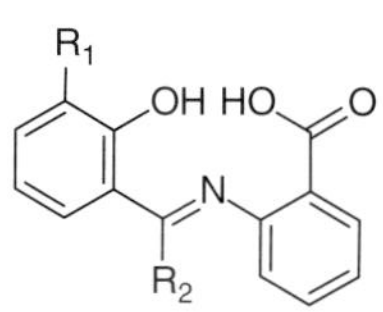

**L37.2a**: $R_1 = R_2 = H$
**L37.2b**: $R_1$ = OMe; $R_2$ = H
**L37.2c**: $R_1$ = H; $R_2$ = Me

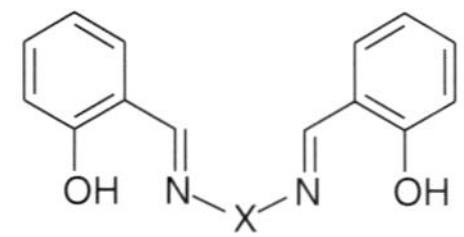

**L38.1a**: X = –$C_6H_4$-$CH_2$-$C_6H_4$–
**L38.1b**: X = (*o*-OMe)-$C_6H_3$-(*o*-OMe)

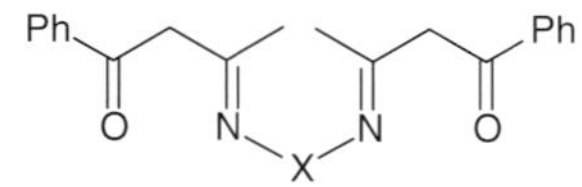

**L38.2a**: X = –$C_6H_4$-$CH_2$-$C_6H_4$–
**L38.2b**: X = (*o*-OMe)-$C_6H_3$-(*o*-OMe)

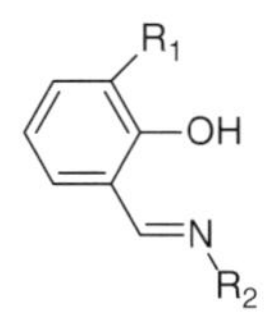

**L39a.1**: $R_1$ = H; $R_2$ = Ph
**L39a.2**: $R_1$ = H; $R_2$ = *o*-Me-$C_6H_4$–
**L39a.3**: $R_1$ = H; $R_2$ = *m*-Me-$C_6H_4$–
**L39a.4**: $R_1$ = H; $R_2$ = *p*-Me-$C_6H_4$–
**L39b.1**: $R_1$ = OMe; $R_2$ = Ph
**L39b.2**: $R_1$ = OMe; $R_2$ = *o*-Me-$C_6H_4$–
**L39b.3**: $R_1$ = OMe; $R_2$ = *m*-Me-$C_6H_4$–
**L39b.4**: $R_1$ = OMe; $R_2$ = *p*-Me-$C_6H_4$–
**L40a**: $R_1$ = H; $R_2$ = Me
**L40b**: $R_1$ = H; $R_2$ = cyclohexyl
**L40c**: $R_1$ = H; $R_2$ = 2-$NH_2$-py-

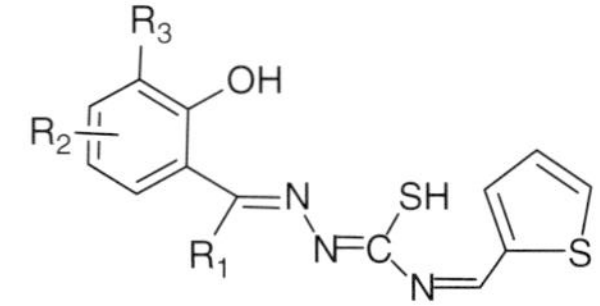

**L41a**: $R_1 = R_2 = R_3 = H$
**L41b**: $R_1$ = Me; $R_2 = R_3$ = H
**L41c**: $R_1 = R_2$ = H; $R_3$ = OMe
**L41d**: $R_1$ = H; $R_2$ = –$C_4H_4$–; $R_3$ = H

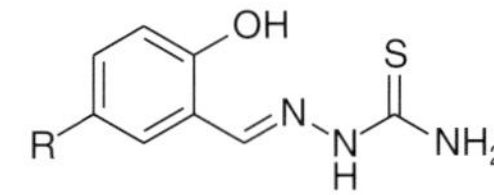

**L42a**: R = Cl
**L42b**: R = $NO_2$

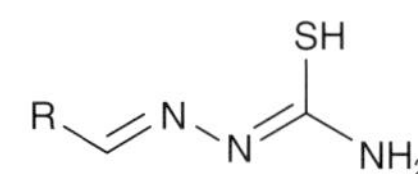

**L43a**: R = 2-furanyl
**L43b**: R = 2-thienyl
**L43c**: R = *p*-OMe-$C_6H_4$–
**L43d**: R = (2,3-$CH_2OCH_2$)-$C_6H_3$–
**L43e**: R = –$(CH_2)_5$–

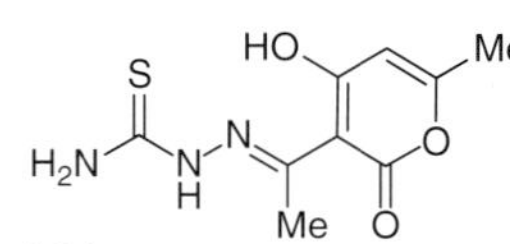

**L44**

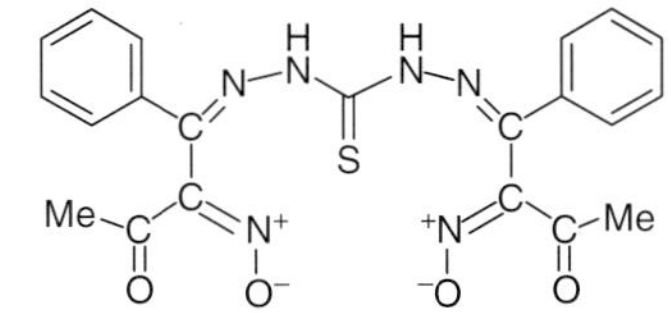

**L45**

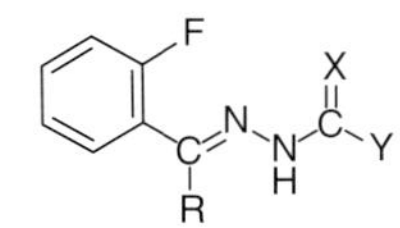

**L46a**: R = H; X = S; Y = $OCH_2Ph$
**L46b**: R = Me; X = O; Y = $NH_2$
**L46c**: R = H; X = O; Y = $NH_2$
**L46d**: R = Me; X = S; Y = $NH_2$
**L46e**: R = H; X = S; Y = $NH_2$

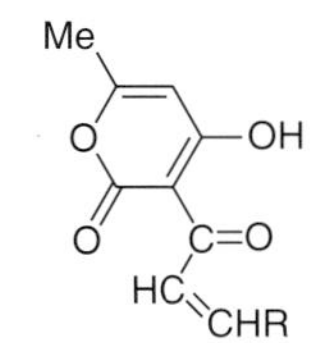

**L47a**: R = *p*-Me-$C_6H_4$
**L47b**: R = *p*-OMe-$C_6H_4$
**L47c**: R = *p*-Cl-$C_6H_4$
**L47d**: R = 3,4-$(OMe)_2$-$C_6H_3$

**Figure 15.1** (*Continued*)

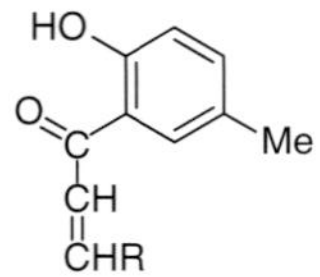

**L48a**: R = $p$-Me-$C_6H_4$–
**L48b**: R = $p$-OMe-$C_6H_4$–
**L48c**: R = $p$-Cl-$C_6H_4$–
**L48d**: R = 3,4-$(OMe)_2$-$C_6H_3$–

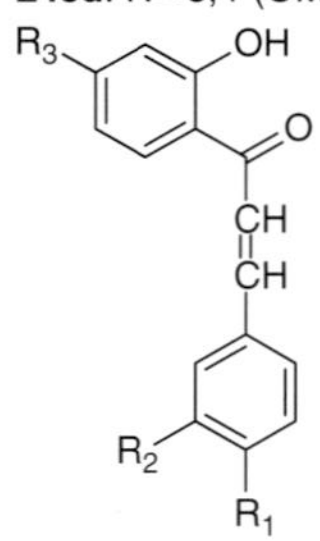

**L49a**: $R_1 = R_2 = R_3 = H$
**L49b**: $R_1$ = OMe; $R_2 = R_3 = H$
**L49c**: $R_1$ = Cl; $R_2 = R_3 = H$
**L49d**: $R_1$ = Me; $R_2 = R_3 = H$
**L49e**: $R_1 = R_2 = R_3$ = OMe
**L49f**: $R_1$ = Me; $R_2$ = H; $R_3$ = OMe
**L49g**: $R_1 = R_3$ = OMe; $R_2$ = H
**L49h**: $R_1$ = Cl; $R_2$ = H; $R_3$ = OMe
**L49i**: $R_1$ = 2-naphthyl; $R_2 = R_3 = H$

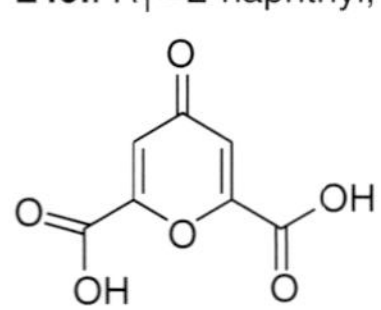

**L50**

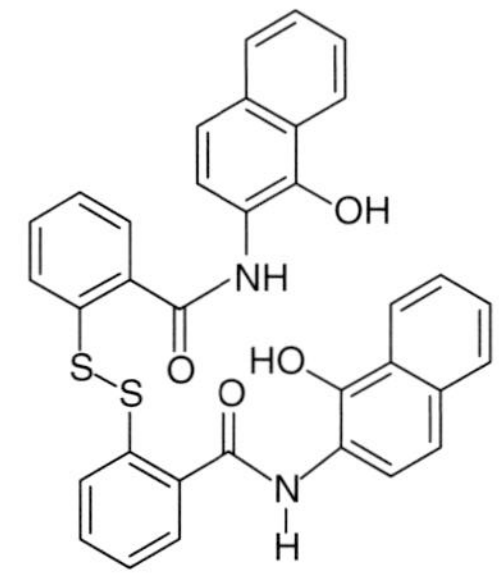

**L51**

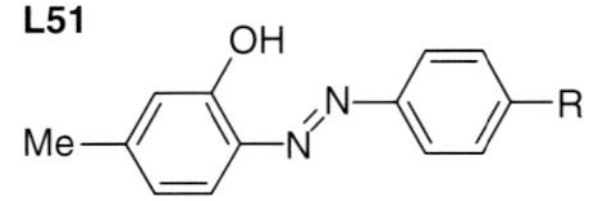

**L52a**: R = H
**L52b**: R = Cl
**L52c**: R = Me
**L52d**: R = OMe
**L52e**: R = OEt

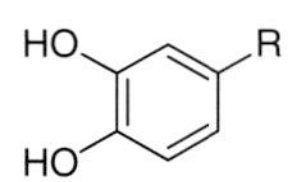

**L53a**: R = $(CH_2)_2CH_2NH_2$
**L53b**: R = $CH(OH)CH_2NHCH(CH_3)_2$
**L53c**: R = $CH(OH)CH_2NH_2$

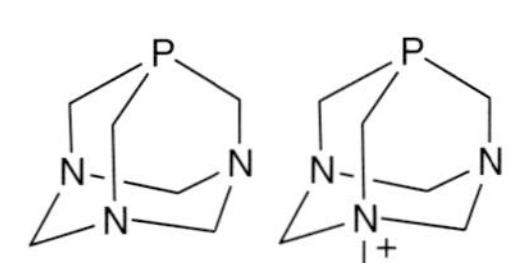

**L54**:- N-Me

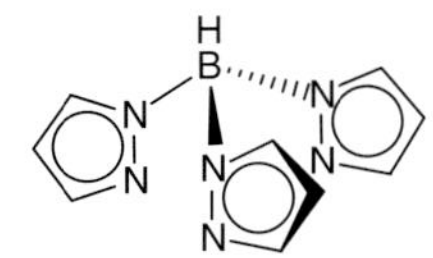

**L55.1**

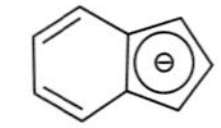

**L55.2**: $\eta^5$-indenyl

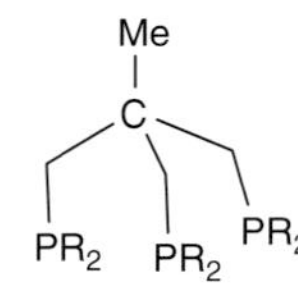

**L56a**: R = $m$-$CF_3$-$C_6H_4$–
**L56b**: R = Ph

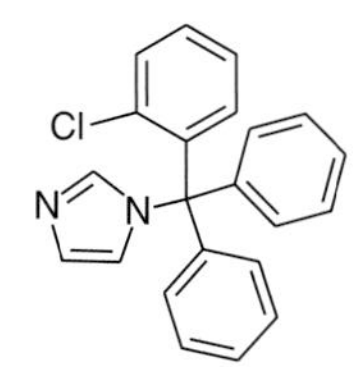

**L57a**

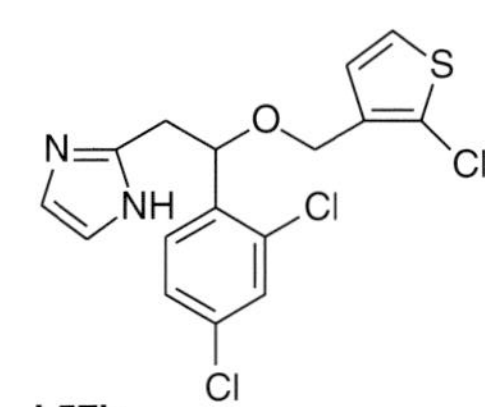

**L57b**

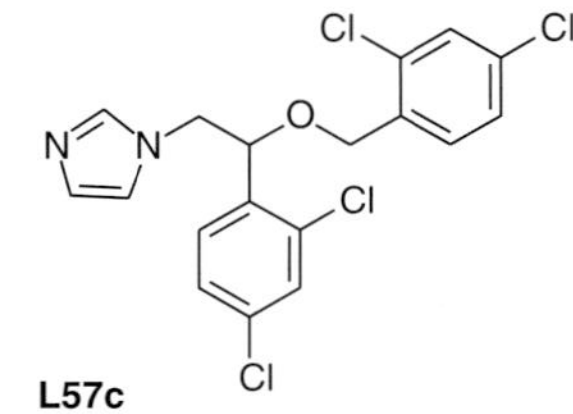

**L57c**

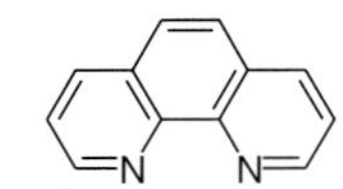

phen

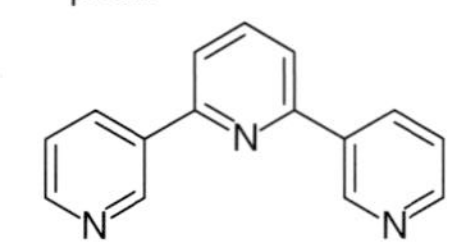

terpy

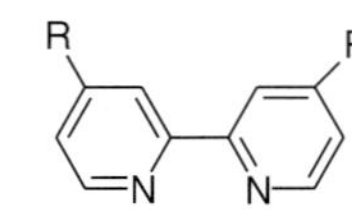

R = H, bipy
R = Me, diMebipy

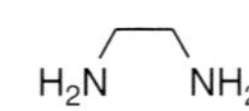

en

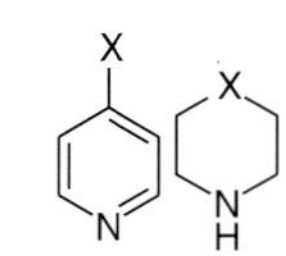

X = H, py X = $CH_2$ pip
X = $NH_2$, ampy X = O mor

**Figure 15.1** (*Continued*)

[24]. The novel generations of triazoles (voriconazole, posaconazole [20b, 25], and ravuconazole [26]) have also presented resistance and cross-resistance [27]. In fact, even the last and the newest class of antifungal agents, the echinocandins (caspofungin, micafungin, and anidulafungin) [28] have presented, although rarely, fungal-resistant cases [29]. Finally, it can be noticed and concluded that the development of new antifungal agents for the treatment of EIDs and IFIs is very important and urgent. Among many novel drug possibilities, metal-based drugs represent a therapeutic alternative against invasive microorganisms [30]. In this way, ruthenium complexes provide a rich platform and suitable building blocks for the design of novel bioactive compounds, once the biological activity of the organic ligand is known, due to the specific properties inherent in the transition metal center [31]. Historically, in 1959, Somers had observed that several different metal ions presented fungitoxicity (against *Alternaria tenuis*), among them ruthenium chloride [32] and, in 1987, the first antifungal activity paper on ruthenium (III) complexes and Schiff bases was published [33]. In this pioneer work, two different types of complexes, **C1.1** [Ru(**L1.1**)$_3$] and **C1.2** ([RuCI(**L1.2**)(H$_2$O)$_2$], were synthesized and their fungicidal activity against *Aspergillus niger* and *Helminthosporium oryzae* was evaluated by the agar plate technique at three concentrations: 1000, 100, and 10 ppm. The complexes **C1.1** were obtained from the coordination with the corresponding monoanionic Schiff bases **L1.1** formed between anthranilic acid and benzaldehyde (**L1.1.1**), acetophenone (**L1.1.2**), *meta*-hydroxyacetophenone (**L1.1.3**), vanillin (**L1.1.4**), and cinnamaldehyde (**L1.1.5**). The **C1.2** analogues were prepared from the bis-deprotonated Schiff bases **L1.2** from anthranilic acid and salicylaldehyde (**L1.2.1**) and *ortho*-hydroxyacetophenone (**L1.2.2**). All these complexes have significant fungitoxicity at 1000 ppm and they presented quite remarkable inhibition against both species of fungus and the complexes are more active (63.3–78.2%) than their corresponding Schiff bases (49.8–56.2%) [33]. Since this first study, almost 100 scientific works about the antifungal activity of ruthenium complexes have been published; and this text shows the reported results in relation to the investigated pathogenic fungus, the most active compounds, the ruthenium oxidation state and the type of the studied ligands. The numbering of the complexes is given according to their appearance in the text and only the investigated pharmacophore ligand is specified; the secondary and common ligands (such as triphenylphosphine, pyridine, and others) are only cited as usual. Table 15.1 summarizes the general formula of all studied complexes, the specification of the investigated fungus, some remarks about the antifungal activity, and presents the related references. Inasmuch, the majority of the studied ruthenium compounds are not structurally characterized by X-ray crystallography and an octahedral geometry is attributed for almost all studied compounds [31a], the general structure of the complexes is similar, and the major difference among them is the presence of the specific pharmacophore ligand. Therefore, only the individual structures of each ligand are shown in Figure 15.1.

**Table 15.1** General formulas of the studied ruthenium complexes with antifungal activity (selected fungus and relevant observations) and references.

| Complex | Ruthenium complexes | Antifungal activity - standard | References |
|---|---|---|---|
| **C1.1, C1.2** | [Ru(**L1.1**)$_3$] (**C1.1**), [RuCI(**L0.2**)($H_2O$)$_2$] (**C1.2**) | *Aspergillus niger, Helminthosporium oryzae* | [33] |
| **C2, C3.1,C.3.2, C4.1, C4.2** | [$Ru_2$Cl(B)$_2$(**L2**)]$Cl_3$ (B = phen, 2,2′,6′,2′-terpy) (**C2**), [$Ru_3$(phen)$_5$(**L3**)$_2$]$Cl_6$ (**C3.1**), [$Ru_3$(phen)$_5$(**L4**)$_2$]$Cl_6$ (**C3.2**), [$Ru_4$(phen)$_6$(**L3**)$_2$]$Cl_8$ (**C4.1**), [$Ru_4$(phen)$_6$(**L4**)$_2$]$Cl_8$ (**C4.2**) | *Alternaria alternata, Aspergillus* sp. *Blastomyces dermatiditis, Candida* sp. *Coccidioides immitis, Cryptococcus neoformans, Epidermophyton floccosum, Histoplasma capsulatum, Malassezia furfur, Microsporum canis, Mucor* sp., *Paracoccidioides brasiliensis, Penicillium marneffei, Pityrosporum ovale, Pneumocystis carinii, Sporothrix schenkii, Trichophyton rubrum, T. interdigitale, Trichosporon beigelii, Rhodotorula* sp. | [34] |
| **C5a-b** | [Ru(phen)$_2$(**L5.1**)]($ClO_4$)$_2$, (**C5a**), [$Ru_2$(phen)$_4$(**L5.2**)]($ClO_4$)$_4$ (**C5b**), [$Ru_3$(phen)$_6$(**L5.3**)]($ClO_4$)$_6$ (**C5c**) | *A. flavus, A. niger, Botrytis cinerea, Curvularia luneta, T. rubrum, T. mentagrophytes* | [35] |
| **C6** | [$RuCl_2$($PPh_3$)$_2$(**L6**)]$_2$[$PdX_2$] (**C6a,C6b**) (**C6a:** X = Cl, **C6b:** X = OAc), [$RuCl_2$($PPh_3$)$_2$($\eta^2$-**L6**)] (**C6c**) | *A. niger, C. albicans, Fusarium oxysporum, Penicillium digitatum* (≅ Nystatin) | [36] |
| **C7** | [$RuX_3$($EPh_3$)(**L7**)$_2$](X = Cl, Br; E = As, P) | *C. albicans, A.niger* | [37] |
| **C8** | [$RuX_2$($EPh_3$)$_2$]$_2$(**L8**)] (X = Cl, Br; E = As, P) | *A. flavus, Fusarium oxysporium, Rhizoctonia solani* | [38] |
| **C9-C17** | [$Ru_2$(**L9-L17**)$_5$] (**L** = $S_2$C-$NR_1R_2$-) | **C9-C17** - *Candida* sp.: *C. albicans* (2 clinical isolates), *C. dubliniensis*(6 clinical isolates), *P. brasiliensis* (7 clinical isolates), *Cryptococcus neoformans, S. schenckii*(MIC ≅ of $10^{-5}$ to $10^{-8}$ mol $ml^{-1}$); **C9-C13** - *Aspergillus* sp. (≅ Fluconazole, < Amphotericin B) | [39, 40] |
| **C18a** | [Ru(**L18a**)(bipy)($H_2O$)]($ClO_4$)$_2$ | *C. albicans, A. alternate, A. niger* ( ≅ Nystatin) | [41] |

| | | | |
|---|---|---|---|
| **C18b, C18c, C18bd, C18be** | $[Ru(\mathbf{L18b})_2](PF_6)_2$, $[Ru(\mathbf{L18c})_2](PF_6)_2$, $[Ru(\mathbf{L18c})(\mathbf{L18d})](PF_6)_2$, $[Ru(\mathbf{L18b})(\mathbf{L18e})](PF_6)_2$ | *Curvularia lunata, F. oxysporum, F. udum, Macrophomina phaseolina, Rhizoctonia solani* (MIC 7-36 μg $ml^{-1}$, Carbendazim (Bavistin) (8-25 μg $ml^{-1}$)) | [42] |
| **C19** | $[Ru(B)_2(\mathbf{L19a\text{-}b})](PF_6)_2 \cdot 2H_2O$ (B = bipy, phen, diMebipy) | *A. niger* (comparable with fluconazole) | [43a] |
| **C20** | $[Ru(B)_2(\mathbf{L20})](PF_6)_2 \cdot 2H_2O$ (B = bipy, phen, diMebipy) | *A. niger* at 1.5 mg $ml^{-1}$ (comparable with fluconazole) | [43b] |
| **C21** | $[Ru(en)_2(\mathbf{L21})](ClO_4)_2$ (en = ethylenediamine) | *Saccharomyces cerevisiae* | [43c] |
| **C22a-c** | $[Ru(\mathbf{L22.1})_2(\mathbf{L22a})]^{2+}$ (**C22a**), $[Ru(\mathbf{L22.1})_2(\mathbf{L22b})]^{2+}$(**C22b**), $[Ru(\mathbf{L22.1})_2(\mathbf{L22c})]^{2+}$(**C22c**) | *Neurospora crassa.* | [43d] |
| **C23a-d** | $[Ru(B)_4(\mathbf{L23a\text{-}b})](ClO_4)_2 \cdot 2H_2O$ (B = py, ampy) | *N. crassa* | [43e] |
| **C24** | $[RuCl_2(\mathbf{L24a\text{-}b})_2]$ | *C. albicans, A. flavus, A. niger* | [44] |
| **C25** | $[Ru(RCO_2)_2(\mathbf{L25})_2)]$ (R = Me, Pr, Ph) | *Agrocyte arvalis, Agaricus bisporus, Actinocorallia herbida.* | [45] |
| **C26b** | $[RuCl_2(\mathbf{L26b})]Cl \cdot 2H_2O$ | *C. albicans* – only the free ligand active against *A. flavus* (≅ Amphotericin B) | [46] |
| **C26 a,c-e** | $[RuCl(\mathbf{L26a,c\text{-}e})(H_2O)]x\ H_2O$ (X = 3,6,7) | *Penicillium vercosum* | [47] |
| **C26f.Ru(III)** | $[RuX(EPPh_3)(\mathbf{L26f})]$ (X = Cl, Br; E = As, P) | *A. niger, Fusarium* sp. | [48] |
| **C26f.Ru(II)** | $[Ru(CO)(B)(\mathbf{L26f})]$ (B = $PPh_3$, py or pip) | *Fusarium* sp | [49] |
| **C27** | $[RuX(EPh_3)(\mathbf{L27})]$ (X = Cl, Br; E = As, P) | *C. albicans, A. niger* (≈ cotrimazine) | [50] |
| **C28** | $[Ru(X)(PPh_3)(\mathbf{L28})]$ (X = Cl, Br) | *A. niger, Botrytis cinerea* | [51] |
| **C29** | $[Ru(X)(EPh_3)_2(\mathbf{L29})]$ (X = Cl, Br; E = As, P) | *A. niger, B. cinerea* (more active than co-trimazole) | [52] |
| **C30** | $[RuCl(CO)(py)(\mathbf{L30})]$ | *A. niger, B. cinerea* (superoxide dismutase activity (SOD) studied) | [53] |
| **C31** | $[Ru(CO)(\mathbf{L31})_2(B)]$ (B = $PPh_3$, $AsPh_3$, py, pip, mor)) | *C. albicans, A. flavus, F. oxysporium, Saccharomyces cerevisiae, R. solani* | [54] |

*(Continued)*

Table 15.1 (Continued)

| Complex | Ruthenium complexes | Antifungal activity - standard | References |
|---|---|---|---|
| **C32** | $[RuCl_2(\mathbf{L32})]Cl$ | *A. niger*, *F. oxysporium* | [55] |
| **C33.1, C33.2** | $[RuCl_2(\mathbf{L33.1})_2]Cl$ (**C33.1**), $[Ru_2(\mu\text{-}Cl)_2Cl_2(\mathbf{L33.2})_2]$ (**C33.2**) | *A. niger*, *F. oxysporium* | [56a] |
| **C34a-h** | $[Ru_2Cl_4(\mathbf{L34})_2]$ | *A. niger*, *F. oxysporium* | [56b] |
| **C35** | $[RuCl_2(\mathbf{L35})]$ | *A. flavus*, *Fusarium* sp. | [57] |
| **C36** | $[RuCl_2(\mathbf{L36})]Cl$ | *Alternaria porri* , *F. oxysporum* | [58] |
| **C37** | $[Ru(CO)(PPh_3)(B)(\mathbf{L37})]$ ($B = PPh_3$, py, pip) | *A. flavus*, *F. oxysporium*, R. solani | [59] |
| **C38** | $[RuX_2(EPh_3)_2(\mathbf{L38})]_2$ ($X = Cl, Br; E = P, As$) | *A. niger*, *F. oxysporium*, *R. solani* | [60] |
| **C39a** | $[RuCl(CO)(\mathbf{L39a})(PPh_3)(B)]$ (B = PPh3, py or pip) | *A. flavus* | [60] |
| **C39b** | $[RuX_2(EPPh_3)_2(\mathbf{L39b})]$ ($X = CI, Br; E = As, P$) | *A. niger*, *Fusarium* sp. | [61] |
| **C40** | $[RuX(EPPh_3)(\mathbf{L40})_2]$ ($X = Cl, Br, E = As, P$) | *A. flavus*, *Fusarium* sp. | [62] |
| **C41** | ($[Ru(CO)(B)(\mathbf{L41a\text{-}d})]$ ($B = PPh_3, AsPh_3$, py) | *C. albicans*, *C. parapsilosis* (eletron donating groups ↑ activity) | [63] |
| **C42** | $[Ru(\mathbf{L42})_2]\cdot 2H_2O$ | *C. albicans*, *A. fumigatus* (≅ echinocandin) | [64] |
| **C43** | $[RuCl(CO)(PPh_3)(\mathbf{L43a\text{-}e})]$ | *A. niger* | [65] |
| **C44** | $[Ru(\mathbf{L44})(CO)(B)(EPh_3)]$ ($B = PPh_3, AsPh_3$, py, pip or mor; $E = As,P$) | *A. niger*, *C. albicans* | [66] |
| **C45** | $[Ru(\mathbf{L44})H_2O]Cl_3$ | *C. albicans*, *T. mentagrophytes*, *T.rubrum* ($MIC > 40\ \mu g\ ml^{-1}$) | [67] |
| **C46** | $[Ru(\mathbf{L46})_3]$ | *Colletotrichum capsici*, *Penicillium notatum*, *Sclerotium rolfsii*(≅ Bavistin) | [68] |

| | | | |
|---|---|---|---|
| **C47** | [RuCl(CO)($EPh_3$)(B)(**L47**)] (BvPPh$_3$, AsPh$_3$, py; E = P, As) | *A. niger*, *Mucor* sp. | [69] |
| **C48** | [RuX$_2$($EPh_3$)$_2$(**L48**)] (X = Cl, Br; E = P, As) | *A. niger*, *Mucor* sp. | [70] |
| **C49.1** | [Ru(CO)(B)(**L49a-d**)$_2$] (B = PPh$_3$, AsPh$_3$, py) | *A. niger* | [71a] |
| **C.49.2** | [RuCl(B)(CO)(EPPh$_3$)(**L49e-h**)] (E = P, As, B = PPh$_3$, AsPh$_3$, py) | *A. niger*, *Mucor* sp. | [71b] |
| **C49.2** | [Ru(B)(CO)(PPPh$_3$)(**L49c,d** or **L49i**)] (B = py, pip) | *A. fumigatus* | [72] |
| **C50a-b** | [Ru(**L50**)(MeCN)(CO)(EPPh$_3$)$_2$] (**C50a**) (**C50a**: E = P, **C50b**: E = As) | *A. niger*, *C. albicans* (MIC values 50-100 mg ml$^{-1}$; comparable to Fluconazole, DNA interactions | [73] |
| **C51** | [Ru$_2$(Cl)$_2$(H$_2$O)(**L51**)] | *A. niger* ($1.6 \times 10^{-4}$ to $8.5 \times 10^{-5}$ mol l$^{-1}$) | [74] |
| **C52a-b** | [RuX(PPh$_3$)$_2$(**L51**)] (**C51a**), [Ru(CO)(PPh$_3$)$_2$(**L51**)] (**C51b**) | *C. albicans* (**C51a**); *C. albicans*, *A. niger* (**C51b**) | [75, 76] |
| **C53** | [Ru(NH$_3$)$_4$(**L52**)]PF$_6$ | *C. albicans* (three isolates), *C. glabrata*, *C. tropicalis* | [77] |
| **C54** | [RuX$_2$($\eta^6$-*p*-cymene)(**L54**)] (**C54a**:X = Cl, **b**:X = Br, **c**:X = I, **d**:X = NCS) | *Trichrophyton mentagrophytes* (**C54a**);*Cladosporium resinae* (**C54d**) | [78] |
| **C55g-i** | [Ru(MeCN)(PPh$_3$)(**L55.1**)(**L54**)$_2$] PF$_6$ (**C55g**), [Ru(X)(PPh$_3$)((**L55.1**)(**L54-N-Me**)]OTf (**C55h**:X = Cl, **C55i**:X = H) | *Rhodosporidium toruloides* | [79] |
| **C56.1–C56.3** | [Ru(CF$_3$CO$_2$)$_2$(**L56a**)] (**C56.1**), [Ru(MeCN)$_3$(**L56a**)]OTf$_2$ (**C56.2**), [Ru(MeCN)$_3$(**L56b**)]OTf (**C56.3**) | *C. albicans*, *C. tropicalis* (125-250 μg ml$^{-1}$) | [80] |
| **C57.1a-c, C57.2a-c** | [RuCl$_2$($\eta^6$-*p*-cymene)(**L57a-c**)] (**C57.1a-c**), [RuCl($\eta^6$-*p*-cymene)(**L57 a-c**)$_2$]Cl (**C57.2a-c**), [Ru[($\eta^6$-*p*-cymene)(**L57a-c**)$_3$](PF$_6$)$_2$ (**C57.3 a-c**) | *Culvularia lunata* (activity order **C57.3 > C57.2 > C57.1** | [81] |

## 15.2 Antifungal Activity Investigations of Ruthenium Complexes

### 15.2.1 Ruthenium Complexes with Activity against Several Pathogenic Fungi Species: Dinuclear, Trinuclear, and Tetranuclear ruthenium Polydentate Polypyridil ligands, Heterotrimetallic di-Ruthenium-Mono-Palladium Complexes, Dinuclear bis-β-Diketones and Pentadithiocarbamate Ligands

The most interesting and broad-spectrum investigation about antimicrobial activity of ruthenium complexes that indicates the high potentiality of ruthenium complexes as antibacterial and antifungal agents is described in a quite recent patent (2013) [34]. In this work, all synthesized complexes $[Ru_2Cl(B)_2(\mathbf{L2})]Cl_3$ (**C2**) (B = 1,10-phenantroline (phen), 2,2′,6′,2″-terpyridine (terpy)), $[Ru_3(phen)_5(\mathbf{L3})_2]Cl_6$ (**C3.1**), $[Ru_3(phen)_5(\mathbf{L4})_2]Cl_6$ (**C3.2**), $[Ru_4(phen)_6(\mathbf{L3})_2]Cl_8$(**C4.1**), and $[Ru_4(phen)_6(\mathbf{L4})_2]Cl_8$ (**C4.2**) were tested for antifungal and antibacterial activity. The inventors reported that **C2, C3** or **C4** ruthenium(II) phenanthroline-bridged di-bipyridine complexes may be used for the treatment and/or prevention of a range of microbial infections, including resistant microbial species. According to the reported data, these complexes present a broad-spectrum action against almost all known clinically relevant fungal species (Table 15.1) and the studied complexes showed high lipophilicity, measured as $\log P$ value. The complexes have low toxicity toward human or animal eukaryotic cells and for the hemolytic activity, the Ru(II) complexes were more concentration dependent than time dependent and exhibited low toxicity to human red blood cells. Another study about the antifungal activity investigation (against *Aspergillus flavus, A. niger, Botrytis cinerea, Curvularia lunata, Trichophyton rubrum,* and *Trichophyton mentagrophytes*) of mono-, di-, and trinuclear ruthenium(II) imidazoyl-substituted phenanthroline complexes $[Ru(phen)_2(\mathbf{L5.1})](ClO_4)_2$ (**C5.1**),$[Ru_2(phen)_4(\mathbf{L5.2})](ClO_4)_4$ (**C5.2**) and $[Ru_3(phen)_6(\mathbf{L5.3})]$ $(ClO_4)_6$ (**C5.3**) was recently published [35].

New ruthenium heterotrinuclear complexes Ru(II)—Pd(II)—Ru(II), $[RuCl_2(PPh_3)_2(\mathbf{L6})]_2[PdX_2]$ (**C6**) (**C6a**: X = Cl, **C6b:** X = OAc), as well the corresponding Ru(II) mononuclear analogues $[RuCl_2(PPh_3)_2(\eta^2\text{-}\mathbf{L6})]$ (**C6c**), were screened against *A. niger, C. albicans, Fusarium oxysporum,* and *Penicillium digitatum,* and the results were quite comparable to the usually antifungal clinical agent Nystatin [36].

There is another paper about antifungal activity of dinuclear Ru(II) complexes of the type $[RuX_3(EPh_3)(\mathbf{L7})_2]$ (**C7**) (X = Cl, Br; E = As, P). All studied compounds presented quite remarkable antifungal activity against *C. albicans* and *A. niger* [37]. In a previous study [38], 12 other analogous hexacoordinated dinuclear ruthenium(III) complexes of the type $[RuX_2(EPh_3)_2]_2(\mathbf{L8})]$ (**C8**) (X = Cl, Br; E = As, P) were synthesized. The antifungal activity for the free ligands and for some of the complexes was screened *in vitro* with 7-day-old cultures of *A. flavus, Fusarium oxysporium,* and *Rhizoctonia solani* showing quite reasonable results. Although there is a sufficient increase in the fungicidal activity of ruthenium complexes as compared to free ligands, it cannot attain the effectiveness of the conventional fungicide bavistin (carbendazim).

Remarkable broad-spectrum antifungal activity was discovered when nine dinuclear ruthenium dithiocarbamate (DTCB) coordination compounds [$Ru_2$(**L9-L17**)$_5$] (**C9-C17**), and their corresponding free ligands, were tested and compared with amphotericin B and fluconazole against five different fungal species with clinical interest and related IFI, such as [39] *Candida* species (*C. albicans,* two clinical isolates, *C. glabrata, C. krusei, C. parapsilosis, C. tropicalis, C. dubliniensis,* six clinical isolates), *P. brasiliensis* (seven clinical isolates), *Cryptococcus neoformans, S. schenckii,* and also against seven different species of *Aspergillus* [40] (*A. clavatus, A. flavus, A. fumigatus, A. niger, A. nomius, A. tamarii,* and *A. terreus*). Almost all compounds showed good antifungal activity results (MIC, minimum inhibitory concentrations, in values on the order of $10^{-5}$ mol $l^{-1}$ to $10^{-8}$ mol $ml^{-1}$) against all fungal species, except for *Candida glabrata,* the less susceptible fungal species. In some cases, the obtained MIC values for the antifungal activity of these complexes were close to or even lower than the obtained MIC value for the classic clinically used antifungal agent fluconazole; besides, the cytotoxic assays ($IC_{50}$) showed that the complexes were not so toxic ($IC_{50}$ values were much higher than MIC values). In fact, the ruthenium dinuclear pentakis-dithiocarbamate complexes can be considered as potential novel antifungal agents with high potency and low cytotoxicity for the development of new drugs to attack the terrible worldwide health problem of IFIs. Sreeramulu and coworkers [82] have also reported the high potentiality as antifungal agent of dithiocarbamate ligands coordinated to ruthenium.

### 15.2.2 Aromatic and Heteroaromatic Ligands in Ru Monometallic Centers (Pyridine, Phenantroline, Terpyridine, Quinoline, and Phenazine)

Among many different types of ligands, the aromatic and heteroaromatic rings are well known and have many different applications, the pyridyl ruthenium complexes usually are an alternative to pharmacologically active novel compounds [83], and, as cited in this text previously, some polymetallic ruthenium complexes are really very promising antifungal agents [33–35].

The antifungal study of ruthenium mononuclear-centered complexes have also been explored and the antifungal study of Ru(II) coordinated with 4-substituted-terpyridine [Ru(bipy)($H_2O$)(**L18a**)]$(ClO_4)_2$ (**C18a**) (bipy = 2,2′-bipyridine) showed that *C. albicans, Alternaria alternata,* and *A. niger* are very susceptible to its presence, even more than to nystatin, and these microorganisms are much less susceptible to the free ligands [41]. Four other 4-substituted-terpyridines **L18b-e** were synthesized and the obtained two symmetrical complexes, [Ru(**L18b**)$_2$]$(PF_6)_2$ (**C18b**) and [Ru(**L18c**)$_2$]$(PF_6)_2$ (**C18c**) and two unsymmetrical [Ru(**L18b**)(**L18d**)]$(PF_6)_2$ (**C18bd**) and [Ru(**L18b**)(**L18e**)]$(PF_6)_2$ (**C18be**), exhibited quite good antifungal activity against most of the tested pathogens and, remarkably, in some cases their activity (MIC values from 7 to 36 μg $ml^{-1}$) was better than the fungicide carbendazim (bavistin) (8–25 μg $ml^{-1}$) [42].

Satyanarayana and coworkers [43] have studied several kinds of azoheteroaromatic ligands: (i) two halo-chromenone-imidazolyl-substituted phenanthroline ligands **L19a** and **L19b** that were used to prepare Ru(II) complexes with general formula $[Ru(B)_2(\mathbf{L19a})](PF_6)_2{\cdot}2H_2O$ **(C19a)** and $[Ru(B)_2(\mathbf{L19b})](PF_6)_2{\cdot}2H_2O$ **(C19b),** (B = bipy, phen and 4,4′-dimethyl-2,2′-bipyridine (diMebipy)). The DNA binding to calf-thymus DNA (CT-DNA) has been investigated; besides, a remarkable antifungal activity similar to fluconazole for all Ru(II) studied complexes was observed, especially for the most lipophilic analogues $[Ru(phen)_2(\mathbf{L19a})]^{2+}$ (**C19a.3**) and $[Ru(phen)_2(\mathbf{L19b})]^{2+}$ (**C19b.3**). It was also observed that under irradiation at 365 nm, the complexes promote the photocleavage of plasmid pBR 322 DNA and that inhibition studies suggest that singlet oxygen plays a significant role in the cleavage mechanism of Ru(II) complexes [43a]. (ii) Other similar ruthenium compounds $[Ru(B)_2(\mathbf{L20})](PF_6)_2{\cdot}2H_2O$ (**C20**) (B = bipy, diMebipy, phen) with substituted phenazine-phenantroline ligand BDPPZ (**L20**), $[Ru(bipy)_2(\mathbf{L20})]^{2+}$ (**C20.1**), $[Ru(diMebipy)_2(\mathbf{L20})]^{2+}$ (**C20.2**), and $[Ru(phen)_2(\mathbf{L20})]^{2+}$ (**C20.3**) showed an appreciable activity against *A. niger* at 1.5 mg $ml^{-1}$ concentration, comparable with the standard drug fluconazole; in addition, as in the previous study, the most lipophilic derivative $[Ru(phen)_2(\mathbf{L20})]^{2+}$ was the most active one and photocleavage studies were also carried out [43b]. (iii) Three condensed phenanthroline-phenazines **L21,** two 7-substituted dipyridophenazines (DPPZ) (**L21a**: 7-Me, **L21b:** 7-$NO_2$), and one benzo-dipyrido-phenazine (**L21c**) and their corresponding ruthenium(II) complexes $[Ru(en)_2(\mathbf{L21})][ClO_4]_2$ (**C21**) (en = ethylenediamine) were also investigated, but the antifungal activity was only demonstrated against *Saccharomyces cerevisiae* [43c]. (iv) Other functionalized phenanthroline derivatives **L22**were studied as mixed ligands to ruthenium(II) coordination compounds $[Ru(\mathbf{L22.1})_2(\mathbf{L22})](ClO_4)_2{\cdot}2H_2O$ (**C22**); and all three complexes $[Ru(\mathbf{L22.1})_2(\mathbf{L22a})]^{2+}$ (**C22a**), $[Ru(\mathbf{L22.1})_2\ (\mathbf{L22b})]^{2+}$ (**C22b**), and $[Ru(\mathbf{L22.1})_2(\mathbf{L22c})]^{2+}$ (**C22c**) demonstrated that they can bind to DNA through the intercalation mode and all the studied compounds showed good activity against *Neurospora crassa* [43d]. (v) Mixed polypyridyl Ru(II) complexes (**C23a-C23d**) $[Ru(B)_4(\mathbf{L23a\text{-}L23b})](ClO_4)_2{\cdot}2H_2O$ (**C23a**: B = 4-$NH_2$-pyridine (ampy), **L23a; C23b:** B = py, **L23a; C23c:** B = ampy, **L23b; C23d:** B = py, **L23b**) also demonstrated inhibition toward *N. crassa* [43e].

Functionalized condensed benzoxa- and benzodiazepine-quinoline derivatives **L24a** and **L24b** were also studied as ligands for obtaining ruthenium(III) complexes $[RuCl_2(\mathbf{L24a})_2]$ (**C24a**) and $[RuCl_2(\mathbf{L24b})_2]$ (**C24b**) and these compounds showed slight inhibition against *C. albicans, A. flavus,* and *A. niger* [44].

Ruthenium carboxylates, such as acetate, butyrate, and benzoate ($RCO_2^-$), can be used as ligands to prepare new compounds, and in this way the reaction with 3-(or *beta-*)hydroxypyridine (3-pyOH) (**L25**) led to the corresponding Ru(II) complexes $[Ru(RCO_2)_2(\mathbf{L25})_2)]$ (**C25**) that showed antifungal activity against *Agrocyte arvalis, Agaricus bisporus,* and *Actinocorallia herbida* [45].

### 15.2.3 Schiff bases, Thiosemicarbazones, and Chalcones

#### 15.2.3.1 Schiff bases (Tetradentate Salen Like, Tridentate, and bidentate)

The imines ($R_2C{=}NR$) are generally well known as Schiff bases since 1864, from the family name of its discoverer, the German and Italian naturalized chemist Hugo Schiff [84]. These nitrogen ligands displayed a broad range of biological activities, and they are also remarkable as "privileged" ligands in the organometallic complexes [43e, 85]. The Salen (contraction for **sal**icylaldehyde and ethyl**en**ediamine) ligand **L26a** was first prepared in 1933 [86], and it is a particular class of Schiff bases obtained when two equivalents of salicylaldehyde are combined with a diamine that presents four coordinating sites and two axial sites open to ancillary ligands. In fact, the Salen type is used in the literature to describe the class of [*O*,*N*,*N*,*O*] tetradentate bis-Schiff-base ligands [87]. Although, Ru(salen) coordination compounds have been prepared before [85a, 88], the antifungal activity investigation of the ruthenium(III) complexes [$RuCl_2$(**L26b**)]Cl·$2H_2O$ (**C26b**) with *O,O′*-dimethyl-Salen-type Schiff bases **L26b** was made more recently [46], and they presented significant antimicrobial activity against *C. albicans* with inhibition comparable to AMB. However, only the free ligand was active against *A. flavus.* Minor cytotoxicity records were reported at the highest concentration level using MTT assay. Besides **L26a,** in another study [47], three other analogous functionalized Salen Schiff bases **L26c-e** were also used as ligands for ruthenium(III) complexes [$RuCl_2$(**L26c-e**)]Cl·$2H_2O$ (**C26c-e**), and all four tested derivatives showed high effect only against *Penicillium verrucosum.* Bis-(*o*-substituted) Salen Schiff bases **L26f,** and also the **L26a** and **L26b** derivatives, as ligands in Ru(III) complexes [RuX($EPPh_3$)(**L26f**)] (**C26f·Ru(III)**) [48] (X = Cl, Br; E = As, P) showed a little larger broad spectrum with antifungal activity against *A. niger* and *Fusarium* sp. However, if Ru(II) complexes are produced [Ru(CO)(B)(**L26f**)] (**C26f·Ru(II)**) [49] (B = $PPh_3$, py or piperidine = pip), the observed activity is only against *Fusarium* sp.

Homologous tetradentate Schiff bases **L27** were used as a bioactive ligand to prepare low-spin Ru(III) octahedral complexes [RuX($EPh_3$)(**L27**)] (**C27**) (X = Cl, Br; E = P, As) that were active against *C. albicans* and *A. niger.* Almost all compounds were active, while the free ligands were not, and in some cases the activity (inhibition diameter zone) was comparable to cotrimazine, the used standard antifungal [50]. Three other analogous Salen-type complexes that presented antifungal activity using Ru(III) and Ru(II) were investigated by Arunachalam and coworkers. In the first study, the complex **C28** with Ru(III) [Ru(X)($PPh_3$)(**L28**)] (X = Cl, Br) showed the *in vitro* cytotoxicity against *A. niger and Botrytis cinerea* at 0.25, 0.50, and 1% concentrations and all ruthenium(III) Schiff-base complexes **C28** are more active than the free ligands, ruthenium(III) precursors, and standard reference (co-trimoxazole) [51]. Other studied Ru(III) complexes [Ru(X)$(EPh_3)_2$(**L29**)] (**C29**) (X = Cl, Br; E = P, As) presented similar *in vitro* cytotoxicity against *A. niger and B. cinerea* [52]. Finally, the diamagnetic Ru(II) complexes of the type [RuCl(CO)(py)(**L30**)] (**C30**) also presented high activity against *A. niger* and *B. cinerea*; in addition, the superoxide dismutase activity (SOD) was examined [89]. It is noteworthy that it is ever mentioned that Salen coordination compounds can have their biological activity related to SOD [53].

Other polyfunctionalized Schiff-base ligands **L31** studied as Ru(II) complexes [Ru(CO)(B)($\mathbf{L31}$)$_2$] (**C31**) (B = $PPh_3$, $AsPh_3$, py, pip, morpholine (mor)) showed antifungal activity against *C. albicans, A. flavus, F. oxysporium, S. cerevisiae,* and *R. solani* [54].

Sharma and Shivastava studied tetradentate Schiff bases **L32** derived from condensation between isatin and α,ω-diamines (1,3-diamino-propane, 1,4-diaminobutane, 1,2-diamino-benzene and 1,3-diamino-benzene) [55] that generated the corresponding Ru(III) isatin tetradentate Schiff-base complexes [$RuCl_2$(**L32**)]Cl (**C32**), which showed quite reasonable growing inhibition against *A. niger* and *F. oxysporium.* The same authors also studied two other different functionalized Schiff bases, the *N*-acyl-hydrazones **L33.1** and **L33.2** [56], as ligands to coordinate with Ru(III) in order to investigate the antifungal activity of the corresponding complexes. The six tridentate 2- and 4-pyridinyl-functionalized ligands generated the corresponding ruthenium complexes **C33.1** [$RuCl_2$($\mathbf{L33.1}$)$_2$]Cl as mononuclear species. On the other hand, with the three tetradentate salicyloyl-substituted ligands the corresponding dinuclear complexes **C33.2** [$Ru_2$(μ-Cl)$_2$$Cl_2$($\mathbf{L33.2}$)$_2$] were obtained [56a]. The antifungal activity was investigated and both fungi *F. oxysporium* and *A. niger* were susceptible to all studied *N*-acyl-hydrazone derivatives more than to the free ligands; but the standard antifungal agent, redomil, was more active in all cases. It can be noticed that the three dinuclear complexes present a slightly higher antifungal activity than the other mononuclear analogues [56a]. The third investigated Schiff base, formed by diacetylmonooxime functionalized ligands **L34,** generated eight dinuclear Ru(III)—Ru(III) coordination compounds **C34a-h** with general formula [$Ru_2Cl_4$($\mathbf{L34}$)$_2$] that presented activity against *A. niger* and *A. alternatum* more than the corresponding free ligands [56b].

Ten of the biologically active tetradentate macrocyclic Schiff-base ligands **L35** produced the corresponding complexes with Ru(II) [$RuCl_2$(**L35**)] (**C35**) that were investigated in antifungal evaluation tests (at 500 and 1000 μg $ml^{-1}$) against *A. flavus* and *Fusarium* sp. [57]. The results showed that the ruthenium chelates are more active against the studied fungus as compared with their free ligands, but it could not reach the effectiveness of the conventional fungicides amphotericin B and bavistin. An analogous macrocyclic Schiff-base 14-membered-dibenzo-tetramethyl-tetraaza [N4] macrocyclic ligand **L36** and the corresponding Ru(III) complex [Ru(**L36**)$Cl_2$]Cl (**C36**) were synthesized and both of them were active against plant pathogenic fungi *Alternaria porri* and *F. oxysporium* [58]. Tridentate dianionic Schiff bases **L37** were also investigated as anionic ligands to prepare Ru(II) complexes **C37** [Ru(CO)($PPh_3$)(B)(**L37**)] (B = $PPh_3$, py or piperidine (pip)) show growth inhibitory activity against *A. flavus, F. oxysporium,* and *R. solani* [59].

Dinuclear ruthenium(III) coordination compounds [$RuX_2$($EPh_3$)$_2$(**L38**)]$_2$ (**C38**) (X = Cl, Br; E = P, As) containing a bidentate Schiff-base ligand **L38** have been prepared and the inhibition antifungal activity of the ligand and their corresponding complexes were investigated against the *A. niger, F. oxysporium,* and *R. solani* [60]. In fact, the same authors [61] have been demonstrating, since 2000, that even the Ru(II) analogue **C39a** [RuCl(CO)(**L39a**)($PPh_3$)(B)] (B = $PPh_3$, py or pip) [61a] and the corresponding Ru(III) [$RuX_2(EPPh_3)_2$(**L39b**)] (**C39b**) (X = CI, Br; E = As, P) are bioactive [61b]. However, *A. flavus* was susceptible only to Ru(II) compounds, while *A. niger* and *Fusarium* sp. were susceptible to the corresponding Ru(III) derivatives. Three analogous non-bicyclic monobasic bidentate Salen ligands **L40** were used to prepare Ru(III) complexes [RuX($EPPh_3$)(**L40**)$_2$] (**C40**) (X = Cl, Br, E = As) that also showed a moderate inhibition toward *A. flavus* and *Fusarium* sp [62]. Ruthenium(II) coordinated with tetradentate bis-functionalized ligands **L41,** both Schiff base and thiosemicarbazone moieties in the same molecule, with general formula [Ru(CO)(B)(**L41a-d**)] (**C41**) (B = $PPh_3$, $AsPh_3$, py) were tested against *C. albicans* and *C. parapsilosis.* The results show higher antifungal activity for the complexes in comparison with the corresponding free ligands and metal precursors, and it is also noticed that the compounds containing electron-donating groups are more active [90].)

#### 15.2.3.2 Thiosemicarbazones

Thiosemicarbazones and semicarbazones have a wide range of bioactivities, among them antifungal, that are generally related to metal ion coordination [63]. Ru(III) complexes [Ru($H_2O$)$_2$(**L42a-b**)$_2$] (**C42a-b**) with anionic *ortho*-hydroxyaryl-thiosemicarbazones have shown antifungal activity against *C. albicans* and *A. fumigatus* and the activities were comparable with the newest standard antifungal echinocandin [64]. Ruthenium(II) compounds [RuX(CO)($PPh_3$)$_2$(**L43a-e**)] (**C43a-e**) (X = H, Cl) were investigated, but only the furyl-ligand derivatives showed antifungal activity against *A. niger* at 0.25%, 0.50% concentrations, but were less effective than bavistin [65]. Ru(II) carbonyl complexes of dehydroacetic acid thiosemicarbazones (**L44**) [Ru(CO)(B)($EPh_3$)(**L44**)] (**C44**) (B = $PPh_3$, $AsPh_3$, py, pip or mor, E = As, P) present antifungal activity for *A. niger* and *C. albicans* [66]. The neutral polyfunctionalized pentadentate thiosemicarbazone, named as bis-(hydroxyisonitrobenzoylacetone) thiocarbohydrazone (**L45**), generates the corresponding pentachelated Ru(III) compound [Ru($H_2O$)(**L45**)]$Cl_3$ (**C45**) which is active against *C. albicans, T. mentagrophytes,* and *T. rubrum* with quite good antifungal activity (MIC values up to 40 $\mu g\,ml^{-1}$), while the free ligand **L45** is less active (MIC values up to 200 $\mu g\,ml^{-1}$) [67]. Another kind of functionalized fluorinated thioimines related to hydrazinecarbodithioic acid can be complexed by Ru(III) producing a novel coordination compound [Ru(**L46**)$_3$] (**C46**). The complex **C46a** derived from the more lipophilic ligand **L46a** showed maximum potential inhibition for *Colletotrichum capsici, Penicillium notatum,* and *Sclerotium rolfsii* comparable to bavistin. The corresponding hydrazinecarbothiomide **C456c** and hydrazinecarboxamide **C46d,e** complexes have lower activity [91].

#### 15.2.3.3 Chalcone Derivatives

Chalcones also present a large variety of pharmacological applications [92], but the investigation of the corresponding complexes and their bioactivities are not so common [68]. In relation to the antifungal activity investigation of chalcones coordinated to ruthenium, Viswanathamurthi and coworkers have isolated new complexes with cyclic chalcone derivatives such as (i) Ru(II) compounds [RuCl(CO)($EPh_3$)(B)(**L47**)] (**C47**) (B = $PPh_3$, $AsPh_3$, py; E = P or As) that presented *in vitro* antifungal activity against *A. niger* and *Mucor* sp. by the usual disc diffusion method [69]; (ii) Ru(III) complexes **C48** coordinated with monoanionic chalconate **L48** [$RuX_2(EPh_3)_2$(**L48**)] (X = Cl, Br; E = P, As) also showed similar antifungal results against *A. niger* and *Mucor* sp [70].; (iii) some bis-coordinated complexes of the type [Ru(CO)(B)(**L49a-d**)$_2$] (**C49.1**) (B = $PPh_3$, $AsPh_3$ or py) with *ortho*-hydroxy-*para*-substituted chalconates **L49a-d** presented low antifungal activity against *A. niger* [71]; however, analogous *para*-methoxy-ortho-hydroxychalcones **L49e-h** when used as ligands generated the corresponding complexes[RuCl(B)(CO)($EPPh_3$)(**L49e-h**)] (**C49.2**) (E = P, As, B = $PPh_3$, $AsPh_3$, py), but only the complexes **C49.2e** derived from the 3,4-dimethoxy ligand **L49e** were active against *A. niger* and *Mucor* sp. [71b].The *ortho*-hydroxy-*para*-methyl- **L49c** and *ortho*-hydroxy-*para*-chloro-**L49d** substituted ligands, as well the 2-naphthyl derivative **L49i,** were used to prepare mono-coordinated chalconate Ru(II) compounds with general formula similar to **C49.2:** [RuCl(B)(CO)(PPh3)(**L49c,d** or **L49i**)] (B = py, pip), but only the complex derived from the ligand **L49c** showed activity against *A. fumigatus* [72].

Two ruthenium complexes **C50a** and **C50b** with 4-oxo-4H-pyran-2,6-dicarboxylic acid (**L50**) presenting a new coordination mode ($\kappa^2CO^3$), [Ru(MeCN)(CO)($EPh_3$)$_2$(**L50**)] (**C50a**: E = P, **C50b:** E = As) showed *in vitro* antimicrobial activity against different species of pathogenic bacteria and fungi. These compounds presented a quite promising inhibition toward *A. niger* and *C. albicans* (MIC values 50–100 $mg\,ml^{-1}$) when compared to fluconazole, the arsenic derivative **C50b** being more active. Besides, DNA–protein interactions of these complexes have been examined by photophysical studies, which revealed that they can bind to DNA through nonintercalation mode and strongly quench the intrinsic fluorescence of bovine serum albumin, through a static process [73].

### 15.2.4 Other ligands (Dithio-Naphtyl-Benzamide, Arylazo, Catecholamine, Organophosphorated, Hydridotris(pyrazolyl)borate and Bioactive Azole Ligands)

Binuclear ruthenium(III) complex [$Ru_2Cl_2(H_2O)_4$(**L51**)] (**C51**) obtained with the tetrachelate ligand 2,2′-dithiobis[*N*-(2-hydroxy-naphth-3-yl)benzamide (**L51**) inhibit the growth of *A. niger* at the concentration of $1.6\times10^{-4}$ – $8.5\times10^{-5}$ $mol\,l^{-1}$, while other complexes were inactive [74].

Arylazo derivatives were used as ligands to prepare ruthenium(III) compounds such as [$RuX(PPh_3)_2$(**L52a-e**)] (**C52a**:X = Cl, Br; **L52** = 2-(arylazo)phenolates) and the corresponding synthesized complexes showed great promise in the growth inhibition of *C. albicans* [75]. The corresponding Ru(II) CO analogues

**C52b** [Ru(CO)($PPh_3$)$_2$(**L52a-e**)] were also investigated and they presented a slightly better activity against *C. albicans* and *A. niger* [76].

Catecholamines with ever described biological activity, such as dopamine (**L53a**), isoproterenol (**L53b**), and noradrenaline (**L53c**) were coordinated to ruthenium(III) in order to yield complexes [Ru($NH_3$)$_4$(**L53**)]$PF_6$ (**C53**) with potentially antifungal activity, and these novel compounds showed strong inhibitory action against the pathogenic yeasts *C. albicans* (three isolates), *C. glabrata,* and *C. tropicalis* [77].

The ruthenium-*para*-cymene coordination compounds **C54** with water-soluble organophosphorated ligand **L54** (1,3,5-triaza-7-phosphatricyclo[3.3.1.1]decane (PTA)) [Ru$X_2$($\eta^6$-*p*-cymene)(**L54**)] (**C54a**: X = Cl, **C54b:** Br, **C54c:** I or **C54d:** NCS), were prepared and a correlation of the activity with the type of X-ligands was observed. The chloride derivative **C54a** inhibited the growth of *T. mentagrophytes* and the NCS **C54d** derivative inhibited *Cladosporium resinae* [78].

Ten half-sandwich ruthenium(II) complexes **C55a-C55j** containing the tridentate k$^3$(*N,N,N*)-hydridotris(pyrazolyl) borate (**L55.1**) as one ligand and the same water-soluble cagelike tertiary phosphane-1,3,5-triaza-7-phosphaadamantane (PTA) (**L54**) as the second ligand were prepared and investigated as possible antifungal agents against *C. albicans, C. parapsilosis,* and *Rhodosporidium toruloides* [79]. They may be classified in five different types: (i) neutral mono-PTA-substituted**C55a-C55d**[Ru(X)($PPh_3$)(**L54**)(**L55.2**)] (**C55a**: X = Cl, **C55b:** X = I, **C55c:**X = NCS, **C55d:**X = H) , (ii) neutral di-PTA-substituted **C55e** [RuCl (**L55.1**)(**L54**)$_2$], (iii) cationic di-PTA-substituted**C55f** [Ru(**L55.1**)(**L54**)$_2$]$PF_6$, (iv) cationic **C55g** [Ru(MeCN)($PPh_3$) (**L55.1**)(**L54**)$_2$]$PF_6$, and (v) N-methylated derivatives, mono-cationic complexes**C55h-i** [Ru(X)($PPh_3$) (**L55.1**)(**L54-N-Me**)] OTf (**C55h** X = Cland **C55i** X = H) and the *bis*-cationic complex **C55j** [RuCl (**L55.1**)(**L54-N-Me**)$_2$]OTf$_2$. In addition, two other PTA-Ru(II) complexes **C55l** and **C55m** containing $\eta^5$-indenyl ligand (**L55.2**), respectively, [Ru(**L55.2**) (**L54**)$_3$]Cl and [RuCl($PPh_3$)(**L55.2**)(**L54**)] were synthesized. *C. albicans* and *C. parapsilosis* were resistant to the 12 investigated ruthenium complexes and only *R. toruloides* were susceptible to the lower water-soluble derivatives **C55g, C55h,** and **C55i,** suggesting that the solubility could be an important factor for antifungal activity. In fact, the DNA-binding properties of the Ru(II) complexes were also investigated, and the tests indicated a disturbance of the DNA structure for almost all complexes. However, the precise mechanism of the biological action must be clarified to facilitate the synthesis of more active bioinorganic antifungal agents [79].

Three ruthenium complexes, also with phosphorated ligands [Ru($CF_3CO_2$)$_2$ (**L56a**)] (**C56.1**), [Ru(MeCN)$_3$(**L56a**)]OTf$_2$ (**C56.2**), and [Ru(MeCN)$_3$(**L56b**)]OTf (**C56.3**), showed antifungal activity against *C. albicans* and *C. tropicalis* with a range of MICs from 125 and 250 μg ml$^{-1}$ [80].

One of the most remarkable studies regarding the antifungal activities of ruthenium complexes have been reported by Turel and coworkers [81]. They have used the azole antifungal agents clotrimazole (ctz), tioconazole (tcz), and miconazole (mcz), as corresponding ligands **L57a-c,** in order to obtain, as usually expected in bioinorganic chemistry, a synergism between the metal ion and the bioactive ligand [93]. Nine organoruthenium complexes with mono-, bis-, and tris-azole ligands [$RuCl_2$($\eta^6$-*p*-cymene)(**L57a-c**)] (**C57.1a-c**), [RuCl($\eta^6$-*p*-cymene)($\mathbf{L57a\text{-}c}$)$_2$]Cl (**C57.2a-c**), and [Ru($\eta^6$-*p*-cymene)($\mathbf{L57a\text{-}c}$)$_3$]($PF_6$)$_2$ (**C57.3a-c**) were prepared and completely characterized, and all nine isolated ruthenium complexes showed antifungal activity at low millimolar concentrations against *Culvularia lunata*. The individual azole ligands **L57a-c** showed more potent antifungal activity in comparison with their corresponding mono-(**C57.2a-c**), and tris-azole (**C57.3a-c**) complexes and, besides, antifungal activities decreased with the increasing number of ligands in these complexes. Tris-azole complexes were less potent at 0.01 mM concentrations, but **C57.3b-c** still showed a 1.5-fold decrease. At 0.5 mM concentrations, the tris-azole analogues **C57.3a-c** still significantly reduced the radial growth rate by 2.3-, 1.9-, and 2.8-fold, as compared to the control, respectively, and also at 0.01 mM concentrations 2.3-, 2,4-, and 2.7-fold decline in the growth rate was observed for **C57.1a-c.** The authors suggested that the antifungal activity mechanism for these azole ruthenium complexes may not be correlated to the usual coordination to lanosterol 14α-demethylase inhibiting ergosterol biosynthesis and fungal growth [94], since the azoles are bound to ruthenium.

## 15.3 Conclusion

Ruthenium complexes really can provide a rich platform and suitable building blocks for the design of novel antifungal agents [31], taking advantage of the specific properties inherent to the transition metal center. Well-known drugs would be successful as a scaffold for miomimetic ruthenium pharmaceuticals to afford new ruthenium complexes able to overcome microbial resistance, although just the last cited work shows the investigation of an antifungal agent used as a possible bioactive ligand for coordination with ruthenium. Nevertheless, ruthenium complexes with various types of coordinating ligands, such as aromatic, heteroaromatic, Schiff bases, thiosemicarbazones, chalcones, catecholamines, half-sandwich, hydridotris(pyrazolyl)borate, and organophosphorated, among others, and even in several kinds of structure, since mononuclear, dinuclear, trinuclear and polymeric species figure prominently with potential antifungal activity. The studies herein reviewed illustrate that almost all complexes are more active than the corresponding free ligands and these data can be explained considering the higher lipophilicity of the complexes [30].[1] A preliminary overall SAR suggests a strong influence from steric and lipophilic parameters in the

1 It is noteworthy that many of the works reported herein cite a presumably well-known "Tweedy's concept or Tweedyás theory": "...*Chelation considerably reduces the polarity of the metal ion because of partial sharing of its positive charge with the donor groups and possible p-electron delocalization over the whole chelate ring. Such chelation could enhance the lipophilic character of central metal atom, which subsequently favours its permeation through the lipid layers of cell membrane*...." [95]. However, for the best of our knowledge, this reference, as cited, does not exist, although it is possible to find two citations by B.G. Tweedy, on the Web of Science®, in Phytopathology (Meeting Abstracts) [96].In fact, some other citations talks about Overtone's theory, that is, the well-known Overton's concept or rule, that was described, in 1900, by Ernst Overton who postulated that the entry of dyes into plant cells depends on the lipophilicity of the compounds [97].

antifungal activity of the presented ruthenium complexes, and many of them have triphenylphosphine, triphenylarsine, aromatic, or heteroaromatic lipophilic moieties. In addition, the general influence of lipophilicity and steric hindrance on the antifungal activity of several known antifungal agents has been shown by a quite recent quantitative structure–activity relationship (QSAR) study [98]. However, these studies herein reported generally used nonstandard antifungal activity assessment methodology and this nonuniformity of results prevents a good comparison of results. In fact, a suitable method to be used would be the disk diffusion, followed by MIC value determination, and all antimicrobial activity essays would have to be performed according to the CLSI guidelines [99], to establish a set of reference data for these compounds; moreover, cytotoxicity tests also have to be carried out. A mechanistic comprehension about the ruthenium complexes activity is also crucial to a clinical success, as well as to rationalize the design of novel bioactive compounds with improved potential. Besides, the candidate drugs need to be demonstrated by *in vitro* studies with targeted biomolecules and tissues, followed by *in vivo* investigations, especially with fungi that have higher clinical interest, as the fungi related to severe and fatal IFI (*Candida* sp. *Aspergillus* sp. *S. schenckii,* and *P. brasiliensis*) and with comparison to more clinically relevant antifungal agents. Anyway, the ruthenium complexes can be considered as potential novel bioactive compounds, since the large variety of ligands with different steric hindrance, electronic, inductive, or lipophilic effects can provide adjustable physicochemical properties (such as triphenylphosphine/arsine to increase lipophilicity in many reported cases herein); and the usefulness of these ligands offers much scope for the design of more elaborate and efficient antifungal agents to attack the terrible worldwide health problem of IFIs and EIDs.

## References


1 Fauci, A.S. (2004) *J. Am. Med. Assoc.*, **292**, 1887.

2 (a) Gellert, G.A. (1994) *Nature*, **370**, 409; (b) Jones, K.E., Patel, N.G., Levy, M.A., Storeygard, A., Balk, D., Gittleman, J.L., and Daszak, P. (2008) *Nature*, **451**, 990.

3 van Doorn, H.R. (2014) *Medicine*, **42**, 60.

4 (a) Janes, C.R., Corbett, K.K., Jones, J.H., and Trostle, J. (2012) *Lancet*, **380**, 1884; (b) Morse, S.S., Mazet, J.A.K., Woolhouse, M., Parrish, C.R., Carroll, D., Karesh, W.B., Zambrana-Torrelio, C., Lipkin, W.I., and Daszak, P. (2012) *Lancet*, **380**, 1956.

5 (a) Rex, J.H., Pfaller, M.A., Rinaldi, M.G., Polak, A., and Galgiani, J.N. (1993) *Clin. Microbiol. Rev.*, **6**, 367; (b) Hemaiswarya, S., Kruthiventi, A.K., and Doble, M. (2008) *Phytomedicine*, **15**, 639.

6 (a) Dryden, M.S., Cooke, J., and Davey, P. (2009) *J. Antimicrob. Chemother.*, **64**, 885; (b) Donnici, C.L., Stoianoff, M.A.R., Lopes, J.C.D., Silva, L.L., Oliveira, S.R., Nogueira, L.J., Montanari, C.A., and Martins, C.V.B. (2012) *Mycoses*, **55-S4**, 111; (c) Donnici, C.L., Stoianoff, M.A.R., Silva, L.L., dos Santos, V.M.R., Montanari, C.A., Lopes, J.C.D., and Moreira, F.O.

(2012) *Mycoses*, **55-S4**, 106; (d) Oliveira, S.R., Nogueira, L.J., Donnici, C.L., Magalhaes, T.F.F., Martins, C.V.B., Montanari, C.A., and Resende, M.A. (2011) *Mycoses*, **54**, e389.

7 Rüping, M.J.G.T., Vehreschild, J.J., and Cornely, O.A. (2008) *Drugs*, **68**, 1941.

8 (a) Pfaller, M.A. and Diekema, D.J. (2004) *J. Clin. Microbiol.*, **42**, 4419; (b) Pfaller, M.A. and Diekema, D.J. (2007) *Clin. Microbiol. Rev.*, **20**, 133.

9 (a) Gudlaugsson, O., Gillespie, S., Lee, K., Berg, J.V., Hu, J., Messer, S., Herwaldt, L., Pfaller, M., and Diekema, D. (2003) *Clin. Infect. Dis.*, **37**, 1172; (b) Lin, S.-J., Schranz, J., and Teutsch, S.M. (2001) *Clin. Infect. Dis.*, **32**, 358; (c) Wilson, L.S., Reyes, C.M., Stolpman, M., Speckman, J., Allen, K., and Beney, J. (2002) *Value Health*, **5**, 26.

10 (a) Blumberg, H.M., Jarvis, W.R., Soucie, J.M., Edwards, J.E., Patterson, J.E., Pfaller, M.A., Rangel-Frausto, M.S., Rinaldi, M.G., Saiman, L., Wiblin, R.T., Wenzel, R.P., and Group, N.S. (2001) *Clin. Infect. Dis.*, **33**, 177; (b) Hajjeh, R.A., Sofair, A.N., Harrison, L.H., Lyon, G.M., Arthington-Skaggs, B.A., Mirza, S.A., Phelan, M., Morgan, J., Lee-Yang, W., Ciblak, M.A., Benjamin, L.E., Sanza, L.T., Huie, S., Yeo, S.F., Brandt, M.E., and Warnock, D.W. (2004) *J. Clin. Microbiol.*, **42**, 1519; (c) Sullivan, D.J., Moran, G.P., Pinjon, E., Al-Mosaid, A., Stokes, C., Vaughan, C., and Coleman, D.C. (2004) *FEMS Yeast Res.*, **4**, 369.

11 Dorocka-Bobkowska, B., Konopka, K., and Düzgüneş, N. (2003) *Arch. Oral Biol.*, **48**, 805.

12 (a) Alexander, B.D. and Pfaller, M.A. (2006) *Clin. Infect. Dis.*, **43**, S15; (b) Hope, W.W., Walsh, T.J., and Denning, D.W. (2005) *Lancet Infect. Dis.*, **5**, 609; (c) Revankar, S.G., Sutton, D.A., and Rinaldi, M.G. (2004) *Clin. Infect. Dis.*, **38**, 206; (d) Martins, C.V.B., da Silva, D.L., Neres, A.T.M., Magalhães, T.F.F., Watanabe, G.A., Modolo, L.V., Sabino, A.A., de Fátima, A., and de Resende, M.A. (2009) *J. Antimicrob. Chemother.*, **63**, 337; (e) Pfaller, M.A., Diekema, D.J., and International Fungal Surveillance Participant Group (2004) *Clin. Microbiol. Infect.*, **10** (Suppl. 1), 11; (f) Nucci, M., Marr, K.A., Queiroz-Telles, F., Martins, C.A., Trabasso, P., Costa, S., Voltarelli, J.C., Colombo, A.L., Imhof, A., Pasquini, R., Maiolino, A., Souza, C.A., and Anaissie, E. (2004) *Clin. Infect. Dis.*, **38**, 1237; (g) Wingard, J.R. and Leather, H. (2004) *Biol. Blood Marrow Transplant.*, **10**, 73; (h) Brown, G.D., Denning, D.W., and Levitz, S.M. (2012) *Science*, **336**, 647.

13 Miceli, M.H., Diaz, J.A., and Lee, S.A. (2011) *Lancet Infect. Dis.*, **11**, 142.

14 (a) Colombo, A.L. and Guimarães, T. (2003) *Rev. Soc. Bras. Med. Trop.*, **36**, 599; (b) Pinho Resende, J.C., De Resende, M.A., and Saliba, J.L. (2002) *Mycoses*, **45**, 306; (c) Álvares, C.A., Svidzinski, T.I.E., and Consolaro, M.E.L. (2007) *J. Bras. Patol. Med. Lab.*, **43**, 319.

15 (a) Nuño, C.G., Alfonso, P.P.P., Vázquez, J.C.R., Gómez, M.M.R., Prats, I.P., and González, J.G. (2000) *Acta Méd.*, **9**, 67; (b) Filho, G.P., Pedroso, E.R.P., Giannetti, A.V., and Darwich, R. (2005) *Arq Neuropsiquiatr*, **63**, 1094; (c) Maertens, J., Theunissen, K., and Boogaerts, M. (2002) *Curr. Med. Chem.*, **1**, 65.

16 (a) Marques, S.A., Franco, S.R.V.S., Camargo, R.M.P., Dias, L.D.F., Haddad Júnior, V., and Fabris, V.E. (1993) *Rev. Inst. Med. Trop. S. Paulo*,

**35**, 327; (b) Neto, R.J.P., Machado, A.A., Castro, G., Quaglio, A.S.S., and Martinez, R. (1999) *Rev. Soc. Bras. Med. Trop.*, **32**, 57.

17 (a) Kosinski, R.M., Axelrod, P., Rex, J.H., Burday, M., Sivaprasad, R., and Wreiole, A. (1992) *J. Clin. Microbiol.*, **30**, 501; (b) Fagundes-Pereyra, W.J., Carvalho, G.T.C., Góes, A.M., Silva, F.C.L., and Sousa, A.A. (2006) *Arq Neuropsiquiatr*, **64**, 269.

18 (a) Yasuda, M.A.S. (2005) *Expert Opin. Pharmacother.*, **6**, 385; (b) Sundriyal, S., Sharma, R.K., and Jain, R. (2006) *Curr. Med. Chem.*, **13**, 1321; (c) Varga, J., Rigó, K., Tóth, B., Téren, J., and Kozakiewicz, Z. (2003) *Food Technol. Biotechnol.*, **41**, 29; (d) Walsh, T.J., Petraitis, V., Petraitiene, R., Field-Ridley, A., Sutton, D., Ghannoum, M., Sein, T., Schaufele, R., Peter, J., Bacher, J., Casler, H., Armstrong, D., Espinel-Ingroff, A., Rinaldi, M.G., and Lyman, C.A. (2003) *J. Infect. Dis.*, **188**, 305.

19 Ghannoum, M.A. and Rice, L.B. (1999) *Clin. Microbiol. Rev.*, **12**, 501.

20 (a) Denning, D.W. and Hope, W.W. (2010) *Trends Microbiol.*, **18**, 195; (b) Cowen, L.E. (2008) *Nat. Rev. Microbiol.*, **6**, 187.

21 Martins, M.D. and Rex, J.H. (1996) *New Horizons*, **4**, 338.

22 Odds, F.C., Brown, A.J.P., and Gow, N.A.R. (2003) *Trends Microbiol.*, **11**, 272.

23 Vandeputte, P., Ferrari, S., and Coste, A.T. (2012) *Int. J. Microbiol.*, Article ID 713687.

24 (a) Espinel-Ingroff, A., Shadomy, S., and Gebhart, R.J. (1984) *Antimicrob. Agents Chemother.*, **26**, 5; (b) Denning, D.W., Venkateswarlu, K., Oakley, K.L., Anderson, M.J., Manning, N.J., Stevens, D.A., Warnock, D.W., and Kelly, S.L. (1997) *Antimicrob. Agents Chemother.*, **41**, 1364.

25 (a) Chiou, C.C., Groll, A.H., and Walsh, T.J. (2000) *Oncologist*, **5**, 120; (b) Sabo, J.A. and Abdel-Rahman, S.M. (2000) *Ann. Pharmacother.*, **34**, 1032; (c) De Sarro, A., La Camera, E., and Fera, M.T. (2008) *J. Chemother.*, **20**, 661.

26 Yamazumi, T., Pfaller, M.A., Messer, S.A., Houston, A., Hollis, R.J., and Jones, R.N. (2000) *Antimicrob. Agents Chemother.*, **44**, 2883.

27 (a) Pfaller, M.A., Diekema, D.J., Messer, S.A., Boyken, L., Hollis, R.J., and Jones, R.N. (2004) *Diagn. Microbiol. Infect. Dis.*, **48**, 101; (b) Pfaller, M.A., Messer, S.A., Boyken, L., Rice, C., Tendolkar, S., Hollis, R.J., and Diekema, D.J. (2004) *J. Clin. Microbiol.*, **42**, 3137.

28 Vanden Bossche, H. (2002) *Expert Opin. Ther. Pat.*, **12**, 151.

29 (a) Fera, M.T., La Camera, E., and De Sarro, A. (2009) *Expert Rev. Anti Infect. Ther.*, **7**, 981; (b) Perlin, D.S. (2007) *Drug Resist. Updat.*, **10**, 121.

30 Thompson, K.H. and Orvig, C. (2003) *Science*, **300**, 936.

31 (a) Ramos, A.I., Braga, T.M., and Braga, S.S. (2012) *Mini Rev. Med. Chem.*, **12**, 227; (b) Clarke, M.J. (2003) *Coord. Chem. Rev.*, **236**, 209.

32 Somers, E. (1959) *Nature*, **184**, 475.

33 Sharma, V.K., Pandey, O.P., and Sengupta, S.K. (1987) *Transition Met. Chem.*, **12**, 509.

34 Mulyana, Y., Feterl, M.L., Keene, F.R., Li, F., Collins, J.G., Warner, J.M., and Heimann, K.R.M. (2013) Use of new and known ruthenium complex for treating microbial infection such as bacterial infection caused by e.g. Escherichia

coli, protozoal infection caused by Plasmodium falciparum, and fungal infection caused by Aspergillus flavus, WO2013091014-A1, US2013172308-A1, AU2011265513-A1, CA2762717-A1.

35 Balakrishnan, S.K., Dass, S.K., Rajendran, H.A.D., and Savrimuthu, I. (2015) *Int. J. Pharm. Pharm. Sci.*, **7**, 317.

36 Al-Noaimi, M., Nafady, A., Warad, I., Alshwafy, R., Husein, A., Talib, W.H., and Hadda, T.B. (2014) *Spectrochim. Acta, Part A*, **122**, 273.

37 Priya, N.P., Arunachalam, S.V., Sathya, N., Chinnusamy, V., and Jayabalakrishnan, C. (2009) *Transition Met. Chem.*, **34**, 437.

38 Karvembu, R., Jayabalakrishnan, C., and Natarajan, K. (2002) *Transition Met. Chem.*, **27**, 574.

39 Donnici, C.L., Nogueira, L.J., Araujo, M.H., Oliveira, S.R., Magalhães, T.F.F., Lopes, M.T.P., Silva, A.C.A., Ferreira, A.M.C., Martins, C.V.B., and Stoianoff, M.A.R. (2014) *Molecules*, **19**, 5402.

40 Nogueira, L.J., de Resende, M.A., Oliveira, S.R., de Araujo, M.H., Magalhaes, T.F.F., de Oliveira, M.B., Martins, C.V.B., Lopes, M.T.P., Silva, A., and Donnici, C.L. (2011) *Mycoses*, **54**, E323.

41 Kiruthika, M., Elayaperumal, R., and Vennila, T. (2012) *Chem. Sin.*, **3**, 1175.

42 Anthonysamy, A., Balasubramanian, S., Shanmugaiah, V., and Mathivanan, N. (2008) *Dalton Trans.*, 2136.

43 (a) Kumar, K.A., Reddy, K.L., and Satyanarayana, S. (2010) *Supramol. Chem.*, **22**, 629; (b) Kumar, K.A., Reddy, K.L., Vidhisha, S., and Satyanarayana, S. (2009) *Appl. Organomet. Chem.*, **23**, 409; (c) Shilpa, M., Nagababu, P., Kumar, Y.P., Latha, J.N.L., Reddy, M.R., Karthikeyan, K.S., Md Gabra, N., and Satyanarayana, S. (2011) *J. Fluoresc.*, **21**, 1155; (d) Devi, C.S., Kumar, D.A., Singh, S.S., Gabra, N., Deepika, N., Kumar, Y.P., and Satyanarayana, S. (2013) *Eur. J. Med. Chem.*, **64**, 410; (e) Devi, C.S. and Satyanarayana, S. (2012) *J. Coord. Chem.*, **65**, 474.

44 Basavaraju, B., Naik, H.S.B., and Prabhakara, M.C. (2007) *E-J. Chem.*, **4**, 32.

45 Fabiyi, F.A.S., Ibironke, O.L., and Lopez, M.L. (2011) *Int. J. Chem. (Toronto, ON, Canada)*, **3**, 3.

46 El-Gamel, N.E.A. and Fekry, A.M. (2015) *Bioelectrochemistry*, **104**, 35.

47 Refat, M.S., El-Korashy, S.A., Kumar, D.N., and Ahmed, A.S. (2008) *Spectrochim. Acta, Part A*, **70**, 898.

48 Viswanathamurthi, P., Dharmaraj, N., Anuradha, S., and Natarajan, K. (1998) *Transition Met. Chem.*, **23**, 337.

49 Viswanathamurthi, P. and Natarajan, K. (1999) *Transition Met. Chem.*, **24**, 638.

50 Priya, N.P., Arunachalam, S., Sathya, N., and Jayabalakrishnan, C. (2010) *J. Coord. Chem.*, **63**, 1440.

51 Arunachalam, S., Priya, N.P., Saravanakumar, C., Jayabalakrishnan, C., and Chinnusamy, V. (2010) *J. Coord. Chem.*, **63**, 1795.

52 Arunachalam, S., Priya, N.P., Boopathi, K., Jayabalakrishnan, C., and Chinnusamy, V. (2010) *Appl. Organomet. Chem.*, **24**, 491.

53 Tovmasyan, A., Reboucas, J.S., and Benov, L. (2014) *Antioxid. Redox Signaling*, **20**, 2416.

54 (a) Thangadurai, T.D. and Ihm, S.-K. (2005) *Synth. React. Inorg. Met.-Org. Chem.*, **35**, 499; (b) Thangadurai, T.D. and Ihm, S.-K. (2004) *Transition Met. Chem.*, **29**, 189.

55 Sharma, V.K., Srivastava, A., and Srivastava, S. (2006) *J. Serb. Chem. Soc.*, **71**, 917–928.

56 (a) Sharma, V.K., Srivastava, S., and Srivastava, A. (2006) *J. Coord. Chem.*, **59**, 1321; (b) Sharma, V.K., Srivastava, S., and Srivastava, A. (2005) *Rev. Roum. Chim.*, **50**, 751.

57 Shanker, K., Rohini, R., Ravinder, V., Reddy, P.M., and Ho, Y.-P. (2009) *Spectrochim. Acta, Part A*, **73**, 205.

58 Rani, S., Kumar, S., and Chandra, S. (2014) *Spectrochim. Acta, Part A*, **118**, 244.

59 Jayabalakrishnan, C., Karvembu, R., and Natarajan, K. (2002) *Transition Met. Chem.*, **27**, 790.

60 Karvembu, R., Jayabalakrishnan, C., Dharmaraj, N., Renukadevi, S.V., and Natarajan, K. (2002) *Transition Met. Chem.*, **27**, 631.

61 (a) Dharmaraj, N., Natarajan, K., and Viswanathamurthi, P. (2001) *Transition Met. Chem.*, **26**, 105; (b) Viswanathamurthi, P., Dharmaraj, N., and Natarajan, K. (2000) *Synth. React. Inorg. Met.-Org. Chem.*, **30**, 1273.

62 Ramesh, R. and Maheswaran, S. (2003) *J. Inorg. Biochem.*, **96**, 457.

63 Beraldo, H. and Gambino, D. (2004) *Mini Rev. Med. Chem.*, **4**, 31.

64 Mehta, B.H. and Shaikh, J.A. (2009) *J. Ind. Council Chem.*, **26**, 1.

65 Muthusamy, G., Viswanathamurthi, P., Muthukumar, M., and Natarajan, K. (2009) *Phosphorus, Sulfur Silicon Relat. Elem.*, **184**, 2115.

66 Kannan, S., Sivagamasundari, M., Ramesh, R., and Liu, Y. (2008) *J. Organomet. Chem.*, **693**, 2251.

67 Patil, M.V. and Malve, S.P. (2004) *J. Indian Chem. Soc.*, **81**, 683.

68 Al-Obaidi, O.H. (2013–2014) *J. Chem. Biol. Phys. Sci. Sec. B*, **4**, 259.

69 Muthukumar, M. and Viswanathamurthi, P. (2010) *J. Coord. Chem.*, **63**, 1263.

70 Muthukumar, M., Viswanathamurthi, P., and Karvembu, R. (2010) *Phosphorus, Sulfur Silicon Relat. Elem.*, **185**, 2201.

71 (a) Muthukumar, M., Viswanathamurthi, P., and Natarajan, K. (2008) *Spectrochim. Acta, Part A*, **70**, 1222; (b) Viswanathamurthi, P. and Muthukumar, M. (2011) *J. Chem. Sci.*, **123**, 567.

72 Kaveri, M.V., Prabhakaran, R., Karvembu, R., and Natarajan, K. (2005) *Spectrochim. Acta, Part A*, **61**, 2915.

73 Kamatchi, T.S., Kalaivani, P., Poornima, P., Padma, V.V., Fronczekc, F.R., and Natarajan, K. (2014) *RSC Adv.*, **4**, 2004.

74 Jhaumeer-Laulloo, S., Bhowon, M.G., and Hosany, A. (2004) *J. Indian Chem. Soc.*, **81**, 547.

75 Venkatachalam, G., Ramesh, R., and Mobin, S.M. (2005) *J. Organomet. Chem.*, **690**, 3937.

76 Naresh Kumar, K., Ramesh, R., and Liu, Y. (2006) *J. Inorg. Biochem.*, **100**, 18.

77 de Lima, R., Lever, A.B.P., Ito, I., and Santana da Silva, R. (2003) *Transition Met. Chem.*, **28**, 272.

78 Allardyce, C.S., Dyson, P.J., Ellis, D.J., Salter, P.A., and Scopelliti, R. (2003) *J. Organomet. Chem.*, **668**, 35.

79 García-Fernández, A., Díez, J., Manteca, Á., Sánchez, J., Gamasa, M.P., and Lastra, E. (2008) *Polyhedron*, **27**, 1214.
80 Sülü, M., Küçükbay, H., Durmaz, R., and Günal, S. (2000) *Microbiologica*, **23**, 73.
81 Kljun, J., Scott, A.J., Lanišnik Rižner, T., Keiser, J., and Turel, I. (2014) *Organometallics*, **33**, 1594.
82 Ahamad, M., Rao, R.M., Md Rafi, M., Mohiddin, G.J., and Sreeramulu, J. (2012) *Arch. Appl. Sci. Res.*, **4**, 858.
83 (a) Antonarakis, E.S. and Emadi, A. (2010) *Cancer Chemother. Pharmacol.*, **66**, 1; (b) Lainé, P.P., Campagna, S., and Loiseau, F. (2008) *Coord. Chem. Rev.*, **252**, 2552; (c) Metcalfe, C., Rajput, C., and Thomas, J.A. (2006) *J. Inorg. Biochem.*, **100**, 1314.
84 Qin, W.L., Long, S., Panunzio, M., and Biondi, S. (2013) *Molecules*, **18**, 12264.
85 (a) Anand, P., Patil, V.M., Sharma, V.K., Khosa, R.L., and Masand, N. (2012) *Int. J. Drug Des. Discovery*, **3**, 851; (b) Kumar, S., Dhar, D.N., and Saxena, P.N. (2009) *J. Sci. Ind. Res.*, **68**, 181.
86 Pfeiffer, P., Breith, E., Lübbe, E., and Tsumaki, T. (1933) *Justus Liebigs Ann. Chem.*, **503**, 84.
87 Cozzi, P.G. (2004) *Chem. Soc. Rev.*, **33**, 410.
88 Katsuki, T. (2003) *Synlett*, 281.
89 Arunachalam, S., Priya, N.P., and Meeran, H.S. (2014) *J. Indian Chem. Soc.*, **91**, 53.
90 Raja, G., Sathya, N., and Jayabalakrishnan, C. (2011) *Synth. React. Inorg. Met.-Org. Chem.*, **41**, 716.
91 Jadon, S.C.S., Saxena, C., and Singh, R.V. (1997) *Transition Met. Chem.*, **22**, 385.
92 (a) Amanaganti, J. and Subhashini, N.J.P. (2013) *Int. J. Chem. Sci.*, **11**, 1335; (b) Donda, H.K., Faldu, S.D., Kapuriya, K.G., Kadam, S.S., and Ganure, A.L. (2013) *Int. J. Drug Des. Discovery*, **4**, 1031; (c) Ni, L., Meng, C.Q., and Sikorski, J.A. (2004) *Expert Opin. Ther. Pat.*, **14**, 1669.
93 Fromtling, R.A. (1988) *Clin. Microbiol. Rev.*, **1**, 187.
94 Sheehan, D.J., Hitchcock, C.A., and Sibley, C.M. (1999) *Clin. Microbiol. Rev.*, **12**, 40.
95 Tweedy, B.G. (1964) *Phytopathology*, **55**, 910.
96 (a) Tweedy, B.G. (1964) *Phytopathology (Meeting Abstract)*, **54**, 910; (b) Tweedy, B.G. and Turner, N. (1965) *Phytopathology (Meeting Abstract)*, **55**, 1080.
97 (a) Overton, E. (1900) *Jahrb. Wiss. Bot.*, **34**, 669; (b) Perouansky, M. (2015) *Br. J. Anaesth.*, **114**, 537.
98 Katritzky, A.R., Slavov, S.H., Dobchev, D.A., and Karelson, M. (2008) *Bioorg. Med. Chem.*, **16**, 7055.
99 National Committee for Clinical Laboratory Standards Institute (2008) *Reference Method for Broth Dilution Antifungal Susceptibility Testing of Filamentous Fungi* Approved Standard, M38-A32, NCCLS, Wayne, PA.

# Index


*a*
A431 carcinoma cells 151–155
acid phosphatases (AcPs) 209
Amphotericin B (AMB) 294
antibacterial (fluoro)quinolones
  HIV integrase inhibitor elvitegravir 245–246
  interactions with metal ions 241
    ruthenium-quinolone complexAFM experiments 242
    $\beta$-ketocarboxylate functionality 244
    and cinoxacin 243
    cysteine cathepsins 244
    cytotoxicity assays 243
    human serum albumin (HSA) 243
    immunoglobulins 245
    nalidixic acid 243, 245
    pharmacological characterization 245
    thionated quinolone 244
antibacterial 8-hydroxyquinolines
  2,9-dimethyl-1,10-phenanthroline 247
  mode-of-action 246–247
  organoruthenium-clioquinol complex 247
anticancer agents
  antimetastatic effects, of NAMI 165
  biological reductants 161
  biotinylated Ru(arene) complex 172
  delivery strategies 162
  enhanced permeability and retention effect 163
  KP1019 164
  levonorgestrel 171
  ligand exchange rates 161
  macromolecular carrier systems
    dendrimers 169
    organometallic compounds 166–168
    polymers and liposomes 168–169
  metallaprisms, metallarectangles and metallacycles 174
  NKP-1339 164
  organometallic hormone-receptor targeting agents 170
  peptides 173–174
  protein binding 163–166
  RAPTA analogues 172
  RAPTA-C 169
  steroid hormones 170
  structures of 162
  vitamins 171
anticancer compounds
  acid phosphatases (AcPs) 209
  cellular uptake 204–205
  cisplatin, side-effects and acquired drug resistance 201
  DNA and DNA-related cellular targets
    glutathione (GSH) levels 205
    KP1019 activity 206
    PARP-1 (Poly(ADP-ribose) polymerase-1) 206
  endosomal/lysosomal system and ER-Golgi network 207
  glycolytic pathways 209–211
  macromolecular ruthenium-conjugates
    EPR effect 212
    RAPTA-C polymeric micelles 211
    RuPMC 214
    NAMI-A and NAMI-A-type complexes 203
  Ru($\eta^6$-arene) scaffold 203
  targeting signaling pathways 207
  thioredoxin reductase (TrxR) 207


*Ruthenium Complexes: Photochemical and Biomedical Applications*, First Edition.
Edited by Alvin A. Holder, Lothar Lilge, Wesley R. Browne, Mark A.W. Lawrence, and Jimmie L. Bullock Jr.

anticancer compounds (*contd.*)
titanocene dichloride and budotitane 202
transition metals 201
antifungal activity investigations
aromatic and heteroaromatic ligands 305–306
arylazo 310–312
catecholamine 310–312
chalcones 310
cytotoxicity tests 313
dithio-naphtyl-benzamide 310–312
organophosphorated ligand 310–312
pathogenic fungi species 300, 304–305
Schiff bases 307–309
structure-activity relationship 312
thiosemicarbazones 309
antimalarial 4-aminoquinolines
mechanism of action 248–249
metallocene analogues, of chloroquine 249–252
aromatic and heteroaromatic ligands 305–306
artificial water oxidation 49, 50
arylazo 310–312

## *b*
bidentate 307–309
bridging ligands 7
binuclear mixed-metal complexes 73
1,4-bis(pyrid-3-yl)benzene (bpb) 57
bioactive azole ligands 310–312
bioactive ligand scaffolds 284–285
bioactive sites
Pt(II) chemotherapeutic agents 140–141
Rh(III) complexes 141, 142
Ru(II) metal centre 149–150
and Co(III) metal centres 151
and Cu, 150–151in nanomedicine 152–155
Pt(II) bioactive sites 143–146
Rh(III) bioactive sites 147
and V(IV) metal centres 151
2,2′-bipyridine ligands 69
Brewer, K.J.
award-recognized educator 4
collaboration with Winkel, Brenda 16–18
early years 4–6
graduate studies and Clemson University 6–11
legacy 20
mentor and role model 4
photochemical $H_2$ production 18–19
postdoctoral research and the University of California 11–12
professional career 19–20
publications per year from 1985–2015 4
study of Metal-DNA interactions 16–18
Virginia Tech 15–16
Washington State University 13–15
$[(bpy)_2Ru(bpy\text{-}4\text{-}CH_3,4'\text{-}CONHCH_2(4\text{-}py)Co(dmgBF_2)_2(OH_2)](PF_6)_2$ 75
$[(bpy)_2Ru(bpy\text{-}4\text{-}CH_3,4'\text{-}CONH(4\text{-}py)Co(dmgBF_2)_2(OH_2)](PF_6)_2$ 75

## *c*
Capillary zone electrophoresis (CZE)-ICP-MS experiments 243
catecholamine 310–312
cathepsin B 209
chemistry of ruthenium(II) and (III) 31, 36
cisplatin 223
Co(III) dihydride 9
$[Co(dmgH)_2(H_2O)_2]$ 72
cobaloxime catalyst 68
cobalt(II)-polypyridyl catalysts 69
cobalt(III)-hydride 69
coordination chemistry 20
element 25–26
stereochemistry and common oxidation states *see* stereochemistry and common oxidation states
cross metathesis reaction 33
cyclometallated systems 228
cytochrome $c_3$-viologen-ruthenium(II) triad complex 78
cytotoxic organoruthenium complexes 285

## *d*
dendrimers 169
Dewar–Chatt–Duncanson model 27
dichloro(*p*-cymene)ruthenium(II) dimer 28
dichlorofluorescin-diacetate (DCFH-DA) 187

diffusion theory 124
4,4′-dimethyl-2,2′-bipyridine ligand 69
dinuclear bis-(-diketones and pentadithiocarbamate ligands 300, 304, 305
dinuclear ruthenium dithiocarbamate (DTCB) 305
dinuclear, trinuclear and tetranuclear ruthenium polydentate polypyridil ligands 300, 304, 305
dithio-naphtyl-benzamide 310–312
DNA binding, ruthenium complexes
  biological consequences
    antimicrobial activity 191
    cellular uptake experiments 190
    DCFH-DA 187
    FACS analysis 189
    fluorometric assay 187
    H3K56A mutant 189
    KP1019 187
    NAMI-A 190
    platinum complexes' abilities 187
    prokaryotes and eukaryotes 190
  intercalation 184–185
  irreversible covalent binding 182–184
  noncovalent binding interactions 185–186
  topoisomerases and telomerase
    COMET assays 195
    DNA replication 192
    eukaryotes 195
    gel electrophoresis 193
    human genome sequence analysis 196
    in vitro DNA binding and viscosity assays 194
    in vitro plasmid relaxation assays 193
    KP1019 and KP418 193
  DNA targetingbinding modes, of small molecules 222–223
  cellular uptake and localisation 231–232
  cyclometalated ligands 230
  cyclometalated systems 228–231
  dinuclear complexes 228
  intercalation 223
  metal complexes 223–224
  mononuclear complexes 226–228
  non-canonical DNA structures 221
  $[Ru(bpy)_2(dppz)]^{2+}$ and DNA light switch effect 224–226

*e*
emerging infectious diseases (EID) 293
energy issue and energy 45–46
enhanced permeability and retention (EPR) 164, 211, 284
$[Et_3NH]BF_4$ 71

*f*
[FeFe] and [NiFe]-hydrogenase 78
ferrocifen 170
fluorescence-activated cell sorting (FACS) analysis 189
fluorometric assay 187

*g*
greenhouse gases 67

*h*
heptacoordinated Ru intermediates 56–57
heterotrimetallic di-ruthenium-mono-palladium complexes 300, 304, 305
high valent Mn-O intermediate 49
higher oxidation states of ruthenium 36–37
Human genome sequence analysis 196
human serum albumin (HSA) 243
hydridotris(pyrazolyl)borate 310–312
hydrogen evolution catalyst (HEC) 71, 77
hydrogenases, with ruthenium(II) complexes 77–84

*i*
Invasive Fungal Infections (IFI) 294
$[Ir(tpp)Cl_3]$ 14
Ir(III) complex $[Ir(dpp)_2Cl_2]^+$ 11

*k*
KP1019/NKP-1339
  activation by reduction 277–278
  clinical development 276
  clinical studies and perspectives 281
  KP1019-loaded micelle-forming PEGylated polymers 276
  mode of action 278–281

KP1019/NKP-1339 (*contd.*)
poly(lactic acid) nanoparticles 276
tumor targeting mediated by plasma proteins 276–277

## l

levonorgestrel 171
ligand photosubstitution reactions with ruthenium compounds
caging and uncaging biologically active ligands 92–96
caging cytotoxic ruthenium complexes with organic ligands 96–100
low energy photosubstitution, *see* low energy photosubstitution
metal-ligand antibonding $e_g$ orbital 91
photodynamic therapy (PDT) 91
light-driven hydrogen ($H_2$) production 77
low energy photosubstitution
phototherapeutic window 100
Ru photophysics 100–105
TPA photosubstitution 109–110
triplet-triplet annihilation upconversion 105–106
upconverting nanoparticles 106–109
low valent ruthenium 32

## m

macrocyclic tetradentate ligands 11
metal-to-ligand charge transfer (MLCT)
electronic spectra and NO-sensor measurements 258
HOMO and LUMO energies 259
tetraazamacrocyclic ligand chelator 259
metronomic chemotherapy 131
Meyer's blue dimer 53, 54
mixed metal Ru(II)-Fe(II) bimetallic and tetrametallic complexes 10
Mixed-Metal/Supramolecular Complexes 139–156
mixed-valence manganese μ-oxo dimers 12
mono-or polydentate ligand scaffolds 282
monodentate bridging ligands 7
monometallic and bimetallic Fe(II) cyano complexes 10
Monte Carlo methods 124
multimetallic complexes 3, 14, 15
$[(\mu\text{-pdt})Fe_2(CO)_5(PPh_2(C_6H_4CCbpy))Ru(bpy)_2]^{2+}$ 81

## n

NAMI-A
antimetastatic activity 274
biotransformation 273
chemical properties 273
clinical studies and perspectives 275–276
mode of action 274–275
[NiFe]-hydrogenases 83
[NiFeSe]-hydrogenase 78
nitric oxide synthase (NOS) pathway 207
nitrogen oxide derivative-ruthenium complexes
metal-to-ligand charge transfer (MLCT)
electronic spectra and NO-sensor measurements 258
HOMO and LUMO energies 259
quantum yield and LUMO energy 261
tetraazamacrocyclic ligand chelator, 259nitrosyl ruthenium complexes
binuclear systems 261
clinical application 263
molecular orbital calculations 265
zinc-phthalocyanine 263
photochemical pathways 258
photovasorelaxation 265–268
*N,N*-chelate ethylenediamine 282

## o

organophosphorated ligand 310–312
organoruthenium(II) compoundsbioactive ligand scaffolds 284–285
cytotoxic organoruthenium complexes 285
enhanced permeability and retention 284
hydrolysis 283
mono-or polydentate ligand scaffolds 282
*N,N*-chelate ethylenediamine 282
piano-stool complexes 282
RAPTA-C 282, 283
RM175 282
Sadler-type compounds 283

Os(II)-based antenna complexes 14
oxo-centered trinuclear ruthenium(III) acetate 34

*p*
PARP-1 (Poly(ADP-ribose) polymerase-1) 206
pathogenic fungi species 300, 304–305
pentaammineruthenium(III) complexes 272
Petersen, J.D. 7
P-glycoprotein 278
phenantroline 305–306
phenazine 305–306
phosphorescence lifetime imaging microscopy (PLIM) 231
photo-stability 128
photobleaching 125, 131
photocatalyst 18
photochemical $H_2$ production 18, 19
photodynamic therapy (PDT) 3, 18, 91
- bioactive sites 140, 152
- biochemical properties 127
- DNA interactions 139
- dose-determining parameter
  - in vitro experiments 122–124
  - in vivo tissue response models 125–126
  - oxygen consumption model 125
- drug development 127
- energy transfers 118
- high quantum energy 127
- immune response 131
- and immunology 126
- light propagation properties 118
- long wavelength activation 128–129
- photo-stability 128
- photon absorption 118
- photophysical and photochemical deactivation pathways 119
- radical production 120
- ROS generation 119
- singlet oxygen production 120
- subcellular localization 130–131
- supramolecular complex 142–143
- thermal effects 127
- type I/II photochemical reactions 119

photolabile Ru-L coordination bond 92
photosynthesis and solar fuels 46–47
photosystem I (PSI) 46
photosystem II (PSII) 46
photovasorelaxation 265–268
phthalocyanine-macrocycle-containing [FeFe] Hase model 83
"piano-stool" complexes 242, 282
poly-3,4-ethylenedioxythiophene/poly-styrenesulfonate (PEDOT/PSS) nanocomposite 77
polyoxometalates (POMs) 57–60
polypyridyl ligands 9
polypyridyl ruthenium(II) complexes 68
proton coupled electron transfer (PCET) 49
pUC18 DNA 97, 144, 147–149
pyridine 305–306
2-pyridinephenyl 171

*q*
Q-absorption band 128
quinolone 305–306

*r*
RAPTA-C 169, 283
Ru(II) species 31
$[Ru^{IV}{}_4(\mu\text{-}O)_4(\mu\text{-}OH)_2(H_2O)_4(\gamma\text{-}SiW_{10}O_{36})_2]^{10-}$ 59
$[Ru(bda)pic_2]$ dimer 58
$[Ru(bpy)_3]Cl_2$ 72
$[Ru(bpy)_3]^{2+}$ cation sensitized $H_2$ photoproduction 82
$[Ru(bpy)_2(\text{L-pyr})Co(dmgH)_2(Cl)]^{2+}$ 72
$[Ru(bpy)_2(\text{L-pyr})Co(dpgBF_2)_2(OH_2)]^{2+}$ 72
$[[Ru(bpy)_2(phen\text{-}NH_2)]^{2+}$ 81
$Ru(bpy)_2(\text{L-pyr})Co(dmgBF_2)_2(OH_2)]^{2+}$ 72
$[RuCI(L1.2)(H_2O)_2]$ 299
$[Ru_3\ (CO)_{12}]$ trimer 30
Ru-Hbpp catalyst 54–55
$[Ru(L1.1)_3]$ 299
RuP/CoPmodified $TiO_2$ 75
$Ru(phen)_2(\text{L-pyr})Co(dmgBF_2)_2(OH_2)]^{2+}$ 72
Ru photophysics 100, 105
$Ru(tpy)(tpp)](PF_6)_2$ 14
$RuO_4$ 37
ruthenium (0) and (-2) 29–30
ruthenium(I) 29
ruthenium(II) 31–34
- photosensitizers 84
- phythalocyanine macrocycles 82
- polypyridyl complexes (RPCs)

ruthenium(II) (*contd.*)
  anticancer therapeutics 232–234
  biological application 225
  intracellular DNA 226
  MLCT luminescence 226
  Raman microscopy 231
ruthenium(III) 34–35
  KP1019/NKP-1339 276, 281
  NAMI-A 273–276
  structural formulas 272
ruthenium(IV), ruthenium(V), and ruthenium(VII) 36–37
ruthenium(VIII) 37
ruthenium-and cobalt-containing complexes for hydrogen production
  bridged systems 70–77
  hydrogenases, with ruthenium(II) complexes 77–84
  nonbridged systems 68–69

***s***
Sadler-type compounds 283
Schiff bases 307–309
single site Ru-WOCs 55–56
$[(SiP^{iPr}{}_3)Ru(N_2)]$ 30
size exclusion chromatography-inductively coupled plasma mass spectrometry (SEC-ICP/MS) 165
sodium-dependent multivitamin transporter (SMVT) system 171
species jump 293
stereochemistry and common oxidation states
  alkenes and alkynes 26
  anhydrous binary oxides 26
  chemistry of ruthenium(II) and (III) 31–36
  higher oxidation states of ruthenium 36–37
  redox processes 26
  ruthenium in low oxidation states 27–30
substitution reactions of ruthenium(II) and (III) 36
supramolecular complex 155
  DNA photomodification agents 142–143

***t***
2,2′:6′,2″-terpyridine 305–306
thioredoxin reductase (TrxR) 207
2,3,5,6-tetrakis(2-pyridyl)pyrazine 14, 15
$(tpy)Ru(tpp)IrCl_3\,](PF_6)_2$ 14
trans-plasma membrane electron transport (tPMET) systems 209
transmission electron microscopy (TEM) 231
tridentate 307–309
triplet-triplet annihilation upconversion 105–106
tris(2,2′-bipyridine)ruthenium(II)-cobalt(II) macrocycle system 69
tumor targeting mediated by plasma proteins 276–277
two-photon absorption (TPA) photosubstitution 109–110

***u***
upconverting nanoparticles (UCNP) 106, 108

***v***
Vanadium 151–152, 201
vascular endothelial growth factor 207
Virginia Tech 15–16

***w***
Warburg effect 209
Washington State University 13–15
water oxidation catalysis
  artificial water oxidation 49–50
  heptacoordinated Ru intermediates 56–57
  Meyer's blue dimer 53–54
  polyoxometalates 57–60
  Ru-Hbpp catalyst 54–55
  ruthenium oxide 50–52
  SC/P/WOC components 61
  single site Ru-WOCs 55–56
solar energy conversion 60

***z***
zero valent ruthenium–cobalt (Ru–Co)-based nanocluster 76